INTRODUCTION TO PROBABILITY

INTRODUCTION TO PROBABILITY
Multivariate Models and Applications

Volume 2

N. Balakrishnan
McMaster University, Hamilton, Ontario, Canada

Markos V. Koutras
University of Piraeus, Piraeus, Greece

Konstadinos G. Politis
University of Piraeus, Piraeus, Greece

This edition first published 2022

Registered Office
John Wiley & Sons, Inc., 111 River Street, Hoboken, NJ 07030, USA

Editorial Office
111 River Street, Hoboken, NJ 07030, USA

For details of our global editorial offices, customer services, and more information about Wiley products visit us at www.wiley.com.

Wiley also publishes its books in a variety of electronic formats and by print-on-demand. Some content that appears in standard print versions of this book may not be available in other formats.

Library of Congress Cataloging-in-Publication Data

Names: Balakrishnan, N., author. | Koutras, Markos V., author. |
        Politis, Konstantinos. G., author.
Title: Introduction to probability : multivariate models and applications /
        N. Balakrishnan, McMaster University, Hamilton, Ontario, Canada, Markos
        V. Koutras, University of Piraeus, Piraeus, Greece, Konstantinos G. Politis,
        University of Piraeus, Piraeus, Greece.
Description: First edition. | Hoboken, NJ : Wiley, 2022. | Includes
        bibliographical references and index.
Identifiers: LCCN 2021031022 (print) | LCCN 2021031023 (ebook) | ISBN
        9781118123331 (hardback) | ISBN 9781118548622 (adobe pdf) | ISBN
        9781118548554 (epub)
Subjects: LCSH: Probabilities–Textbooks.
Classification: LCC QA273.B254727 2022 (print) | LCC QA273 (ebook) | DDC
        519.2–dc23
LC record available at https://lccn.loc.gov/2021031022
LC ebook record available at https://lccn.loc.gov/2021031023

Cover Design: Wiley
Cover Image: © Mike Jack/Getty Images

Set in 10/12.5pt NimbusRomNo9L by Straive, Chennai, India

CONTENTS

PREFACE

This book is a follow-up of the book *Introduction to Probability: Models and Applications*, written by the same authors (which is referred throughout as Volume I). It thus forms the second volume for teaching probability theory.

In the first volume, we discussed the basic rules and concepts of probability, introduced the notion of conditional probability and independence, presented several combinatorial methods for probabilistic computations and finally went on to the introduction and study of discrete and continuous random variables; besides the presentation of general properties of random variables, we also discussed some well-known discrete and continuous distributions. These topics, in our view, form the core for an introduction to probability. The present volume discusses more advanced topics such as joint distributions, measures of dependence, multivariate random variables, some well-known multivariate discrete and continuous distributions, generating functions, laws of large numbers and the central limit theorem, which are the key topics for a second course on probability.

The form and structure of each chapter are similar to those in Volume I. In the beginning of each chapter, we provide a brief historical account of some pioneers in Probability who made exemplary contributions to the topic discussed in that chapter. This is done in order to provide students with a feeling of the history of Probability Theory and an appreciation of the vital contributions made by some renowned probabilists. Of course, books on the history of Probability and Statistics, would provide more elaborate details on their lives and contributions!

At the end of each section, the exercises have been classified into two groups, Group A and Group B. Exercises in Group A are usually routine extensions of the theory, involve simple calculations based on the theoretical tools presented in that section and are meant to consolidate the knowledge gained by the reader. Exercises in Group B are more advanced and would require both critical thinking and an ability to understand and use in an appropriate way the corresponding theory. In addition to regular exercises, we have also provided in each chapter, a long set of True/False questions and a set of multiple-choice questions. In our view, these will not only be useful for students to practice with (and assess their progress) but also for instructors to conduct regular in-class quizzes.

Special effort has been made to give the theoretical results in their simplest forms, so that they can be understood easily by the reader. In an effort to offer additional means for understanding the concepts presented, intuitive approaches as well as illustrative graphical representations/figures have been used throughout.

In each chapter, we have also included a section with some examples/problems for which the use of a computer is necessary. We demonstrate, through ample examples, how one can make effective use of computers for understanding probability concepts and carrying out various probability calculations. For these examples, the use of computer algebra software, such as Mathematica, Maple, and Derive, is recommended. These programs provide excellent tools for creating graphs in an easy way as well as for performing mathematical operations such as derivative, summation, integration, etc.; most importantly, one can handle symbols and variables without having to replace them with specific numerical values. To assist in this process, an example set of Mathematica commands is given each time (analogous commands can be assembled for other programs mentioned above as well). These commands may be used to perform a specific task and then several other similar tasks are set forth in the form of exercises. No effort is made to present the most effective Mathematica program for tackling the suggested problem and no detailed description of the Mathematica syntax is provided; the interested reader may refer to the Mathematica Instruction Manual (Wolfram Research) to check the, virtually unlimited, commands available in this software package (or any other computer algebra software) and use them for creating several alternative instruction sets for the suggested exercises.

Moreover, as in the first volume, at the end of each chapter, we have included a section detailing a case study (application) through which we demonstrate the usefulness of the results and concepts discussed in that chapter for a real-life problem.

In the first volume, we focused on situations wherein we assigned probabilities to events with regard to a single random variable X. In particular, we distinguished between discrete and continuous random variables and discussed the most important probability distributions in both cases. Even though there were examples in which more than one variable can be defined within the same random experiment, we were only interested in how each of these variables varied individually. However, there are many instances wherein our interest may be to study the way two or more variables vary simultaneously and, as a result, our primary objective will be to formulate, and answer, probability statements concerning more than one variable. For example, suppose a company wishes to forecast its annual turnover, along with its total expenses, for next year. Let X denote the company's turnover and Y the company's expenses; then, our interest may be on probability statements of the form

$$P(X \in A, Y \in B),$$

for some intervals (A and B) on the positive half-line. If the company sets as a target the turnover to be at least x and the expenses to be at most y (x and y are specific values), then the probability that both these targets of the company will be met is

$$P(X \geq x, Y \leq y).$$

When x and y are treated as fixed, problems of this type have already been considered in Volume I. To be specific, let A be the event $\{X \geq x\}$ and B be the event $\{Y \leq y\}$. Then, we simply want to find the probability that A and B occur at the same time, which is simply $P(A \cap B)$. However, and in analogy with the case of a single variable (i.e. univariate case), the company for its financial planning may want to see how the probability in the above equation varies for *different values* of x and y. This is accomplished with the concept of *joint distribution* of X and Y, which would explain how the two variables vary together. The central role played by probability distributions in the first volume is now extended to joint probability distributions of two or more variables, referred to as *multivariate probability distributions*. The particular case of two variables (known as the *bivariate case*) deserves special consideration and is therefore treated separately. This makes the transition from the univariate to the multivariate case smoother, and helps one to get a better understanding about the main changes needed to pass from the case of one variable to the multivariate case. For this reason, we have adhered to this structure, and in the first two chapters, we discuss (discrete and continuous) bivariate random variables and distributions, while in the third chapter, we treat the multivariate case, along with the fundamental concept of independence between random variables. In Chapter 4, we present techniques for obtaining the distribution of transformations of random variables, and then we use them to derive three additional distributions that are of special importance in Statistics.

The presentation of the concept of independence between random variables in Chapter 3, is restricted to the consideration whether two or more variable are independent or not. However, if two random variables are not independent, we would naturally be interested to introduce a measure to *quantify* their relationship. In Chapter 5, we discuss two such measures, called *covariance* and *correlation*. In Chapter 6, we present in detail some of the most important multivariate distributions. The last two chapters of the book deal with two areas of probability theory that provide essential tools for handling a wide range of applications: generating functions (Chapter 7) and limit theorems (Chapter 8). Their common appeal is in situations when one is interested in sums or averages of random variables. For example, when we want to estimate the mean from a (potentially large) population, we draw a random sample from that population. In this case, essentially no statistical knowledge is needed for one to imagine that our best "guess" (i.e. *estimate*) for the population mean will be the sample average, that is, the arithmetic mean of the data collected. The theoretical justification for this (intuitively appealing) guess is provided by the results of Chapter 8, with the use of the so-called *laws of large numbers*. The central limit theorem, one of the fundamental results in probability theory, is also presented in the same chapter; it goes further to consider the difference between the sample average and the theoretical mean and highlights the importance of the normal distribution in statistical theory and practice.

This book is intended as a textbook for a second course in probability and can be used by majors in Mathematics, Statistics, Physics, Computer Science, Engineering, Actuarial Science, Business, and Operations Research. Even though we have tried to make the material as much self-contained as possible, it is inevitable that many concepts, methods, and results discussed here, especially with regard to multivariate distributions, depend on analogous issues relating to univariate distributions. It is, therefore, expected that the reader is well-acquainted with basic probability theory and univariate distributions (in other words, the coverage in Volume I), while some knowledge of multivariate calculus (such as performing multiple integration and differentiation) is also required.

The material of this book emerged from a similar book (*Introduction to Probability: Theory and Applications: Part II*, Stamoulis Publications) written by one of us (MVK) in Greek, which is being used as a textbook for many years in several Greek Universities. Of course, we have expanded and transformed the material presented in that book to make it broader and to be suitable for an international audience.

It is a pleasure to thank our friends and colleagues Apostolos Bozikas, Christina Koutropoulou, and Sotitios Losidis, who read parts of the book, offered many useful comments and corrections and, in some cases, provided assistance with Mathematica. Also, special thanks go to our students who attended the classes and made several insightful remarks and suggestions through the years.

In a book of this size and content, it is inevitable that there are some typographical errors and mistakes (that have clearly escaped several pairs of eyes). If you do notice any of them, please bring them to our attention so that we can make necessary corrections/changes in future editions of this book.

It is our sincere hope that instructors find this textbook to be easy-to-use for teaching an intermediate course on probability, while the students find the book to be user-friendly with easy and logical explanations, plethora of examples, and numerous exercises (including computational ones) that they could practice and learn with!

April 2021

NB, MVK, KGP
Hamilton, Ontario, Canada
Piraeus, Greece

ACKNOWLEDGMENTS

It is a pleasure to thank the entire Wiley production team, and especially Ms. Kimberly Monroe-Hill, Ms. Kathleen Pagliaro, and Ms. Sarah Keegan, for their immense help and cooperation during the entire course of this book project. We are also grateful to Sundaramoorthy Balasubramani, who has been of great help in the final stages of the production of this book.

Apart from the books listed in the bibliography, *Wikipedia* has been used in some cases as a source for biographical details of the authors at the beginning of each chapter. The sources for the authors' photos are as follows: adoc-photos/Contributor/Getty Images (Chapter 1), https://commons.wikimedia.org/wiki/File:Bravais2.gif (Chapter 2), https://wikivisually.com/wiki/File:Linnik.jpg (Chapter 3), Creative Commons Attribution (Chapters 4, 6, and 7), Smith Archive/Alamy Stock Photo (Chapter 5), and Konrad Jacobs [https://commons.wikipedia.org.wiki] (Chapter 8).

CHAPTER 1

TWO-DIMENSIONAL DISCRETE RANDOM VARIABLES AND DISTRIBUTIONS

Francis Galton (Birmingham 1822 – Haslemere, England 1911)

Sir Francis Galton, a half-cousin of Charles Darwin, was born on 16 February 1822, in Birmingham, England, and died on 17 January 1911, at the age of 88, in Haslemere, England. He was a renowned statistician, sociologist, psychologist, and anthropologist, in addition to being a tropical explorer and geographer. Galton published numerous papers and books. He was very much interested in the joint behavior of quantities, created the statistical concept of correlation, and promoted regression toward the mean, a method which is currently of widespread use in all areas of statistical applications.

Introduction to Probability: Multivariate Models and Applications, Volume 2, First Edition.
N. Balakrishnan, Markos V. Koutras, and Konstadinos G. Politis.
© 2022 John Wiley & Sons, Inc. Published 2022 by John Wiley & Sons, Inc.

1.1   INTRODUCTION

Quite often, in order to study a random experiment, it is not enough to observe a single characteristic (i.e. a random variable); we might be interested in the study of two or more (usually numerical) characteristics. For example:

- when rolling two dice, we focus on both the indication of the first and the second die;

- when studying the operation of a gas station it makes sense to look at the number of cars waiting to be served in each of the gas pumps of the station.

In these cases, apart from the study of each random variable separately, the determination of the behavior of one in relation to the behavior of the others might be of particular interest. In this chapter, we will deal systematically with the case where we have two discrete random variables (as in the first example above), while the case of two continuous random variables will be considered in Chapter 2. The joint behavior of more than two random variables will be covered later on in Chapter 3.

When one studies a discrete variable, two functions which play a prominent role are the probability function and the (cumulative) distribution function. In this chapter, we study the simultaneous behavior, from a probabilistic viewpoint, of two discrete random variables and we extend the definition of these functions to cover the bivariate case, that is the case of two variables. In such a case, one speaks of a *joint probability function* and a *joint distribution function*, respectively. We shall also consider the expectation for a function of these two random variables (which is a random variable itself) and the conditional expectation of one variable given the other.

1.2   JOINT PROBABILITY FUNCTION

We begin our discussion with the simple case wherein, on the same random experiment, we are interested simultaneously in two discrete random variables. In the next chapter, we will discuss the case of two continuous random variables, while in Chapter 3, we will consider the case of more than two variables.

We shall start with an introductory example to illustrate some of the ideas involved.

Example 1.2.1 A box contains three balls numbered 1 to 3. We select two balls at random, without replacement, and let us define

X: the number on the first ball drawn,
Y: the largest number on the two balls selected.

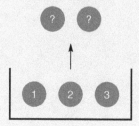

As in the case of a single variable, it is convenient to define a sample space Ω whose elements are equally likely. Because the experiment involves selecting two balls, we consider the elementary events (i, j) associated with the experiment, in which i denotes the number on the first ball and j the number on the second ball. There are clearly three choices for the number on the first ball, so that i can take any value in the set $\{1, 2, 3\}$. However, as the sampling is without replacement, j cannot be the same as i, which means that for each value of $i = 1, 2, 3$, the number j on the second ball can take a value from the set $\{1, 2, 3\} - \{i\}$. We thus arrive at the following sample space

$$\Omega = \{(1, 2), (1, 3), (2, 1), (2, 3), (3, 1), (3, 2)\}.$$

This contains six elementary events which are clearly equally likely, i.e. $P(\omega) = 1/6$ for any $\omega \in \Omega$ (in familiar terminology, the elementary events in our sample space are equiprobable). Suppose, in a realization of this experiment, we get the outcome $\omega = (1, 2)$. Then, form the definition of the variables X and Y, we obtain $X = X(\omega) = 1$ and $Y = Y(\omega) = 2$. Instead, if we get the outcome $\omega = (3, 2)$, then the variables X and Y take the values $X = X(\omega) = 3$ and $Y = Y(\omega) = 3$. Table 1.1 presents all possible outcomes from this experiment, along with the values of X and Y in each case:

Looking at the last row of this table, it is apparent that the outcome $X = 1, Y = 2$ may occur only with the outcome $\omega = (1, 2)$ and thus corresponds to the elementary event

$$A = \{\omega \in \Omega : X(\omega) = 1, Y(\omega) = 2\} = \{(1, 2)\}.$$

The probability of the event A is $1/6$, and so we see that the probability the pair of variables (X, Y) takes the value $(1, 2)$ is $1/6$. We denote this probability by $P(X = 1, Y = 2)$, so that we have

$$P(X = 1, Y = 2) = \frac{1}{6}.$$

Consider now the case when both X and Y take the value 3. From Table 1.1, we observe that this occurs if **either** of the outcomes $\omega = (3, 1)$, $\omega = (3, 2)$ occur. Hence, the result $X = 3, Y = 3$ corresponds to the event

$$B = \{\omega \in \Omega : X(\omega) = 3, Y(\omega) = 3\} = \{(3, 1), (3, 2)\},$$

and so the probability of event B is $2/6$, i.e.

$$P(X = 3, Y = 3) = \frac{2}{6}.$$

$\omega \in \Omega$	$(1, 2)$	$(1, 3)$	$(2, 1)$	$(2, 3)$	$(3, 1)$	$(3, 2)$
$P(\omega)$	$1/6$	$1/6$	$1/6$	$1/6$	$1/6$	$1/6$
$X = X(\omega)$	1	1	2	2	3	3
$Y = Y(\omega)$	2	3	2	3	3	3
(X, Y)	$(1, 2)$	$(1, 3)$	$(2, 2)$	$(2, 3)$	$(3, 3)$	$(3, 3)$

Table 1.1 Values of the pair (X, Y) with their respective probabilities.

The quantity $P(X = x, Y = y)$, in the above example, expresses the joint allocation of probabilities for values that X, Y may take on together. This generalizes the concept of a probability function for a single variable, $P(X = x)$, and is therefore called the **joint probability function** of the pair (X, Y).

Definition 1.2.1 Let Ω be the sample space for a chance experiment, and X, Y be two discrete random variables defined on it. Denote by $R_{X,Y}$ the set of all possible values for the pair (X, Y). Then, the function defined by

$$f(x, y) = P(X = x, Y = y) = P(\{\omega \in \Omega : X(\omega) = x \text{ and } Y(\omega) = y\}),$$

for any $(x, y) \in R_{X,Y}$, is called the joint probability function of the variables X, Y or, equivalently, the probability function of the two-dimensional random variable (or, the random pair) (X, Y). The distribution of (X, Y) is referred to as a bivariate (or two-dimensional) discrete distribution.

The set $R_{X,Y}$ is referred to as **the range of values** for the pair (X, Y). It is obvious that if (x, y) belongs to $R_{X,Y}$, then x must belong to the range of values of X, say R_X, and y to the range of values of Y, say R_Y. Thus, we have

$$R_{X,Y} \subseteq R_X \times R_Y \subseteq \mathbb{R}^2,$$

where $\times$ denotes the Cartesian product between two sets; moreover,

$$R_X = \{x \in \mathbb{R} : (x, y) \in R_{X,Y}\} \subseteq \mathbb{R}, \quad R_Y = \{y \in \mathbb{R} : (x, y) \in R_{X,Y}\} \subseteq \mathbb{R}.$$

In the univariate case, two main properties of the probability function for a variable X are that it takes nonnegative values, and the sum of its values over the entire range of values of X is equal to one. These properties also hold in the bivariate case, when we consider the probability function $f(x, y)$ associated with the pair (X, Y), as we explain below.

First, in Definition 1.2.1, we see that the function $f(x, y)$ is defined as a probability, and so its values cannot be negative; i.e. we have $0 \leq f(x, y) \leq 1$ for any x, y. Next, if we consider all pairs (x, y) in the set $R_{X,Y}$, we see that the events

$$\{\omega \in \Omega : X(\omega) = x, Y(\omega) = y\}$$

form a *partition* of the sample space Ω (try to explain why!). This, in particular, means that if we add all the values of $f(x, y)$ for all pairs (x, y), we obtain the probability of the entire sample space, i.e.

$$\sum_{(x,y) \in R_{X,Y}} f(x, y) = 1.$$

We state these properties in the following proposition, along with another property of the function f which is evident.

Proposition 1.2.1 *The joint probability function $f(x, y)$ of a random pair (X, Y), with range $R_{X,Y}$, satisfies the following properties:*

1. $f(x, y) = 0$ for any $(x, y) \notin R_{X,Y}$,

2. $f(x, y) \geq 0$ *for any* $(x, y) \in R_{X,Y}$,

3. $\displaystyle\sum_{(x,y)\in R_{X,Y}} f(x, y) = 1.$

We should mention at this point that the converse of Proposition 1.2.1 is also true. To be specific, if a real function $f(x, y)$ of two variables satisfies the three properties given in Proposition 1.2.1, then there exists a pair (X, Y) of random variables such that f is the joint probability function of (X, Y).

Example 1.2.1 *(Continued)*
From Table 1.1, we can present the joint probability function of the pair (X, Y) in Example 1.2.1 in a concise from, as in Table 1.2:

x \ y	2	3
1	1/6	1/6
2	1/6	1/6
3	–	2/6

Table 1.2 The joint probability function $f(x, y) = P(X = x, Y = y)$ for the pair (X, Y).

In this example, we have the range of values of (X, Y) as

$$R_{X,Y} = \{(1, 2), (1, 3), (2, 2), (2, 3), (3, 3)\},$$

from which we observe

$$R_{X,Y} \subseteq R_X \times R_Y = \{1, 2, 3\} \times \{2, 3\}.$$

Moreover, it is easy to verify that

$$\sum_{(x,y)\in R_{X,Y}} f(x, y) = f(1, 2) + f(1, 3) + f(2, 2) + f(2, 3) + f(3, 3)$$

$$= \frac{1}{6} + \frac{1}{6} + \frac{1}{6} + \frac{1}{6} + \frac{2}{6} = 1.$$

Example 1.2.2 Find the value of the constant c in each of the two cases below, so that $f(x, y)$ is the joint probability function of a random pair (X, Y):

(i) $f(x, y) = cxy$, where $x = 1, 2, 3$ and $y = 1, 2, 3$

(ii) $f(x, y) = \dfrac{c}{2^x 4^y}$, where $x, y = 0, 1, 2, \dots$.

SOLUTION   In each case, we find the value of c such that the function f satisfies

$$f(x, y) \geq 0 \quad \text{for any } (x, y) \in R_{X,Y}$$

and

$$\sum_{(x,y) \in R_{X,Y}} f(x, y) = 1$$

(as in the case of one variable, we do not have to check the condition $f(x, y) = 0$ for $(x, y) \notin R_{X,Y}$, as this is implicit by the statement of the example).

(i) The first condition above is satisfied whenever $c \geq 0$. The second condition gives

$$\sum_{(x,y) \in R_{X,Y}} f(x, y) = \sum_{x=1}^{3} \sum_{y=1}^{3} f(x, y) = 1.$$

The joint probability function of the pair (X, Y) is as follows:

Using the formula for $f(x, y)$ given in the statement, we get

$$1 = \sum_{x=1}^{3} \sum_{y=1}^{3} f(x, y) = c + 2c + 3c + 2c + 4c + 6c + 3c + 6c + 9c = 36c,$$

which implies that the required value of the constant c, so that $f(x, y)$ is a proper joint probability function, is $c = 1/36$ (Table 1.3).

x \ y	1	2	3
1	c	$2c$	$3c$
2	$2c$	$4c$	$6c$
3	$3c$	$6c$	$9c$

Table 1.3   The joint probability function $f(x, y)$ of the pair (X, Y) in Part (i) of Example 1.2.2.

(ii) The condition $f(x, y) \geq 0$ leads to $c \geq 0$ again. In order to calculate the (double) sum over all x, y for $x = 0, 1, 2 \ldots$ and $y = 0, 1, 2, \ldots$, we observe first that

$$\sum_{x=0}^{\infty} \sum_{y=0}^{\infty} f(x, y) = \sum_{x=0}^{\infty} \sum_{y=0}^{\infty} \frac{c}{2^x 4^y} = c \sum_{x=0}^{\infty} \left(\tfrac{1}{2}\right)^x \left\{ \sum_{y=0}^{\infty} \left(\tfrac{1}{4}\right)^y \right\}.$$

Now, upon recalling the (infinite) sum of a geometric series

$$\sum_{k=0}^{\infty} a^k = \frac{1}{1-a}, \quad \text{for } |a| < 1,$$

and applying this twice in the previous formula, we obtain

$$\sum_{x=0}^{\infty} \sum_{y=0}^{\infty} f(x,y) = c \sum_{x=0}^{\infty} \left(\frac{1}{2}\right)^x \cdot \frac{1}{1-\frac{1}{4}} = \frac{4c}{3} \sum_{x=0}^{\infty} \left(\frac{1}{2}\right)^x = \frac{4c}{3} \cdot \frac{1}{1-\frac{1}{2}} = \frac{8c}{3}.$$

Thus, for f to be a valid joint probability function in this case, we must have $8c/3 = 1$, which yields immediately that $c = 3/8$.

For a two-dimensional random variable (X, Y), if we know its joint probability function $f(x,y)$, we can, at least in principle, calculate any probability associated with these variables. Specifically, let A be a subset of the range $R_{X,Y}$ for the pair (X, Y). Then, the probability of the event that (X, Y) takes some value in the set A can be found as

$$P((X,Y) \in A) = \sum_{(x,y) \in A} P(X = x, Y = y) = \sum_{(x,y) \in A} f(x,y).$$

Using this formula, for the set

$$A = \{(s,t) \in R_{X,Y} : s \le x \text{ and } t \le y\},$$

where x, y are any given real numbers we obtain

$$P(X \le x, Y \le y) = \sum_{s \le x} \sum_{t \le y} P(X = s, Y = t) = \sum f(s,t), \tag{1.1}$$

where the last summation extends over all pairs (s,t) in the set $R_{X,Y}$ such that $s \le x$ and $t \le y$.

The function

$$F(x,y) = P(X \le x, Y \le y), \quad x, y \in \mathbb{R}, \tag{1.2}$$

is called the **joint (cumulative) distribution function** of the two-dimensional random variable (X, Y).

Example 1.2.3 A shop has two cashiers. Let X denote the number of customers waiting for service by the first cashier at 11 : 00am on a given day, and Y denote the number of customers waiting for service by the second cashier, at the same time instant. Suppose the shop manager has available data from the previous five months and (interpreting a probability as the limit of the corresponding relative frequency, as is done in the case of single variable), she has determined the joint probabilities $P(X = x, Y = y)$, for X and Y, as in Table 1.4.

x \ y	0	1	2
0	0.10	0.04	0.02
1	0.08	0.20	0.06
2	0.06	0.14	0.30

Table 1.4   The joint probability function $f(x, y) = P(X = x, Y = y)$ for the pair (X, Y) in Example 1.2.3.

To check that the function $f(x, y)$ given in Table 1.4 is indeed a valid joint probability function, we note first that

$$f(x, y) \geq 0 \quad \text{for any } (x, y) \in R_{X,Y} = \{(0, 0), (0, 1), \dots, (2, 2)\}.$$

Also, adding all probabilities in the table, we find that

$$\sum_{(x,y) \in R_{X,Y}} f(x, y) = \sum_{x=0}^{2} \sum_{y=0}^{2} f(x, y) = f(0, 0) + f(0, 1) + \cdots + f(2, 2)$$

$$= 0.10 + 0.04 + \cdots + 0.30 = 1.$$

Thus, for example, the probability that at 11 : 00am there is one person waiting for service by each cashier is

$$P(X = 1, Y = 1) = f(1, 1) = 0.20,$$

while the probability that there are two persons waiting for service by the first cashier and none waiting to be served by the second cashier is

$$P(X = 2, Y = 0) = f(2, 0) = 0.06.$$

Next, the probability that there is at most one person waiting for service by each cashier is the probability that (X, Y) takes values in the shaded area A_1 in the graph below and this can happen if the value of (X, Y) belongs to the set $\{(0, 0), (0, 1), (1, 0), (1, 1)\}$. Thus,

$$P(X \leq 1, Y \leq 1) = F(1, 1) = P((X, Y) \in A_1) = \sum_{x=0}^{1} \sum_{y=0}^{1} f(x, y)$$

$$= f(0, 0) + f(0, 1) + f(1, 0) + f(1, 1) = 0.42.$$

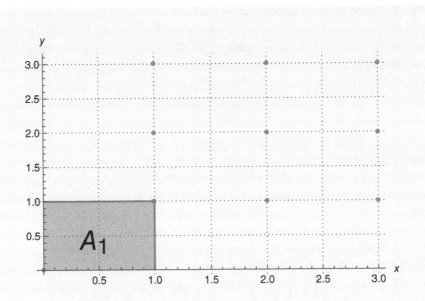

If we want the probability that the number of persons waiting for service is the same for both cashiers, this is simply

$$P(X = Y) = P\left((X, Y) \in \{(0,0), (1,1), (2,2)\}\right) = f(0,0) + f(1,1) + f(2,2)$$
$$= 0.60,$$

while the probability that the *total number* of persons waiting for service by the two cashiers is equal to 3 is

$$P(X + Y = 3) = P\left((X, Y) \in \{(1,2), (2,1)\}\right) = f(1,2) + f(2,1)$$
$$= 0.06 + 0.14 = 0.20.$$

Finally, the probability that there is one more person waiting for service by the first cashier than by the second one is equal to

$$P(X = Y + 1) = P\left((X, Y) \in \{(1,0), (2,1)\}\right) = f(1,0) + f(2,1)$$
$$= 0.08 + 0.14 = 0.22.$$

We now list some properties of a bivariate distribution function in the discrete case, most of which parallel and generalize in a straightforward way the properties of a distribution function in the univariate case. If we have a single variable X having distribution F, then we saw in Chapter 4 of Volume I that

$$\lim_{t \to -\infty} F(t) = 0, \quad \lim_{t \to \infty} F(t) = 1.$$

The analogous property in the bivariate case is that

$$\lim_{\substack{x \to -\infty \\ y \to -\infty}} F(x, y) = 0$$

and

$$\lim_{\substack{x \to \infty \\ y \to \infty}} F(x, y) = 1;$$

from the last relation, we see that $F(x, y)$ tends to one when *both* x and y tend to infinity.

Next, when there is only one variable X, its distribution function, $F(t)$, has a single argument t and is nondecreasing in that argument. For the case when two variables X, Y have a joint distribution given in (1.2), we see that for any real x_1, x_2 with $x_1 \leq x_2$ and $y \in \mathbb{R}$,

$$F(x_1, y) = P(X \leq x_1, Y \leq y) \leq P(X \leq x_2, Y \leq y) = F(x_2, y).$$

Similarly, if $y_1 \leq y_2$, we then obtain for any real x that

$$F(x, y_1) = P(X \leq x, Y \leq y_1) \leq P(X \leq x, Y \leq y_2) = F(x, y_2).$$

We thus see that, in the bivariate case, $F(x, y)$ is nondecreasing in *each of its arguments* x and y.

Finally, recall that a univariate discrete distribution function $F(t)$ changes value (more precisely, it has an upward jump) only at the points belonging to the range of the corresponding random variable. The same is true in the bivariate case; in particular, $F(x, y)$ defined in (1.2) changes values *only* at the points (x, y) belonging to the range $R_{X,Y}$.

The following example illustrates all these aspects.

Example 1.2.4 Assume that the random variables X and Y can take on only the values 0 and 1, and that

$$P(X = 0, Y = 0) = 1/3, \quad P(X = 0, Y = 1) = 1/6, \quad P(X = 1, Y = 0) = 1/9.$$

Obtain the joint cumulative distribution function of X and Y.

SOLUTION First, we observe that the joint probability function is not completely specified in the statement. If we add the probabilities of the three events $\{X = 0, Y = 0\}$, $\{X = 0, Y = 1\}$ and $\{X = 1, Y = 0\}$ we do not get 1, but rather

$$\frac{1}{3} + \frac{1}{6} + \frac{1}{9} = \frac{11}{18}.$$

This is due to the fact that in the statement we are only given the probabilities for three values of the pair (X, Y), while (since both X and Y assume two values each) it is clear that there are four possibilities. The probability which is *not given* is that of the event

$\{X = 1, Y = 1\}$ and since we must have the sum of values of a joint probability function over its range to be 1, we readily find

$$P(X = 1, Y = 1)$$
$$= 1 - [P(X = 0, Y = 0) + P(X = 0, Y = 1) + P(X = 1, Y = 0)]$$
$$= 1 - \frac{11}{18} = \frac{7}{18}.$$

Now, in order to find the joint distribution function of X and Y, we first note that when either $x < 0$ or $y < 0$, then $F(x, y) = 0$. To see what happens in the remaining cases, we need to consider only the four pairs

$$(0, 0), \ (0, 1), \ (1, 0), \ (1, 1)$$

in the range of (X, Y). Specifically, we find the following:

- For $0 \le x < 1, 0 \le y < 1$, we have (see area Π_1 in the graph)

$$F(x, y) = P(X \le x, Y \le y) = \sum_{(s,t) \in R_{X,Y} \,:\, s \le x, t \le y} f(s, t) = f(0, 0) = \frac{1}{3};$$

- For $0 \le x < 1, y \ge 1$, we get (area Π_2 in the graph)

$$F(x, y) = P(X \le x, Y \le y) = \sum_{(s,t) \in R_{X,Y} \,:\, s \le x, t \le y} f(s, t)$$
$$= f(0, 0) + f(0, 1) = \frac{1}{3} + \frac{1}{6} = \frac{1}{2};$$

- For $x \ge 1, 0 \le y < 1$, which corresponds to area Π_3 in the graph, we have that

$$F(x, y) = P(X \le x, Y \le y) = \sum_{(s,t) \in R_{X,Y} \,:\, s \le x, t \le y} f(s, t)$$
$$= f(0, 0) + f(1, 0) = \frac{1}{3} + \frac{1}{9} = \frac{4}{9};$$

- Finally, for $x \ge 1$ and $y \ge 1$ (area Π_4 in the graph), we obtain similarly that

$$F(x, y) = P(X \le x, Y \le y) = \sum_{(s,t) \in R_{X,Y} \,:\, s \le x, t \le y} f(s, t)$$
$$= f(0, 0) + f(0, 1) + f(1, 0) + f(1, 1) = \frac{1}{3} + \frac{1}{6} + \frac{1}{9} + \frac{7}{18} = 1.$$

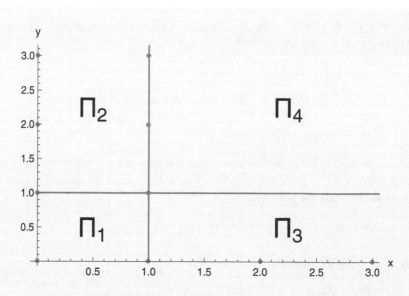

Summarizing the above, we can express the joint distribution function of the random vector (X, Y) as

$$F(x,y) = P(X \leq x, Y \leq y) = \begin{cases} 0, & \text{if } x < 0 & \text{or} & y < 0 \\ 1/3, & \text{if } 0 \leq x < 1 & \text{and} & 0 \leq y < 1 \\ 4/9, & \text{if } x \geq 1 & \text{and} & 0 \leq y < 1 \\ 1/2, & \text{if } 0 \leq x < 1 & \text{and} & y \geq 1 \\ 1, & \text{if } x \geq 1 & \text{and} & y \geq 1. \end{cases}$$

EXERCISES

Group A

1. At a gas station, there are four dispensing pumps, two of which are for gas and the other two are for diesel. Let X be the number of gas pumps which are in use at a particular time during the day, and Y be the number of diesel pumps in use at the same time. The joint probability function, $f(x, y) = P(X = x, Y = y)$, of the variables X and Y is as given in the following table:

x \ y	0	1	2
0	0.13	0.07	0.02
1	0.15	0.19	0.21
2	0.04	0.12	0.07

(i) Calculate the probabilities $P(X = 1, Y = 1)$ and $P(X \leq 1, Y \leq 1)$.

(ii) Explain in words what the event $\{X \neq 0, Y \neq 0\}$ represents, and find the probability of that event.

(iii) Obtain the probability for each of the following events:

(a) the number of gas pumps in use is the same as the number of diesel pumps in use;

(b) the number of gas pumps in use is smaller than the number of diesel pumps in use;

(c) the total number of pumps in use is 3;

(d) the total number of pumps in use is at least 3.

2. A mini market has three cashier points. The first of these (Cashier Point I) serves only customers who have purchased at most five different products from the mini market, while the other two serve customers who have purchased more than five different products. Let X be the number of persons for service at Cashier Point I and Y be the total number of customers who have purchased more than five products and are waiting at the other two cashier points. We assume that the joint probability function, $f(x, y) = P(X = x, Y = y)$, of X and Y is known to be as follows:

x \ y	0	1	2	3
0	0.11	0.06	0.02	0.00
1	0.13	0.09	0.04	0.02
2	0.06	0.10	0.05	0.02
3	0.06	0.05	0.04	0.02
4	0.01	0.02	0.04	0.06

(i) Write down the range, $R_{X,Y}$, of the joint distribution of X and Y. Do we have $R_{X,Y} = R_X \times R_Y$?

(ii) Calculate the probabilities $P(X = 1, Y = 2)$ and $P(X = 1, Y \geq 2)$.

(iii) Explain in words what each of the following events means and find the associated probability:

$$\{X = Y\}, \ \{X = Y + 2\}, \ \{X \geq Y + 2\}, \ \{X \geq 1 \text{ and } 2X \leq Y\}.$$

3. Nick has a red and a green dice and he rolls both of them. Let X be the outcome of the green die and Y be the sum of the two outcomes.

(i) What is the range of values for each of the variables X and Y?

(ii) Write down the range of the two-dimensional random variable (X, Y).

(iii) Find the joint probability function of X and Y, and check that it satisfies the conditions of Proposition 1.2.1.

4. The joint probability function of the variables X and Y is

$$f(x, y) = P(X = x, Y = y) = \begin{cases} \dfrac{cx}{y}, & \text{if } x = 1, 2 \text{ and } y = 1, 2, \\[2mm] 0, & \text{otherwise.} \end{cases}$$

(i) Find the value of c.

(ii) Write down the joint distribution function of X and Y. Thence, calculate $P(X \neq Y)$.

5. Obtain the joint distribution function of the variables X and Y in Example 1.2.1.

6. Find the value of $c \in \mathbb{R}$ such that, in each of the cases below, the function f will be a valid joint probability function of a two-dimensional random variable:

   (i) $f(x, y) = cxy$, for $x = 1, 2$ and $y = 2, 4, 6$;

   (ii) $f(x, y) = c$, for $x = 1, 2, \ldots, m$ and $y = 1, 2, \ldots, n$, where m, n are two given positive integers;

   (iii) $f(x, y) = c(x + y)$, for $x = 0, 1, 2$ and $y = 0, 1, 2$;

   (iv) $f(x, y) = c(x^2 + y^2)$, for $x = -1, 0, 1, 2$ and $y = -1, 2, 3$;

   (v) $f(x, y) = \dfrac{c}{3^x 7^y}$, for $x = 0, 1, 2, \ldots$ and $y = 0, 1, 2, \ldots$.

7. Two discrete variables X and Y have a joint probability function as

$$f(x, y) = \begin{cases} c(x^2 + y^2), & \text{if } x = 1, 2 \text{ and } y = 0, 1, 2, \\ 0, & \text{otherwise.} \end{cases}$$

Calculate the following probabilities:

$$P(X = Y), \ P(X > Y), \ P(X < Y), \ P(X + Y = 2), \ P(X + Y > 2).$$

Group B

8. In the experiment of tossing two dice, let X be the number of times that a six appears and Y be the number of times an even integer appears.

   (i) Derive the joint probability function of X and Y.

   (ii) Calculate the following probabilities:

$$P(X \leq 2, Y = 2), \ P(X > 0, Y < 2), \ P(Y > X), \ P(Y = 2X).$$

9. In a certain city, there are three candidates – Smith, Jones, and Allen – running for Mayor in the forthcoming election. It has been estimated by a recent poll that 45% of eligible voters support Mr. Smith, 30% support Ms. Jones, and the remaining 25% support Mr. Allen. We select at random a person who is eligible to vote, and define the following random variables:

$$X = \begin{cases} 1, & \text{if the person supports Mr. Smith,} \\ 0, & \text{otherwise,} \end{cases}$$

and

$$Y = \begin{cases} 1, & \text{if the person supports Ms. Jones,} \\ 0, & \text{otherwise.} \end{cases}$$

(i) Find the joint probability function of the random variable (X, Y).

(ii) Find $P(X + Y = 0)$ and $P(X + Y = 1)$.

10. The random vector (X, Y) has its joint probability function as follows:

x \ y	0	1	2
2	0.20	0.15	0.05
3	0.08	0.12	0.08
4	0.07	0.13	0.12

(i) Write down, in a similar tabular form, the joint probability function of the random pair (Z, W), where

$$Z = X + Y, \quad W = X - Y,$$

after specifying the range of values for (Z, W).

(ii) Derive the joint distribution function of (Z, W), and then use it to calculate $P(Z \le 4, W \le 1)$.

11. Let (X, Y) be a random vector with range

$$R_{X,Y} = \{(x, y) : x = 0, \pm 1, \pm 2, \ldots, \pm k \text{ and } y = 0, \pm 1, \pm 2, \ldots, \pm k\}$$

for a given positive integer k. Express each of

$$P(X > Y), \quad P(X > |Y|), \quad P(|X| \ge Y), \quad P(X > Y + 1)$$

in terms of the joint probability function

$$f(x, y) = P(X = x, Y = y)$$

of X and Y.

1.3 MARGINAL DISTRIBUTIONS

For an experiment in which we consider two variables X and Y, even if we know their joint probability function, we may still be interested in the probability function of just X and/or that of just Y. More generally, having studied both one- and two-dimensional discrete random variables, two questions that arise in relation to a given pair of variables X, Y are:

- If we know the joint probability function of X, Y, can we obtain the probability function for each of X and Y separately?

- If we know the probability function for each of X, Y separately, can we derive from them their joint probability function?

The answer to the second question is negative, while for the first question it is affirmative, as we explain below.

In Table 1.1, we have the joint probability function for the pair (X, Y) in Example 1.2.1. Can we deduce from Table 1.1, for example, the probability function of X alone? A look at the columns of Table 1.1 shows that, for the pair (X, Y), there are five possible outcomes, namely

$$(1, 2), (2, 2), (1, 3), (2, 3), (3, 3),$$

and that the first four of these have a probability of $1/6$, while the event $\{(X, Y) = (3, 3)\}$ has a probability of $1/3$. Suppose now we are interested in the probability of the event $\{X = 1\}$. Upon observing that this event is equivalent to the union of the events $\{X = 1, Y = 2\}$ and $\{X = 1, Y = 3\}$, them being disjoint, we have

$$f_X(1) = P(X = 1) = P(\{X = 1, Y = 2\} \text{ or } \{X = 1, Y = 3\})$$

$$= P(X = 1, Y = 2) + P(X = 1, Y = 3)$$

$$= f(1, 2) + f(1, 3) = \frac{1}{6} + \frac{1}{6} = \frac{1}{3}.$$

The same can be done for the events $\{X = 2\}$ and $\{X = 3\}$, so that we obtain

$$P(X = i) = \frac{1}{3}, \quad \text{for } i = 1, 2, 3.$$

We now present the following general result.

Proposition 1.3.1 *Let (X, Y) be a jointly discrete random variable with joint probability function $f(x, y)$, with range $R_{X,Y}$. Then, the probability function for each of X and Y can be found as*

$$f_X(x) = P(X = x) = \sum_{y:(x,y) \in R_{X,Y}} f(x, y) \quad \text{for any } x \in R_X,$$

and

$$f_Y(y) = P(Y = y) = \sum_{x:(x,y) \in R_{X,Y}} f(x, y) \quad \text{for any } y \in R_Y.$$

Proof: Since (X, Y) have joint probability function $f(x, y)$, we have, for (x, y) in the range of (X, Y),

$$f(x, y) = P(X = x, Y = y).$$

Now, let $x \in R_X$ be fixed. For all values y in the range of Y, consider the events

$$\{X = x, Y = y\}.$$

Taking the union of these events over all $y \in R_Y$, we see that this is simply

$$\bigcup_{y \in R_Y} \{X = x, Y = y\} = \{X = x\}.$$

Further, for different values of $y \in R_Y$, the events on the left-hand side above are mutually exclusive, which implies that

$$P(X = x) = P\left(\bigcup_{y \in R_Y} \{X = x, Y = y\} \right) = \sum_{y \in R_Y} P(X = x, Y = y) = \sum_{y \in R_Y} f(x, y).$$

We thus see that by adding all probabilities $f(x, y)$ with respect to y, we arrive at the probability function of just X, i.e. when it is considered regardless of Y. In a similar way, we obtain

$$\bigcup_{x \in R_X} \{X = x, Y = y\} = \{Y = y\}$$

which implies, using again that the events on the left-hand side are mutually exclusive, that

$$P(Y = y) = P\left(\bigcup_{x \in R_X} \{X = x, Y = y\} \right) = \sum_{x \in R_X} P(X = x, Y = y) = \sum_{x \in R_X} f(x, y). \qquad \square$$

The key point to remember from the above discussion is that if we want the probability function of just X, then we sum over y, while if we want the probability function of just Y, we sum over all values of x.

The distribution of a random variable X alone, when it arises from the joint distribution of X and some other variable, is usually referred to as the **marginal distribution** of X. Similarly, the probability functions f_X and f_Y in Proposition 1.3.1 are called **marginal probability functions**.

Example 1.3.1 At a coffee shop in New York, customers may order coffee and hot chocolate in each of three sizes: Small (S), Medium (M), and Large (L). The percentages of customers, who order either coffee or hot chocolate, according to the size of the drink, is as given in the table. We assume, as usual, that these percentages (which are in fact relative frequencies) are based on a large amount of data so that each relative frequency can be viewed as a probability.

	Small (%)	Medium (%)	Large (%)
Coffee	14	23	19
Hot chocolate	11	20	13

Let X be the variable denoting the drink that a customer orders. For convenience, we assign numerical values to X, and set $X = 0$ if the customer orders coffee and $X = 1$ if the customer orders hot chocolate. Similarly, let Y be the size of the drink and we then set $Y = 0, 1$, and 2 according to the size being S, M, and L, respectively. Then, we see from the table that the joint probability function of X and Y is given by

$$f(0, 0) = 0.14, \quad f(0, 1) = 0.23, \quad f(0, 2) = 0.19,$$

$$f(1, 0) = 0.11, \quad f(1, 1) = 0.20, \quad f(1, 2) = 0.13.$$

It should be clear that the choice of values for X, Y here is arbitrary, and that we could have used, for example, $X = 5$ and $X = 10$ for the events "choice of coffee" and "choice of hot chocolate".

Suppose now we are interested in the probability function of X alone. As X takes only two values, we simply want to find (based on the table of percentages) the percentage of customers who select coffee for their drink and the percentage of those who select hot chocolate. We then see that the former equals $14\% + 23\% + 19\% = 56\%$, i.e. a probability of 0.56 that a randomly selected customer orders coffee. The percentage of those who choose hot chocolate is simply the sum of the percentages in the second row of the table, i.e. $11\% + 20\% + 13\% = 44\%$, or a probability of 0.44. Thus, the marginal probability function of X is

$$f_X(0) = 0.56, \quad f_X(1) = 0.44.$$

Working in the same way for the variable Y, we see that we now have to add the values *in each column*, to arrive at

$$f_Y(0) = 0.25, \quad f_Y(1) = 0.43, \quad f_Y(2) = 0.32.$$

The use of the term *marginal probability function* for f_X and f_Y above comes from the fact the values of these functions are frequently inserted at the margins of the table along with the joint probability function, as they correspond to the row and column totals of that table, as shown below.

	Small (%)	Medium (%)	Large (%)	
Coffee	14	23	19	56%
Hot chocolate	11%	20%	13%	44%
	25%	43%	32%	100%

Example 1.3.2 Let us reconsider Example 1.2.3 regarding the number of persons waiting to be served at two cashier points of the shop. For the number of customers, X, waiting to be served by the first cashier, we find

$$f_X(0) = P(X = 0) = \sum_{y=0}^{2} f(0, y) = f(0, 0) + f(0, 1) + f(0, 2) = 0.16,$$

$$f_X(1) = P(X = 1) = \sum_{y=0}^{2} f(1, y) = f(1, 0) + f(1, 1) + f(1, 2) = 0.34,$$

$$f_X(2) = P(X = 2) = \sum_{y=0}^{2} f(2, y) = f(2, 0) + f(2, 1) + f(2, 2) = 0.50.$$

In an analogous manner, we find the (marginal) probability function of the number of customers, Y, awaiting service by the second cashier, to be

$$f_Y(0) = P(Y = 0) = \sum_{x=0}^{2} f(x,0) = f(0,0) + f(1,0) + f(2,0) = 0.24,$$

$$f_Y(1) = P(Y = 1) = \sum_{x=0}^{2} f(x,1) = f(0,1) + f(1,1) + f(2,1) = 0.38,$$

$$f_Y(2) = P(Y = 2) = \sum_{x=0}^{2} f(x,2) = f(0,2) + f(1,2) + f(2,2) = 0.38.$$

Inserting these two probability functions at the margins of Table 1.4, we simply obtain the table below. Note that the value (in bold) in the far right-hand cell of the last row in this table is 1 (or 100%, if expressed as a percentage, as in the previous example), and this is always true since this is the sum of the probabilities $f_X(x)$ (or $f_Y(y)$), while it is evident that

$$\sum_{x \in R_X} f_X(x) = \sum_{y \in R_Y} f_Y(y) = 1.$$

x \ y	0	1	2	f_X
0	0.10	0.04	0.02	0.16
1	0.08	0.20	0.06	0.34
2	0.06	0.14	0.30	0.50
f_Y	0.24	0.38	0.38	**1.00**

The (cumulative) distribution functions of the random variables X and Y alone when these arise from the joint distribution of the pair (X, Y) are called the **marginal distribution functions**, and are denoted by F_X and F_Y, respectively.

Note that, upon using

$$F_X(x) = P(X \le x) = P(X \le x \text{ and } (X, Y) \in R_{X,Y}),$$

we simply obtain

$$F_X(x) = \sum_{s \le x} \left(\sum_{y:(s,y)\in R_{X,Y}} f(s,y) \right) = \sum_{s \le x} f_X(s).$$

Similarly, for $F_Y(y)$, we have

$$F_Y(y) = \sum_{t \le y} f_Y(t).$$

Observe that this is exactly the same relation between the probability function and the distribution function of a discrete variable in the univariate case, as seen in Volume I (see Chapter 4).

Further, we should mention that the marginal distribution functions can be found directly from the joint distribution function as

$$F_X(x) = \lim_{y \to \infty} F(x, y), \quad F_Y(y) = \lim_{x \to \infty} F(x, y). \tag{1.3}$$

To justify these expressions, note first that

$$F_X(x) = P(X \le x) = P\left(\lim_{n \to \infty} \{X \le x, Y \le n\}\right)$$

and by an appeal to the continuity property of probability, we get

$$F_X(x) = \lim_{n \to \infty} P(X \le x, Y \le n) = \lim_{n \to \infty} F(x, n) = \lim_{y \to \infty} F(x, y).$$

The proof of the second expression in (1.3) follows in an analogous manner.

Example 1.3.3 Let (X, Y) be a two-dimensional discrete random variable having the joint probability function as

$$f(x, y) = \frac{6}{3^x 4^y}, \quad x = 1, 2, \dots \text{ and } y = 1, 2, \dots$$

(i) Find the marginal probability functions of the variables X and Y.

(ii) Calculate the probability $P(X \ge 3 \text{ and } Y \ge 4)$.

SOLUTION

(i) For the marginal probability function of X, we obtain from Proposition 1.3.1 that

$$f_X(x) = \sum_{y=1}^{\infty} f(x, y) = \sum_{y=1}^{\infty} \frac{6}{3^x 4^y} = \frac{6}{3^x \cdot 4} \sum_{y=1}^{\infty} \left(\frac{1}{4}\right)^{y-1}$$

$$= \frac{6}{3^x \cdot 4} \cdot \frac{1}{1 - \frac{1}{4}} = \frac{2}{3^x}, \quad \text{for } x = 1, 2, \dots$$

Similarly, for the variable Y, we find

$$f_Y(y) = \sum_{x=1}^{\infty} f(x, y) = \sum_{x=1}^{\infty} \frac{6}{3^x 4^y} = \frac{6}{3 \cdot 4^y} \sum_{x=1}^{\infty} \left(\frac{1}{3}\right)^{x-1}$$

$$= \frac{2}{4^y} \cdot \frac{1}{1 - \frac{1}{3}} = \frac{3}{4^y}, \quad \text{for } y = 1, 2, \dots$$

(ii) Let us denote the events $\{X < 3\}$ and $\{Y < 4\}$ by A and B, respectively. Then, we see that

$$P(X \geq 3, Y \geq 4) = 1 - P(A \cup B) = 1 - P(A) - P(B) + P(AB),$$

which readily gives

$$P(X \geq 3, Y \geq 4) = 1 - P(X \leq 2) - P(Y \leq 3) + P(X \leq 2, Y \leq 3)$$
$$= 1 - F_X(2) - F_Y(3) + F(2, 3),$$

where $F_X(x)$ and $F_Y(y)$ are the marginal distribution functions of X and Y, and $F(x, y)$ is the joint distribution function of the pair (X, Y). Next, we find

$$F_X(2) = P(X \leq 2) = P(X = 1) + P(X = 2) = f_X(1) + f_X(2) = \frac{2}{3} + \frac{2}{3^2} = \frac{8}{9}$$

and

$$F_Y(3) = P(Y \leq 3) = f_Y(1) + f_Y(2) + f_Y(3) = \frac{3}{4} + \frac{3}{4^2} + \frac{3}{4^3} = \frac{63}{64},$$

using f_X and f_Y found in Part (i). In addition, we find

$$F(2, 3) = P(X \leq 2, Y \leq 3)$$
$$= f(1, 1) + f(1, 2) + f(1, 3) + f(2, 1) + f(2, 2) + f(2, 3)$$
$$= \frac{6}{3 \cdot 4} + \frac{6}{3 \cdot 4^2} + \frac{6}{3 \cdot 4^3} + \frac{6}{3^2 \cdot 4} + \frac{6}{3^2 \cdot 4^2} + \frac{6}{3^2 \cdot 4^2} = \frac{7}{8},$$

using the expression of $f(x, y)$ given in the statement. Thus, we arrive at

$$P(X \geq 3, Y \geq 4) = 1 - \frac{8}{9} - \frac{63}{64} + \frac{7}{8} = \frac{1}{576}.$$

EXERCISES

Group A

1. The joint probability function of a random vector (X, Y) is

$$f(x, y) = c(x^2 + y^2), \quad \text{for } x = 0, 1, 2 \text{ and } y = 1, 2, 3,$$

for some real constant c.

  (i) Find the value of c.

  (ii) Obtain the marginal probability functions of X and Y.

2. The joint probability function of a pair (X, Y) is

$$f(x, y) = \frac{x + y}{c}, \quad \text{for } x = 0, 1, 2, 3 \text{ and } y = 1, 2, 3,$$

for some $c \in \mathbb{R}$.

   (i) What is the value of c?

   (ii) Find the marginal probability functions of X and Y.

   (iii) Calculate

$$P(X \geq 1), \quad P(Y \leq 2), \quad P(X \geq 1, Y \leq 2), \quad P(X \geq 1 \text{ or } Y \leq 2).$$

3. Let X and Y be two random variables, each of which assumes only the values 0 and 1. It is known that their joint probability function, $f(x, y)$, satisfies

$$f(0, 1) + f(1, 1) = 2 \{ f(0, 0) + f(1, 0) \}.$$

   (i) Obtain the marginal probability function of Y, and thence find $E(Y)$ and $Var(3Y - 2)$.

   (ii) Is the information given above sufficient to determine the marginal distribution of X? Explain why or why not!

4. Suppose (X, Y) has the joint probability function as

$$f(x, y) = c \left(\frac{x}{y^2} \right), \quad x = 1, 2, 3 \text{ and } y = 1, 2.$$

After finding the value of the constant $c \in \mathbb{R}$, derive the marginal probability functions of X and Y, and thence find $E(X)$ and $E(Y)$.

5. For the pair (X, Y) of random variables in Example 1.3.3, having a joint probability function $f(x, y)$ as in the statement of the example, find a general expression for

$$P(X \geq x, Y \geq y), \quad \text{for } x = 1, 2, \dots \quad \text{and } y = 1, 2, \dots.$$

Then, verify that $P(X \geq 3, Y \geq 4)$ is as found in Part (ii) of Example 1.3.3.

6. The range of values for the two-dimensional random variable (X, Y) is the set

$$R_{X,Y} = \{ (2, 4), (3, 4), (2, 5), (3, 5) \}.$$

It is known that the following hold for the joint probability function $f(x, y)$:

$$f(3, 4) = 3 \cdot f(3, 5), \quad f(2, 4) = f(2, 5) + 2 \cdot f(3, 5), \quad f(3, 4) = \frac{f(2, 5) + f(3, 5)}{2}.$$

   (i) Obtain the values $f(x, y)$, for all $(x, y) \in R_{X,Y}$.

   (ii) Find the marginal probability functions of X and Y, and thence find their expected values.

7. An airline offers its customers, for a certain trip, travel insurance which costs either US\$20 or US\$35 (depending on the coverage of the policy). Further, the airline offers car rental for one week, which costs US\$75, US\$100 or US\$130, depending on the type of the car chosen. Let X denote the amount that a traveling passenger pays for car rental and Y be the amount that she pays for insurance (X and Y may also take a zero value if one does not opt for the corresponding services). The joint probability function of X and Y is as given in the following table:

x \ y	0	20	35
0	0.21	0.06	0.05
75	0.11	0.09	0.04
100	0.12	0.10	0.06
130	0.04	0.05	0.07

(i) Find the marginal probability functions and the marginal distribution functions of X and Y.

(ii) Find
$$P(X > 0), \quad P(X > 0 \text{ and } Y > 0), \quad P(X \geq 4Y).$$

(iii) What is the probability that a traveler pays at least

(a) US\$80,

(b) US\$100

in total for the two services combined (travel insurance and car rental)?

8. In the case of a single variable X, the events $\{X \leq a\}$ and $\{X > a\}$ are complementary for any real a; consequently, their probabilities add up to one. When we have two variables X and Y, the events

$$\{X \leq a, Y \leq b\} \text{ and } \{X > a, Y > b\}$$

are no longer complementary (explain why!).

(i) Show that

$$P(\{X > a, Y > b\}) = 1 - P(\{X \leq a\} \text{ or } \{Y \leq b\})$$

for any real a, b.

(ii) Let F be the joint distribution function of X and Y, and F_X and F_Y be the marginal distribution functions of X and Y, respectively. Use the result in Part (i) to establish that

$$P(\{X > a, Y > b\}) = 1 - F_X(a) - F_Y(b) + F(a, b).$$

(Note that the arguments needed to solve this exercise have already been used in Example 1.3.3, with the only difference being that the "$\geq$" signs in the example have been replaced by strict inequalities above.)

Group B

9. Nick tosses a coin three times. Let X be the number of heads that appear in the first toss of the coin and Y be the total number of heads in the three tosses.

   (i) What is the range, $R_{X,Y}$, of the random vector (X, Y)?

   (ii) Show that the joint probability function of (X, Y) is

   $$f(x, y) = \frac{1}{8} \binom{2}{y - x}, \quad \text{for } (x, y) \in R_{X,Y}.$$

   (iii) Find the marginal probability functions of X and Y. What do you observe?

10. The joint probability function of a random pair (X, Y) is

    $$f(x, y) = c, \quad \text{for } x = 1, 2, \ldots, y \quad \text{and } y = 1, 2, \ldots, n,$$

    where n is a given positive integer.

    (i) Show that

    $$c = \frac{2}{n(n + 1)}.$$

    (ii) Find the marginal probability functions of X and Y, and then find their expected values.

11. A pair (X, Y) of discrete random variables has joint probability function

    $$f(x, y) = \binom{n}{x} \binom{n - x}{y} p^x q^y r^{n-x-y}, \quad (x, y) \in R_{X,Y},$$

    where n is a positive integer, $p, q, r \geq 0$ are such that $p + q + r = 1$, and

    $$R_{X,Y} = \{(x, y) : x, y = 0, 1, \ldots, n \text{ and } x + y \leq n\}.$$

    Establish that the marginal distributions of X and Y are binomial with parameters (n, p) and (n, q), respectively. (The bivariate distribution corresponding to $f(x, y)$ above is called a **trinomial distribution**, and will be studied in detail in Chapter 6.)

12. The joint probability function of the pair (X, Y) is given by

    $$f(x, y) = e^{-\lambda} \frac{(\lambda p)^x}{x!} \cdot \frac{(\lambda q)^{y-x}}{(y - x)!}, \quad \text{for } x = 0, 1, 2, \ldots, y \text{ and } y = 0, 1, 2, \ldots,$$

    where $\lambda > 0$, and $p, q \in (0, 1)$ with $p + q = 1$. Show that the marginal distributions of the variables X and Y are Poisson with parameters λp and λ, respectively.

1.4 EXPECTATION OF A FUNCTION

In many situations when the joint distribution of two random variables X and Y is known, we may be interested in calculating probabilities associated with a function of these variables. For example, when X denotes the number of boys and Y denotes the number of girls in a randomly selected family from a population, we may be interested in

- the distribution, or the expected value, of the number of children in the family,

- the probability that there are more girls than boys in the family,

and so on. In the former case, we want to find the distribution, or the expectation of the random variable $X + Y$, while in the latter we want to calculate the probability $P(X - Y < 0)$.

In general, let (X, Y) be a two-dimensional random variable with joint probability function $f(x, y)$, with $(x, y) \in R_{X,Y}$, and let $h(x, y)$ be a real function of two variables. Then, $Z = h(X, Y)$ is a (one-dimensional) random variable. Finding the distribution of Z, from the joint distribution of (X, Y), may not be a straightforward task and we will take this up in Chapter 4 of the book. However, there is always an explicit expression for $E(Z)$ in terms of the joint probability function f.

To provide another motivating example, let us reconsider Example 1.2.3, wherein X and Y correspond to the numbers of customers awaiting service at the two cashier points. Then:

- the variable $Z = X + Y$ represents the total number of customers awaiting service,

- the variable $Z = \max\{X, Y\}$ corresponds to the number of waiting customers at the cashier point that has the largest queue,

- the variable $Z = \min\{X, Y\}$ corresponds to the number of waiting customers at the cashier point that has the shortest queue.

As $Z = h(X, Y)$ is a univariate discrete random variable, its expectation (provided it exists) can be found readily as

$$E[h(X, Y)] = E(Z) = \sum_{z \in R_Z} z P(Z = z) = \sum_{z \in R_Z} z f_Z(z),$$

where $f_Z(z) = P(Z = z)$ denotes the probability function of Z. However, this expression assumes knowledge of $f_Z(z)$, and so we usually employ the following result, which is a direct extension of the result corresponding to a function of a single random variable.

Proposition 1.4.1 *Let (X, Y) be a pair of discrete random variables with joint probability function $f(x, y)$, and $h(x, y)$ be a real function of two variables. Then, the expected value of the random variable $h(X, Y)$ is given by*

$$E[h(X, Y)] = \sum_{(x,y) \in R_{X,Y}} h(x, y) f(x, y),$$

assuming that the sum on the right-hand side converges absolutely (to a finite value).

Proof: The proof is omitted. □

Example 1.4.1 A pair of discrete random variables (X, Y) has range

$$R_{X,Y} = \{(1, 1), ((1, 2), (2, 1), (2, 2)\},$$

while, for the joint probability function $f(x, y)$, it is known that

$$f(2, 2) = 2f(1, 2), \quad f(1, 2) = 2f(2, 1), \quad f(2, 1) = f(1, 1).$$

(i) Calculate the values of the function $f(x, y)$, for all $(x, y) \in R_{X,Y}$, and find the marginal probability functions of X and Y.

(ii) Find, using two different ways, each of $E(X + Y)$ and $E(XY)$.

(iii) Examine which, if any, of the following hold:

$$E(X + Y) = E(X) + E(Y), \quad E(XY) = E(X)E(Y).$$

SOLUTION

(i) Let us set, for convenience, $f(1, 1) = c$. Then, from the expressions given in the statement, we have

$$f(2, 1) = c, \quad f(1, 2) = 2c, \quad f(2, 2) = 4c,$$

so that the condition

$$\sum_{x=1}^{2} \sum_{y=1}^{2} f(x, y) = 1$$

gives $8c = 1$, which means $c = 1/8$. The joint probability function and the marginal probability functions of X and Y are as given in the following table:

x \ y	1	2	f_X
1	1/8	2/8	3/8
2	1/8	4/8	5/8
f_Y	2/8	6/8	1

Table 1.5   The joint probability function of (X, Y) in Example 1.4.1

(ii) From Proposition 1.4.1, we have

$$E(X + Y) = \sum_{(x,y) \in R_{X,Y}} (x + y)f(x, y)$$

$$= (1 + 1)f(1, 1) + (1 + 2)f(1, 2) + (2 + 1)f(2, 1) + (2 + 2)f(2, 2)$$

$$= 2 \cdot \frac{1}{8} + 3 \cdot \frac{2}{8} + 3 \cdot \frac{1}{8} + 4 \cdot \frac{4}{8} = \frac{27}{8}.$$

Similarly, we have

$$E(XY) = \sum_{(x,y) \in R_{X,Y}} xyf(x, y)$$

$$= (1 \cdot 1)f(1,1) + (1 \cdot 2)f(1,2) + (2 \cdot 1)f(2,1) + (2 \cdot 2)f(2,2)$$

$$= 1 \cdot \frac{1}{8} + 2 \cdot \frac{2}{8} + 2 \cdot \frac{1}{8} + 4 \cdot \frac{4}{8} = \frac{23}{8}.$$

Another way to find the expectations above is to determine the probability function for each of the random variables $Z = X + Y$ and $W = XY$, and then use them to find the corresponding means using the formula for a single variable. For this purpose, we see that the ranges of values of Z and W are $R_Z = \{2,3,4\}$ and $R_W = \{1,2,4\}$, respectively. Concerning the probability functions of Z and W, we find from Table 1.5, first for variable Z, that

$$f_Z(2) = P(Z = 2) = P(X + Y = 2) = P(X = 1, Y = 1) = \frac{1}{8},$$

$$f_Z(3) = P(Z = 3) = P(X + Y = 3) = P(X = 1, Y = 2) + P(X = 2, Y = 1) = \frac{3}{8},$$

$$f_Z(4) = P(Z = 4) = P(X + Y = 4) = P(X = 2, Y = 2) = \frac{4}{8}.$$

Similarly, for the variable W we obtain

$$f_W(1) = P(W = 1) = P(XY = 1) = P(X = 1, Y = 1) = \frac{1}{8},$$

$$f_W(2) = P(W = 2) = P(XY = 2) = P(X = 1, Y = 2) + P(X = 2, Y = 1) = \frac{3}{8},$$

$$f_W(4) = P(W = 4) = P(XY = 4) = P(X = 2, Y = 2) = \frac{4}{8}.$$

We then find, using the formula for the expectation of a discrete univariate random variable, that

$$E(Z) = \sum_{z \in R_Z} z f_Z(z) = 2 \cdot \frac{1}{8} + 3 \cdot \frac{3}{8} + 4 \cdot \frac{4}{8} = \frac{27}{8}$$

and

$$E(W) = \sum_{w \in R_W} w f_W(w) = 1 \cdot \frac{1}{8} + 2 \cdot \frac{3}{8} + 4 \cdot \frac{4}{8} = \frac{23}{8}.$$

Clearly, these agree with the values obtained earlier.

(iii) Because the row and column totals of Table 1.5 give the probability functions of X and Y, it is easy to find their expectations. In particular, we have

$$E(X) = \sum_{x \in R_X} x f_X(x) = 1 \cdot f_X(1) + 2 \cdot f_X(2) = 1 \cdot \frac{3}{8} + 2 \cdot \frac{5}{8} = \frac{13}{8},$$

$$E(Y) = \sum_{y \in R_Y} y f_Y(y) = 1 \cdot f_Y(1) + 2 \cdot f_Y(2) = 1 \cdot \frac{2}{8} + 2 \cdot \frac{6}{8} = \frac{14}{8}.$$

It is now apparent that the equality

$$E(X + Y) = E(X) + E(Y)$$

holds; but, we do find

$$E(XY) = \frac{23}{8} \neq \frac{13}{8} \cdot \frac{14}{8} = E(X) \cdot E(Y).$$

Proposition 1.4.2 *Let $h_1(X, Y)$ and $h_2(X, Y)$ be two functions of the two-dimensional random variable (X, Y). Then, we have*

$$E\left[h_1(X, Y) + h_2(X, Y)\right] = E\left[h_1(X, Y)\right] + E\left[h_2(X, Y)\right]$$

provided the expectations on the right-hand side exist.

Proof: Let $f(x, y)$, for $(x, y) \in R_{X,Y}$, be the joint probability function of (X, Y). Let us introduce the function

$$h(x, y) = h_1(x, y) + h_2(x, y).$$

In view of Proposition 1.4.1, we can then write

$$
\begin{aligned}
E\left[h_1(X, Y) + h_2(X, Y)\right] &= E\left[h(X, Y)\right] \\
&= \sum_{(x,y) \in R_{X,Y}} h(x, y) f(x, y) \\
&= \sum_{(x,y) \in R_{X,Y}} [h_1(x, y) + h_2(x, y)] f(x, y) \\
&= \sum_{(x,y) \in R_{X,Y}} [h_1(x, y) f(x, y) + h_2(x, y) f(x, y)] \\
&= \sum_{(x,y) \in R_{X,Y}} h_1(x, y) f(x, y) + \sum_{(x,y) \in R_{X,Y}} h_2(x, y) f(x, y) \\
&= E\left[h_1(X, Y)\right] + E\left[h_2(X, Y)\right],
\end{aligned}
$$

which yields the desired result.   □

With reference to the question in Part (iii) of Example 1.4.1, the last proposition shows that $E(X + Y) = E(X) + E(Y)$ is always true (just put $h_1(X, Y) = X, h_2(X, Y) = Y$ in the proposition), meaning that

the expected value of the sum of two variables is equal to the sum of their expected values.

In contrast, and as we demonstrated in Example 1.4.1, this is not true when we consider products instead of sums for random variables.

In addition, by applying Proposition 1.4.2 to the special case when $h_1(X, Y) = aX$ and $h_2(X, Y) = bY$, with $a, b \in \mathbb{R}$, we readily obtain the following corollary.

Corollary 1.4.1 (*Expectation for a linear combination of random variables*) *For any random variables X and Y and any real numbers a and b, we have*

$$E(aX + bY) = aE(X) + bE(Y).$$

EXERCISES

Group A

1. Assume that (X, Y) has a joint probability function

$$f(x, y) = c \ 2^{x-y+2}, \quad x = 1, 2 \text{ and } y = 1, 2.$$

   (i) Find the value of c and then obtain the marginal probability functions of X and Y.

   (ii) Calculate $E(X)$, $E(Y)$ and $E(X/Y)$, and then verify that $E(X/Y) \neq E(X)/E(Y)$.

2. An examination consists of two parts, each of which has three multiple-choice questions. Each question carries 10 points. Let X be the number of points that a student gets in the first part of the exam and Y be the number of points that the same student receives in the second part of the exam.

   Suppose the joint probability function of (X, Y) is given as follows:

$$P(X = x, Y = y)$$

x \ y	0	10	20	30
0	0.02	0.06	0.02	0.10
10	0.02	0.15	0.10	0.06
20	0.01	0.10	0.14	0.01
30	0.02	0.05	0.10	0.04

   (i) What is the average total mark achieved by a student in this exam (i.e. when we add the points from both parts of the exam)?

   (ii) Find the average mark achieved by a student in the exam, if the first part has a weight of 60% and the second part has a weight of 40%.

   (iii) Suppose a student's final mark is formed by the points the student receives for only one of the two parts of the exam, which is the one in which the student performed the best. What is the expected value of the student's final mark in this case?

3. In a small ship that carries vehicles, the cost per trip for a regular car is US$20, while the cost for a truck is US$75. Let X be the number of passenger cars that the ship carries during a trip and Y be the number of trucks on the same trip. The joint probability function of X and Y is as follows:

$$P(X = x, Y = y)$$

x \ y	0	1	2	3
0	0.000	0.005	0.015	0.025
1	0.010	0.020	0.030	0.040
2	0.020	0.030	0.050	0.055
3	0.040	0.050	0.065	0.075
4	0.040	0.035	0.045	0.045
5	0.030	0.040	0.045	0.050
6	0.035	0.035	0.040	0.030

Calculate the expected amount of money that the company owning the ship would make during a trip.

4. Calculate the expectations $E(XY), E(X + Y), E(X^2 + 3Y^2 - XY)$ for each of the joint probability functions for (X, Y) below:

   (i) $f(x, y) = \dfrac{x + y}{21}, x = 1, 2$ and $y = 1, 2, 3$;

   (ii) $f(x, y) = \dfrac{1}{6}, x = 1, 2, \ldots, y$ and $y = 1, 2, 3$;

   (iii) $f(x, y) = \dfrac{1}{9}, x = 2, 3, 4$ and $y = -1, 1, 2$;

   (iv) $f(x, y) = \dfrac{x^2 + y^2}{32}, (x, y) \in \{(1, 1), (1, 4), (2, 3)\}$.

5. Let X be a random variable which takes the values $-1, 0$, and 1, each with probability $1/3$, i.e.

$$P(X = -1) = P(X = 0) = P(X = 1) = \frac{1}{3},$$

   and Y be another variable defined as

$$Y = \begin{cases} 0, & \text{if } X \neq 0, \\ 1, & \text{if } X = 0. \end{cases}$$

   Check that $E(XY) = E(X)E(Y)$.

6. Let X and Y be two discrete random variables with joint probability function

$$f(x, y) = \begin{cases} \dfrac{1}{4}, & \text{if } (x, y) \in \{(1, 2), (-1, 2), (1, 3), (-1, 3)\}, \\ 0, & \text{otherwise.} \end{cases}$$

   Show that, for these variables, $E(XY) = E(X)E(Y)$.

7. The joint probability function of X and Y is given by

$$f(0,0) = f(1,0) = f(0,1) = \frac{1}{3}.$$

   (i) Find the marginal probability functions of X and Y.

   (ii) Calculate

$$E(X), \ E(Y), \ E(X+Y), \ E(XY), \ E\left(\frac{X+1}{Y+1}\right), \ E\left(\frac{Y+1}{X+1}\right).$$

8. Suppose, for the random variables X and Y, it is known that $E(X) = 1, E(Y) = 3$, $E(XY) = 2$, $Var(X) = 2$ and $Var(Y) = 4$. Obtain the expectation for each of the following random variables:

$$X^2 + Y^2, \ (X - Y)^2, \ 4X^2 - 2Y^2, \ (X+Y)(X-Y), \ (X+Y-2)(X-2Y+1).$$

Group B

9. Nick tosses two dice. Let X be the outcome of the first die and Y be the outcome of the second one.

   (i) Find the distribution for each of

$$Z = \min\{X, Y\} \text{ and } W = \max\{X, Y\}.$$

   (ii) Obtain $E(Z)$ and $E(W)$ using two different methods.

   (iii) Verify that
$$E(Z) + E(W) = E(X) + E(Y).$$

   Can you interpret the last result intuitively?

10. (Continuation of the previous exercise) Find the joint probability function of Z and W. Thence, calculate $E(ZW)$. Is it true in this case that $E(ZW) = E(Z)E(W)$?

11. The joint probability function of X and Y is given by

$$f(x, y) = \frac{3}{2^{x+1}4^{y+1}}$$

   for $x = 0, 1, 2, \ldots$ and $y = 0, 1, 2, \ldots$.

   (i) Find the marginal probability functions of X and Y, and then calculate their expected values.

   (ii) Obtain the expected value of the product XY. What do you observe?

12. Six persons sit at random around a circular table. When person A wants to pass a written message to person B, the transfer of the message is done by the direction in which A and B are closer to one another, and let Z be the number of persons sitting between A and B in that direction. Then, find $E(Z)$.

   [Hint: Let X and Y be the places of the two persons, and then $Z = h(X, Y)$. Record in a two-way table (with respect to X and Y) the values that Z can take on, and then calculate the quantity $E(Z) = E[h(X, Y)]$.]

1.5   CONDITIONAL DISTRIBUTIONS AND EXPECTATIONS

When we consider probabilities of events in a sample space, we have seen in Volume I that the concept of conditional probability arises naturally. There are many occasions wherein we want to find the probability that an event A occurs *knowing* that another event, B, has already occurred. In this case, the conditional probability of A, given B has occurred, is given by

$$P(A|B) = \frac{P(A \cap B)}{P(B)}$$

provided $P(B) > 0$.

Suppose now we have two random variables, X and Y, with joint probability function $f(x, y)$. If we know that Y has taken the value $Y = y$, we may then be interested in calculating the conditional probabilities

$$P(X = x|Y = y)$$

for any value of X in its range. This is provided in the following definition.

Definition 1.5.1 Let (X, Y) be a pair of random variables with joint probability function $f(x, y)$ and marginal probability functions $f_X(x)$ and $f_Y(y)$. Then, the **conditional probability function** of X, given $Y = y$ (here, y is a fixed value in the range of Y such that $P(Y = y) > 0$), is denoted by $f_{X|Y}(x|y)$ and is given by

$$f_{X|Y}(x|y) = P(X = x|Y = y) = \frac{P(X = x, Y = y)}{P(Y = y)} = \frac{f(x, y)}{f_Y(y)},$$

where x takes any value such that $(x, y) \in R_{X,Y}$. Similarly, the conditional probability function of Y, given $X = x$ (here, x is in the range of X such that $P(X = x) > 0$), is given by

$$f_{Y|X}(y|x) = P(Y = y|X = x) = \frac{P(X = x, Y = y)}{P(X = x)} = \frac{f(x, y)}{f_X(x)},$$

where y takes any value such that $(x, y) \in R_{X,Y}$.

It is easy to check that the function $f_{X|Y}(x|y)$ is a (discrete, univariate) probability function. In fact, it is obvious that it takes only nonnegative values. Moreover, we recall from the definition of a marginal probability function that

$$f_Y(y) = \sum_{x:(x,y) \in R_{X,Y}} f(x, y),$$

which immediately yields

$$\sum_{x:(x,y) \in R_{X,Y}} f_{X|Y}(x|y) = \sum_{x:(x,y) \in R_{X,Y}} \frac{f(x, y)}{f_Y(y)} = 1$$

showing that $f_{X|Y}(x|y)$ is a valid probability function. It is important to keep in mind that we view this as a function of x alone, while the value of y is kept fixed (and is often assumed known, like $y = 3$). In a similar fashion, we can show that for a fixed $x \in R_X$, the function $f_{Y|X}(y|x)$ is a valid probability function.

Example 1.5.1 Let us consider again Table 1.4 in Example 1.2.3, concerning the numbers of customers waiting to be served by two cashiers in a shop. The joint probability function of the variables X and Y, representing the customers awaiting service by the first and second cashiers, respectively, is given below (this is similar to Table 1.4, but also contains the marginal probability functions of X and Y):

x \ y	0	1	2	$f_X(x)$
0	0.10	0.04	0.02	0.16
1	0.08	0.20	0.06	0.34
2	0.06	0.14	0.30	0.50
$f_Y(y)$	0.24	0.38	0.38	1.00

Suppose now we know that $Y = 0$ (no customer is waiting to be served by the second cashier). Then, according to Definition 1.5.1, the conditional probability function of X, given $y = 0$, is given by

$$f_{X|Y}(x|0) = \frac{f(x,0)}{f_Y(0)} = \begin{cases} \dfrac{0.10}{0.24} = 0.42, & x = 0, \\ \dfrac{0.08}{0.24} = 0.33, & x = 1, \\ \dfrac{0.06}{0.24} = 0.25, & x = 2, \end{cases}$$

which can be written concisely as in the following table:

x	$f_{X|Y}(x|0)$
0	0.42
1	0.33
2	0.25

This has the following interpretation. If we know that there is no one waiting to be served by the second cashier, then the probability that the queue at the first cashier is also empty has probability 42%, while there is a probability of 33% that exactly one person is waiting to be served, and a probability of 25% that exactly two persons are waiting for service.

In a similar way, we find the conditional probability function of Y, given $X = 2$, as

$$f_{Y|X}(y|2) = \frac{f(2,y)}{f_X(2)} = \begin{cases} \dfrac{0.06}{0.50} = 0.12, & y = 0, \\ \dfrac{0.14}{0.50} = 0.28, & y = 1, \\ \dfrac{0.30}{0.50} = 0.60, & y = 2, \end{cases}$$

which can be written concisely as in the following table:

y	$f_{Y|X}(y|2)$
0	0.12
1	0.28
2	0.60

We have found above the conditional probability function of X, given $Y = 0$. It should be noted that, if we instead know that $Y = 1$, then the conditional probability function of X is

$$f_{X|Y}(x|1) = \frac{f(x, 1)}{f_Y(1)} = \begin{cases} \dfrac{0.04}{0.38} = 0.105, & x = 0, \\[2mm] \dfrac{0.20}{0.38} = 0.526, & x = 1, \\[2mm] \dfrac{0.14}{0.38} = 0.369, & x = 2. \end{cases}$$

We thus observe that the conditional probability function of X is different, depending on whether $Y = 0$ or $Y = 1$. Intuitively, this means that the number of customers that wait to be served by the first cashier depends on the number of persons waiting for service by the second cashier. A formal detailed discussion of the concept of *independence* between two random variables is made in Chapter 3.

Once we have defined conditional probability functions, we may proceed to consider conditional distribution functions. Since the function

$$g(y) = f_{Y|X}(y|x) = \frac{f(x, y)}{f_X(x)} \tag{1.4}$$

is, for a given $x \in R_X$, a probability function, it corresponds to a (cumulative) distribution function

$$G(y) = \sum_{t \leq y} g(t).$$

This is called the **conditional (cumulative) distribution function** of Y, given $X = x$, and is denoted by $F_{Y|X}(y|x)$. Similarly,

$$\sum_{y:(x,y) \in R_{X,Y}} y \, g(y)$$

defines the expectation corresponding to the probability function in (1.4), and is called the **conditional expectation** of Y, given $X = x$. We denote it by $E(Y|X = x)$. Putting together all the notation above, we have the following:

$$F_{Y|X}(y|x) = P(Y \leq y|X = x) = \sum_{t \leq y} f_{Y|X}(t|x)$$

and

$$E(Y|X = x) = \sum_{y:(x,y)\in R_{X,Y}} y f_{Y|X}(y|x). \tag{1.5}$$

Of course, we can define analogously the cumulative distribution function of X, given $Y = y$ (written as $F_{X|Y}(x|y)$) and the expectation of X, given $Y = y$, denoted by $E(X|Y = y)$.

Finally, as with the usual (i.e. unconditional) expectation, the conditional expectation of a function $h(Y)$, given $X = x$, is simply given by

$$E[h(Y)|X = x] = \sum_{y:(x,y)\in R_{X,Y}} h(y) f_{Y|X}(y|x).$$

Example 1.5.2 In Example 1.5.1 we found, based on the joint probability function in Table 1.4 (people waiting to be served at two cashier points), the conditional distribution of Y, given $X = 2$, as

y	$f_{Y|X}(y|2)$
0	0.12
1	0.28
2	0.60
	1.00

In contrast with formula (1.5), which looks rather complicated, when a conditional probability function is available (as in the table above), it is very simple to work out the corresponding conditional expectation. We simply multiply each value of y in the table by the associated probability (in the same row of the table) and then sum up the values. Thus, we have

$$E(Y|X = 2) = 0 \cdot (0.12) + 1 \cdot (0.28) + 2 \cdot (0.60) = 1.48.$$

It may be noted that (in analogy with the unconditional case), even though the variable Y is integer-valued, its conditional expectation need not be so. Here we have found that, if we know that there are 2 persons awaiting service by the first cashier, the expected number of persons waiting to be served by the second cashier is 1.48.

In Example 1.5.1, we also found the conditional probability functions of X, given $Y = 0$ and $Y = 1$. From these functions, we find immediately

$$E(X|Y = 0) = 0 \cdot (0.42) + 1 \cdot (0.33) + 2 \cdot (0.25) = 0.83,$$
$$E(X|Y = 1) = 0 \cdot (0.10) + 1 \cdot (0.53) + 2 \cdot (0.37) = 1.27.$$

We observe that these two expectations are unequal, that is,

$$E(X|Y = 0) \neq E(X|Y = 1).$$

In the examples we have considered so far, we have assumed that the range of values for a random pair (X, Y) is known in advance, that is, before the random experiment is carried out. But, in several instances, this may not be the case, in particular when the range of values of one variable depends on the value that the other variable takes. The following example illustrates this issue, as well as the use of conditional expectations for making decisions.

Example 1.5.3 Suppose Karen, who is participating in a TV quiz show, has reached the final stage of the show. At this stage, she has the option to choose whether she wants to answer one, two, or three questions. If she answers all questions correctly, she wins

- US\$500 if she has chosen to answer one question,

- US\$1000 if she has chosen to answer two questions, or

- US\$2000 if she has chosen to answer three questions.

If she gives at least one wrong answer (or does not give an answer to a question at all), she receives nothing.

The probability that Karen answers a given question correctly is 0.80. What is the best strategy for Karen to maximize her expected profit from the show?

SOLUTION   Let us set X to take the values 1, 2, and 3 if Karen decides to choose one, two, and three questions, respectively. As we are interested in the amount of money she receives from the show, let us also define a variable Y representing her earnings. The ranges of values for X and Y are

$$R_X = \{1, 2, 3\}, \quad R_Y = \{0, 500, 1000, 2000\};$$

but, it is clear that for (X, Y), some combinations are not possible (for example, the pairs of values $(1, 1000)$ and $(3, 500)$ cannot occur).

The best strategy for Karen is the one that maximizes

$$E(Y|X = i)$$

for $i = 1, 2, 3$. So, we need to calculate all three conditional expectations.

For $i = 1$, the probability function of Y is given by

$$P(Y = 500|X = 1) = 0.80, \quad P(Y = 0|X = 1) = 0.20,$$

and so

$$E(Y|X = 1) = 0 \cdot P(Y = 0|X = 1) + 500 \cdot P(Y = 500|X = 1)$$

$$= 0 + 500 \cdot (0.80) = 400.$$

For $i = 2$, the values that Y can take are $Y = 0$ (if Karen does not answer both questions correctly) and $Y = 1000$. The probability that she wins US\$1000 in this case is $(0.8)^2$ (assuming implicitly that the events corresponding to her answering correctly the two questions are independent, which seems to be a reasonable assumption here). Consequently,

$$E(Y|X = 2) = 0 \cdot P(Y = 0|X = 2) + 1000 \cdot P(Y = 1000|X = 2)$$

$$= 0 + 1000 \cdot (0.80)^2 = 640.$$

Finally, for $i = 3$, the possibilities are $Y = 2000$ or $Y = 0$ depending on whether Karen answers all three questions correctly or not. The former event has probability $(0.80)^3$, assuming again that the events corresponding to her answering correctly successive questions are independent. This gives

$$E(Y|X = 3) = 0 \cdot P(Y = 0|X = 3) + 2000 \cdot P(Y = 2000|X = 3)$$
$$= 0 + 2000 \cdot (0.80)^3 = 1024.$$

We thus see that the best strategy, to maximize her expected profit, is to choose three questions to answer.

An important point to note here is that, given $X = x$, the *conditional distribution* of the variable W, denoting the number of correct answers that Karen gives, is binomial with parameters $n = x$ and $p = 0.80$. See Exercises 8 and 11 at the end of this section for variants of the present problem, where this remark could prove useful!

EXERCISES

Group A

1. The basketball teams of two Universities, A and B, have reached the Final Four of the NCAA tournament. The two teams are drawn to play in different semifinals, and so if they both win their first games, they will play against each other in the final. Suppose each team has a 60% chance of winning their semifinal game, and further suppose if one or both teams reach the final, there is an even chance for the finalists to win the trophy.

   Let X and Y be the number of wins that the teams from Universities A and B, respectively, achieve in the Final Four.

   (i) Write down the range of values for the pair (X, Y) and its joint probability function.

   (ii) Derive the marginal probability functions of X and Y.

   (iii) Obtain the conditional probability function and the conditional expectation of X, given that the second team does not qualify for the final.

2. The joint probability function of (X, Y) is

$$f(x, y) = c(x + y), \quad x = 1, 2, 3 \text{ and } y = 5, 10.$$

   Find the value of c, and then calculate the conditional expectations $E(X|Y = 5)$ and $E(X|Y = 10)$. Are these two expectations equal?

3. A box contains two balls marked "0" and two balls marked "1". Suppose we select two balls from the box without replacement, and let X and Y be the numbers on the first and second balls drawn, respectively.

   (i) Give the probability function of X.

(ii) Calculate the conditional probability functions

$$f_{Y|X}(y|0) \quad \text{and} f_{Y|X}(y|1), \ y = 0, 1,$$

and write down the joint probability function of X and Y in the form of a two-way table.

(iii) Find the conditional probability functions of X, given $Y = 0$ and $Y = 1$. Thence, calculate $E(X|Y = 0)$ and $E(X|Y = 1)$ and verify that

$$E(X|Y = 0) = 2E(X|Y = 1).$$

4. Suppose four paintings by a famous painter are on sale at a forthcoming auction. The company organizing the auction is particularly interested in the number of paintings that will be sold at a price which is at least three times as much as the asking price for a painting. Let X be the total number of paintings that will be sold at the auction and Y be the number of those that will fetch at least three times the asking price. Suppose the joint probability function of X and Y is as given in the following table (the organizers are certain that at least one painting will be sold, and so the event $\{X = 0\}$ has probability zero):

x \ y	0	1	2	3	4
1	0.03	0.02	0	0	0
2	0.05	0.05	0	0	0
3	0.10	0.15	0.17	0.10	0
4	0.07	0.08	0.10	0.05	0.03

(i) If exactly one painting is sold at the auction, what is the probability that it is sold at a price at least three times the asking price?

(ii) Let W be the number of paintings that will be sold, but will not fetch at least three times the asking price. Then, find the expected value of W.

(iii) Obtain the conditional probability function of Y, given $X = 3$, and thence find $E(Y|X = 3)$.

5. From an urn containing n tickets numbered $1, 2, \ldots, n$, suppose we select a ticket at random, and denote by X the number on that ticket. If $X = x$, we select a number Y randomly from the set $\{1, 2, \ldots, x\}$.

(i) Write down the probability function of X and the conditional probability function of Y, given $X = x$.

(ii) Obtain the joint probability function of (X, Y).

(iii) Give an expression, in terms of a sum, for the marginal distribution function of Y.

6. Consider the joint probability function of (X, Y) (see Example 1.3.3) given by

$$f(x, y) = \frac{6}{3^x 4^y}, \quad x = 1, 2, \ldots \text{ and } y = 1, 2, \ldots$$

In Example 1.3.3, we found the marginal probability functions f_X and f_Y. Using these:

(i) Calculate $E(X)$, $E(Y)$ and $E(3X - 2Y)$.

(ii) For a given $y \in \{1, 2, \ldots\}$, obtain the conditional probability function $f_{X|Y}(x|y)$. Does this depend on the value of y chosen?

(iii) For a given $x \in \{1, 2, \ldots\}$, obtain the conditional probability function $f_{Y|X}(y|x)$. Does this depend on the value of x chosen?

(iv) Obtain $E(X|Y = y)$ and $E(Y|X = x)$. What do you observe?

(Hint: You can answer most parts of this question without any calculations, by simply identifying the marginal distributions of X and Y to a well-known univariate distribution. The same also applies to the conditional probability functions you will find in Parts (ii) and (iii), hence all expectations can be found immediately by appealing to known properties of that distribution.)

7. Suppose the joint probability function of (X, Y) is as given in the following table:

x \ y	2	3	4
1	0.06		0.10
2	0.04	0.07	
3		0.15	0.25

(i) Fill the missing entries in this table if it is known that

$$f_{X|Y}(2|2) = 0.2 \quad \text{and} \quad f_{Y|X}(2|1) = 0.2.$$

(ii) Obtain $E(Y|X = x)$ for $x = 1, 2, 3$.

8. Consider the following modification of Example 1.5.3:

- If Karen decides to answer one question, she receives US$500 if she answers correctly, otherwise she receives nothing.

- If she decides to answer two questions, she receives US$300 for each correct answer she gives (that is, she gets US$0, US$300 or US600 when she gives 0, 1 and 2 correct answers, respectively).

- If she decides to answer three questions, she receives US$200 for each correct answer she gives.

Find the number of questions that Karen should choose under this scenario in order to maximize her expected earnings from the show.

9. The joint probability function of (X, Y) is given by

x \ y	1	2
5	c	$2c$
10	$3c$	$1/3$

  (i) After finding the value of c, derive the marginal probability functions of X and Y.

  (ii) Calculate the conditional probabilities $f_{X|Y}(5|y)$ for $y = 1, 2$.

  (iii) Obtain $E(X|Y = 1)$, $E(X|Y = 2)$.

10. The joint probability function of (X, Y) is given by

y x	1	2
0	a	b
1	$3a$	$1/8$
2	$2b$	$1/3$

for some suitable constants a and b.

Find the values of a and b if it is known that

$$E(X|Y = 1) = E(Y|X = 2).$$

Group B

11. Consider Example 1.5.3 again, with X denoting the number of questions that Karen chooses to answer and W denoting the number of correct answers she gives. Obtain the joint probability function of (X, W), and thence find the marginal probability function of W.

12. The number of fish caught by a fisherman in an hour is a random variable X taking values $0, 1, 2, 3, 4$ with probabilities $0.1, 0.2, 0.3, 0.25, 0.15$, respectively. A percentage of 20% among the fish caught have a weight exceeding 5 lb.

Let Y be the number of fish whose weight exceeds 5 lb caught by the fisherman.

  (i) Verify that the conditional distribution of Y, given $X = 4$, is the binomial distribution with parameters $(n = 4, p = 0.2)$. Thence, calculate $P(X = 4, Y = 3)$.

  (ii) Using arguments similar to those in Part (i), obtain the joint probability function of (X, Y).

  (iii) Derive the marginal probability function of Y.

13. The random variables X and Y have joint probability function as

$$f(x, y) = \frac{1}{e^2 y!(x - y)!}, \quad x = 0, 1, 2, \ldots, \quad y = 0, 1, 2, \ldots, x.$$

  (i) Find the marginal probability function of X and the conditional probability function of Y, given $X = x$, for $x = 0, 1, 2, \ldots$;

  (ii) Thence, obtain

$$E(Y|X = x), \quad E(Y^2|X = x), \quad x = 0, 1, 2, \ldots.$$

1.6   BASIC CONCEPTS AND FORMULAS

Joint probability function for a pair of discrete random variables	$f(x, y) = P(X = x, Y = y)$ $= P(\{\omega \in \Omega : X(\omega) = x \text{ and } Y(\omega) = y\})$
Properties of the joint probability function	• $f(x, y) = 0$ if $(x, y) \notin R_{X,Y}$; • $f(x, y) \geq 0$ for any $(x, y) \in R_{X,Y}$; • $\displaystyle\sum_{(x,y) \in R_{X,Y}} f(x, y) = 1$
Joint (cumulative) distribution function	$F(x, y) = \displaystyle\sum_{s \leq x, t \leq y} f(s, t)$
Calculation of probabilities using joint probability function	$P((X, Y) \in A) = \displaystyle\sum_{(x,y) \in A} f(x, y)$
Marginal probability functions	$f_X(x) = \displaystyle\sum_{y : (x,y) \in R_{X,Y}} f(x, y),$ $f_Y(y) = \displaystyle\sum_{x : (x,y) \in R_{X,Y}} f(x, y)$
Expectation of function of two random variables	$E[h(X, Y)] = \displaystyle\sum_{(x,y) \in R_{X,Y}} h(x, y) f(x, y)$
Expectation of a sum and linear combination	$E\left[h_1(X, Y) + h_2(X, Y)\right]$ $= E\left[h_1(X, Y)\right] + E\left[h_2(X, Y)\right]$ $E(aX + bY) = aE(X) + bE(Y)$
Conditional probability function of X, given $Y = y$	$f_{X|Y}(x|y) = \dfrac{f(x, y)}{f_Y(y)}, \quad \text{for } f_Y(y) > 0$
Conditional probability function of Y, given $X = x$	$f_{Y|X}(y|x) = \dfrac{f(x, y)}{f_X(x)}, \quad \text{for } f_X(x) > 0$
Conditional (cumulative) distribution functions	$F_{X|Y}(x|y) = P(X \leq x | Y = y) = \displaystyle\sum_{t \leq x} f_{X|Y}(t|y),$ $F_{Y|X}(y|x) = P(Y \leq y | X = x) = \displaystyle\sum_{s \leq y} f_{Y|X}(s|x)$
Conditional expectations	$E(X|Y = y) = \displaystyle\sum_{x : (x,y) \in R_{X,Y}} x f_{X|Y}(x|y);$ $E(h(X)|Y = y) = \displaystyle\sum_{x : (x,y) \in R_{X,Y}} h(x) f_{X|Y}(x|y);$ $E(Y|X = x) = \displaystyle\sum_{y : (x,y) \in R_{X,Y}} y f_{Y|X}(y|x);$ $E(h(Y)|X = x) = \displaystyle\sum_{y : (x,y) \in R_{X,Y}} h(y) f_{Y|X}(y|x)$

1.7 COMPUTATIONAL EXERCISES

1. Let us assume we have a random vector (X, Y) with joint probability function

$$f(x, y) = c(x^2 + y^2), \quad x = -5, -3, -1, 1, 3, 5, 7, \quad y = -1, 0, 1, 2, 3,$$

for a suitable real constant c.

The following set of commands in Mathematica enables us to find the value of c, to write the joint probability function in a tabular form, and then to calculate the expectation of $U = h(X, Y)$, where $h(x, y) = 3x^2y^3$.

```
In[1]:= s1:= Sum[x^2 + y^2, {x, -5, 7, 2}, {y, -1, 3}];
c = 1/s1;
f[x_, y_]:= c*(x^2 + y^2);
t = Table[f[x, y], {x, -5, 7, 2}, {y, -1, 3}];
Print["c= ", c];
Print[MatrixForm[N[t]]]; h[x_, y_]:= 3 x^2 y^3;
m:= 0; m2:= 0;
Do[m = m + f[x, y]*h[x, y]; m2 = m2 + f[x, y]*h[x, y]^2,
{x, -5, 7, 2}, {y, -1, 3}];
Print["Mean= ", N[m]];
Print["Variance= ", N[m2 - m^2]]
```

```
Out[1]= c= 1/700
(0.0371429  0.0357143  0.0371429  0.0414286  0.0485714
 0.0142857  0.0128571  0.0142857  0.0185714  0.0257143
 0.0028571  0.0014285  0.0028571  0.0071428  0.0142857
 0.0028571  0.0014285  0.0028571  0.0071428  0.0142857
 0.0142857  0.0128571  0.0142857  0.0185714  0.0257143
 0.0371429  0.0357143  0.0371429  0.0414286  0.0485714
 0.0714286  0.07       0.0714286  0.0757143  0.0828571)
Mean= 712.5
Variance= 1.3637*10^6
```

By an appropriate modification of the above, in each of the following cases find the value of c, then write the joint probability function in the form of a table, and finally find the expectation of XY:

(i) $f(x, y) = cxy, \ x = 1, 2, 3, 5, 7, \ y = 4, 8, 12, 16, 20;$

(ii) $f(x, y) = \dfrac{cx^2}{y^3}, \ x = 1, 2, 3, \ y = 1, 2, 3;$

(iii) $f(x, y) = \dfrac{x + y}{c}, \ x = 1, 2, 3, \ y = 1, 2, 3;$

(iv) $f(x, y) = c(x + y + 1)^2, \ x = -3, -2, -1, 0, 1, 2, 3, \ y = 0, 1, 2, 3, 4.$

2. In the following cases, $f(x, y)$ is the joint probability function of X and Y taking values in the specified range for each of the two variables. Use Mathematica to calculate the value of c appearing in $f(x, y)$ below:

(i) $f(x, y) = \dfrac{c}{4^{x+1}7^y}$, $x = 0, 1, 2, \ldots$, $y = 0, 1, 2, \ldots$;

(ii) $f(x, y) = \dfrac{c}{5^{x+1}9^{y+1}}$, $x, y = 0, 1, 2, \ldots$;

(iii) $f(x, y) = \dfrac{c}{5^{x+1}9^{y+4}}$, $x, y = 0, 1, 2, \ldots$;

(iv) $f(x, y) = c\,\dfrac{5^x 2^y}{x! y!}$, $x, y = 0, 1, 2, \ldots$;

(v) $f(x, y) = \dfrac{c}{x!(x-y)!}$, $x = 0, 1, 2, \ldots$, $y = 0, 1, 2, \ldots, x$.

3. A pair of discrete random variables (X, Y) has joint probability function as

$$f(x, y) = \frac{x + y + xy}{c}, \quad x = 1, 2, \ldots, 10, \quad y = 5, 6, \ldots, 25,$$

for a suitable real constant c. Identify the value of c, and then calculate

$$E(X), \ E(Y), \ E(XY), \ E\left(\frac{X}{Y}\right).$$

Compare the last two quantities with the values of $E(X)E(Y)$ and $E(X)/E(Y)$, respectively. What do you observe?

4. The joint probability function of X and Y is given by

$$f(x, y) = c(x + y), \quad x = 1, 2, \ldots, 12, \ y = 1, 2, \ldots, 12,$$

for some suitable constant c. After finding the value of c, use Mathematica to calculate

   (i) the marginal distributions of X and Y;

   (ii) the expected values of X and Y;

   (iii) the value of $E(X + Y)$ and verify that $E(X + Y) = E(X) + E(Y)$.

5. The joint probability function of X and Y in Example 1.2.3 is given in Table 1.4. Use Mathematica to

   (i) present the values of this function in a table format, as in Table 1.4;

   (ii) obtain the marginal probability functions of X and Y;

   (iii) find the conditional probability functions $f_{X|Y}(x|0)$ and $f_{Y|X}(y|2)$, and verify that the results you obtain agree with those in Examples 1.2.3 and 1.5.1.

6. Let $f(x, y)$ be the probability function of (X, Y), wherein X takes values in the set $\{1, 2, \cdots, n\}$ and for a given $X = x$, the range of Y is the set $\{1, 2, \cdots, x\}$. Write a program to calculate $E\big[g(X, Y)\big]$ for a function g of the two variables. As an application,

   (i) solve Exercise 5 of Section 1.5;

(ii) calculate

$$E(3X^2 + 2Y^4), \quad E(X/Y), \quad E(X - \sqrt{Y}), \quad E(\ln(X + Y))$$

when the joint probability function of (X, Y) is given by

$$f(x, y) = c, \quad x = 1, 2, \dots, 10, \ y = 1, 2, \dots, x$$

(you need to find the value of c first).

7. We toss a coin 10 times, and denote by X the number of times the sequence HT occurs and by Y the number of times the sequence THH occurs (here, H stands for "Heads" and T stands for "Tails", as usual).

   (i) In Mathematica, create a set having all 2^{10} outcomes for this experiment (it might be convenient to code H and T as 0 and 1, respectively).

   (ii) After identifying the range of values, $R_{X,Y}$, for (X, Y), find their joint probability function.

   (iii) Thence, calculate

   $$P(X = 1 | Y = 2) \text{ and } P(Y = 0 | X = 3).$$

8. The table below gives the joint distribution (that is, the joint probability function) of the pair (X, Y).

x \ y	0	1	2
0	2/21	3/21	4/21
1	3/21	4/21	5/21

With the following commands, we obtain the marginal probability functions of X and Y as well as all conditional distributions of X given Y and those of Y given X.

```
In[1]:= n = 2; m = 3;
t = {{2/21, 3/21, 4/21}, {3/21, 4/21, 5/21}};
Print[MatrixForm[t]]
Print["Marginal probability function of X"]
Do[Xmarg[i] = Sum[t[[i, j]], {j, 1, m}]; Print[Xmarg[i]], {i, 1, n}]
Print["Marginal probability function of Y"]
Do[Ymarg[j] = Sum[t[[i, j]], {i, 1, n}]; Print[Ymarg[j]], {j, 1, m}]
Print["Marginal probability functions of X given Y"]
Do[Print ["------"]
Do[Print[t[[i, j]]/Ymarg[j]], {i, 1, n}], {j, 1, m}]
Print["Marginal probability functions of Y given X"]
Do[Print ["------"]
Do[Print[t[[i, j]]/Xmarg[i]], {j, 1, m}], {i, 1, n}]
```

```
Out[1]=
(2/21    1/7      4/21
 1/7     4/21     5/21)
Marginal probability function of X
3/7
4/7
Marginal probability function of Y
5/21
1/3
3/7
Marginal probability functions of X given Y
-------
2/5
3/5
-------
3/7
4/7
-------
4/9
5/9
Marginal probability functions of Y given X
-------
2/9
1/3
4/9
-------
1/4
1/3
5/12
```

By a suitable modification of this program and adding some new commands as appropriate, calculate the following for the pair (X, Y) with joint probability function given in the table below:

  (i) the marginal probability functions as well as the expectation and variance for each of X and Y;

 (ii) all the conditional distributions of X given Y and those of Y given X;

(iii) all the conditional expectations and variances of X given Y and those of Y given X.

x \ y	0	1	2	3
0	7/370	6/185	17/370	11/185
1	9/370	7/185	19/370	12/185
2	11/370	8/185	21/370	13/185
3	13/370	9/185	23/370	14/185
4	3/74	2/37	5/74	3/37

1.8   SELF-ASSESSMENT EXERCISES

1.8.1   True–False Questions

1. When we toss a coin twice, let X be the number of heads and Y be the number of tails observed. Then, $P(X + Y = 2) = 1$.

2. If (X, Y) has joint probability function $f(x, y)$, then

$$\sum_{x:(x,y)\in R_{X,Y}} f(x, y) = 1.$$

3. Let (X, Y) be a discrete, two-dimensional random variable with range $R_{X,Y}$ and $F_X(x)$ be the marginal distribution function of X. Then,

$$F_X(x) = \sum_{y:(x,y)\in R_{X,Y}} P(X \le x, Y = y).$$

   Questions 4–12 below refer to a pair of random variables (X, Y) with joint probability function

$$f(0, 0) = f(0, 1) = \frac{1}{4}, \quad f(1, 0) = \frac{1}{8}, \quad f(1, 1) = \frac{3}{8}.$$

4. $P(X = 1) = \dfrac{1}{2}$.

5. $P(X = Y) = \dfrac{5}{8}$.

6. $f_{X|Y}(1|0)$ cannot be defined.

7. $P(X = 1|Y = 1) = \dfrac{3}{5}$.

8. X and Y have the same expectation.

9. The conditional expectation of X, given $Y = 1$, is $\dfrac{1}{3}$.

10. $E(X^2|Y = 0) = \dfrac{1}{3}$.

11. $E(XY) = \dfrac{3}{8}$.

12. Let $F_Y(y)$ be the marginal distribution function of Y. Then, $F_Y(1/2) = \dfrac{1}{2}$.

13. Let (X, Y) have joint probability function $f(x, y)$ and denote the marginal probability function of X by $f_X(x)$. Then,

$$\sum_{y:(x,y)\in R_{X,Y}} \frac{f(x, y)}{f_X(x)} = 1.$$

14. If X and Y are discrete random variables, then

$$E(2X^2 + 3Y^2) = 2[E(X)]^2 + 3[E(Y)]^2.$$

15. For any pair of discrete variables (X, Y),

$$E(2X - Y) + E(X + Y) = 3E(X).$$

16. An urn contains 10 black balls and 8 red balls. We select from it two balls without replacement, and let X and Y denote the number of black and red balls selected, respectively. Then,

$$E(X|Y = 1) = 1.$$

17. For a random pair (X, Y), let $F_{X|Y}(x|y)$ be the conditional distribution function of X, given $Y = y$ (assuming that $P(Y = y) > 0$). Then, the following holds for $(x, y) \in R_{X,Y}$:

$$P(X > x|Y = y) = 1 - F_{X|Y}(x|y).$$

1.8.2 Multiple Choice Questions

1. Assume that the joint probability function of (X, Y) is

$$f(x, y) = \frac{c(x + y)}{9}, \quad x = 1, 2, \ y = 2, 3, 4.$$

Then, the value of c is

(a) 3   (b) $\dfrac{1}{3}$ (c) $\dfrac{1}{27}$ (d) $\dfrac{5}{9}$ (e) $\dfrac{9}{5}$

2. Assume that the pair (X, Y) has joint probability function

$$f(1, 1) = f(1, 2) = \frac{1}{6}, \ f(2, 1) = \frac{1}{4}, \ f(2, 2) = \frac{5}{12}.$$

Then, the marginal probability function of Y is

(a) $f_Y(1) = \dfrac{2}{7}, \ f_Y(2) = \dfrac{5}{7}$

(b) $f_Y(1) = \dfrac{5}{12}, \ f_Y(2) = \dfrac{7}{12}$

(c) $f_Y(1) = \dfrac{1}{3}, \ f_Y(2) = \dfrac{2}{3}$

(d) $f_Y(1) = \dfrac{2}{3}, \ f_Y(2) = \dfrac{1}{3}$

(e) $f_Y(1) = \dfrac{2}{5}, \ f_Y(2) = \dfrac{3}{5}$

3. The joint probability function of (X, Y) is

$$f(10, 20) = f(10, 40) = \frac{1}{5}, \ f(20, 20) = \frac{1}{2}, \ f(20, 40) = \frac{1}{10}.$$

Then, $E(X + Y)$ is equal to

(a) 22.5   (b) 36   (c) 40   (d) 42   (e) 45

4. Nick tosses a die twice, and let X and Y denote the outcomes of the first and second tosses, respectively. Then, $P(X \leq 2|Y = 4)$ is

(a) $\dfrac{1}{18}$ (b) $\dfrac{1}{3}$ (c) $\dfrac{1}{6}$ (d) $\dfrac{1}{2}$ (e) $\dfrac{1}{9}$

5. Assume that each of the discrete random variables X and Y takes only the values 1 and 2, while their joint probability function is

$$f(1, 1) = \frac{4}{15}, \ f(1, 2) = \frac{1}{6}, \ f(2, 1) = \frac{1}{5}, \ f(2, 2) = \frac{11}{30}.$$

Then, $P(X = 1|Y = 2)$ is equal to

(a) $\dfrac{5}{11}$ (b) $\dfrac{5}{17}$ (c) $\dfrac{1}{6}$ (d) $\dfrac{5}{16}$ (e) $\dfrac{11}{17}$

6. Assume that the joint probability function of (X, Y) is

$$f(x, y) = \frac{x^2 + y^2}{c}, \quad x = 2, 3, \ y = 2, 3,$$

for a suitable constant c. Then, $P(Y = 3)$ equals

(a) $\dfrac{9}{31}$ (b) $\dfrac{3}{8}$ (c) $\dfrac{9}{13}$ (d) $\dfrac{31}{48}$ (e) $\dfrac{31}{52}$

7. Wendy has a green and a blue dice and she tosses them once. Let X be the outcome of the blue die and Y be the sum of the two outcomes. Then, $P(X = 3|Y = 8)$ is

(a) $\dfrac{1}{5}$ (b) $\dfrac{1}{6}$ (c) $\dfrac{1}{4}$ (d) $\dfrac{5}{36}$ (e) $\dfrac{1}{9}$

8. A box contains 5 blue chips and 4 red ones. We select two chips at random without replacement. Let X be the number of blue chips and Y be the number of red chips selected. The marginal probability function of Y is

(a) $f_Y(0) = \dfrac{25}{81}, \quad f_Y(1) = \dfrac{40}{81}, \quad f_Y(2) = \dfrac{16}{81}$

(b) $f_Y(0) = \dfrac{1}{6}, \quad f_Y(1) = \dfrac{5}{9}, \quad f_Y(2) = \dfrac{5}{18}$

(c) $f_Y(0) = \dfrac{5}{18}, \quad f_Y(1) = \dfrac{5}{9}, \quad f_Y(2) = \dfrac{1}{6}$

(d) $f_Y(0) = \dfrac{1}{3}, \quad f_Y(1) = \dfrac{5}{9}, \quad f_Y(2) = \dfrac{1}{9}$

(e) $f_Y(0) = \dfrac{5}{9}, \quad f_Y(1) = \dfrac{1}{3}, \quad f_Y(2) = \dfrac{1}{9}$

9. The joint probability function of (X, Y) is

$$f(x, y) = \frac{1}{5^x 2^{y-2}}, \quad x = 1, 2, \ldots, \ y = 1, 2, \ldots.$$

The marginal distribution of X is (for $x = 1, 2, \ldots$)

(a) $f_X(x) = \dfrac{4}{5^x}$

(b) $f_X(x) = \dfrac{2}{5^x}$

(c) $f_X(x) = \dfrac{1}{5^x}$

(d) $f_X(x) = \dfrac{2}{5^{x+1}}$

(e) $f_X(x) = \dfrac{4}{5^{x+1}}$

Questions 10–15 refer to a random pair (X, Y) whose joint probability function is as follows:

x \ y	1	2	3
1	0.10	0.15	0.15
2	0.12	0.08	0.05
3	0.06	0.14	0.15

10. $f_{X|Y}(3|2)$ is

    (a) $\dfrac{1}{5}$ (b) $\dfrac{35}{37}$ (c) $\dfrac{14}{37}$ (d) $\dfrac{2}{5}$ (e) $\dfrac{3}{7}$

11. $P(X > 1, Y \leq 2)$ equals

    (a) 0.32   (b) 0.43   (c) 0.45   (d) 0.40   (e) 0.60

12. The value of the conditional (cumulative) distribution function $F_{X|Y}(x|y)$ for $x = y = 2$ is

    (a) $\dfrac{13}{20}$ (b) $\dfrac{8}{37}$ (c) $\dfrac{23}{37}$ (d) $\dfrac{9}{20}$ (e) $\dfrac{9}{13}$

13. $E(X + Y)$ equals

    (a) 2.00   (b) 2.07   (c) 1.70   (d) 4.02   (e) 3.90

14. $E(Y|X = 1)$ equals

    (a) 0.357   (b) 1.857   (c) 2.125   (d) 2.050   (e) 2.750

15. $E(X^2 - Y^2)$ equals

    (a) 0.36   (b) 1.30   (c) -0.0824 (d) 9.46   (e) -0.36

1.9 REVIEW PROBLEMS

1. We toss a coin n times and denote by X the number of heads in the first toss and by Y the total number of heads in the n tosses.

    (i) Find the joint probability function of (X, Y).

    (ii) Derive the marginal probability functions of X and Y.

    (iii) Show that the conditional probability function of Y, given $X = x$, is a (univariate) binomial distribution, while the conditional distribution of X, given $Y = y$, is a (univariate) hypergeometric distribution.

    [Hint: For Parts (i) and (ii), generalize the arguments used in Exercise 9 of Section 1.3.]

2. Let X be a random variable having the Bernoulli distribution with success probability p, so that
$$P(X = 1) = p = 1 - P(X = 0).$$

    Define the variables $Y = X^2$ and $W = 1 - X$.

    (i) Obtain the joint probability functions of (X, Y) and (X, W).

    (ii) Calculate the means of XY and X^3 and verify that

$$E(XY) = E(X^3)$$

    (a) by using the joint distribution of (X, Y) found in Part (i), and Proposition 1.4.1 for the expectation of a function of two variables;

    (b) by finding the third moment around zero of a Bernoulli distribution.

3. Let X be the outcome in a single throw of a die. If $X = x$, we toss a coin x times, and denote the number of heads we observe in them by Y.

    (i) Obtain the joint probability function of (X, Y).

    (ii) Derive the marginal probability function of Y.

    (iii) If we know that the number of heads observed in this experiment is 2, what is the expected outcome of the die?

    [Hint: Use the distribution of X and the fact that given $X = x$, the conditional distribution of Y is binomial with parameters x and $1/2$.]

4. **(Bivariate hypergeometric distribution)** A box contains a white, b black and c red balls and we take successively, without replacement, n balls out of the box. Let X be the number of white balls that were selected and Y be the number of black balls selected. Find the joint probability function of (X, Y). The distribution of (X, Y) is a generalization of the familiar (univariate) hypergeometric distribution; this distribution will be further discussed in Chapter 6.

Application: A medical drawer contains 4 boxes with aspirins, 2 boxes with sleeping pills, and 3 boxes with multivitamins. We select two boxes at random (without replacement), and let X and Y be the numbers of boxes with aspirins and with sleeping pills, respectively, that we selected. Then, find

$$P(X + Y = 1), \quad P(X \leq 1), \quad P(X \leq 1, Y \leq 1).$$

5. Let (X, Y) be a pair of discrete random variables whose range is

$$R_{X,Y} = \{(x, y) : x \in \{x_0, x_1, \dots\}, \text{ and } y \in \{y_0, y_1, \dots\}\}.$$

Show that the joint probability function of X and Y, denoted by $f(x, y)$, is related to the joint distribution function, $F(x, y)$, of X and Y through the formula

$$f(x_i, y_j) = \begin{cases} F(x_0, y_0), & i = j = 0, \\ F(x_0, y_j) - F(x_0, y_{j-1}), & i = 0, j \geq 1, \\ F(x_i, y_0) - F(x_{i-1}, y_0), & i \geq 1, j = 0, \\ F(x_i, y_j) - F(x_{i-1}, y_j) \\ \quad -F(x_i, y_{j-1}) + F(x_{i-1}, y_{j-1}), & i \geq 1, j \geq 1. \end{cases}$$

Application: Assume that the joint distribution function of (X, Y) is

$$F(x, y) = \begin{cases} 0, & x < 0 \text{ or } y < 0, \\ 1/3, & 0 \leq x, y < 1, \\ 1/2, & 0 \leq x < 1 \text{ and } y \geq 1, \\ 2/3, & x \geq 1 \text{ and } 0 \leq y < 1, \\ 1, & x, y \geq 1. \end{cases}$$

Obtain the joint probability function of (X, Y), and thence calculate

$$P(X = 1, Y = 1), \ P(X = 1), \ P(X + Y = 1), \ P(X^2 - Y^2 = -1).$$

6. Let X and Y be two discrete random variables with ranges R_X and R_Y, respectively. If (X, Y) has joint probability function $f(x, y)$ and range $R_{X,Y}$, verify that the marginal distribution of X satisfies

$$f_X(x) = \sum_{y:(x,y) \in R_{X,Y}} f_{X|Y}(x|y) f_Y(y), \ (x, y) \in R_{X,Y}.$$

Application 1: A random variable N follows the Poisson distribution with parameter $\lambda > 0$, while for another variable X it is known that

(i) when the variable N takes the value n, then X can only take one of the values $0, 1, 2, \dots, n$,

(ii) $P(X = x | N = n) = \binom{n}{x} p^x (1 - p)^{n-x}, \ x = 0, 1, \dots, n$,

for some $p \in (0, 1)$, that is, the conditional distribution of X, given $N = n$, is binomial with parameters (n, p).

Show that the distribution of X is Poisson with parameter λp.

Application 2: The number of persons who are treated at an outpatient ward in a hospital has the Poisson distribution with parameter $\lambda = 15$. It has been estimated that 90% of the patients who are treated in the ward leave within three hours of their admittance. Find

(i) the probability that, on a given day, there are at least three patients who stay for more than three hours in the ward;

(ii) the expected number of persons treated daily and *not staying* in the ward for more than three hours.

7. Suppose in a certain town, 15% of married couples have no children, 25% have one child, 35% have two children, and 25% have three children. We assume further that male and female children are equally likely in each family and, after choosing at random a family from that population, let X and Y be the numbers of boys and girls in this family, respectively.

(i) Obtain the joint probability function of (X, Y).

(ii) What is the probability that the family does not have the same number of boys and girls?

(iii) Calculate the expected numbers of boys, girls and children in the family.

8. The joint probability function of (X, Y) is

$$f(x, y) = \frac{x + y}{21}, \quad x = 1, 2, 3, \ y = 1, 2.$$

(i) Calculate $P(X \leq 2, Y = 1)$, $P(X = Y)$, $P(X + 2Y = 5)$.

(ii) Write down the marginal probability functions of X and Y.

(iii) Find the conditional probability functions $f_{X|Y}(x|y)$ and $f_{Y|X}(y|x)$, and thence the values of $P(X = 2|Y = 2)$, $P(Y = 1|X = 1)$.

(iv) Calculate $E(X|Y = 1)$ and $E(Y|X = 3)$.

9. Nigel and Garry, who like playing chess, agree to play a series of four chess games. Let X be the number of games that Nigel wins and Y be the number of games that Garry wins. Assume that, in each game, each player has a probability of 30% to win and that there is a probability of 40% of the game ending in a draw.

Calculate the probability that

(i) there are exactly two draws in the series of four games;

(ii) Gary wins two games and Nigel wins one;

(iii) each player wins at least one game;

(iv) the series of four games ends in a draw (that is, X and Y take the same value).

Can you write down the marginal distribution of X without doing any calculations?

10. In a city there are two stores selling cars of a certain make. Let X and Y denote the numbers of cars sold weekly in the first and second stores, respectively. The joint probability function of X and Y is as follows:

x \ y	0	1	2	3
0	0.07	0.06	0.05	0.03
1	0.05	0.06	0.05	0.04
2	0.07	0.08	0.10	0.05
3	0.03	0.05	0.07	0.05
4	0	0.02	0.04	0.03

(i) What is the probability that, in a given week, the number of cars sold by the two stores is the same?

(ii) Find the probability that the first store sells exactly two cars more than the second one.

(iii) Calculate the probability that, in a given week, the total number of cars sold together by the two stores is (a) exactly 5, and (b) at least 3.

(iv) Obtain the marginal probability functions of X and Y, and thence find $E(X), E(Y)$ and $E(X + Y)$.

(v) What is the expected number of cars sold by the second store in a week when the first store has sold two cars?

11. The joint probability function of (X, Y) is

$$f(1,2) = f(1,3) = \frac{1}{7}, \quad f(2,2) = \frac{2}{7}, \quad f(2,3) = \frac{3}{7}.$$

(i) Find the marginal probability functions of X and Y.

(ii) Calculate

$$E(X), \ E(Y), \ E(X + Y), \ E(3X - 2Y), \ E(XY), \ E(X/Y).$$

(iii) Obtain the conditional probability functions of X, given Y, and of Y, given X. Thence, calculate

$$E(X|Y = 2), \quad E(X|Y = 3), \quad E(Y|X = 1), \quad E(Y|X = 2).$$

12. The random variables X and Y have joint probability function as

$$f(x, y) = \frac{1}{2^{x-1}3^y}, \quad x = 1, 2, \dots \text{ and } y = 1, 2, \dots .$$

(i) Calculate $P(X \le 2, Y > 3)$.

(ii) Show that

$$E(Y|X = x) = \frac{3}{2}$$

for all $x = 1, 2, \dots .$

13. The joint probability function of (X, Y) is given by

x \ y	10	20
2	a	$2/9$
3	$a - b$	a
4	$2a$	$2b$

for some constants a and b.

(i) Find the values of a and b if it is known that

$$2f(3, 10) + 3f(4, 20) = 2f(2, 20).$$

(ii) Calculate $E(X|Y = 10)$ and $E(X|Y = 20)$. Are these two expectations equal?

(iii) Find the value of $E(X^2 + Y)$.

1.10   APPLICATIONS

1.10.1   Mixture Distributions and Reinsurance

In certain insurance portfolios, especially those associated with catastrophic risks, e.g. due to national disasters, the insurance company may not afford to undertake the accumulated risk of potentially huge losses; in these cases, it is quite common practice the primary insurance company to transfer part of this risk to another company by *reinsurance*. One of the most common types of reinsurance treaties is *excess of loss reinsurance*; under this agreement, the (primary) insurance company pays up to a certain amount, say z_0, to the policyholder who made the claim. If a claim exceeds that amount, the excess over z_0 is paid by the reinsurer. Apparently, if a claim is less than or equal to z_0 monetary units, it does not reach the reinsurer.

Suppose now we examine the following problem, from the reinsurer's point of view: we want to know the probability that within a fixed time period, say a month, there will be at least k claims that the reinsurance company has to pay for. Let N denote the number of claims that will be made to the primary insurer and X be the number, among them, that will be above z_0 so that they reach the reinsurer. We use the generic symbol Z for the size of a claim and assume that the distribution function for the sizes of these claims is F, so that an arbitrary claim reaches the reinsurer with probability

$$p = P(Z > z_0) = 1 - F(z_0).$$

The distribution F may be either discrete or continuous, but we focus first on the variables N, X which are both discrete. The problem may then be formulated as follows. We wish to calculate the probability

$$P(X \geq k)$$

so that the reinsurance company makes provisions that this probability does not exceed a fixed level, say α, for some small value of $\alpha \in (0, 1)$.

A common assumption in this context, which is supported by empirical evidence, is that the distribution of N is Poisson (recall that the Poisson is called the distribution for *rare events*). Write λ for the parameter of this distribution. Can we use this fact to obtain the distribution of X? In order to accomplish this we first note that, given that $N = n$, the distribution of X is binomial with parameters n and p (each claim exceeds the value z_0, independently of the other claims, with the same probability p). Using the result of Exercise 6 in Section 1.9, we then obtain that the distribution of X is Poisson with parameter λp. A distribution which arises in this form is called a **mixture distribution**, while the distribution of N and the conditional distribution of X, given N, are the *components* of that mixture.

Suppose now the reinsurer poses the condition $P(X \geq k) < \alpha$, or equivalently,

$$P(X < k) \geq 1 - \alpha.$$

Since the distribution of X is Poisson with parameter λp we get

$$\sum_{x=0}^{k-1} e^{-\lambda p} \frac{(\lambda p)^x}{x!} \geq 1 - \alpha. \tag{1.6}$$

For catastrophic risks, it is typical that the distribution of individual claims is heavy-tailed. A distribution with such behavior which is used often as a model for insurance claims is the Pareto distribution, with density

$$f_Z(z) = \frac{cb^c}{(z+b)^{c+1}}, \qquad z \geq 0.$$

Here $b > 0, c > 0$ are the parameters of the distribution. Is is easy to see that the distribution function associated with this density is

$$F_Z(z) = 1 - \left(\frac{b}{z+b} \right)^c, \qquad z \geq 0.$$

For a portfolio with Pareto claims, as above, the probability that a claim reaches the reinsurer is simply

$$p = 1 - F_Z(z) = \left(\frac{b}{z+b} \right)^c,$$

and now substituting the value of p in (1.6) we arrive at the formula

$$\sum_{x=0}^{k-1} \exp\left[-\lambda \left(\frac{b}{z+b} \right)^c \right] \frac{\lambda^x \left((b/(z_0+b)) \right)^{cx}}{x!} \geq 1 - \alpha.$$

We write for simplicity $g(z_0)$ for the quantity on the left-hand side above. It is easy to check that g is a increasing function of z_0, so that the condition

$$g(z_0) \geq 1 - \alpha$$

is equivalent to the condition that z_0 exceeds a certain value, say z^*. Figure 1.1 shows the function $g(z_0)$ for $\lambda = 40, b = 2, c = 4, k = 4$.

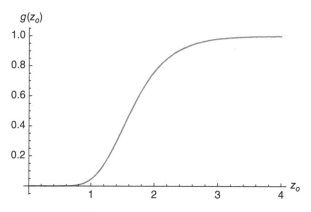

Figure 1.1 The function $g(z_0)$.

Finally, the following table gives the value z^*, which is the minimum value for z_0 so that (1.6) is satisfied, for $b = 2, c = 4$ and for different values of λ, k, and α.

α	λ	k	z^*
0.001	40	2	8.9
0.001	40	4	4.3
0.001	80	2	11.0
0.001	80	4	5.4
0.005	40	2	6.9
0.005	40	4	3.6
0.005	80	2	8.6
0.005	80	4	4.7
0.01	40	2	6.2
0.01	40	4	3.3
0.01	80	2	7.7
0.01	80	4	4.3

KEY TERMS

bivariate (or two-dimensional) distribution

conditional cumulative distribution function

conditional distribution

conditional expectation

conditional probability function

joint cumulative distribution function

joint distribution

joint probability function

marginal distribution

marginal probability function

mixture distribution

random pair (of discrete variables); or two-dimensional (discrete) random variable

CHAPTER 2

TWO-DIMENSIONAL CONTINUOUS RANDOM VARIABLES AND DISTRIBUTIONS

Auguste Bravais (Annonay, Ardèche 1811 – Haslemere, Le Chesnay, France 1863)

Auguste Bravais was born on 23 August 1811, in Annonay, Ardéche, and died on 30 March 1863, in LeChesnay, France. He was well-known for his work on crystallography, the concept of Bravais lattices and Bravais law. He also carried out work on meteorology, astronomy, statistics, and hydrography. He studied at the Collège Stanislas in Paris and then joined the École Polytechnique in 1829. In a pioneering paper in 1846, he introduced his study on correlation, while studying the existence of two errors simultaneously, and arrived at a definition of the *correlation coefficient* before Karl Pearson. In this paper, Bravais mathematically found the equation of the normal surface for the frequency of error. Making use of analytic and geometric methods, he essentially found the mathematical equations for what would later become *regression line*.

Introduction to Probability: Multivariate Models and Applications, Volume 2, First Edition.
N. Balakrishnan, Markos V. Koutras, and Konstadinos G. Politis.
© 2022 John Wiley & Sons, Inc. Published 2022 by John Wiley & Sons, Inc.

2.1   INTRODUCTION

In the preceding chapter, we discussed the joint distribution of two discrete random variables. In this chapter, our focus is on the case of two continuous random variables. We extend the concepts and tools from the univariate case to the situation when two variables vary simultaneously. In many instances, the transition from the discrete to the continuous case is straightforward and intuitively obvious; many mathematical formulas for the continuous case can be directly derived from the respective formulas for discrete variables by simply replacing the summation operators with integrals. Although it is possible, and may be of interest in some applications, that one of the two variables is discrete and the other is continuous, this is still somewhat less common and would require a careful analysis. We do not discuss this case in this book.

2.2   JOINT DENSITY FUNCTION

For the allocation of probabilities in the case of a discrete random variable, we used a joint probability function. The continuous analogue of it is a *joint density function*. Recall that, in the case of a single variable, a nonnegative real-valued function f is a density function for a continuous variable X if, for any subset A of the real line which is either an interval or can be expressed in terms of intervals via the usual set operations (unions, intersections, and complements), we have

$$P(X \in A) = \int_A f(x)dx.$$

The following definition is a natural generalization of this concept to two variables.

Definition 2.2.1 Let X and Y be two random variables defined on the same sample space. We say that X and Y have a joint continuous distribution if there exists a nonnegative function of two variables, $f : \mathbb{R} \times \mathbb{R} \mapsto [0, \infty)$, such that for any region $\Gamma (\Gamma \subseteq \mathbb{R} \times \mathbb{R})$ which can be expressed using a finite or infinitely countable number of set operations (unions, intersections, and complements) between rectangles, we have

$$P((X, Y) \in \Gamma) = \int_\Gamma \int f(x, y)dx \, dy. \tag{2.1}$$

Then, the function f is said to be the **joint density function** of the random variables X and Y, or the joint density function of the two-dimensional random variable (X, Y).

When X and Y have a joint continuous distribution, we say that the pair (X, Y) is a continuous two-dimensional (bivariate) random variable. Moreover, if X and Y have a joint continuous distribution, then each of X and Y is a continuous random variable, in the usual univariate sense. Their densities (when each variable is considered separately) can, at least in theory, be obtained from their joint density f, as discussed in Section 2.3.

Note that the double integration in (2.1) is taken over the region Γ, which can in principle be "almost' any subset of the plane. In the important special case when Γ can be written in the form of a Cartesian product

$$\Gamma = A \times B = \{(a, b) : a \in A, b \in B\},$$

where A and B are any sets that can be formed using the usual set operations between intervals of the real line, the integral in (2.1) becomes

$$P((X, Y) \in \Gamma) = \int_B \int_A f(x, y) dx\, dy = \int_A \int_B f(x, y) dy\, dx \qquad (2.2)$$

(since the integrand is nonnegative, the order of integration is immaterial). Thus, for example, if $A = (-1, 1)$ and $B = (2, 4)$, (2.2) reduces to a double Riemann integral written in a familiar form as

$$P((X, Y) \in \Gamma) = \int_2^4 \int_{-1}^1 f(x, y) dx\, dy.$$

Further, by taking $A = (-\infty, \infty)$, and $B = (-\infty, \infty)$ in (2.2), we obtain immediately that

$$1 = \int_{-\infty}^\infty \int_{-\infty}^\infty f(x, y) dx\, dy.$$

Note that this property and the nonnegativity of f are characteristic properties of a density function. Specifically, if a function f satisfies

D1. $f(x, y) \geq 0$ for any $x, y \in \mathbb{R}$,

D2. $\int_{-\infty}^\infty \int_{-\infty}^\infty f(x, y) dx\, dy = 1$,

then f is indeed a density of a continuous pair of variables (X, Y).

Taking the other extreme case in (2.2) when both A and B are singletons, say $A = \{a\}$ and $B = \{b\}$, we obtain

$$P(X = a, Y = b) = \int_b^b \int_a^a f(x, y) dx\, dy = 0. \qquad (2.3)$$

This simply means that the event the pair (X, Y) takes a specific value (a, b) is zero, which is analogous to the univariate case, namely, $P(X = x) = 0$ for any real x if X is continuous.

More generally, taking one of A and B to be a singleton, one verifies easily that

$$P(X = a, Y \in B) = P(X \in A, Y = b) = 0.$$

An important consequence of the above is that, for any $a_1 \leq a_2$ and $b_1 \leq b_2$, the value of the probability

$$P(a_1 \leq X \leq a_2, b_1 \leq Y \leq b_2)$$

remains the same if one, or more, of the "$\leq$" signs above is replaced by "$<$". Thus, we have

$$P(a_1 \leq X \leq a_2, b_1 \leq Y \leq b_2) = P(a_1 \leq X < a_2, b_1 \leq Y \leq b_2)$$
$$= P(a_1 \leq X \leq a_2, b_1 < Y < b_2)$$
$$= P(a_1 < X < a_2, b_1 < Y < b_2),$$

and so on. Similarly, the sign "$\geq$" might be replaced by "$>$" when calculating probabilities associated with two continuous variables without affecting the values of the probabilities.

There is also a natural geometric interpretation of the fact that the probability $P(X = a, Y = b)$ in (2.3) is equal to zero. Recall that the integral of the density f over a set E in the univariate case is the area under the curve $(x, f(x))$ when x takes values in E.

Similarly, the double integral on the right-hand side of (2.2) represents the *volume* under the three-dimensional surface containing the points $(x, y, f(x, y))$ when the pair (x, y) takes values in Γ. Thus, if this surface is one- or two-dimensional, it has volume zero. From this viewpoint we see that, if at least one of the sets A and B in (2.2) is finite or countably infinite (e.g. the set of positive integers), then the associated probability is zero. Furthermore, the same geometric argument shows that when X and Y have a joint continuous distribution, we have

$$P(X + Y = 1) = 0, \quad P(X^2 + Y^2 = 4) = 0,$$

and so on.

Example 2.2.1 The pair of variables (X, Y) has a joint continuous distribution with density function

$$f(x, y) = \begin{cases} cx^2y, & 0 \leq x \leq 1, 0 \leq y \leq 1, \\ 0, & \text{otherwise,} \end{cases}$$

for some constant c. Find the value of c and thence calculate $P(X > Y)$.

SOLUTION   For finding the value of c, we employ the fact that f, being a probability density, satisfies Properties D_1 and D_2. The first one implies that c must be nonnegative, while from the second we obtain,

$$1 = \int_0^1 \int_0^1 cx^2y \, dx \, dy = \int_0^1 cy \left[\frac{x^3}{3}\right]_0^1 dy = c \int_0^1 \frac{y}{3} dy = c \left[\frac{y^2}{6}\right]_0^1 = \frac{c}{6}.$$

This implies that $c = 6$. Next, to find the probability $P(X > Y)$, we express it as

$$P(X > Y) = P((X, Y) \in \Gamma),$$

where $\Gamma = \{(x, y) : 0 \leq y < x \leq 1\}$, as shown in the graph below.

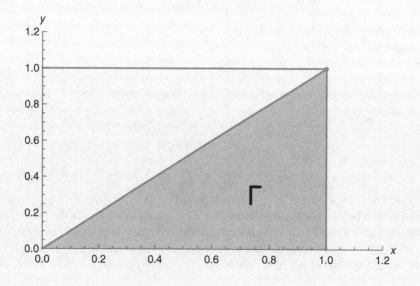

Then, we obtain

$$P(X > Y) = \int_0^1 \int_y^1 6x^2 y \, dx \, dy = \int_0^1 6y \left[\frac{x^3}{3} \right]_y^1 dy = \int_0^1 (2y - 2y^4) dy$$

$$= \left[y^2 - \frac{2y^5}{5} \right]_0^1 = 1 - \frac{2}{5} = \frac{3}{5}.$$

Note that the same result is obtained if the order of integration is reversed. More explicitly,

$$P(X > Y) = \int_0^1 \int_0^x 6x^2 y \, dy \, dx = \int_0^1 6x^2 \left[\frac{y^2}{2} \right]_0^x dx$$

$$= \int_0^1 3x^4 \, dx = \left[\frac{3x^5}{5} \right]_0^1 = \frac{3}{5}.$$

Example 2.2.2 The joint density function of X and Y is given by

$$f(x, y) = \begin{cases} cxy, & 0 < x < y < 1, \\ 0, & \text{otherwise,} \end{cases}$$

for a suitable constant c. Calculate the probabilities

$$P(X \le 1/3, \ Y \ge 3/4) \quad \text{and} \quad P(X > 1/4, \ 1/2 < Y \le 1).$$

SOLUTION Here, by using Property D_2 again, we have

$$\int_{-\infty}^{\infty} f(x, y) \, dx \, dy = \int_0^1 \int_0^y cxy \, dx \, dy = 1.$$

In order to carry out the above double integration, we note that $f(x, y)$ is nonzero only is the shaded area in the graph below, that is when $(x, y) \in A$ there.

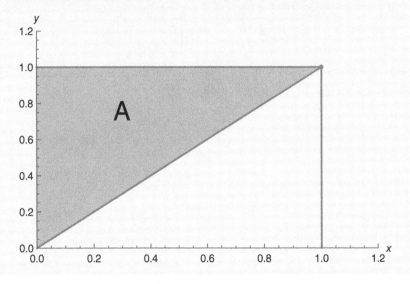

We consequently obtain that

$$\int_0^1 \int_0^y cxy \, dx \, dy = \int_0^1 cy \left(\int_0^y x \, dx \right) dy = \int_0^1 cy \left[\frac{x^2}{2} \right]_0^y dy = c \int_0^1 \frac{y^3}{2} dy$$

$$= c \left[\frac{y^4}{8} \right]_0^1 = \frac{c}{8}.$$

As this must equal one, we get $c = 8$.

In order to calculate $P(X \leq 1/3, Y \geq 3/4)$, we first observe that this is the same as $P((X, Y) \in B)$, where B is the dark-shaded rectangle shown below.

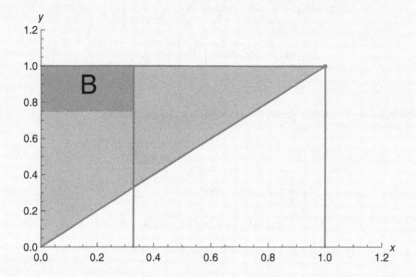

Thus, we get

$$P(X \leq 1/3, Y \geq 3/4) = \int_{3/4}^1 \int_0^{1/3} 8xy \, dx \, dy = \int_{3/4}^1 8y \left[\frac{x^2}{2} \right]_0^{1/3} dy$$

$$= \int_{3/4}^1 \frac{4y}{9} dy = \left[\frac{2y^2}{9} \right]_{3/4}^1 = \frac{7}{72}.$$

Similarly, for $P(X > 1/4, 1/2 < Y \leq 1)$, we note that the event in the brackets corresponds to the dark-shaded area Δ in the following graph.

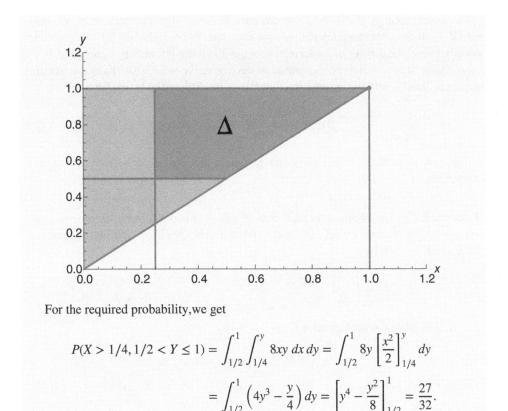

For the required probability, we get

$$P(X > 1/4, 1/2 < Y \le 1) = \int_{1/2}^{1} \int_{1/4}^{y} 8xy \, dx \, dy = \int_{1/2}^{1} 8y \left[\frac{x^2}{2} \right]_{1/4}^{y} dy$$

$$= \int_{1/2}^{1} \left(4y^3 - \frac{y}{4} \right) dy = \left[y^4 - \frac{y^2}{8} \right]_{1/2}^{1} = \frac{27}{32}.$$

When we consider "cumulative" probabilities of the form $P(X \le x, Y \le y)$, the Definition (1.1) of the joint distribution function applies in the same way as in the discrete case. Specifically, if X and Y are two continuous random variables, then the function $F : \mathbb{R} \times \mathbb{R} \mapsto [0, 1]$ defined by

$$F(x, y) = P(X \le x, Y \le y)$$

is said to be the **joint (cumulative) distribution function** of X and Y. It is apparent that

$$F(x, y) = P(X \in (-\infty, x], Y \in (-\infty, y]) = \int_{-\infty}^{y} \int_{-\infty}^{x} f(s, t) \, ds \, dt, \qquad (2.4)$$

a relation that parallels (1.1), bearing in mind that sums have now been replaced by integrals.

The above relation shows how we can pass from the density function of a random pair (X, Y) to the corresponding distribution function. We can also do the reverse; in the univariate case, the density is the derivative of the distribution function. Here, since $F(x, y)$ is a function of two variables, the notion of derivative we need is the *partial derivative*. More specifically, with the assumption that F is twice differentiable, we have

$$f(x, y) = \frac{\partial^2 F(x, y)}{\partial x \partial y} = \frac{\partial^2 F(x, y)}{\partial y \partial x} \tag{2.5}$$

(this shows, in particular, that the order in which the partial derivatives are taken is immaterial).

Example 2.2.3 The lifetimes, in thousands of hours, of two electronic components are represented by random variables, X and Y, respectively. The joint distribution function of X and Y is given by

$$F(x, y) = \begin{cases} (1 - e^{-2x})(1 - e^{-3y}), & x, y > 0, \\ 0, & \text{otherwise.} \end{cases}$$

(i) Calculate the probabilities

$$P(X \le 3/2, Y \le 2) \quad \text{and} \quad P(X > 1 \text{ or } Y > 2).$$

(ii) Obtain the joint density function of X and Y.

(iii) Suppose a system, consisting of these two devices, would only work if both components work. Then, find the probability that the system will work for more than 800 hours.

SOLUTION

(i) In terms of the joint distribution function, we simply have

$$P(X \le 3/2, Y \le 2) = F(3/2, 2) = (1 - e^{-2 \cdot 3/2})(1 - e^{-3 \cdot 2})$$

$$= (1 - e^{-3})(1 - e^{-6}) \cong 0.948.$$

For the second probability which is required, we argue as follows:

$$P(X > 1 \text{ or } Y > 2) = 1 - P(X \le 1, Y \le 2) = 1 - F(1, 2)$$

$$= 1 - (1 - e^{-2})(1 - e^{-6}) \cong 1 - 0.863 = 0.137.$$

(ii) First, we differentiate $F(x, y)$ with respect to x to obtain

$$\frac{\partial}{\partial x} F(x, y) = (1 - e^{-3y}) \frac{\partial}{\partial x} (1 - e^{-2x}) = (1 - e^{-3y})(2e^{-2x}).$$

Next, we differentiate the obtained expression with respect to y to obtain from (2.5) that

$$f(x,y) = \frac{\partial^2 F(x,y)}{\partial x \partial y} = \frac{\partial}{\partial y}\left[(1 - e^{-3y})(2e^{-2x})\right]$$

$$= 2e^{-2x}\frac{\partial}{\partial y}\left(1 - e^{-3y}\right) = 6e^{-2x}e^{-3y}.$$

(iii) Since the variables X and Y are measured in thousands of hours, the probability that the system will work for more than 800 hours is

$$P(X > 4/5, Y > 4/5),$$

so that each component still works after $(4/5) \cdot 1000 = 800$ hours. An important point to note here (used implicitly also in Part (i) above) is that the event inside the brackets is **not** the complement of the event $\{X \leq 4/5, Y \leq 4/5\}$ (because one of the components might be working after 800 hours, while the other one has failed), so that the above probability is not the same as $1 - F(4/5, 4/5)$ (for further details on this, see Exercise 8 of Section 1.3 and, for an extension, Exercise 8 at the end of the present section). Instead, in order to find the required probability, we integrate the density found in Part (ii) over the range $(4/5, \infty) \times (4/5, \infty)$, which yields

$$P(X > 4/5, Y > 4/5) = \int_{4/5}^{\infty}\int_{4/5}^{\infty} f(x,y)\,dx\,dy = \int_{4/5}^{\infty}\int_{4/5}^{\infty} 6e^{-2x}e^{-3y}\,dx\,dy$$

$$= \int_{4/5}^{\infty} e^{-3y}\left(\int_{4/5}^{\infty} 6e^{-2x}\,dx\right)dy$$

$$= \int_{4/5}^{\infty} e^{-3y}\left[-3e^{-2x}\right]_{4/5}^{\infty}dy$$

$$= \int_{4/5}^{\infty} 3e^{-3y}e^{-8/5}dy = e^{-8/5}\left[-e^{-3y}\right]_{4/5}^{\infty} = e^{-8/5}e^{-12/5}$$

$$= e^{-4}.$$

Example 2.2.4 Examine whether the function

$$F(x,y) = 1 - \lambda e^{-\lambda(x+y)}, \qquad\qquad x > 0,\ y > 0,$$

can be, for some value of $\lambda > 0$, the distribution function of a two-dimensional random variable.

SOLUTION   If, for some positive λ, the function $F(x, y)$ is a distribution function, then the function

$$f(x, y) = \frac{\partial^2 F(x, y)}{\partial x \partial y}$$

should be the density function for a pair of random variables. However, differentiating $F(x, y)$ with respect to x and y successively, we find

$$\frac{\partial^2 F(x, y)}{\partial x \partial y} = -\lambda^3 e^{-\lambda(x+y)} < 0.$$

This cannot be a joint density function, and so there does not exist a pair of variables (X, Y) with joint distribution function F as given.

Example 2.2.5   Let X and Y be the market shares of two competing products, a and b, of cereals. We regard X and Y as (continuous) random variables due to their random fluctuations from day to day. The joint density function of X and Y is given by

$$f(x, y) = \begin{cases} 27(x + y), & 0 < x < 1/3,\ 0 < y < 1/3, \\ 0, & \text{otherwise.} \end{cases}$$

   (i) Derive the joint distribution function of (X, Y).

   (ii) Calculate the probability that at least one of the two products has a market share of at least 25%.

SOLUTION   (i) As the function f is zero if at least one of its arguments is negative, it is clear from (2.4), that $F(x, y) = 0$ if $x < 0$ or $y < 0$. This is depicted by the shaded area in the following graph.

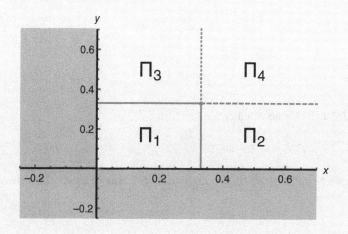

Now, let (x, y) be a point in the area Π_1 on the graph, that is, $0 \leq x < 1/3$ and $0 \leq y < 1/3$. Then, we have

$$F(x, y) = P(X \leq x, Y \leq y) = \int_{-\infty}^{x} \int_{-\infty}^{y} f(s, t) \, dt \, ds$$

$$= \int_{0}^{x} \int_{0}^{y} 27(s + t) \, dt \, ds = \frac{27xy}{2}(x + y).$$

Next, for (x, y) such that $x \geq 1/3$ and $0 \leq y < 1/3$ (area Π_2 on the graph), it is easy to see that

$$F(x, y) = P(X \leq x, Y \leq y) = \int_{0}^{1/3} \int_{0}^{y} 27(s + t) \, dt \, ds = \frac{3y(3y + 1)}{2}.$$

Similarly, for (x, y) such that $0 \leq x < 1/3$ and $y \geq 1/3$ (area Π_3 on the graph), we have

$$F(x, y) = P(X \leq x, Y \leq y) = \int_{0}^{x} \int_{0}^{1/3} 27(s + t) \, dt \, ds = \frac{3x(3x + 1)}{2}.$$

Finally, for (x, y) such that $x \geq 1/3$ and $y \geq 1/3$, which is depicted by area Π_4, we have

$$F(x, y) = P(X \leq x, Y \leq y) = \int_{0}^{1/3} \int_{0}^{1/3} 27(s + t) \, dt \, ds = 1,$$

as expected.

Combining all these, we have $F(x, y)$ as follows:

$$F(x, y) = \begin{cases} 0, & \text{if } x < 0 \text{ or } y < 0, \\[2mm] \dfrac{27xy(x + y)}{2}, & \text{if } 0 \leq x < 1/3 \text{ and } 0 \leq y < 1/3, \\[2mm] \dfrac{3y(y + 1)}{2}, & \text{if } x \geq 1/3 \text{ and } 0 \leq y < 1/3, \\[2mm] \dfrac{3x(x + 1)}{2}, & \text{if } 0 \leq x < 1/3 \text{ and } y \geq 1/3, \\[2mm] 1, & \text{if } x \geq 1/3 \text{ and } y \geq 1/3. \end{cases}$$

Observe that $F(x, y)$ is an increasing function in each of the variables x and y, and this is true for any bivariate distribution function.

(ii) The event we are interested in, expressed in words, contains the phrase "at least one", and so it is more convenient to work with the complementary event. Let us define the event

A : both products have a market share of less than 25%.

This has its probability as

$$P(A) = P(X < 1/4, Y < 1/4) = P(X \leq 1/4, Y \leq 1/4) = F(1/4, 1/4);$$

from the above distribution function $F(x, y)$, we then find

$$P(A) = F(1/4, 1/4) = \frac{27 \cdot \frac{1}{4} \cdot \frac{1}{4} \left(\frac{1}{4} + \frac{1}{4} \right)}{2} = \frac{27}{64}.$$

Hence, the desired probability that at least one product has a market share of at least 25% is

$$P(A') = 1 - P(A) = 1 - \frac{27}{64} = \frac{37}{64} \cong 57.81\%.$$

As already mentioned, for any continuous pair of random variables (X, Y), we have $P(X = a, Y = b) = 0$ for all real a, b. It is then evident that the value of the density function $f(x, y)$ at the point $x = a, y = b$ **does not express** the probability $P(X = a, Y = b)$ as one might be tempted to think, following the interpretation for bivariate discrete random variables. In fact, just as in (continuous) univariate case, $f(x, y)$ does not represent *any* probability and, in particular, can take values greater than one.

However, large values of $f(x, y)$ are in fact proportional to the probability that X takes values close to the point x and Y takes values close to y. In other words,

> The larger the value of the quantity $f(a, b)$, the larger is the probability that the pair (X, Y) takes values in a small area around the point (a, b).

To get a better insight on the above statement, let ϵ and δ be two small positive real numbers. Then,

$$P\left(a - \frac{\epsilon}{2} < X < a + \frac{\epsilon}{2}, b - \frac{\delta}{2} < Y < b + \frac{\delta}{2} \right) = \int_{a-\epsilon/2}^{a+\epsilon/2} \int_{b-\delta/2}^{b+\delta/2} f(x, y) dy dx.$$

The left-hand side gives the probability that X lies in an interval of length ϵ centered at a and Y lies in an interval of length δ centered at b. On the other hand, the right-hand side of the equation gives the volume of the rectangle

$$\left(a - \frac{\epsilon}{2}, a + \frac{\epsilon}{2} \right) \times \left(b - \frac{\delta}{2}, b + \frac{\delta}{2} \right)$$

below the surface defined by the function $f(x, y)$.

For infinitesimal $\epsilon > 0$ and $\delta > 0$ (that is, when $\epsilon \to 0$ and $\delta \to 0$), we may assume (approximately) that the values of the function f are essentially equal to $f(a, b)$. We thus arrive at the approximate formula

$$P\left(a - \frac{\epsilon}{2} < X < a + \frac{\epsilon}{2}, b - \frac{\delta}{2} < X < b + \frac{\delta}{2} \right) \cong \int_{a-\epsilon/2}^{a+\epsilon/2} \int_{b-\delta/2}^{b+\delta/2} f(a, b) dy dx$$

$$= f(a, b) \int_{a-\epsilon/2}^{a+\epsilon/2} \int_{b-\delta/2}^{b+\delta/2} dy dx$$

$$= \epsilon \delta f(a, b). \tag{2.6}$$

An alternative interpretation for $f(a, b)$ emerges if we write, in view of (2.6), that

$$f(a, b) \cong \frac{P\left(a - \frac{\epsilon}{2} < X < a + \frac{\epsilon}{2}, \; b - \frac{\delta}{2} < X < b + \frac{\delta}{2}\right)}{\epsilon \cdot \delta}$$

which readily shows that

> The value of the function f at the point $x = a, y = b$ is (approximately) equal to the ratio of the probability that the value (X, Y) falls inside a rectangle with dimensions ϵ and δ centered at the point (a, b), divided by the area of that rectangle.

Exercises

Group A

1. The value X (in dollars) and the total sales, Y (in tens of thousand units) for a product have a joint density function as

$$f(x, y) = \begin{cases} 5xe^{-xy}, & 0.2 < x < 0.4, y > 0, \\ 0, & \text{otherwise.} \end{cases}$$

   (i) Show that this satisfies Properties D_1 and D_2 for a joint density function.

   (ii) Calculate the probability that the value of the product is less than US0.3 and the total sales are at least 15000 units.

   (iii) Find the probability that the value of the product is between US0.25 and US0.3, while the total sales of the product are less than 12000 units.

2. Let (X, Y) be a random pair with joint density function $f(x, y)$. Express, in terms of $f(x, y)$, each of the following probabilities:

$$P(X > Y), \quad P(X \le |Y|), \quad P(X - Y > 1), \quad P(X + Y \le 4).$$

3. With reference to Example 2.2.5, calculate the probability that

   (i) product a has a market share of less than 25% while product b has a market share higher than 15%;

   (ii) both products have a market share higher than 20%.

4. Find the value of the constant $c \in \mathbb{R}$, such that each of the following functions is the joint density function of a bivariate continuous random variable:

   (i) $f(x, y) = \begin{cases} c(y^2 - x^2)e^{-y}, & \text{if } y \ge 0 \text{ and } |x| \le y, \\ 0, & \text{otherwise;} \end{cases}$

   (ii) $f(x, y) = \begin{cases} cx(x - y), & \text{if } 0 < x < 1 \text{ and } x > |y|, \\ 0, & \text{otherwise.} \end{cases}$

5. The joint distribution function of a bivariate random variable (X, Y) is

$$F(x, y) = \begin{cases} \left[1 - \exp(-2x^3)\right] \left[1 - \exp(-4y^4)\right], & x > 0, y > 0, \\ 0, & \text{otherwise.} \end{cases}$$

(i) Find

$$P(X \leq 1, Y \leq 2), \quad P(X = 1), \quad P(X > 1 \text{ or } Y > 3/2).$$

(ii) Calculate the probability

$$P(1 \leq X \leq 2 \text{ and } 1 \leq Y \leq 3).$$

(iii) Obtain the joint density function of X and Y.
[Hint: For Part (ii), consider the events

$$A : X \leq 2 \text{ and } Y \leq 3, \qquad B : X \leq 2 \text{ and } Y < 1,$$
$$C : X < 1 \text{ and } Y \leq 3, \qquad D : 1 \leq X \leq 2 \text{ and } 1 \leq Y \leq 3,$$

and observe that $P(A) = P(D) + P(B \cup C)$.]

6. Examine whether the function

$$F(x, y) = \begin{cases} 1 - e^{-(2x - y)}, & x > 0, y > 0, \\ 0, & \text{otherwise,} \end{cases}$$

is a joint distribution function for a pair of random variables.

Group B

7. The joint density function of a bivariate random variable (X, Y) is

$$f(x, y) = \begin{cases} 3e^{-3x - y}, & x > 0, y > 0, \\ 0, & \text{otherwise.} \end{cases}$$

Calculate

$$P(X + Y \leq 2), \quad P(X \leq Y) \quad \text{and} \quad P(X < Y^2).$$

8. For any jointly continuous pair of variables (X, Y) with joint distribution function F, show that

$$P(v < X \leq x, w < Y \leq y) = F(x, y) + F(v, w) - F(v, y) - F(x, w)$$

whenever $v < x$, $w < y$.

[A similar result appears in Part (ii) of Exercise 8 of Section 1.3. In fact, both results are valid for (X, Y) being *either* a two-dimensional continuous or a discrete variable.]

2.3 MARGINAL DISTRIBUTIONS

In Chapter 1, we saw that the joint probability function of two random variables also provides the marginal probability functions. As we will see, the same is true for the case of two continuous variables, X and Y (replacing now probability functions by density functions). The problem here is how, using the joint density function of X and Y, we can determine the (usual) univariate densities of X and Y, which we denote by $f_X(x)$ and $f_Y(y)$, respectively.

To begin with, let A be a subset of the real line, which can be expressed as a union of (finite or infinite) intervals. Then, since f_X is the density of X, we have

$$P(X \in A) = \int_A f_X(x)dx. \tag{2.7}$$

However,

$$P(X \in A) = P(X \in A, Y \in (-\infty, \infty)),$$

so that, using the definition of a joint probability density function, we obtain

$$P(X \in A) = \int_A \int_{-\infty}^{\infty} f(x,y)dy\, dx = \int_A \left(\int_{-\infty}^{\infty} f(x,y)dy \right)\, dx.$$

Upon comparing this expression with (2.7), we see that the density, f_X, of X can be expressed through the joint density f of X and Y as

$$f_X(x) = \int_{-\infty}^{\infty} f(x,y)dy, \qquad\qquad x \in \mathbb{R}.$$

In a similar manner, for the density, f_Y, of Y we have

$$f_Y(y) = \int_{-\infty}^{\infty} f(x,y)\, dx, \qquad\qquad y \in \mathbb{R}.$$

The above relations are similar to those given in Proposition 1.3.1 for the discrete case. The difference is that sums have now been replaced by integrals. An easy way to remember the last two results is to keep in mind that, for a given joint density f,

- the density of X is found by integrating f over all possible values of y, and

- the density of Y is found by integrating f over all possible values of x.

As in the discrete case, when f_X and f_Y are found from the joint density function as above, they are called the **marginal density functions** of X and Y, respectively.

We thus have the following proposition.

Proposition 2.3.1 *Let (X, Y) be jointly continuous random variables with joint density $f(x, y)$. Then, the density functions of the variables X and Y are given by the formulas*

$$f_X(x) = \int_{-\infty}^{\infty} f(x, y) dy, \qquad\qquad x \in \mathbb{R},$$

and

$$f_Y(y) = \int_{-\infty}^{\infty} f(x, y) \, dx, \qquad\qquad y \in \mathbb{R}.$$

Example 2.3.1 Jim, who is a first year student, is taking a Calculus course and a Probability course for his studies. The *proportion* of time that he spends on studying for these courses during the semester, represented by X (for Calculus) and Y (for Probability), have a joint density

$$f(x, y) = \begin{cases} 12(x + y^2), & 0 \le x \le 1/2, \ 0 \le y \le 1/2, \\ 0, & \text{otherwise.} \end{cases}$$

The probability that he spends no more than 25% of his time studying each of these two courses is then

$$F(1/4, 1/4) = P\left(X \le \frac{1}{4}, Y \le \frac{1}{4}\right) = 12 \int_0^{1/4} \int_0^{1/4} (x + y^2) \, dx \, dy$$

$$= 12 \int_0^{1/4} \left(\int_0^{1/4} x \, dx\right) dy + 12 \int_0^{1/4} \left(\int_0^{1/4} dx\right) y^2 dy$$

$$= 12 \int_0^{1/4} \left[\frac{x^2}{2}\right]_0^{1/4} dy + 12 \int_0^{1/4} \frac{1}{4} y^2 dy$$

$$= 12 \int_0^{1/4} \frac{1}{32} dy + 3 \left[\frac{y^3}{3}\right]_0^{1/4} = \frac{7}{64} \cong 11\%.$$

The marginal density functions of X and Y are obtained as

$$f_X(x) = \int_{-\infty}^{\infty} f(x, y) dy = \int_0^{1/2} 12(x + y^2) dy = 6x + \frac{1}{2}, \quad 0 \le x \le 1/2,$$

and

$$f_Y(y) = \int_{-\infty}^{\infty} f(x, y) \, dx = \int_0^{1/2} 12(x + y^2) \, dx = 6y^2 + \frac{3}{2}, \quad 0 \le y \le 1/2.$$

Thus, the probability that Jim devotes at least 25% of his time to study Calculus is

$$P(X \ge 1/4) = \int_{1/4}^{1/2} \left(6x + \frac{1}{2}\right) dx = \frac{11}{16} \cong 69\%,$$

while the probability that he spends no more than 25% of his studying time on Probability is

$$P(Y \leq 1/4) = \int_0^{1/4} \left(6y^2 + \frac{3}{2}\right) dy = \frac{13}{32} \cong 41\%.$$

Finally, by combining the results above, we find the probability that Jim devotes less than 25% of his time in at least one of the two subjects.

$$P\left(X \leq \frac{1}{4} \text{ or } Y \leq \frac{1}{4}\right) = P\left(X \leq \frac{1}{4}\right) + P\left(Y \leq \frac{1}{4}\right) - P\left(X \leq \frac{1}{4} \text{ and } Y \leq \frac{1}{4}\right)$$

$$= \left(1 - \frac{11}{16}\right) + \frac{13}{32} - \frac{7}{64} = \frac{39}{64} \cong 61\%.$$

As in Chapter 1, the cumulative distribution functions of X and Y arising from the joint distribution function, $F(x, y)$, of the pair (X, Y) are called **marginal distribution functions**.
It is clear that

$$F_X(x) = P(X \leq x) = P(X \leq x, -\infty < Y < \infty)$$

$$= \int_{-\infty}^x \left(\int_{-\infty}^\infty f(s, y) dy\right) ds = \int_{-\infty}^x f_X(s) ds$$

and

$$F_Y(y) = P(Y \leq y) = P(-\infty < X < \infty, Y \leq y)$$

$$= \int_{-\infty}^y \left(\int_{-\infty}^\infty f(x, t) \, dx\right) dt = \int_{-\infty}^y f_Y(t) dt,$$

which are in fact the well-known formulas for the univariate case (see, e.g. Chapter 6 in Volume I).
The marginal distribution functions can also be obtained from the joint distribution function as

$$F_X(x) = \lim_{y \to \infty} F(x, y), \qquad F_Y(y) = \lim_{x \to \infty} F(x, y).$$

These can be shown to be true by arguing exactly as in the discrete case (see the discussion before Example 1.3.3).

Example 2.3.2 John and Paula are waiting at a bus stop for different buses. Let X and Y be the time (in hours) until John's and Paula's buses arrive, respectively. Suppose they have a joint distribution function

$$F(x, y) = \left(1 - e^{-5x}\right)\left\{1 - (6y + 1)e^{-6y}\right\}, \qquad x \geq 0, y \geq 0,$$

while $F(x, y) = 0$ if $x < 0$ or $y < 0$.

(i) Find the marginal distribution functions of X and Y.

(ii) Find the probability that both wait for more than 10 minutes.

(iii) Obtain the marginal densities of X and Y. What is the joint density function of (X, Y)?

SOLUTION

(i) The marginal distribution functions $F_X(x)$ and $F_Y(y)$ are easily found from $F(x, y)$ by taking the limits as $y \to \infty$ and $x \to \infty$, respectively. Thus, we get

$$F_X(x) = \lim_{y \to \infty} F(x, y) = \lim_{y \to \infty} \left[\left(1 - e^{-5x} \right) \left\{ 1 - (6y + 1)e^{-6y} \right\} \right] = 1 - e^{-5x},$$

$x \geq 0$,

and

$$F_Y(y) = \lim_{x \to \infty} F(x, y) = \lim_{x \to \infty} \left[\left(1 - e^{-5x} \right) \left\{ 1 - (6y + 1)e^{-6y} \right\} \right]$$
$$= 1 - (6y + 1)e^{-6y}, \qquad\qquad y \geq 0.$$

(ii) Let us define the events A and B

A: John waits for 10 minutes at most until his bus arrives;
B: Paula waits for 10 minutes at most until her bus arrives.

Then, the event that they both wait for more than 10 minutes can be expressed as $A'B'$, which gives (as time is measured in hours, a 10-minute period corresponds to $x = y = 1/6$),

$$P(A'B') = P\left((A \cup B)'\right) = 1 - P(A) - P(B) + P(AB)$$
$$= 1 - F_X(1/6) - F_Y(1/6) + F(1/6, 1/6)$$
$$= 1 - \left(1 - e^{-5/6} \right) - \left(1 - 2e^{-1} \right) + \left(1 - e^{-5/6} \right)\left(1 - 2e^{-1} \right)$$
$$= 2e^{-11/6} \cong 0.3198,$$

that is, there is about a 32% chance that they both wait for more than 10 minutes for their buses to arrive.

(iii) The marginal densities of X and Y are simply found by differentiating the corresponding marginal distribution functions. Thus, we get the marginal density of X as

$$f_X(x) = F'_X(x) = 5e^{-5x}, \qquad\qquad x \geq 0,$$

and that of Y as

$$f_Y(y) = F'_Y(y) = 36ye^{-6y}, \qquad\qquad y \geq 0.$$

We observe the following to be true in this example:

$$f(x,y) = f_X(x)f_Y(y)$$

and

$$F(x,y) = F_X(x)F_Y(y)$$

for all real x, y. Neither of them are true in general; they are properties for a special case of two-dimensional random variables, which will be discussed in detail in the next chapter.

Exercises

Group A

1. Obtain the marginal density functions for the following joint density functions for a pair (X, Y):

   (i) $f(x,y) = \begin{cases} 4x^2 e^{-2x-y}, & x, y > 0, \\ 0, & \text{otherwise}; \end{cases}$

   (ii) $f(x,y) = \begin{cases} (xy)^{-2}, & \text{if } x, y \geq 1, \\ 0, & \text{otherwise}. \end{cases}$

2. The joint density function of (X, Y) is

   $$f(x,y) = \begin{cases} 2e^{-x-2y}, & x, y > 0, \\ 0, & \text{otherwise}. \end{cases}$$

   (i) Find the marginal density functions of X and Y.

   (ii) Calculate $P(X > 1)$, $P(Y \leq 2)$, $P(X > 1, \ Y < 2)$ and $P(X > 1 \text{ or } Y < 2)$.

3. Consider the following two joint density functions of (X, Y):

   (i) $f(x,y) = \begin{cases} 2(x+y-2xy), & 0 \leq x \leq 1, \quad 0 \leq y \leq 1, \\ 0, & \text{otherwise}; \end{cases}$

   (ii) $f(x,y) = \begin{cases} 1, & 0 \leq x \leq 1, \quad 0 \leq y \leq 1, \\ 0, & \text{otherwise}. \end{cases}$

   Show that, in both cases, the marginal distributions of X and Y are uniform distributions over the interval $[0, 1]$. What can we conclude from this result? (See also Exercise 8 in this section).

4. The joint density function of (X, Y) is given by

   $$f(x,y) = \begin{cases} \dfrac{3x(x+y)}{5}, & 0 < x < 1 \text{ and } 0 < y < 2, \\ 0, & \text{otherwise}. \end{cases}$$

(i) Find the probability

$$P(0 < X < 1/2,\ 1 < Y < 2).$$

(ii) Derive the marginal density functions $f_X(x)$ and $f_Y(y)$.

(iii) Obtain the joint distribution function $F(x, y)$.

(iv) Find the marginal distribution functions $F_X(x)$ and $F_Y(y)$.

5. The joint distribution function of a continuous random pair (X, Y) is

$$F(x, y) = \left(1 - e^{-2x^4}\right)\left(1 - e^{-5y^4}\right), \qquad\qquad x > 0, y > 0.$$

Find the marginal distribution functions and the marginal density functions of X and Y.

6. The joint distribution function of lifetimes X and Y for two lightbulbs, in thousands of hours, is given by

$$F(x, y) = \begin{cases} \left[1 - \exp(-x^3/3)\right]\left[1 - \exp(-2y^4)\right], & x > 0, y > 0, \\ 0, & \text{otherwise.} \end{cases}$$

(i) Obtain the marginal distribution functions of X and Y.

(ii) What is the probability that

    a. both light bulbs last for at least 500 hours?

    b. at least one of the light bulbs works for at least 500 hours?

(iii) Find the marginal density functions of X and Y. What is the joint density function of (X, Y)?

Group B

7. The joint distribution function of lifetimes X and Y (in days) of two bacteria found in the human body is

$$f(x, y) = \begin{cases} e^{-x(y+1)}, & x \geq 0,\ 0 \leq y \leq b, \\ 0, & \text{otherwise,} \end{cases}$$

for a suitable constant b.

(i) Find the value of b and show that $E(X) = 1 - e^{-1}$.

(ii) Does

$$f(x, y) = f_X(x)f_Y(y)$$

hold for all (x, y) in the range of (X, Y)?

(iii) Find

$$P(1 \leq X \leq 2),\quad P(X \leq 1,\ Y \leq 1),\quad P(X > 1 | Y \leq 1).$$

(iv) Does

$$F(x, y) = F_X(x)F_Y(y)$$

hold for all (x, y) in the range of (X, Y)?

8. Suppose g and h are two (univariate) density functions with corresponding distribution functions G and H, respectively, and let $\alpha \in \mathbb{R}$ be such that $|\alpha| \leq 1$. Let

$$f_\alpha(x, y) = g(x)h(y) \left[1 + \alpha\{2G(x) - 1\}\{2H(y) - 1\} \right], \qquad x, y \in \mathbb{R}.$$

(i) Verify that f_α is the joint density function of a two-dimensional random variable (X, Y).

(ii) Show that the marginal density functions of X and Y are g and h, respectively.

Observe that we have formed a family of joint densities

$$f_\alpha(x, y), \quad -1 \leq \alpha \leq 1,$$

so that each member of this family has the same marginal densities g and h.

This family of bivariate distributions is called the *Farlie–Gumbel–Morgenstern family*; see Kotz et al. (2000).

2.4 EXPECTATION OF A FUNCTION

As in the discrete case there are many instances where, given two continuous variables X and Y, we may be interested in the expected value of a function of these variables. For example, if X and Y represent the financial returns from two investments A and B, we may wish to know what the expected profit (or loss) from this joint venture will be. The total return (profit if it is positive, or loss if it is negative) is $X + Y$, and so we are interested in $E(X + Y)$ in this case. Similarly, if X denotes the income a company makes in a given month, and Y is the company's total expenses for that month, then $X - Y$ represents the profit that the company makes and $E(X - Y)$ is the expected profit for that month. As another example, let X and Y be the future lifetimes of two electronic devices in a system. Then, assuming that the system works as long as at least one of the two devices operates, the lifetime of the system is $Z = \max\{X, Y\}$, and the expected life of the system is $E(Z) = E(\max\{X, Y\})$.

Now, suppose we want to find the expectation of a function $h(X, Y)$ of two variables X and Y. The following result, which is the continuous analogue of Proposition 1.4.1, shows that all we need to know is the joint density function, $f(x, y)$, of (X, Y).

Proposition 2.4.1 *Let (X, Y) be a pair of continuous random variables with joint density function $f(x, y)$, and let $h(x, y)$ be a real function of the two variables. Then, the expectation of the random variable $h(X, Y)$ is given by*

$$E[h(X, Y)] = \int_{-\infty}^{\infty} \int_{-\infty}^{\infty} h(x, y)f(x, y) \, dx \, dy$$

provided that the double integral converges absolutely.

Proof: The proof of this result is omitted. $\qquad\qquad\qquad\qquad\qquad\qquad\qquad\qquad\square$

Note that this formula enables us to find the expected value of $h(X, Y)$ without knowing its probability density function. In fact, it is generally not easy to find the density of $h(X, Y)$ from the joint density $f(x, y)$, especially when h is a complicated function. For simple functions, and in particular, for $h(X, Y) = X + Y$, we will discuss the problem further in Chapter 4.

Example 2.4.1 Thelma and Louise make an appointment to have lunch together at 1pm on the following day. The earliest each may arrive at the appointment is $12 : 50$pm. Assume that the time, in hours after $12 : 50$pm, that each arrives at the appointment is a random variable, with X for Thelma and Y for Louise, and that the joint density function of X and Y is

$$f(x, y) = 192x^2 y, \qquad\qquad 0 \leq x \leq 1/2,\ 0 \leq y \leq 1/2.$$

Calculate the expected time (in hours) that whoever arrives first will have to wait for the other.

SOLUTION   The required quantity is

$$E(|X - Y|) = \int_{-\infty}^{\infty} \int_{-\infty}^{\infty} |x - y| f(x, y)\, dx\, dy.$$

For finding this, we split the range of integration for one of the variables, say x, into $[0, y]$ and $(y, 1/2)$ since the sign of the difference $x - y$ depends on which of these intervals x belongs to. Thus, we obtain

$$E(|X - Y|) = \int_0^{1/2} \int_0^{1/2} |x - y| 192 x^2 y\, dx\, dy$$

$$= \int_0^{1/2} \left\{ \int_0^y (-x + y) 192 x^2 y\, dx + \int_y^{1/2} (x - y) 192 x^2 y\, dx \right\} dy$$

$$= \int_0^{1/2} \left\{ \left[-48 x^4 \right]_0^y y + \left[64 x^3 \right]_0^y y^2 + \left[48 x^4 \right]_y^{1/2} y - \left[64 x^3 \right]_y^{1/2} y^2 \right\} dy$$

$$= \int_0^{1/2} \left(-48 y^5 + 64 y^5 + 3y - 48 y^5 - 8 y^2 + 64 y^5 \right) dy$$

$$= \int_0^{1/2} \left(32 y^5 + 3y - 8 y^2 \right) dy$$

$$= \left[\frac{32 y^6}{6} + \frac{3 y^2}{2} - \frac{8 y^3}{3} \right]_0^{1/2} = \frac{1}{8}.$$

As time is measured in hours here, this means that the expected waiting time for the person who arrives first at the meeting is $60/8 = 7.5$ minutes.

Proposition 1.4.2, about the expectation for the sum of two functions for a bivariate random variable, is also valid for continuous variables; that is, if X and Y are two continuous variables, then for any functions h_1 and h_2, we have

$$E\left[h_1(X, Y) + h_2(X, Y)\right] = E\left[h_1(X, Y)\right] + E\left[h_2(X, Y)\right].$$

For proving this, repeat the steps in the proof of Proposition 1.4.2 by replacing the (double) sums with (double) integrals. A special case of this, which is of particular interest, is when $h_1(X, Y) = aX, h_2(X, Y) = bY$, for some $a, b \in \mathbb{R}$. Then, we arrive at

$$E(aX + bY) = aE(X) + bE(Y).$$

Exercises

Group A

1. The joint density function of (X, Y) is

$$f(x, y) = \begin{cases} c(x^2 + y^4), & 0 \le x \le 1 \text{ and } 0 \le y \le 1, \\ 0, & \text{otherwise.} \end{cases}$$

   (i) What is the value of c?

   (ii) Find the marginal density functions of X and Y.

   (iii) Find $E(X)$, $E(3X + 1)$, $E(X^2 + Y)$, and $E(XY)$.

2. The random variables X and Y have joint density function

$$f(x, y) = \begin{cases} ce^{-2x-4y}, & x, y > 0, \\ 0, & \text{otherwise.} \end{cases}$$

   (i) Find the value of c.

   (ii) Obtain the marginal density functions of X and Y.

   (iii) Verify that
$$E(XY) = E(X)E(Y).$$

3. The dimensions of a rectangular plot are random variables, X and Y, with joint density function

$$f(x, y) = \begin{cases} cx^2y^2, & 0 \le x \le 1 \text{ and } 0 \le y \le 1, \\ 0, & \text{otherwise.} \end{cases}$$

   Find $E(X), E(Y)$, $E(XY)$ and $E(X^2Y^2)$, and thence find the expected value of the perimeter and the area of the rectangle.

4. Let the joint density of (X, Y) be

$$f(x, y) = \begin{cases} cx(x + y), & 0 \le x \le 2, \ 0 \le y \le 2, \\ 0, & \text{otherwise.} \end{cases}$$

(i) Show that $c = 3/28$.

(ii) Find the probability $P(X < Y)$.

(iii) Does $E(XY) = E(X)E(Y)$ hold?

5. A certain mixture of copper contains two ingredients, A and B. Let X and Y be two continuous variables representing the weights, in grams, of the substances A and B, respectively, found in 1 kg of copper. The joint density of (X, Y) is

$$f(x,y) = \begin{cases} c\,\sqrt{x}(x+y), & 0 \le x \le 3,\ 0 \le y \le 5, \\ 0, & \text{otherwise.} \end{cases}$$

(i) Show that $c = \left(43\sqrt{3}\right)^{-1}$.

(ii) Find $P(X < Y)$.

(iii) Find $E(X + Y)$ and $E(2X - 3Y^2)$.

Group B

6. The random variables X and Y have joint density

$$f(x,y) = \begin{cases} 1, & |x| < y < 1, \\ 0, & \text{otherwise.} \end{cases}$$

(i) Find the marginal densities of X and Y.

(ii) Examine whether it holds that $E(XY) = E(X)E(Y)$.

7. Assume that (X, Y) have joint density

$$f(x,y) = \begin{cases} cxe^{-7y}, & 0 < x < a,\ 0 < y < \infty, \\ 0, & \text{otherwise,} \end{cases}$$

for some suitable constants a and c (which do not depend on x, y).

(i) Find an expression that a and c must satisfy so that the function f is a valid joint density function.

(ii) Show that $E(XY) = E(X)E(Y)$.

(iii) Determine the values of a and c if it is known that $E(X - 2Y) = 0$.

2.5   CONDITIONAL DISTRIBUTIONS AND EXPECTATIONS

Let (X, Y) be a pair of jointly continuous random variables with joint density function $f(x, y)$. Assume that the marginal density of X is $f_X(x)$, and let x be a real number such that $f_X(x) \ne 0$. Writing the familiar formula

$$f_X(x) = \int_{-\infty}^{\infty} f(x, y)dy$$

in the form

$$\int_{-\infty}^{\infty} \frac{f(x, y)}{f_X(x)} dy = 1,$$

we observe that the function

$$g(y) = \frac{f(x, y)}{f_X(x)}, \qquad y \in \mathbb{R},$$

is a density function of a continuous random variable. Analogous to the discrete case, the distribution having this density is called the **conditional distribution** of Y, given $X = x$. The difference with the discrete case that we have to keep in mind, is that none of the three functions in the above expression represents a probability, and so our intuitive understanding in this case is somewhat more subtle.

Even though the right-hand side depends on both x and y, g is defined as a function of y alone. This is because, conditioning on the event $\{X = x\}$, we regard the value of x as fixed. The standard mathematical notation for the conditional density of X, given $Y = y$, is

$$f_{X|Y}(x|y) = \frac{f(x, y)}{f_Y(y)}, \qquad x \in \mathbb{R},$$

which makes it explicit that the considered density is that of X, given $Y = y$. In a similar way, we may define the conditional density of Y, given $X = x$. We thus have the following definition.

Definition 2.5.1 Let (X, Y) be a pair of continuous random variables with joint density $f(x, y)$ and marginal densities $f_X(x)$ and $f_Y(y)$, respectively. For a given value y such that $f_Y(y) \neq 0$, the **conditional density function** of X, given $Y = y$, is defined as

$$f_{X|Y}(x|y) = \frac{f(x, y)}{f_Y(y)}, \qquad x \in \mathbb{R}.$$

Similarly, for a given x such that $f_X(x) \neq 0$, the conditional density function of Y, given $X = x$, is defined as

$$f_{Y|X}(y|x) = \frac{f(x, y)}{f_X(x)}, \qquad y \in \mathbb{R}.$$

Example 2.5.1 (Marginal normal distributions)
Let us consider the pair (X, Y) of continuous variables with joint density function

$$f(x, y) = \frac{1}{4\pi} \exp\left[-\frac{1}{2}\left(x^2 + \frac{y^2}{4} \right) \right], \qquad x, y \in \mathbb{R}.$$

The marginal density function of X is given by

$$f_X(x) = \int_{-\infty}^{\infty} f(x, y) dy = \frac{1}{4\pi} e^{-x^2/2} \int_{-\infty}^{\infty} e^{-y^2/8} dy.$$

We now observe that the function

$$h(y) = \frac{1}{2\sqrt{2\pi}} \exp\left(-\frac{1}{2} \cdot \frac{y^2}{2^2} \right) = \frac{1}{2\sqrt{2\pi}} e^{-y^2/8}, \qquad y \in \mathbb{R},$$

is the density function of a normal random variable with mean $\mu = 0$ and variance $\sigma^2 = 4$. Thus, we have

$$f_X(x) = \frac{1}{4\pi} e^{-x^2/2} \int_{-\infty}^{\infty} 2\sqrt{2\pi} \; h(y) dy = \frac{1}{\sqrt{2\pi}} e^{-x^2/2} \int_{-\infty}^{\infty} h(y) dy$$

$$= \frac{1}{\sqrt{2\pi}} e^{-x^2/2} \cdot 1,$$

that is,

$$f_X(x) = \frac{1}{\sqrt{2\pi}} e^{-x^2/2}, \qquad\qquad x \in \mathbb{R},$$

which means that $X \sim N(0, 1)$. Consequently, the conditional density of Y, given that $X = x$ is

$$f_{Y|X}(y|x) = \frac{f(x,y)}{f_X(x)} = \frac{\frac{1}{4\pi} \exp\left[-\frac{1}{2}\left(x^2 + \frac{y^2}{4}\right)\right]}{\frac{1}{\sqrt{2\pi}} \exp\left(-\frac{x^2}{2}\right)}$$

$$= \frac{1}{2\sqrt{2\pi}} e^{-y^2/8}, \qquad\qquad y \in \mathbb{R}.$$

This means that the conditional density of Y, given $X = x$, is normal, $N(\mu, \sigma^2)$, with parameters $\mu = 0$ and $\sigma = 2$. (Observe, in particular that this density does not depend on x.) In a similar way, we can show that the marginal distribution of Y is $N(0, 2^2)$ while the conditional density of X, given $Y = y$, is $N(0, 1)$.

We observe, in this example, that

$$f_{X|Y}(x|y) = f_X(x) \text{ and } f_{Y|X}(y|x) = f_Y(y),$$

something that is generally not true, as seen in the following example.

Example 2.5.2   A real number Y is selected randomly from the interval $(0, 1)$. For $Y = y$, we select a second number X randomly from the interval $(0, y)$. Find the density function of X.

SOLUTION   Let $f(x, y)$ be the joint density of X and Y. Then from, Proposition 2.3.1, we have

$$f_X(x) = \int_{-\infty}^{\infty} f(x,y) dy, \qquad\qquad x \in \mathbb{R}, \qquad (2.8)$$

while from Definition 2.5.1, we see that, for any real x,

$$f_{X|Y}(x|y) = \frac{f(x,y)}{f_Y(y)}.$$

Consequently, we can write

$$f_X(x) = \int_{-\infty}^{\infty} f_{X|Y}(x|y) f_Y(y) dy.$$

Since Y is randomly selected from the interval $(0, 1)$, the marginal distribution of Y is uniform in $(0, 1)$ with density function

$$f_Y(y) = 1, \qquad\qquad 0 < y < 1.$$

Further, for given $Y = y$, the conditional distribution of X is uniform in the interval $(0, y)$, and so

$$f_{X|Y}(x|y) = \frac{1}{y}, \qquad\qquad 0 < x < y < 1.$$

Using these two expressions in (2.8), we find

$$f_X(x) = \int_x^1 \frac{1}{y} dy = \left[\ln y\right]_x^1 = -\ln x, \qquad x \in (0, 1).$$

We have thus shown that the marginal density function of X is

$$f_X(x) = \begin{cases} -\ln x, & \text{if } x \in (0, 1), \\ 0, & \text{otherwise.} \end{cases}$$

It is clear that in this example we have $f_{X|Y}(x|y) \neq f_X(x)$.

We have seen that the function

$$g(y) = f_{Y|X}(y|x) = \frac{f(x, y)}{f_X(x)}, \qquad\qquad y \in \mathbb{R},$$

for a given $x \in \mathbb{R}$ such that $f(x) \neq 0$, is a density function. We can then define the corresponding (cumulative) distribution function as

$$G(y) = \int_{-\infty}^{y} g(t) \, dt,$$

and the expected value associated with density $g(y)$ as

$$\int_{-\infty}^{\infty} t g(t) \, dt.$$

These two quantities are called the **conditional (cumulative) distribution function** and the **conditional expected value** of Y, given $X = x$, respectively. Using notation similar to that for the discrete case, we write

$$F_{Y|X}(y|x) = P(Y \leq y | X = x) = \int_{-\infty}^{y} f_{Y|X}(t|x) \, dt$$

for the conditional distribution function of Y, given $X = x$, and

$$E(Y|X = x) = \int_{-\infty}^{\infty} y f_{Y|X}(y|x) dy$$

for the conditional expected value of Y, given $X = x$.

In a similar way, we can define the conditional distribution function of X, given $Y = y$, denoted by $F_{X|Y}(x|y)$, and the conditional expected value of X, given $Y = y$, denoted by $E(X|Y = y)$. Note that these are defined for any y such that $f_Y(y) \neq 0$.

Moreover, as with the usual (i.e. unconditional) expected value, the conditional expectation of a function $h(Y)$, given $X = x$, is given by

$$E[h(Y)|X = x] = \int_{-\infty}^{\infty} h(y) f_{Y|X}(y|x) dy.$$

Example 2.5.3 Suppose that the conditional density function of the random variable X, given $Y = y$, is given by

$$f_{X|Y}(x|y) = \frac{16x + 4y}{y + 1} e^{-4x}, \qquad x > 0,$$

for any $y > 0$. Then, for conditional distribution function of X, given $Y = y$, is obtained by integration by parts as

$$F_{X|Y}(x|y) = P(X \le x|Y = y) = \int_{-\infty}^{x} f_{X|Y}(s|y)\, ds = \int_{0}^{x} \frac{16s + 4y}{y + 1} e^{-4s}\, ds$$

$$= \frac{1}{y + 1} \left\{ \left[-4se^{-4s}\right]_0^x + \int_0^x 4e^{-4s} ds + y\left[-e^{-4s}\right]_0^x \right\}$$

$$= \frac{1 - (4x + 1)e^{-4x} + y(1 - e^{-4x})}{y + 1}, \qquad x > 0.$$

It is then clear that, for any positive y,

$$\lim_{x \to \infty} F_{X|Y}(x|y) = 1,$$

in accordance with the properties of a distribution function.

We also have

$$P(X < 1|Y = 3) = F_{X|Y}(1|3) = 1 - 2e^{-4} \cong 96.3\%.$$

Note that, although we know the conditional density function and the conditional distribution function of X, given Y, it is not possible to find the marginal (i.e. unconditional) density function, or distribution function of X, because we do not have any information about the joint distribution of X and Y.

Example 2.5.4 Let X and Y be the total distance, in miles, run by two soccer players A and B during a soccer game. The joint density of (X, Y) is given by

$$f(x, y) = \begin{cases} \dfrac{1}{120}(2x + y)e^{-x/3}e^{-y/4}, & x, y > 0, \\ 0, & \text{otherwise.} \end{cases}$$

(i) Find the value of the conditional distribution function $F_{X|Y}(x|y)$ for $x = 3$ and $y = 2$.

(ii) Explain whether the result in Part (i) is the same as

$$P(X \le 3 | Y \le 2).$$

SOLUTION (i) We have to find the conditional density $f_{X|Y}(x|y)$. As $f(x, y)$ is given, we see from Definition 2.5.1 that we also need the marginal density $f_Y(y)$. For this, we integrate the joint density with respect to x over the positive half-line, to obtain

$$f_Y(y) = \int_{-\infty}^{\infty} f(x, y)\, dx = \int_0^{\infty} \frac{1}{120} \cdot (2x + y)e^{-x/3}e^{-y/4}\, dx$$

$$= \frac{1}{120} \cdot e^{-y/4} \int_0^{\infty} 2x e^{-x/3}\, dx + \frac{1}{120} \cdot y e^{-y/4} \int_0^{\infty} e^{-x/3}\, dx. \qquad (2.9)$$

For the second integral on the right-hand side of (2.9), we note that it is three times the integral of an exponential density with parameter $1/3$ over $(0, \infty)$. Thus, we have

$$\int_0^{\infty} e^{-x/3}\, dx = 3 \underbrace{\int_0^{\infty} \frac{1}{3}e^{-x/3}\, dx}_{=1} = 3.$$

The first integral on the right-hand side of (2.9) can be calculated easily by integration by parts. But, a quicker way is to use the fact that this integral is a multiple of the expected value of an exponential random variable with parameter $1/3$. Using the fact that this expected value is the reciprocal of the parameter of the exponential distribution, we get

$$\int_0^{\infty} 2x e^{-x/3}\, dx = 6 \underbrace{\int_0^{\infty} x \cdot \frac{1}{3}e^{-x/3}\, dx}_{=3} = 18.$$

Substituting these two into (2.9), we find

$$f_Y(y) = \frac{1}{120}e^{-y/4}(18 + 3y), \qquad y > 0.$$

We then have the conditional density of X, given $Y = y$, to be

$$f_{X|Y}(x|y) = \frac{f(x, y)}{f_Y(y)} = \frac{\dfrac{1}{120}(2x + y)e^{-x/3}e^{-y/4}}{\dfrac{1}{120}e^{-y/4}(18 + 3y)} = \frac{(2x + y)e^{-x/3}}{18 + 3y}.$$

The conditional distribution function $F_{X|Y}(x|y)$ can now be found by integrating the above function as follows:

$$F_{X|Y}(x|y) = \int_0^x f_{X|Y}(s|y)\, ds = \int_0^x \frac{(2s + y)e^{-s/3}}{18 + 3y}\, ds$$

$$= \frac{1}{18 + 3y} \int_0^x 2s\, e^{-s/3}\, ds + \frac{y}{18 + 3y} \int_0^x e^{-s/3}\, ds. \qquad (2.10)$$

It is easy to see that the second integral on the right-hand side of (2.10) is

$$\int_0^x e^{-s/3}\, ds = 3\left(1 - e^{-x/3}\right),$$

while for the first integral on the right-hand side of (2.10), using integration by parts, we obtain

$$\int_0^x 2s\, e^{-s/3}\, ds = \left[-6s\, e^{-s/3}\right]_0^x + \int_0^x 6\, e^{-s/3}\, ds$$

$$= 18 - 6(x + 3)e^{-x/3}.$$

Upon substituting these two expressions in (2.10), we obtain

$$F_{X|Y}(x|y) = \frac{18 - 6(x + 3)e^{-x/3}}{18 + 3y} + \frac{3y\left(1 - e^{-x/3}\right)}{18 + 3y}.$$

For $x = 3$ and $y = 2$, we get the

$$F_{X|Y}(x|y) = \frac{18 - 6(3 + 3)e^{-3/3}}{18 + 3 \cdot 2} + \frac{3 \cdot 2\left(1 - e^{-3/3}\right)}{18 + 3 \cdot 2}$$

$$= \frac{\left(18 - 36e^{-1}\right) + \left(6 - 6e^{-1}\right)}{24} = 1 - \frac{7}{4e} \cong 0.3562.$$

(ii) First, note that while the probability we found in Part (i) is

$$F_{X|Y}(3|2) = P(X \le 3|Y = 2),$$

we are seeking here the probability $P(X \le 3|Y \le 2)$, which is *not the same* as the quantity above. The probability $P(X \le 3|Y \le 2)$ can be found directly from the definition of conditional probabilities as

$$P(X \le 3|Y \le 2) = \frac{P(X \le 3,\, Y \le 2)}{P(Y \le 2)} = \frac{F(3, 2)}{F_Y(2)}.$$

A rather long calculation of the numerator and the denominator above verifies the fact that the two probabilities are indeed different (see Exercise 4).

Exercises

Group A

1. The joint density function of (X, Y) is

$$f(x, y) = \begin{cases} cxy(x^2 + y), & 0 \le x \le 2, \ 0 \le y \le 1, \\ 0, & \text{otherwise,} \end{cases}$$

for a suitable constant $c \in \mathbb{R}$.

(i) Find the value of c, and thence obtain the conditional density functions $f_{X|Y}(x|y)$ and $f_{Y|X}(y|x)$.

(ii) Find $E(X|Y = 1/2)$ and $E(Y|X = 1)$.

2. For the pair (X, Y) of random variables in Example 2.5.1, find

(i) $P(X > 2)$;
(ii) $P(X > 2|Y = 1)$;
(iii) $P(Y \le 3/2|X = 2)$;
(iv) $P(Y \le 3/2)$.

What do you observe?

3. Linda takes a psychology test consisting of two parts, A and B. The scores that she achieves in these parts (out of 10) are continuous random variables, X and Y, respectively, with joint density function

$$f(x, y) = \begin{cases} cx^2y(10 - x)(10 - y), & 0 \le x \le 10, \ 0 \le y \le 10, \\ 0, & \text{otherwise.} \end{cases}$$

(i) Find the value of c.

(ii) Obtain the marginal densities of X and Y, and the conditional densities $f_{X|Y}(x|y)$ and $f_{Y|X}(y|x)$.

(iii) Find
$$P(X > 7), \quad P(X > 7, \ Y > 9) \quad \text{and} \quad P(X > 7, \ Y \le 9).$$

(iv) Calculate $E(X|Y = 5)$.

4. For the random pair (X, Y) in Example 2.5.4,

(i) calculate the joint distribution function $F(x, y)$ of (X, Y);

(ii) derive the marginal distribution function $F_Y(y)$;

(iii) obtain the conditional probability $P(X \le 3|Y \le 2)$, and then verify that this is not the same as the result in Part (i) of the example.

5. Assume that the conditional density function of the random variable X, given $Y = y$, is given by
$$f_{X|Y}(x|y) = \frac{x + y}{1 + y} e^{-x}, \qquad x > 0,$$

for any $y > 0$.

(i) Find the conditional distribution function of X, given $Y = y$, and thence calculate

$$P(X < 3|Y = 2).$$

(ii) Show that

$$E(X|Y = 2) = \frac{4}{3}.$$

6. The market shares of two companies A and B specializing in the same area, as it varies through time, can be represented by two continuous random variables, X and Y, respectively. The joint density of X and Y is given by

$$f(x, y) = 24xy, \qquad\qquad 0 < x < 1,\ 0 < y < 1,\ 0 < x + y < 1,$$

while $f(x, y) = 0$ for all other values of x, y.

(i) Calculate the marginal densities of X and Y, and thence find their expected values.

(ii) Find the conditional density function and the conditional distribution function of Y, given $X = 1/4$.

(iii) What is the expected share of Company A in the market if we know that the market share for Company B is 25%?

Group B

7. A number Y is drawn randomly from the interval $(0, 1)$. Given that $Y = y$, another number X is selected at random from the interval $(y, 1)$. Find the density function of the variable X and the joint density function of X and Y.

8. A restaurant offers a special type of ice cream that consists of mocha, pistachio, and nuts. Let X be the proportion of mocha flavor in the weight of an ice cream and Y be the proportion of pistachio flavor in the weight of the ice cream (the remaining proportion, $1 - X - Y$, is the weight of nuts). Suppose the joint density of X and Y is

$$f(x, y) = \begin{cases} \dfrac{16}{3}(x + 2y), & 0 < x < 1/2,\ 0 < y < 1/2, \\ 0, & \text{otherwise.} \end{cases}$$

Find

(i) the expected values of X and Y;

(ii) the probability that the proportion of mocha flavor in an ice cream is less than 25% if we know that the proportion of pistachio is exactly 30%;

(iii) the expected value of the proportion of mocha flavor in the weight of an ice cream if we know that the proportion of pistachio is exactly 40%.

9. The joint density function of (X, Y) is given by

$$f(x, y) = \begin{cases} 6, & \text{if } 0 \le x \le 1 \text{ and } x^2 \le y \le x, \\ 0, & \text{otherwise.} \end{cases}$$

(i) Show that the conditional distribution of Y, given $X = x$, is the uniform distribution over the interval $[x^2, x]$.

(ii) Calculate $E(Y|X = x)$ and the unconditional expectation $E(Y)$.

10. Let (X, Y) be a pair of continuous random variables with joint density $f(x, y)$, marginal densities $f_X(x)$ and $f_Y(y)$ for X and Y, respectively, while the corresponding conditional densities are $f_{X|Y}(x|y)$ and $f_{Y|X}(y|x)$. Show that

$$f_X(x) = \int_{-\infty}^{\infty} f_{X|Y}(x \mid y) f_Y(y) dy, \qquad x \in \mathbb{R}$$

and

$$f_Y(y) = \int_{-\infty}^{\infty} f_{Y|X}(y \mid x) f_X(x) dy, \qquad y \in \mathbb{R}.$$

Applications:

(i) Suppose Y has the uniform distribution over the interval $[0, 1]$ and that the conditional distribution of X, given $Y = y$, is also uniform, over the interval $[0, \sqrt{y}]$. Then, show that

$$f_X(x) = 2(1 - x), \qquad\qquad 0 \le x \le 1.$$

(ii) The claim amounts in a motor insurance portfolio follow the exponential distribution with parameter λ. Suppose the inflation rate for the following year is a random variable Y with density function

$$f_Y(y) = y^{-2} \exp\left(-y^{-1}\right), \qquad\qquad y > 0.$$

Let X be the amount of money that an insurance company will have to pay for a claim arriving at a motor insurance portfolio, during the *following year*. If it is assumed that, for a given $Y = y$, the conditional distribution of X is exponential with mean λy, calculate the distribution function of X.

2.6 GEOMETRIC PROBABILITY

We have already seen a situation wherein a real number is selected randomly from an interval, such as $(0, 1)$. Problems of this type are of importance in simulation techniques. When we select a number X from the interval $(0, 1)$, a typical assumption is that X follows the uniform distribution over that interval, so that in particular the probability that X belongs to any interval of width δ depends only on the magnitude of δ and not on the location of that interval within $(0,1)$. The problem can be generalized to dimensions higher than one, so that a point is drawn at random from a subset of a plane, a sphere, etc. Suppose on a plane of Cartesian coordinates, we have a square S with vertices at the points $(0, 0)$, $(0, 2)$, $(2, 2)$, and $(2, 0)$, and that we select completely at random a point with coordinates (X, Y) within the interior of the square. It is evident that, as in the univariate case, the probability that the random variable takes a specific value, say $(X, Y) = (1/2, 1)$, is zero. However, we might be interested in the probability that the point selected lies in a certain sub-area within the square. To illustrate this, let us try to calculate the probability that the selected point lies within the interior of the square with vertices at the points $(1, 1)$, $(1, 2)$, $(2, 1)$, and $(2, 2)$ as seen in the figure below.

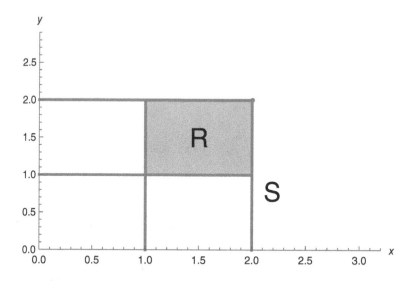

It is clear that the coordinates X and Y are (continuous) random variables, each having range as $(0, 2)$. Since the point (X, Y) is drawn at random, the probability that the point we select is "close" to a given point (a, b) (that is, it belongs to a small area around (a, b)) is the same for all choices of (a, b) in the square. Let $f(x, y)$ be the joint density function of (X, Y). As we described earlier in Section 2.2, the value of $f(a, b)$ is proportional to the probability that (X, Y) is "close" to (a, b). Consequently, as in the case of one variable, the value of $f(x, y)$ must be the same for any (x, y) inside the rectangle, and so the joint density function is

$$f(x, y) = \begin{cases} c, & \text{if } 0 < x < 2 \text{ and } 0 < y < 2, \\ 0, & \text{otherwise.} \end{cases}$$

Then, the condition

$$\int_{-\infty}^{\infty} \int_{-\infty}^{\infty} f(x, y) \, dx \, dy = 1$$

yields

$$1 = \int_0^2 \int_0^2 c \, dx \, dy = 4c,$$

and so $c = 1/4$, and consequently the joint density of X and Y is

$$f(x, y) = \begin{cases} 1/4, & \text{if } 0 < x < 2 \text{ and } 0 < y < 2, \\ 0, & \text{otherwise.} \end{cases}$$

We can now calculate easily the probability that the point (X, Y) lies within $R = \{(x, y) : 1 < x < 2, 1 < y < 2\}$ as

$$p = \int_R \int f(x, y) \, dx \, dy = \frac{1}{4} \int_1^2 \int_1^2 dx \, dy = \frac{1}{4}.$$

This can also be seen, perhaps in a more natural way, by a geometric argument as the double integral $\int_R \int dx\, dy$ expresses algebraically the *area* of the rectangle R. Therefore, we see that the probability a point randomly selected from S belongs to R is the ratio of the area of R to the area of S, i.e.

$$p = \frac{\text{Area}(R)}{\text{Area}(S)} = \frac{1}{4}.$$

This can be generalized to the case when R and S are any regions in the plane (or in higher dimensions, with the obvious interpretation; for example, in three dimensions, area is replaced by volume). Suppose we have a region S in the plane and we pick randomly a point from that region. If (X, Y) are the coordinates of this point, the joint density of (X, Y) is

$$f(x, y) = \begin{cases} c, & \text{if } (x, y) \in S, \\ 0, & \text{otherwise.} \end{cases}$$

The double integral of $f(x, y)$ over x and y must equal 1, and so

$$c \int_S \int dx\, dy = 1 \quad \text{or} \quad c \cdot \text{Area}(S) = 1,$$

which means that c is the reciprocal of the area of S.

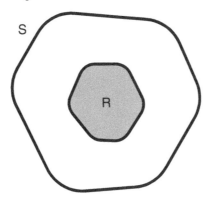

As done above, the probability that the point (X, Y) belongs to a region R, lying within S, is then given by

$$P((X, Y) \in R) = c \int_R \int dx\, dy,$$

and so once again

$$P((X, Y) \in R) = \frac{\text{Area}(R)}{\text{Area}(S)}.$$

The last expression, which is intuitively appealing, is often taken as the definition of the probability that a point drawn completely at random from a region S belongs to a subregion R. In such cases, we are speaking of a **geometric probability**. An alternative expression, which generalizes the univariate case, is to say that the pair (X, Y) has the **uniform distribution** over the region S. Finally, if the region R does not lie entirely within S, the probability $P((X, Y) \in R)$ can be given as

$$P((X, Y) \in R) = \frac{\text{Area}(R \cap S)}{\text{Area}(S)}.$$

Example 2.6.1 Two numbers X and Y are selected at random from the unit interval $(0, 1)$. Find the probabilities $P(Y \leq X)$, $P(X^2 + Y^2 \leq 1)$ and $P(Y \leq X, \ X^2 + Y^2 \leq 1)$.

SOLUTION The range of the bivariate random variable (X, Y) is the rectangle S whose vertices are the points $(0, 0)$, $(0, 1)$, $(1, 1)$, and $(1, 0)$. The probability $P(Y \leq X)$ corresponds to the shaded area R_1 in the figure below, and so

$$P(Y \leq X) = P((X, Y) \in R_1) = \frac{\text{Area}(R_1)}{\text{Area}(S)} = \frac{1}{2} = 50\%.$$

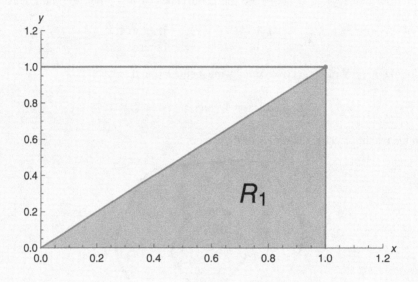

Similarly,

$$P(X^2 + Y^2 \leq 1) = P((X, Y) \in R_2) = \frac{\text{Area}(R_2)}{\text{Area}(S)}$$

where R_2 is the shaded region in the following graph

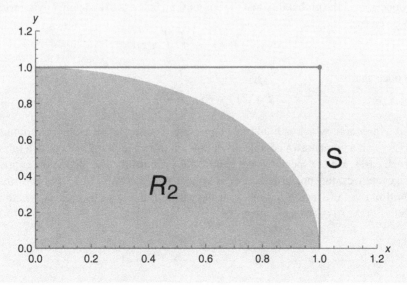

and

$$P(Y \le X,\ X^2 + Y^2 \le 1) = P((X, Y) \in R_3) = \frac{\text{Area}(R_3)}{\text{Area}(S)},$$

where R_3 is the shaded region in the figure below.

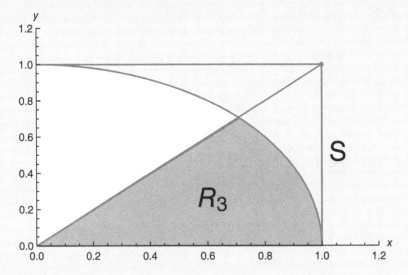

It is easy to see that the areas of R_2 and R_3 are $\pi/4$ and $\pi/8$, respectively, and since S has unit area, we get

$$P(X^2 + Y^2 \le 1) = \frac{\pi/4}{1} = \frac{\pi}{4} \cong 78.5\%$$

and

$$P(Y \le X, X^2 + Y^2 \le 1) = \frac{\pi/8}{1} = \frac{\pi}{8} \cong 39.3\%.$$

Example 2.6.2 Mike and Hannah make an appointment to meet outside a restaurant at $8:30$pm in order to have dinner together. Assuming that, due to traffic delays, each will arrive between $8:25$ and $8:40$pm and the arrival time is completely random in that interval, what is the probability that

(i) Mike waits more than 5 minutes for Hannah?

(ii) the person who arrives first does not have to wait for more than 5 minutes until the other arrives?

SOLUTION   Let X and Y be the times, in minutes, after 8.25 that Mike and Hannah, respectively, arrive at the restaurant (so that both X and Y have a uniform distribution over the interval $(0, 15)$).

(i) The probability that Mike waits for Hannah more than 5 minutes is

$$P(X + 5 < Y) = P((X, Y) \in R_1) = \frac{\text{Area}(R_1)}{\text{Area}(S)},$$

where

$$S = \{(x, y) : 0 \le x \le 15, \ 0 \le y \le 15\},$$
$$R_1 = \{(x, y) : 0 \le x \le 15, \ 0 \le y \le 15 \text{ and } x + 5 \le y\}.$$

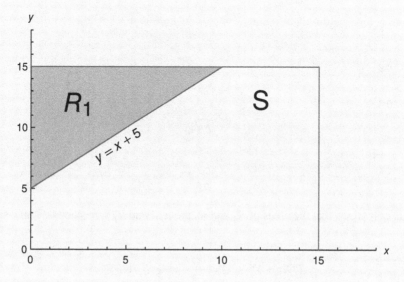

It is clear that the area of S is $15 \cdot 15 = 225$, while that of R_1 is $(10 \cdot 10)/2 = 50$, and so

$$P(X + 5 < Y) = \frac{50}{225} = \frac{2}{9} \cong 22.2\%.$$

(ii) The event that the person who arrives first waits for at most 5 minutes can be expressed as $|X - Y| \le 5$, and the probability that this happens corresponds to the region R_2 in the figure below.

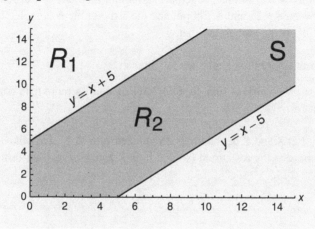

The area of this region is

$$\text{Area}(R_2) = 15 \cdot 15 - 2\left(\frac{1}{2} \cdot 10 \cdot 10\right) = 125,$$

which gives that the required probability as

$$P(|X - Y| \leq Y) = P((X, Y) \in R_2) = \frac{\text{Area}(R_2)}{\text{Area}(S)} = \frac{125}{225} = \frac{5}{9} \cong 55.6\%.$$

Exercises

Group A

1. A shooter aims at a specific target that is centered at the origin and has radius r. Suppose the point hit at the target is chosen randomly from the circle, and let X and Y be its coordinates.

   (i) Verify that the joint density function of (X, Y) is given by

   $$f(x, y) = \begin{cases} \dfrac{1}{\pi r^2}, & \text{for } x^2 + y^2 \leq r^2, \\ 0, & \text{otherwise.} \end{cases}$$

   (ii) What is the probability that the distance between the point that the shooter hits inside the target and the center of the target is less than $r/2$?

   (iii) What is the probability that each of the coordinates has a value less than $r/2$?

   (iv) Find the marginal distributions of X and Y, and check that they are the same.

2. We select at random a point from the rectangle defined on a set of Cartesian coordinates by the vertices $(1, 1), (1, 3), (5, 3)$, and $(5, 1)$. Let (X, Y) be the coordinates of the point selected.

   (i) Find $P(X \leq 4, Y \leq 2)$.

   (ii) What is the probability that $Y - X \leq 1$?

3. The two-dimensional random variable (X, Y) is uniformly distributed over the triangle defined on a plane of Cartesian coordinates by its vertices, located at the points $(-1, 0), (1, 0)$, and $(0, 1)$. Show that

   $$E(XY) = E(X)E(Y).$$

4. We select randomly a point from the rectangle with vertices $(0, 0), (1, 0), (1, 2)$, and $(0, 2)$. If X and Y are the coordinates of this point, find $E(X|Y = y)$ and $E(Y|X = x)$.

5. A point is drawn randomly from the square with vertices at the points $(0, 0), (0, 1), (1, 1)$, and $(1, 0)$. Find the probability that the selected point lies in the region R defined by the equations $y = x$ and $y = x^2$ (see the graph below).

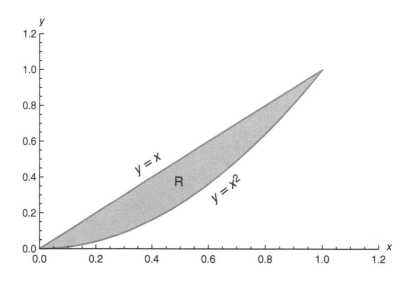

Group B

6. Let (X, Y) have a uniform distribution over the triangle enclosed by the line $x + y = 4$ and the first quadrant of the plane $(x, y \geq 0)$.

    (i) Find $P(X + Y \leq 3)$ and $P(X^2 + Y^2 \leq 4)$.

    (ii) Obtain the marginal density function of Y.

    (iii) Find the conditional density of Y, given $X = x$, and $E(Y|X = x)$.

7. Andrew and Ariel, who do not know each other, visit a coffee shop each morning to have a coffee break from work. Andrew's arrival time at the coffee shop is uniformly distributed between 10 : 45am and 11 : 15am, and he stays in the shop for 15 minutes. Ariel arrives any time between 11 : 05am and 11 : 20am, and stays in the shop for 10 minutes.

    (i) Find the probability that, on a given day, Ariel arrives at the shop at least 5 minutes before Andrew.

    (ii) What is the probability that the first person to arrive at the coffee shop leaves before the other comes in?

    (iii) Find the probability that, in a given week (5 working days), Ariel arrives at the shop before Andrew at least twice.

2.7   BASIC CONCEPTS AND FORMULAS

Properties for a joint density function of a two-dimensional random variable	$f(x, y) \geq 0$ for any $x \in \mathbb{R}, y \in \mathbb{R}$; $\int_{-\infty}^{\infty} f(x, y) \, dx \, dy = 1$
Joint (cumulative) distribution function	$F(x, y) = P(X \leq x, Y \leq y), x \in \mathbb{R}, y \in \mathbb{R}$

Relationship between the joint distribution function and the joint density function	$F(x, y) = \int_{-\infty}^{x} \int_{-\infty}^{y} f(s, t) \, dt \, ds$ $f(x, y) = \dfrac{\partial^2 F(x, y)}{\partial x \partial y} = \dfrac{\partial^2 F(x, y)}{\partial y \partial x}$
Calculation of probabilities using the joint density function	$P((X, Y) \in \Gamma) = \int_{\Gamma} \int f(x, y) \, dx \, dy$
Marginal density functions	$f_X(x) = \int_{-\infty}^{\infty} f(x, y) dy; \; f_Y(y) = \int_{-\infty}^{\infty} f(x, y) \, dx$
Expectation of a function of two random variables	$E[h(X, Y)] = \int_{-\infty}^{\infty} \int_{-\infty}^{\infty} h(x, y) f(x, y) \, dx \, dy$
Expectation of sum and linear combination	$E\left[h_1(X, Y) + h_2(X, Y)\right] =$ $E\left[h_1(X, Y)\right] + E\left[h_2(X, Y)\right];$ $E(aX + bY) = aE(X) + bE(Y)$
Conditional density function of X, given $Y = y$	$f_{X|Y}(x|y) = \dfrac{f(x, y)}{f_Y(y)}, \; \text{for} f_Y(y) > 0$
Conditional density function of Y, given $X = x$	$f_{Y|X}(y|x) = \dfrac{f(x, y)}{f_X(x)}, \; \text{for} f_X(x) > 0$
Conditional (cumulative) distribution functions	$F_{X|Y}(x|y) = P(X \leq x | Y = y) = \int_{-\infty}^{x} f_{X|Y}(s|y) \, ds;$ $F_{Y|X}(y|x) = P(Y \leq y | X = x) = \int_{-\infty}^{y} f_{Y|X}(t|x) \, dt$
Conditional expectations	$E(X|Y = y) = \int_{-\infty}^{\infty} x f_{X|Y}(x|y) \, dx;$ $E(h(X)|Y = y) = \int_{-\infty}^{\infty} h(x) f_{X|Y}(x|y) \, dx;$ $E(Y|X = x) = \int_{-\infty}^{\infty} y f_{Y|X}(y|x) dy;$ $E(h(Y)|X = x) = \int_{-\infty}^{\infty} h(y) f_{Y|X}(y|x) dy$

2.8 COMPUTATIONAL EXERCISES

The syntax for calculating double or, more generally, multiple integrals in Mathematica, is similar to that for the integration with respect to one variable. For definite integrals, like those we often see in the present context, the only thing that changes from the univariate case is that we now have to specify the range of integration for each variable. For example, the command

```
Integrate[c*x^2*y, {x, 0, 1}, {y, 0, 1}]
```

calculates the double integral of the density $f(x, y)$ in Example 2.2.1, over its entire range, and thus we may easily find the value of the constant c. In the following output, we perform the following:

(a) calculate the probability $P(X > Y)$;

(b) calculate $P(0 < X < 1/2$ and $1/3 < Y < 2/3)$;

(c) find the expected value of $Z = h(X, Y)$, where $h(x, y) = xy$;

(d) make a 3-D graphical display of the joint density $f(x, y)$ of (X, Y);

(e) make a Contour Plot of the joint density. The curves on this plot show the pairs (x, y) for which $f(x, y)$ takes a specific value (they correspond, from left to right, to the values $f(x, y) = 1, 2, 3, 4, 5)$.

```
f1[x_, y_] := x^2*y
a = Integrate[f1[x, y], {x, 0, 1}, {y, 0, 1}];
c = 1/a;
Print["c=", c];
f[x_, y_] = c*f1[x, y]
prob1 = Integrate[f[x, y], {x, 0, 1}, {y, 0, x}]
Print["P(X>Y)=", prob1]
prob2 = Integrate[f[x, y], {x, 0, 1/2}, {y, 1/3, 2/3}];
Print["P(0<X<1/2, 1/3<Y<2/3)=", prob2];
h[x_, y_] := x*y;
meanH = Integrate[h[x, y]*f[x, y], {x, 0, 1}, {y, 0, 1}];
Print["E[h(X,Y)]=", meanH];
Print["Joint Density of X, Y :"]
Plot3D[f[x, y], {x, 0, 1}, {y, 0, 1}]
Print["Contour Plot of the Joint Density for X, Y :"]
ContourPlot[f[x, y], {x, 0.4, 1}, {y, 0.2, 1}]
```

```
Out[1]:= c=6
P(X>Y)=3/5
P(0<X<1/2, 1/3<Y<2/3)=1/24
E[h(X,Y)]=1/2
Joint Density of X, Y :
```

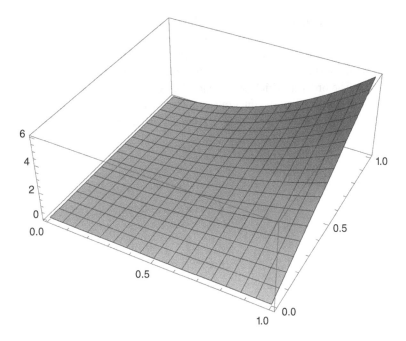

Contour Plot of the Joint Density for X, Y :

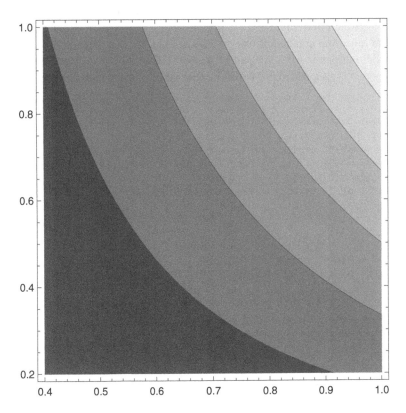

When performing multiple integrations in Mathematica, it is important to keep in mind **the order** in which we declare the ranges of the variables to integrate. For the density in Example 2.2.1 above, the range of values for each variable does not depend on the value of the other variable. In such cases, the order is not important. Consider instead the density in Example 2.2.2. The order in which we declare the ranges of the variables in Mathematica is **not** the same as the one that the integration takes place. In fact, first we declare the outer range of integration, then the inner range, so that for instance the command

```
Integrate[4*x*y, {y, 0, 1}, {x, 0, y}]
```

finds the integral

$$\int_0^1 \int_0^y 4xy \, dx \, dy,$$

while the command

```
Integrate[4*x*y, {x, 0, y}, {y, 0, 1}]
```

calculates the integral

$$\int_0^y \int_0^1 4xy dy \, dx$$

(check that the latter is a function of y, not a constant, and so the two results are different; we have to be watchful of this!).

With the following commands, we do the calculations that we carried out in Example 2.2.3. In addition, we make a 3D plot of the joint distribution function $F(x, y)$, the joint density function $f(x, y)$, and also we make a contour plot of the joint density.

```
In[2]:=F[x_, y_] := (1 - Exp[-2*x])*(1 - Exp[-3*y])
Print["Joint Distribution Function of X,Y: ", F[x, y]]
Plot3D[F[x, y], {x, 0, 2}, {y, 0, 2}]
FX[x_] := Limit[F[x, y], y -> Infinity]
Print["Marginal Distribution Function of X: ", FX[x]]
FY[y_] := Limit[F[x, y], x -> Infinity]
Print["Marginal Distribution Function of Y: ", FY[y]]
prob1 := F[3/2, 2];
Print["P(X<=3/2,Y<=2)=", prob1, "=", N[prob1]];
prob2 := 1 - F[1, 2];
Print["P(X>1 or Y>2)=", prob2, "=", N[prob2]];
f[x_, y_] := D[F[x, y], x, y]
Print["Joint Density Function of X, Y :", f[x, y]]
prob3 := Integrate[f[x, y], {y, 4/5, Infinity}, {x, 4/5, Infinity}]
Print["P(X>4/5, Y>4/5)=", prob3, "=", N[prob3]];


Out[2]:=Joint Distribution Function of X,Y: (1-Exp[-2 x]) (1-Exp[-3 y])
```

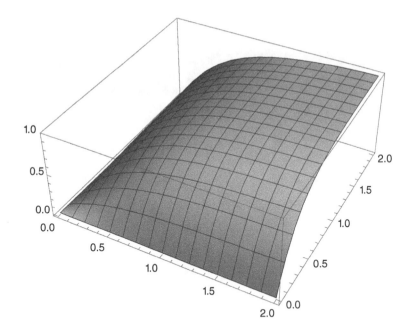

```
Marginal Distribution Function of X: 1-Exp(-2 x)
Marginal Distribution Function of Y: 1-Exp(-3 y)
P(X<=3/2,Y<=2)=(1-1/Exp[6]) (1-1/Exp[3])=0.947858
P(X>1 or Y>2)=1-(1-1/Exp[6]) (1-1/Exp[2])=0.137479
Joint Density Function of X, Y :6 Exp[-2 x-3 y]
P(X>4/5, Y>4/5)=1/Exp[4]=0.0183156

In[3]:=Plot3D[6*Exp[-2*x - 3*y], {x, 0, 2}, {y, 0, 2}]
Print["Contour Plot of the Joint Density for X, Y :"]
ContourPlot[6*Exp[-2*x - 3*y], {x, 0, 2}, {y, 0, 2}]
```

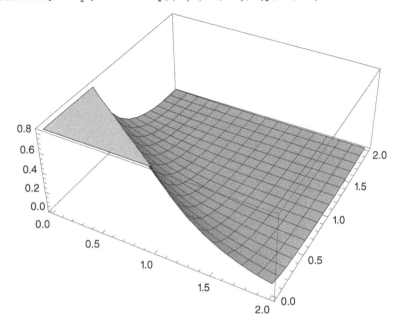

Contour Plot of the Joint Density for X, Y :

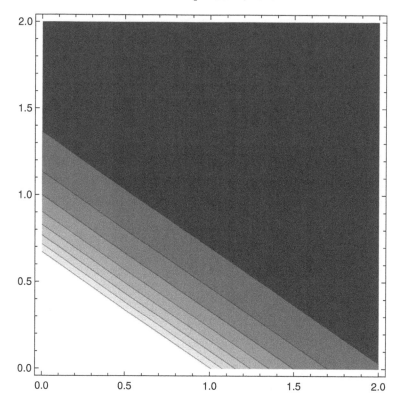

1. The joint density function of (X, Y) is

$$f(x, y) = c\, x^5 y^3 e^{-5x-4y}, \qquad\qquad x > 0, y > 0.$$

(i) Find the value of c, and obtain the marginal densities of X and Y.

(ii) Check whether

$$E(XY) = E(X)E(Y)$$

holds.

(iii) Find the conditional density functions

$$f_{X|Y}(x|y) \ \text{ and } \ f_{Y|X}(y|x),$$

and examine whether these are the same as the marginal densities of X and Y found in Part (i).

2. In each of the following cases, determine the value of c so that, for $i = 1, 2$, $f_i(x, y)$ is the joint density for a pair of random variables (X, Y). Then, find the marginal densities of X and Y in each case:

$$f_1(x, y) = \begin{cases} cx^4(y - x)^3 e^{-y}, & 0 < x < y < \infty, \\ 0, & \text{elsewhere}, \end{cases}$$

and

$$f_2(x,y) = \begin{cases} cx^2 \ln x \ (y+2)^4 e^{-3y}, & 1 < x < 2, \ 0 < y < \infty, \\ 0, & \text{elsewhere.} \end{cases}$$

3. The joint density of X and Y is

$$f(x,y) = \frac{c}{\left(1 + x^2 + y^2\right)^{3/2}}, \qquad x \in \mathbb{R}, y \in \mathbb{R},$$

for some real constant c.

(i) Use Mathematica to determine the value of c.

(ii) Find the marginal density functions of X and Y, and then plot these functions. [The marginal distribution of both X and Y is the **Cauchy distribution**, for which it is known that the mean does not exist; verify this.]

(iii) Find

$$P(X < 0), \quad P(Y \le 2), \quad P(X < 0, \ Y \le 2) \text{ and } P(X < 0 \text{ or } Y \le 2).$$

4. For Exercise 2 of Section 2.6,

(i) use the *Plot* command in Mathematica to draw the rectangle defined in that exercise on a Cartesian set of coordinates;

(ii) solve both parts of the exercise with a simple set of Mathematica commands, and then check that your results agree with those obtained by hand.

5. The joint distribution function of (X, Y) is

$$F(x,y) = \begin{cases} 1 - e^{-x^2} - e^{-y^3} + e^{-x^2 - y^3 - x^2 y^3/10}, & x \ge 0, \ y \ge 0, \\ 0, & \text{otherwise.} \end{cases}$$

(i) Make a *three*-dimensional plot of the function $F(x, y)$ as well as the corresponding contour plot.

(ii) Find the value of x for which $P(X \le x, Y \le 1) = \frac{1}{2}$.

(iii) Find the value of y for which $P(X \le 2, Y \le y) = \frac{1}{2}$.

(iv) Obtain the set of pairs (x, y) for which $P(X \le x, Y \le y) = \frac{1}{2}$.

(v) Derive the marginal (cumulative) distribution functions, $F_X(x)$ and $F_Y(y)$, of the variables X and Y, respectively.

(vi) Find the joint density function, $f(x, y)$, of (X, Y).

(vii) Calculate

$$P(X \le 1, Y \le 2), \quad P(X \le 1 \text{ or } Y \le 2) \text{ and } P(X \le 2, Y > 2).$$

6. The position, (X, Y), of a particle on the positive quadrant of the plane has density function as

$$f(x, y) = cxe^{-x/2}(1 + y)^{-4}, \qquad\qquad x > 0, \ y > 0.$$

Find

$$P(X < Y) \quad \text{and} \quad P(X + Y \leq 4).$$

7. The joint density function of (X, Y) is given by

$$f(x, y) = \begin{cases} c(x^2 + 2y^3 + 3xe^{-3x}), & 0 < x, y < 1, \\ 0, & \text{otherwise.} \end{cases}$$

(i) Find the value of c.

(ii) Make a three-dimensional plot of the function $f(x, y)$.

(iii) Derive the marginal density functions of X and Y.

(iv) Obtain the conditional densities $f_{X|Y}(x|y)$ and $f_{Y|X}(y|x)$.

(v) Find $P(X \leq 0.5|Y = 0.25)$ and $P(Y \leq 0.5|X = 0.25)$.

(vi) Find the value of the limit

$$\lim_{h \to 0} P(X \leq 0.5|0.25 \leq Y \leq 0.25 + h)$$

and compare it to $P(X \leq 0.5|Y = 0.25)$ found in Part (v).

8. Find the value of $c \in \mathbb{R}$ for which

$$f(x, y) = \begin{cases} c\, y^2 e^{-5y} \left[2 + \cos\left(\sqrt{x}\right)\right] e^{-x^3}, & 0 \leq x \leq \pi^2, y \geq 0, \\ 0, & \text{elsewhere,} \end{cases}$$

is a density function for (X, Y). Then, find the marginal densities, the mean and variance of X and Y.

9. Two substances, a and b, take part in a chemical reaction. The times, in seconds, until these substances begin to melt is represented by two random variables, X and Y (for a and b, respectively). The joint density function of X and Y is

$$f(x, y) = \begin{cases} \frac{64(x^2 + y^2)}{5} e^{-2x - 4y}, & x, y \geq 0, \\ 0, & \text{elsewhere.} \end{cases}$$

(i) Find the joint distribution function $F(x, y)$ of X and Y, and verify that

$$\lim_{x \to \infty} \lim_{y \to \infty} F(x, y) = \lim_{y \to \infty} \lim_{x \to \infty} F(x, y) = 1.$$

(ii) Obtain the marginal densities of X and Y, and thence calculate $E(X)$ and $E(Y)$.

(iii) Calculate $E(XY)$, $E(X/Y)$ and examine whether

$$E\left(\frac{X}{Y}\right) = \frac{E(X)}{E(Y)}.$$

(iv) Check whether

$$E\left(\frac{1}{XY}\right) = \frac{1}{E(XY)}$$

holds or not.

2.9 SELF-ASSESSMENT EXERCISES

In the exercises below, $f(x, y)$ denotes the joint density of (X, Y), while $F(x, y)$ is the (cumulative) joint distribution function of (X, Y). The marginal densities of X and Y are $f_X(x)$ and $f_Y(y)$, and their marginal distribution functions are $F_X(x)$ and $F_Y(y)$, respectively. If the values of a function are not given in a certain range, they are assumed to be zero in that range.

2.9.1 True–False Questions

1. For any $x, y \in \mathbb{R}$, we have

$$P(X = x, Y = y) = 0.$$

2. For any real x,

$$f_X(x) = \int_{-\infty}^{\infty} f(x, y) \, dx.$$

3. For any real x and y, we have $f(x, y) \leq 1$.

4. $P(X > a)$ is given by

$$\int_a^{\infty} \int_{-\infty}^{\infty} f(x, y) \, dx \, dy.$$

5. For any real a, we have

$$P(X \leq a) = \int_{-\infty}^{a} f_X(x) \, dx.$$

6. For any real x, we have

$$F_X(x) = \int_{-\infty}^{\infty} F(x, y) dy.$$

7. Let $f(x, y) = 1$ if $0 \leq x, y \leq 1$, and $f(x, y) = 0$ otherwise. Then,

$$P(X - Y = 1/2) = 1/2.$$

8. The joint density of X and Y is given by

$$f(x, y) = 16e^{-4x-4y}, \qquad\qquad x \geq 0, y \geq 0.$$

Then, the marginal distributions of X and Y are the same.

9. For all $a, b \in \mathbb{R}$,

$$P(X > a, Y > b) = 1 - \int_{-\infty}^{a} \int_{-\infty}^{b} f(x, y) dy \, dx.$$

10. For any x in the range of X, we have $F_X(x) = \lim_{y \to \infty} F(x, y)$.

11. For any x in the range of X, we have $f_X(x) = \lim_{y \to \infty} f(x, y)$.

12. The joint density function of (X, Y) is

$$f(x, y) = \begin{cases} c(x^2 + y), & 0 \leq x \leq 1, \ 0 \leq y \leq 2, \\ 0, & \text{otherwise.} \end{cases}$$

Then, the value of c is $1/3$.

13. The joint density function of (X, Y) is

$$f(x, y) = \begin{cases} ce^{-(x+y)}xy^2, & x > 0, \ y > 0, \\ 0, & \text{otherwise.} \end{cases}$$

Then, the value of c is $1/2$.

14. If the joint density of (X, Y) is

$$f(x, y) = xe^{-x-y}, \qquad\qquad x, y \geq 0,$$

then we have

$$f(x, y) = f_X(x)f_Y(y).$$

15. Assuming that (X, Y) have joint density

$$f(x, y) = \begin{cases} 8xy, & 0 < x < y < 1, \\ 0, & \text{otherwise,} \end{cases}$$

then, for any real x and y, we have

$$F(x, y) = F_X(x)F_Y(y).$$

16. Let the joint distribution function of (X, Y) be

$$F(x, y) = (1 - e^{-x^3})\left(1 - e^{-y/3}\right), \qquad\qquad x \geq 0, y \geq 0.$$

Then, $f_{X|Y}(x|3) = 3x^2e^{-x^3}$, for any $x \geq 0$.

17. The joint density of (X, Y) is

$$f(x, y) = cxy(x + y), \qquad 0 \leq x \leq 2, \ 0 \leq y \leq 2,$$

for a suitable constant c. Then,

$$E(Y|X = 1) = 1.$$

2.9.2 Multiple Choice Questions

1. Assume that the joint density function of (X, Y) is

$$f(x, y) = \begin{cases} c(3x + 5y), & 1 \leq x \leq 3, \ 0 \leq y \leq 2, \\ 0, & \text{otherwise.} \end{cases}$$

Then, the value of c is

(a) 11   (b) $\dfrac{1}{44}$ (c) $\dfrac{1}{11}$ (d) $\dfrac{3}{308}$ (e) $\dfrac{1}{52}$

2. Let the joint function of the pair (X, Y) be

$$f(x, y) = \begin{cases} cxy^2, & 0 \leq x \leq 1, \ 0 \leq y \leq 1, \\ 0, & \text{otherwise,} \end{cases}$$

for some suitable c. Then, $E(X^2)$ is equal to

(a) $1/2$ (b) $1/3$ (c) $1/4$ (d) $2/3$ (e) $3/4$

3. Assume that the joint density function of (X, Y) is

$$f(x, y) = 6e^{-2x}e^{-3y}, \qquad x > 0, y > 0.$$

Let $\mathbb{N}$ be the set of positive integers and B be the interval $(0, 2)$. Then, $P(X \in \mathbb{N}, Y \in B)$ is equal to

(a) $1 - e^{-4}$ (b) 0   (c) $1 - e^{-6}$ (d) $1 - e^{-2}$ (e) $2\left(1 - e^{-4}\right)$

4. The joint density function of X and Y is given by

$$f(x, y) = 192x^2y, \qquad 0 \leq x \leq 1/2, \ 0 \leq y \leq 1/2.$$

Then, $E(XY)$ is equal to

(a) 16   (b) 2   (c) $1/4$ (d) $3/8$ (e) $1/8$

5. Which of the following is *always* true for a two-dimensional continuous random variable (X, Y) (assuming that all expectations exist):

   (a) $f(x, y) = f_X(x)f_Y(y)$ $(x, y \in \mathbb{R})$ (d) $E(XY) = E(X)E(Y)$
   (b) $F(x, y) = F_X(x)F_Y(y)$ $(x, y \in \mathbb{R})$ (e) $E(X/Y) = E(X)/E(Y)$
   (c) $E(X + Y) = E(X) + E(Y)$

6. Let (X, Y) have joint density as

$$f(x, y) = \begin{cases} c(x^2 + y), & 0 < x < y < 1, \\ 0, & \text{otherwise.} \end{cases}$$

Then:

   (a) $P(X \le Y) = 1,$ $P(X \le Y - 1) = 1$
   (b) $P(X \le Y) = 0,$ $P(X \le Y - 1) = 1$
   (c) $P(X \le Y) = 1,$ $P(X \le Y - 1) = 0$
   (d) $P(X \le Y) = 1,$ $P(X \le Y - 1) = 1/2$
   (e) $P(X \le Y) = 1/2,$ $P(X \le Y - 1) = 0$

7. (X, Y) has its joint distribution function as

$$F(x, y) = \left(1 - e^{-6x^2}\right)\left(1 - e^{-5y}\right), \qquad x > 0, \ y > 0.$$

Then, the marginal density function of X is (for $x > 0$)

   (a) $f_X(x) = 1 - e^{-6x^2}$ (b) $f_X(x) = e^{-12x}$ (c) $f_X(x) = 60e^{-6x^2}$
   (d) $f_X(x) = 12e^{-6x^2}$ (e) $f_x(x) = 12xe^{-6x^2}$

Questions 8–16 refer to (X, Y) having joint density function

$$f(x, y) = \begin{cases} 2xe^{-4y}, & 0 < x \le 2, 0 < y < \infty, \\ 0, & \text{otherwise.} \end{cases}$$

8. The marginal density function of X is (for $0 < x \le 2$)

   (a) $f_X(x) = 1$ (b) $f_X(x) = 1/2$ (c) $f_X(x) = x$
   (d) $f_X(x) = x/2$ (e) $f_X(x) = x^2/4$

9. The marginal density function of Y is (for $y > 0$)

   (a) $f_Y(y) = e^{-4y}$ (b) $f_Y(y) = 2e^{-4y}$ (c) $f_Y(y) = 2ye^{-4y}$
   (d) $f_Y(y) = 4e^{-4y}$ (e) $f_Y(y) = 16ye^{-4y}$

10. $E(X)$ equals

   (a) $4/3$ (b) 1 (c) $1/2$ (d) $1/3$ (e) $1/4$

11. $E(X^2Y)$ equals

      (a) $1/2$ (b) 2 (c) 4 (d) $1/4$ (e) $1/6$

12. Given that $X = 3$, the conditional density of Y is (for $y > 0$)

    (a) $f_Y(y) = e^{-12y}$ (b) $f_Y(y) = 3e^{-3y}$ (c) $f_Y(y) = 12e^{-12y}$
    (d) $f_Y(y) = 4e^{-4y}$ (e) $f_Y(y) = 12e^{-4y}$

13. $E(X|Y = 5)$ equals

      (a) 1 (b) $4/3$ (c) $20/3$ (d) $1/3$ (e) 5

14. The conditional variance $Var(X|Y = 5)$ is equal to

      (a) 2 (b) 1 (c) $2/9$ (d) $1/3$ (e) $1/9$

15. $P(X > 1, Y > 3)$ equals

    (a) e^{-4} (b) $\dfrac{3e^{-4}}{2}$ (c) e^{-12} (d) $\dfrac{3e^{-12}}{4}$ (e) $\dfrac{e^{-12}}{2}$

16. The value of the conditional (cumulative) distribution $F_{X|Y}(x|y)$, for $x = 1$ and $y = 2$, is

    (a) $\dfrac{1}{4}$ (b) $\dfrac{1}{8}$ (c) $\dfrac{1 - e^{-8}}{2}$ (d) $\dfrac{3}{4}$ (e) $\dfrac{e^{-8}}{4}$

2.10 REVIEW PROBLEMS

1. The joint distribution function of (X, Y) is

$$F(x, y) = \left(1 - e^{-3x^4}\right)\left(1 - e^{-5y^3}\right), \qquad\qquad x > 0, y > 0.$$

    (i) Find the joint density function of X and Y.
    (ii) Find

$$P(X < 1, \ Y < 2), \quad P(X < 1, \ Y \geq 2) \text{ and } P(X > 2, \ Y > 1).$$

2. Let h be the density function of a continuous random variable. Let a new function f, defined on $\mathbb{R}^2$, be

$$f(x, y) = h(x)h(y), \qquad\qquad x, y \in \mathbb{R}.$$

(i) Show that f is the joint density function of a random pair (X, Y).

(ii) Obtain the marginal density functions of X and Y.

(iii) Prove that $P(X \geq Y) = 0.5$.

(iv) Verify that $E(XY) = E(X)E(Y)$.

Application: The joint density function of (X, Y) is given by

$$f(x, y) = \frac{1}{8\pi^2} \exp\left\{-\frac{1}{8\pi}(x^2 + y^2)\right\}, \qquad x, y \in \mathbb{R}.$$

Then, find the following quantities:

$$P(X \leq Y), \quad P(X \leq 1), \quad E(X), \quad E(Y), \text{ and } E(XY).$$

[Hint: For Part (iii), show that

$$P(X \geq Y) = \int_{-\infty}^{\infty} h(x)H(x)\, dx = \left[\frac{H^2(x)}{2}\right]_{-\infty}^{\infty},$$

where H is the cumulative distribution function associated with h.]

3. The joint density function of (X, Y) is

$$f(x, y) = \begin{cases} c(x + y), & 0 \leq x \leq a \text{ and } 0 \leq y \leq b, \\ 0, & \text{otherwise,} \end{cases}$$

where c is a real constant.

(i) Prove that

$$c = \frac{2}{ab(a + b)}.$$

(ii) Obtain the joint distribution function $F(x, y)$ and the marginal density functions of X and Y.

(iii) For $n = 1, 2, \ldots$, find $E(X^n)$, $E(Y^n)$ and $E[(X + Y)^n]$. What do you observe?

4. The joint density function of (X, Y) is

$$f(x, y) = \begin{cases} 1, & 0 \leq x \leq 1 \text{ and } |y| \leq x, \\ 0, & \text{otherwise.} \end{cases}$$

Show that the conditional expectation of Y, given $X = x$ is a linear function of x, and that the conditional expectation of X, given $Y = y$, is a linear function of y.

5. Let (X, Y) have a joint density function

$$f(x, y) = \begin{cases} e^{-y}, & \text{if } y > 0 \text{ and } 0 < x < 1, \\ 0, & \text{otherwise.} \end{cases}$$

For $n = 1, 2, \ldots$, find

$$E(X^n | Y = y).$$

6. Show that any bivariate distribution function $F(x, y)$ associated with (X, Y), either discrete or continuous, satisfies

$$F(x_2, y_2) - F(x_1, y_2) - F(x_2, y_1) + F(x_1, y_1) \geq 0$$

for any real numbers $x_1 < x_2$, $y_1 < y_2$.

[Hint: the left-hand side expresses the probability of an event A, which you should try to identify.]

7. In Example 2.4.1, find the probability that

   (i) Thelma arrives first and waits for less than ten minutes until Louise arrives;

   (ii) given that Thelma arrives first, she has to wait for at most ten minutes until Louise arrives.

8. A shooter aims at a certain target which has radius r. We assume that the point hit at the target is chosen randomly from the circle centered at the center of the target, O, with radius r. If the point A which is hit at the target has a distance less than $r/4$ from O, then the shooter wins a prize of US50. *If that distance is between r/4 and r/2, he wins US20*, and otherwise he receives nothing. Find the shooter's expected profit from that game.

   [Hint: You may find the result of Exercise 1 in Section 2.6 to be useful here.]

9. The joint density function of (X, Y) is

$$f(x, y) = \begin{cases} cx(y - x)e^{-y}, & 0 < x \leq y < \infty, \\ 0, & \text{otherwise.} \end{cases}$$

   (i) Find the value of c.

   (ii) Obtain the marginal density functions of X and Y.

   (iii) Show that

$$f_{Y|X}(y|x) = (y - x)e^{x-y}, \qquad 0 < x \leq y < \infty,$$

   and check that this is a proper density function so that, in particular, it integrates to one.

   (iv) Deduce that, for a given value of $X = x$, the conditional expectation of Y is

$$E(Y|X = x) = x + 2.$$

10. A pair of variables (X, Y) is uniformly distributed over the range $R_{X,Y}$, which is defined by the triangle with vertices at the points with coordinates $(0, 0)$, $(2, 0)$ and $(2, 4)$.

    (i) Obtain the joint density function $f(x, y)$ of (X, Y) and the marginal densities $f_X(x)$ and $f_Y(y)$.

    (ii) Find the quantities $E(X)$, $E(Y)$, $Var(X)$ and $Var(Y)$.

    (iii) Calculate the quantities $E(X|Y = y), E(Y|X = x), \quad Var(X|Y = y)$ and $Var(Y|X = x)$.

11. The joint density function of (X, Y) is given by

$$f(x, y) = \begin{cases} c(x+2)^2 e^{-7y}, & 0 < x < 1 \text{ and } 0 < y < \infty, \\ 0, & \text{otherwise,} \end{cases}$$

for a suitable constant c.

(i) Show that c is $c = 21/19$.

(ii) Find the marginal densities of X and Y.

(iii) Derive the conditional density $f_{X|Y}(x|y)$.

(iv) Calculate the value of the conditional (cumulative) distribution function $F_{X|Y}(x|y)$, for $x = 1/2, y = 2$.

2.11 APPLICATIONS

2.11.1 Modeling Proportions

In many real-life applications, there is a need that arises naturally for modeling two random variables which account for proportions. In these cases, the bivariate density function $f(x, y)$ to be used should have a support described by $x, y \geq 0$ and $x + y \leq 1$. For example:

- when studying the outcome of an electoral contest one wishes to model, as random variables, the proportions of the electoral body who vote for each of the two most popular candidates. These proportions might add to less than one due to the presence of other candidates that are of no interest in our study.

- if we are interested in modeling the buying behavior of consumers, we need a bivariate probability distribution to describe the joint distribution of brand shares; we assume here that there exist only three different brands in the market, so that describing the stochastic behavior of the shares for two of them determines implicitly the share of the third one.

- in compositional data, we are interested in the proportions of each component, which should add up to unity, or to less than one when considering only $n - 1$ among the n components. The percentage composition of different minerals in rocks, the percentages of n nutrients in foods (as seen in nutrition facts labels), the percentage of household expenditures spent on n different groups of commodities are examples of compositional data. In all these cases, a bivariate distribution with density $f(x, y)$ and with support $x \geq 0, y \geq 0$ and $x + y \leq 1$ will be needed when $n = 3$.

The most popular bivariate distribution used for modeling percentages is the bivariate Beta distribution. Its density function reads

$$f(x, y) = \frac{\Gamma(a+b+c)}{\Gamma(a)\Gamma(b)\Gamma(c)} x^{a-1} y^{b-1} (1 - x - y)^{c-1}, \quad x, y \geq 0 \text{ and } x + y \leq 1 \qquad (2.11)$$

where $a, b, c > 0$ are the parameters of the distribution. The marginal density of X is given by

$$f_X(x) = \int_0^1 f(x, y) dy = \frac{\Gamma(a+b+c)}{\Gamma(a)\Gamma(b)\Gamma(c)} x^{a-1} \int_0^1 y^{b-1} (1 - x - y)^{c-1} dy$$

and substituting $y = (1 - x)t$ we obtain

$$f_X(x) = \frac{\Gamma(a+b+c)}{\Gamma(a)\Gamma(b)\Gamma(c)} \, x^{a-1}(1-x)^{b+c-1} \int_0^1 t^{b-1}(1-t)^{c-1}dt.$$

The integral equals $B(b,c) = \Gamma(b)\Gamma(c)/\Gamma(b+c)$, and thus $f(x,y)$ reduces to

$$f_X(x) = \frac{\Gamma(a+b+c)}{\Gamma(a)\Gamma(b+c)} \, x^{a-1}(1-x)^{b+c-1}, \qquad x \geq 0.$$

The last formula indicates that X follows a univariate Beta distribution with mean and variance given by (see, e.g. Section 7.4.2 in Volume I)

$$E(X) = \frac{a}{a+b+c}, \quad Var(X) = \frac{a(b+c)}{(a+b+c)^2(a+b+c+1)}.$$

In a similar way, we may deduce the marginal density of Y as

$$f_Y(y) = \frac{\Gamma(a+b+c)}{\Gamma(b)\Gamma(a+c)} \, y^{b-1}(1-y)^{a+c-1}, \qquad y \geq 0,$$

that is, Y also follows a Beta distribution with

$$E(Y) = \frac{b}{a+b+c}, \quad Var(Y) = \frac{b(a+c)}{(a+b+c)^2(a+b+c+1)}.$$

Using the joint density $f(x,y)$ in (2.11), the marginal density $f_Y(y)$ found above and the formula $f_X(x|Y=y) = f(x,y)/f_Y(y)$, we easily arrive at the conditional density of X given that $Y = y$, namely

$$f_X(x|Y=y) = \frac{\Gamma(a+c)}{\Gamma(a)\Gamma(c)} \frac{x^{a-1}(1-x-y)^{c-1}}{(1-y)^{a+c-1}}, \qquad 0 < x < 1-y.$$

The expectation associated with this density is

$$E(X|Y=y) = \frac{a}{a+c}(1-y), \qquad 0 \leq y \leq 1.$$

Similar formulas can be established for the conditional distribution of Y given that $X = x$.

Let us next present an illustration of the above model in a practical example. Assume that X, Y stand for the percentages of household expenditures, spent on two groups of commodities, A and B, respectively. From past data, it has been observed that the mean expenditure percentage for A was 40% while for B it was 20%. Assuming that the market is stationary, we may use these figures as estimates for $E(X)$ and $E(Y)$, respectively.

Using (2.11) as an appropriate bivariate model for the description of the stochastic behavior of the random variables X, Y, we should have

$$E(X) = \frac{a}{a+b+c} = \frac{2}{5}, \quad E(Y) = \frac{b}{a+b+c} = \frac{1}{5};$$

these conditions yield $a = c = 2b$ and we thus have at hand a whole family of bivariate distributions (for each $b > 0$ we have a different density) that could be used for modeling

X and Y. The parameter b could be chosen according to additional information extracted from past data, e.g. the variance of X or Y.

For simplicity in the calculations that will be carried out, let us use $b = 1$, in which case we have $a = c = 2$ and the joint density is

$$f(x, y) = 24x(1 - x - y), \qquad x, y \geq 0 \text{ and } x + y \leq 1.$$

The marginal densities of X and Y are now given by

$$f_X(x) = 12x(1 - x)^2, \quad 0 \leq x \leq 1$$

and

$$f_Y(y) = 4(1 - y)^3, \quad 0 \leq y \leq 1$$

while the conditional expectations are

$$E(X|Y = y) = \frac{1}{2}(1 - y), \qquad 0 \leq y \leq 1$$

and

$$E(Y|X = x) = \frac{1}{3}(1 - x), \qquad 0 \leq x \leq 1.$$

It is of interest to note that, in view of the last two formulas,

- if $x < E(X) = \frac{2}{5}$, then $E(Y|X = x) > \frac{1}{3}\left(1 - \frac{2}{5}\right) = \frac{1}{5} = E(Y)$,

- if $y < E(Y) = \frac{1}{5}$, then $E(X|Y = y) > \frac{1}{2}\left(1 - \frac{1}{5}\right) = \frac{2}{5} = E(X)$,

i.e. given that we have spent in one of the commodities a percentage smaller than its mean, the expected percentage spent on the other one exceeds its (unconditional) mean.

Let us now answer the following questions:

(a) what is the probability that the household expenditure for each commodity is less than 25%?

(b) what is the probability that the household expenditure for at least one commodity is less than 25%?

(c) given that we have spent 25% for commodity B, what is the probability of spending less than 25% for commodity A as well?

(d) what is the probability that the household expenditure for commodity A exceeds that for commodity B?

For question (a) we have (see also the figure below), after a little algebra that

$$P(X < 1/4, Y < 1/4) = \int_0^{1/4} \int_0^{1/4} 24x(1 - x - y)dx\, dy = \frac{17}{128} \cong 0.1328.$$

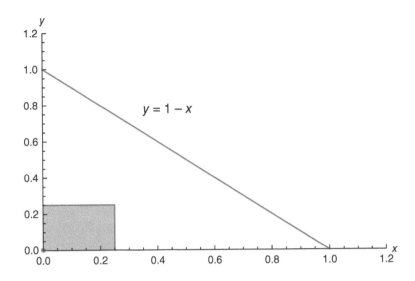

For (b), we may write

$$P\left(X < \frac{1}{4} \text{ or } Y < \frac{1}{4}\right) = P\left(X < \frac{1}{4}\right) + P\left(Y < \frac{1}{4}\right) - P\left(X < \frac{1}{4}, Y < \frac{1}{4}\right)$$

and making use of (a) and the marginal probabilities

$$P\left(X < \frac{1}{4}\right) = \int_0^{1/4} 12x(1 - x)^2 dx = \frac{67}{256} \cong 0.2617$$

and

$$P\left(Y < \frac{1}{4}\right) = \int_0^{1/4} 4(1 - y)^3 dy = \frac{175}{256} \cong 0.6836$$

we get that

$$P\left(X < \frac{1}{4} \text{ or } Y < \frac{1}{4}\right) = \frac{67}{256} + \frac{175}{256} - \frac{17}{128} = \frac{13}{16} = 0.8125.$$

Question (c) requires the conditional density

$$f_X(x|Y = y) = \frac{f(x, y)}{f_Y(y)} = \frac{24x(1 - x - y)}{4(1 - y)^3} = \frac{6x(1 - x - y)}{(1 - y)^3}$$

for $y = 1/4$, that is

$$f_X(x|Y = 1/4) = \frac{6x(3/4 - x)}{(3/4)^3} = \frac{32x(3 - 4x)}{9}, \quad 0 \le x \le 3/4.$$

Thus, the required conditional probability is

$$P\left(X < \frac{1}{4} \mid Y = \frac{1}{4}\right) = \int_0^{1/4} f_X(x|Y = 1/4)dx = \int_0^{1/4} \frac{32x(3 - 4x)}{9} dx = \frac{7}{27}.$$

Note that this probability does not coincide with the unconditional probability $P(X \leq 1/4)$ found earlier.

Finally, for Question (d) we have (see also the following figure)

$$P(X > Y) = \int_0^{1/2} \int_y^{1-y} f(x, y)dx \, dy = \int_0^{1/2} 4(y + 1)(2y - 1)^2 dy = 3/4,$$

after some straightforward calculations.

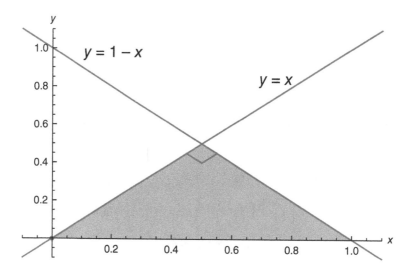

KEY TERMS

conditional cumulative distribution function

conditional distribution

conditional expectation

conditional probability function

geometric probability

joint cumulative distribution function

joint density function

joint distribution

marginal distribution

marginal density function

random pair (of continuous variables); or two-dimensional (continuous) random variable

CHAPTER 3

INDEPENDENCE AND MULTIVARIATE DISTRIBUTIONS

Yuri Linnik (Bila Tserkva 1915 – Leningrad 1972)

Yuri Vladimirovich Linnik was born on 8 January 1915, in Bila Tserkva (in present-day Ukraine) and died on 30 June 1972, in Leningrad (in present-day St. Petersburg, Russia). He studied in St. Petersburg University under the supervision of Professor Vladimir Tartakovski, and subsequently worked in that University and in the Steklov Institute.

He is known for Linnik's large sieve, Linnik's theorem in analytic number theory, and Linnik's ergodic method. He developed many fine results on infinite divisible distributions, which resulted in the book *Decomposition of Random Variables and Vectors*, collaboratively with I.V. Ostrovskii. He also established numerous characterization results for statistical distributions, many of which are based on independence of certain statistics, and these resulted in the book *Characterization Problems in Mathematical Statistics*, jointly with A.M. Kagan and C.R. Rao.

For his fine work and accomplishments, he received both the State Prize and the Lenin Prize of the Soviet Union.

3.1 INTRODUCTION

The concept of stochastic independence of random variables is one of the most important concepts in probability theory. In the case when two random variables are independent, knowledge of the distribution for each of them is sufficient to determine their common behavior. On the other hand, the existence of independence between the random variables we are interested in, greatly simplifies the calculation of probabilities related to both, as well as their marginal and conditional distributions.

In this chapter, we will also see how we can study the common behavior of more than two random variables. We shall introduce the concept of random sample and briefly discuss the distribution of random variables resulting from a random sample when placing them in ascending or descending order (ordered sample).

3.2 INDEPENDENCE

When we consider more than one variable, the concept of independence becomes fundamental in much the same way as it is when we study multiple events defined on the same sample space. Recall that two events A and B on a sample space Ω are independent if

$$P(AB) = P(A)P(B), \tag{3.1}$$

that is, the probability that they both occur is the product of the probabilities that each of them occurs. Intuitively, this means that knowledge of whether one of the two events occurs or not does not affect the probability of the other one occurring. This can be seen more clearly from the expressions

$$P(A|B) = P(A), \quad P(B|A) = P(B),$$

which are equivalent to (3.1) provided $P(A) \neq 0$ and $P(B) \neq 0$. These equations reveal that the conditional probability of event A (respectively, B) occurring, given event B (respectively, A) has occurred, is the same as the unconditional probability of A (respectively, B) occurring.

The idea of independent random variables is essentially a generalization of this notion of independent events. Suppose we throw two dice and denote by X the outcome of the first die and by Y the outcome of the second die. Then, it seems reasonable to assume that knowing the value of X would not affect the probability function of the second outcome, Y. More generally, we can present the following definition with regard to this concept.

Definition 3.2.1 Let X and Y be two random variables defined on the same sample space. Then, we say that X and Y are independent if

$$P(X \in A, Y \in B) = P(X \in A) \, P(Y \in B) \tag{3.2}$$

for any events A and B on that space.

As a special case of the above definition, if we take $A = (-\infty, x]$ and $B = (-\infty, y]$, we see that

$$F(x, y) = F_X(x)F_Y(y), \tag{3.3}$$

i.e. the joint distribution function of X and Y is equal to the product of the marginal distribution functions. Note that, in general, the joint distribution function F contains a lot more information than the marginal distribution functions F_X and F_Y, as it explains how X and Y vary *together*. However, when X and Y are independent, then knowing F is the same as knowing both F_X and F_Y.

What we have shown above is that if X and Y are independent, then (3.3) holds for any x and y. The converse of this result is also true, and is given in the following proposition (the proof of this is rather difficult and is not given here as it is beyond the scope of the present book).

Proposition 3.2.1 *Let X and Y be two random variables with joint distribution function $F(x, y)$ and marginal distribution functions $F_X(x)$ and $F_Y(y)$, respectively. Then, X and Y are independent if and only if (3.3) holds for any $x, y \in \mathbb{R}$.*

Recall that distribution functions, and joint distribution functions, are defined in the same way for both discrete and continuous random variables. Therefore, Proposition 3.2.1 holds for both discrete and continuous cases.

Suppose X and Y are discrete with joint probability function $f(x, y)$ and marginal probability functions $f_X(x)$ and $f_Y(y)$, respectively. Let (x, y) be an arbitrary pair in the range $R_{X,Y}$ of (X, Y). By taking

$$A = \{x\}, \quad B = \{y\}$$

in (3.2), we see that

$$f(x, y) = f_X(x) f_Y(y).$$

Conversely, assume that this holds for any x and y, and let $A \subseteq R_X$, $B \subseteq R_Y$. Then, we obtain

$$P(X \in A, Y \in B) = \sum_{x \in A} \sum_{y \in B} f(x, y) = \sum_{x \in A} \sum_{y \in B} f_X(x) f_Y(y)$$

$$= \sum_{x \in A} f_X(x) \left(\sum_{y \in B} f_Y(y) \right)$$

$$= \sum_{x \in A} f_X(x) P(B) = P(B) \sum_{x \in A} f_X(x) = P(X \in A) P(Y \in B),$$

which shows that X and Y are independent.

Thus, we have the following proposition.

Proposition 3.2.2 *Let X and Y be two discrete random variables with joint probability function $f(x, y)$ and marginal probability functions $f_X(x)$ and $f_Y(y)$, respectively. Then, X and Y are independent if and only if*

$$f(x, y) = f_X(x) f_Y(y)$$

for any $(x, y) \in R_{X,Y}$.

An analogous result holds for continuous random variables as well and is given in the following proposition.

Proposition 3.2.3 *Let X and Y be two random variables with joint density function $f(x,y)$ and marginal density functions $f_X(x)$ and $f_Y(y)$, respectively. Then, X and Y are independent if and only if*

$$f(x,y) = f_X(x)f_Y(y)$$

for any $x, y \in \mathbb{R}$.

Proof: Assume that X and Y are independent. Then (3.3) holds, and by differentiating with respect to x, we get

$$\frac{\partial}{\partial x}F(x,y) = F_X'(x)F_Y(y) = f_X(x)F_Y(y).$$

Upon differentiating with respect to y, we get

$$\frac{\partial}{\partial y}\left(\frac{\partial}{\partial x}F(x,y)\right) = \frac{\partial}{\partial y}\left(f_X(x)F_Y(y)\right) = f_X(x)F_Y'(y) = f_X(x)f_Y(y),$$

that is,

$$f(x,y) = f_X(x)f_Y(y).$$

The proof of the converse proceeds by following the same steps as in the proof of Proposition 3.2.2 by simply replacing the sums by integrals.   □

Summarizing the above results, in order to show that two variables X and Y are independent, it is sufficient to verify one of the following:

(a) $P(X \in A, Y \in B) = P(X \in A)P(Y \in B)$ for any events A and B;

(b) $F(x,y) = F_X(x)F_Y(y)$ for any (x,y);

(c) $f(x,y) = f_X(x)f_Y(y)$ for any (x,y).

Among these three conditions, the ones which are usually easier to check are (b) and (c). Note that in (c), f, f_X, and f_Y are interpreted as density functions in the continuous case, and as probability functions in the discrete case. Thus, we would typically use either (b) or (c) to ascertain that two variables are independent, and then employ (a) to calculate probabilities of interest associated with (X,Y).

Example 3.2.1 Two variables X and Y have joint distribution function

$$F(x,y) = \begin{cases} 1 - e^{-x} - e^{-y} + e^{-x}e^{-y}, & x, y \geq 0, \\ 0, & \text{otherwise.} \end{cases}$$

It is easy to see that the marginal distribution functions of X and Y are given by

$$F_X(x) = \lim_{y \to \infty} F(x,y) = 1 - e^{-x}, \qquad x \geq 0,$$

and

$$F_Y(y) = \lim_{x \to \infty} F(x, y) = 1 - e^{-y}, \qquad y \geq 0.$$

Therefore, for any x and y, we observe that

$$F(x, y) = F_X(x)F_Y(y),$$

and so X and Y are independent. Observe that X and Y both have the standard exponential distribution (i.e. with a unit mean) in this case.

Now, if we want to compute the probability $P(X > 2, Y \leq 3)$, using the independence of X and Y and setting $A = (2, \infty)$, $B = (-\infty, 3]$ in (3.2), we obtain that

$$P(X > 2, Y \leq 3) = P(X > 2)P(Y \leq 3) = \left(1 - F_X(2)\right) F_Y(3) = e^{-2}(1 - e^{-3}).$$

Example 3.2.2 The lifetimes, in thousands of hours, for two mechanical parts in a machine are continuous random variables, denoted by X and Y, respectively; they have a joint density function as

$$f(x, y) = \begin{cases} xe^{-x(1+y)}, & \text{if } x \geq 0 \text{ and } y \geq 0, \\ 0, & \text{otherwise.} \end{cases}$$

Examine whether X and Y are independent.

SOLUTION The marginal density function of X is

$$f_X(x) = \int_{-\infty}^{\infty} f(x, y)dy = \int_0^{\infty} xe^{-x(1+y)} \, dy = xe^{-x} \int_0^{\infty} e^{-xy} \, dy$$

$$= xe^{-x} \left[-\frac{e^{-xy}}{x} \right]_{y=0}^{\infty} = e^{-x}, \qquad x \geq 0,$$

while the marginal density of Y is

$$f_Y(y) = \int_{-\infty}^{\infty} f(x, y)dx = \int_0^{\infty} xe^{-x(1+y)} \, dx = \int_0^{\infty} x \left(-\frac{e^{-x(1+y)}}{1+y} \right)' dx,$$

where the derivative above is considered with respect to x. Integration by parts then yields

$$f_Y(y) = \left[-\frac{x}{1+y} e^{-x(1+y)} \right]_{x=0}^{\infty} + \frac{1}{1+y} \int_0^{\infty} e^{-x(1+y)} \, dx$$

$$= \frac{1}{1+y} \left[-\frac{e^{-x(1+y)}}{1+y} \right]_{x=0}^{\infty} = \frac{1}{(1+y)^2}, \qquad y \geq 0.$$

Because $f(x, y) \neq f_X(x)f_Y(y)$ in general, we see that X, Y are not independent in this case.

Proposition 3.2.3 shows that if two variables X and Y have a joint continuous distribution and their joint density is the product of the marginal densities, then X and Y are independent. The following result is more general; it shows that X and Y are independent whenever the joint density $f(x, y)$ can be factorized as a product of a function of x alone and a function of y alone.

Proposition 3.2.4 *Let X and Y be two random variables having a joint continuous distribution with density $f(x, y)$. If there exist two functions $f_1(x)$ and $f_2(y)$, of x alone and of y alone respectively, such that*

$$f(x, y) = f_1(x) f_2(y), \qquad \text{for any } x, y \in \mathbb{R}, \qquad (3.4)$$

then X and Y are independent random variables.

SOLUTION   *From (3.4), we have the marginal density functions of X and Y as*

$$f_X(x) = \int_{-\infty}^{\infty} f(x, y) dy = f_1(x) \int_{-\infty}^{\infty} f_2(y) dy = c_1 f_1(x), \quad x \in \mathbb{R}$$

and

$$f_Y(y) = \int_{-\infty}^{\infty} f(x, y) \, dx = f_2(y) \int_{-\infty}^{\infty} f_1(x) \, dx = c_2 f_2(y), \quad y \in \mathbb{R},$$

where we have set

$$c_1 = \int_{-\infty}^{\infty} f_2(y) dy, \quad c_2 = \int_{-\infty}^{\infty} f_1(x) dx.$$

But, $f(x, y)$ is a joint density function, and so we must have

$$1 = \int_{-\infty}^{\infty} \int_{-\infty}^{\infty} f(x, y) dx \, dy = \int_{-\infty}^{\infty} \left(\int_{-\infty}^{\infty} f_1(x) dx \right) f_2(y) dy = \int_{-\infty}^{\infty} c_2 f_2(y) dy$$

$$= c_2 \int_{-\infty}^{\infty} f_2(y) dy = c_2 c_1.$$

This gives

$$f_X(x) f_Y(y) = \left(c_1 f_1(x) \right) \left(c_2 f_2(y) \right) = \left(c_1 c_2 \right) f_1(x) f_2(y) = f_1(x) f_2(y) = f(x, y)$$

for any $(x, y) \in \mathbb{R}$. Therefore, we see from Proposition 3.2.3 that X and Y are independent random variables.

It is important to note that in order to apply Proposition 3.2.4, we have to check that (3.4) is valid for any real x and y *without* any restriction involving both x and y, such as $x < y$, $x^2 \geq y$ and so on (when we have restrictions of this form, typically X and Y are not independent). The following example illustrates this point.

Example 3.2.3 In each of the two cases below, examine whether the variables X and Y, having joint density f, are independent:

(i) $f(x,y) = \begin{cases} 4xy, & 0 \le x \le 1, \ 0 \le y \le 1, \\ 0, & \text{otherwise;} \end{cases}$

(ii) $f(x,y) = \begin{cases} 8xy, & 0 \le x \le y \le 1, \\ 0, & \text{otherwise.} \end{cases}$

SOLUTION

(i) If we define the functions $f_1(x) = 4x$, $0 \le x \le 1$, and $f_2(y) = y$, $0 \le y \le 1$, with f_1 and f_2 zero outside the interval $[0, 1]$, it is evident from Proposition 3.2.4 that X and Y are independent. Observe, in particular, that the functions f_1 and f_2 in the proposition are not unique; for instance, we could have instead used the functions $f_1(x) = 2x$, $0 \le x \le 1$, and $f_2(y) = 2y$, $0 \le y \le 1$.

(ii) In this case, Proposition 3.2.4 cannot be used since we have the condition $x \le y$ in the definition of the function $f_2(y)$. So, we have to find the marginal densities of X and Y, which are given by

$$f_X(x) = \int_{-\infty}^{\infty} f(x,t)dt = \int_{x}^{1} 8xt\, dt = 8x \left[\frac{t^2}{2}\right]_{x}^{1} = 4x(1 - x^2), \quad 0 \le x \le 1,$$

and

$$f_Y(y) = \int_{-\infty}^{\infty} f(s,y)ds = \int_{0}^{y} 8sy\, ds = 8y \left[\frac{s^2}{2}\right]_{0}^{y} = 4y^3, \quad 0 \le y \le 1.$$

It is clear that $f(x,y) \ne f_X(x)f_Y(y)$, and so X and Y are not independent.

It is worth noting that Proposition 3.2.4 also holds for discrete random variables (provided we replace density functions by probability functions).

We mentioned in the beginning of this chapter that our intuitive notion of independence between two events is that the outcome of one of them does not have any effect on the outcome of the other. Precisely the same can be said for random variables. To make this more formal, we recall the definition of conditional probability and conditional density functions from the preceding chapters. Let us consider the case of two discrete variables X and Y. Then, the conditional probability function of X given $Y = y$, is

$$f_{X|Y}(x|y) = \frac{f(x,y)}{f_Y(y)}, \qquad x \in R_X,$$

provided $f_Y(y) \ne 0$. Under this last assumption, the relation $f(x,y) = f_X(x)f_Y(y)$ in Proposition 3.2.3 takes the following equivalent form

$$f_{X|Y}(x|y) = f_X(x), \tag{3.5}$$

that is, the conditional probability function of X, given $Y = y$, is for any value of y, the same as the unconditional (i.e. the marginal) probability function of X. In other words, knowing the value of Y does not add any knowledge to our uncertainty about X. The same obviously holds if the roles of X and Y are interchanged; but this argument works in the continuous case as well. Thus, we have proved the following.

> **Corollary 3.2.1** *Let X and Y be two discrete random variables with marginal probability functions f_X and f_Y, respectively, and f be their joint probability function. Then X and Y are independent if and only if one of the following conditions hold:*
>
> (i) $f_{X|Y}(x|y) = f_X(x)$ *for any $x, y \in \mathbb{R}$ such that $f_Y(y) \neq 0$;*
>
> (ii) $f_{Y|X}(y|x) = f_Y(y)$ *for any $x, y \in \mathbb{R}$ such that $f_X(x) \neq 0$.*
>
> *The result also holds if X and Y have a joint continuous density function.*

We note that results analogous to Proposition 3.2.4 and Corollary 3.2.1 also hold for the case of distribution functions. In particular, it is useful to have in mind that we can check independence, both for discrete and continuous random variables, by examining whether the joint cumulative distribution function, $F(x, y)$, of (X, Y) can be written in the form

$$F(x, y) = F_1(x)F_2(y) \tag{3.6}$$

for all $x, y \in \mathbb{R}$.

Further, analogous to Corollary 3.2.1, for any pair of independent random variables X and Y, we have

(i) $F_{X|Y}(x|y) = F_X(x)$ for any $x, y \in \mathbb{R}$ such that $f_Y(y) \neq 0$;

(ii) $F_{Y|X}(y|x) = F_Y(y)$ for any $x, y \in \mathbb{R}$ such that $f_X(x) \neq 0$.

Finally, if two probability (or density) functions are the same, then the associated expectations must coincide. Thus, from Corollary 3.2.1, we also obtain that for X and Y independent, the conditional expectation of X, given $Y = y$, is the same as the unconditional expectation of X. The same is also true for the expectation of Y, and we thus have

(i) $E(X|Y = y) = E(X)$ for any $y \in R_Y$ such that $f_Y(y) \neq 0$;

(ii) $E(Y|X = x) = E(Y)$ for any $x \in R_X$ such that $f_X(x) \neq 0$.

Although (3.4) and (3.6) are useful in checking whether two variables are independent or not, we only need to use them when we have some *reason to doubt* the assumption of independence for X and Y. In many cases, especially when X and Y are related to different chance experiments, and the outcomes of these experiments would not depend on each other (a typical case will be when X and Y relate to successive repetitions of the same experiment), it is natural to take independence for granted. For instance,

- when we throw two dice, the variables

  X: the outcome of the first die,

  Y: the outcome of the second die

  are independent random variables;

- if X and Y represent the heights (or weight, gender, age, etc.) of two persons selected at random from a population, it would seem reasonable to assume that the values of X would not affect those of Y and vice versa, so that X and Y are independent;

- if X stands for the turnover of a shop in a given year and Y is the turnover of another shop located in a different area for the same year, we can probably assume that X and Y are independent;

- if X denotes the number of accidents on a motorway in California during a given month and Y is the total amount of rainfall in Baltimore for next year, it is clear that the value of X would not have any bearing on the value of Y and vice versa, and so X and Y can be assumed again to be independent variables.

Example 3.2.4 Among the units produced in a production line, the percentage of those that are defective is 5%. What is the probability that a first batch of 10 units selected contains exactly two defectives, while a second batch of 20 units contains at most one defective? We assume that both batches are selected from a very large set of produced units.

SOLUTION   We consider the random variables

X: number of defective units in the first batch,
Y: number of defective units in the second batch.

Since both batches are selected from a very large set of produced units, X and Y may be assumed to be independent (otherwise, if the total number of units is small, the value of X would affect the value of Y). The distribution of X is binomial with parameters $n = 10, p = 0.05$, and so

$$f_X(x) = P(X = x) = \binom{10}{x}(0.05)^x(0.95)^{10-x}, \qquad x = 0, 1, \dots, 10.$$

Similarly, the distribution of Y is binomial with $n = 20, p = 0.05$, and so

$$f_Y(y) = P(Y = y) = \binom{20}{y}(0.05)^y(0.95)^{20-y}, \qquad y = 0, 1, \dots, 20.$$

As X and Y are independent, we have $P(X \in A,\ Y \in B) = P(X \in A)P(Y \in B)$, and so by choosing $A = \{2\}, B = \{0, 1\}$, we obtain

$$P(X = 2,\ Y \le 1) = P(X = 2)P(Y \le 1) = P(X = 2)\left[P(Y = 0) + P(Y = 1)\right]$$

$$= \binom{10}{2}(0.05)^2(0.95)^8\left[(0.95)^{20} + \binom{20}{1}(0.05)(0.95)^{19}\right]$$

$$= (0.075)\left[0.358 + 0.377\right] = 0.055.$$

Thus, the event of interest has probability 5.5%.

Example 3.2.5 The amount of time Sandra needs to write an email to one of her friends is exponentially distributed with a mean value of one minute. Today, she has to write two messages to her friends, Vicky and Christina. Let X be the time (in minutes) it will take her to write to Vicky and Y be the time she needs to write to Christina. Let X and Y be independent.

(i) Find the joint density function and the joint distribution function of X and Y.

(ii) What is the probability that Sandra will need less than a minute to write each message?

(iii) What is the probability that the total time she needs to write the two messages will be more than one but no more than three minutes?

SOLUTION

(i) For the density functions and distribution functions of X and Y, we have

$$f_X(x) = e^{-x}, \qquad F_X(x) = 1 - e^{-x}, \qquad\qquad x \geq 0,$$
$$f_Y(y) = e^{-y}, \qquad F_Y(y) = 1 - e^{-y}, \qquad\qquad y \geq 0,$$

and so due to the independence of X and Y, we immediately have

$$f(x, y) = f_X(x) f_Y(y) = e^{-x-y}, \qquad\qquad x, y \geq 0,$$

and

$$F(x, y) = F_X(x) F_Y(y) = (1 - e^{-x})(1 - e^{-y}), \qquad\qquad x, y \geq 0.$$

(ii) The required probability is

$$P(X < 1, \ Y < 1) = P(X \leq 1, \ Y \leq 1) = F(1, 1) = F_X(1) F_Y(1) = \left(1 - e^{-1}\right)^2.$$

(iii) Here, we require the probability $P(1 < X + Y \leq 3)$. For finding this, we first observe that the required probability can be expressed as the difference $P(X + Y \leq 3) - P(X + Y \leq 1)$; so, we calculate each of these two probabilities separately. For the first one, we have

$$P(X + Y \leq 3) = \int_{x+y\leq3} \int f(x,y) dx\, dy = \int_0^3 \int_0^{3-y} e^{-x} e^{-y}\, dx\, dy$$

$$= \int_0^3 e^{-y} \left(\int_0^{3-y} e^{-x}\, dx \right) dy = \int_0^3 e^{-y} \left[-e^{-x} \right]_0^{3-y} dy$$

$$= \int_0^3 e^{-y} \left(1 - e^{y-3} \right) dy = \int_0^3 \left(e^{-y} - e^{-3} \right) dy = 1 - 4e^{-3}.$$

In a similar manner, we find

$$P(X + Y \leq 1) = \int_0^3 \int_0^{1-y} e^{-x} e^{-y} \, dx \, dy = 1 - 2e^{-1},$$

so that the probability that Sandra will need more than one but no more than three minutes to write her emails is

$$P(X + Y \leq 3) - P(X + Y \leq 1) = \left(1 - 4e^{-3}\right) - \left(1 - 2e^{-1}\right)$$

$$= 2e^{-1} - 4e^{-3} \cong 53.7\%.$$

Finally, in view of the results of this section, we can revisit the various two-dimensional random variables considered in the first two chapters and determine whether they are independent random variables or not (interested readers are encouraged to do this!). For example, we find that

- the variables X and Y in Examples 1.2.2, and 1.3.3 are independent discrete random variables;

- the discrete variables X and Y in Example 1.5.1 (regarding the number of people waiting for service by two cashiers) are not independent;

- the variables X and Y having a joint continuous distribution in Examples 2.2.2, 2.2.3, and 2.3.2 are independent;

- the variables X and Y having a joint continuous distribution in Examples 2.3.1 and 2.5.2 are not independent.

EXERCISES

Group A

1. Find the marginal probability functions of X and Y and check whether they are independent for each of the joint probability functions

$$f(x, y) = P(X = x, \ P(Y = y)$$

given below:

(i)

$x \backslash y$	-1	0	1
-1	$1/16$	$3/16$	$1/16$
0	$3/16$	0	$3/16$
1	$1/16$	$3/16$	$1/16$

(ii)

x \ y	0	1000	2000
200	0.15	0.20	0.05
400	0.10	0.30	0.20

(iii)

x \ y	0	1	2
0	0.10	0.06	0.04
2	0.35	0.21	0.14
4	0.05	0.03	0.02

2. For each of the bivariate probability functions below, examine whether X and Y are independent:

(i) $f(x,y) = \dfrac{1}{30}\left(x^2 + y^2\right)$, $x = 0, 1, 2$ and $y = 0, 1, 2$;

(ii) $f(x,y) = \begin{cases} \dfrac{xy}{5}, & \text{if } (x,y) = (1,1), (1,2), (2,1), \\ 0, & \text{otherwise;} \end{cases}$

(iii) $f(x,y) = \dfrac{x+y}{315}\left(x^2 + y^2\right)$, $x = 1, 2, 3$ and $y = 2, 3$.

Then, calculate the probability $P(X \geq 2, \ Y \leq 2)$ for each of these cases.

3. We draw two cards at random without replacement from an ordinary deck of 52 cards. Let X be the number of kings drawn and Y be the number of diamonds drawn. Check whether X and Y are independent.

4. For the random variables X and Y in Example 3.2.2,

(i) calculate the probabilities

$$P(X > 2), \quad P(Y \leq 2), \quad P(X > 2, Y \leq 2),$$

and examine whether the last one is equal to the product of the first two;

(ii) examine whether $E(XY) = E(X)E(Y)$.

5. Consider two random variables X and Y having marginal distribution functions as

$$F_X(x) = \begin{cases} 0, & x < 2, \\ 1/3, & 2 \leq x < 3, \\ 5/6, & 3 \leq x < 4, \\ 1, & x \geq 4, \end{cases}$$

and

$$F_Y(y) = \begin{cases} 0, & y < 0, \\ 1 - \dfrac{e^{-y/2}}{2} - \dfrac{e^{-y/4}}{2}, & y \geq 0, \end{cases}$$

respectively. Assuming that X and Y are independent,

(i) find the joint distribution function of (X, Y);

(ii) calculate the probabilities

$$P(X > 5/2), \quad P(Y \leq 3), \quad P(P(X > 5/2, \ 1 \leq Y \leq 3).$$

(Note here that X is discrete while Y is continuous, but this presents no problem as we are dealing with distribution functions.)

6. Let X and Y be two discrete random variables with joint probability function $f(x, y)$ given by

$$f(1, 2) = \frac{1}{4}, \quad f(1, 4) = \frac{1}{6}, \quad f(2, 2) = \frac{1}{4}, \quad f(2, 4) = \frac{1}{3}.$$

(i) Find the marginal probability functions of X and Y.

(ii) Obtain the expectations of X, Y and their product, XY.

(iii) Find the conditional probability function $f_{X|Y}(x|y)$ for each y in the range of Y.

(iv) Explain whether or not X and Y are independent.

(v) Examine whether the variables X^2 and $Y - 1$ are independent.

7. The joint probability function of a bivariate random variable (X, Y) is given by

$$f(x, y) = c, \qquad \text{for } x = 1, 2, \dots, y \text{ and } y = 1, 2, \dots, 10.$$

(i) Find the value of c.

(ii) Derive the marginal probability functions of X and Y, and examine whether any of the following relations hold:

$$f_{X|Y}(x|y) = f_X(x) \quad \text{and} \quad f_{Y|X}(y|x) = f_Y(y).$$

Are X and Y independent?

8. For each of the bivariate density functions below, show that the variables X and Y are independent without finding their marginal density functions (a and b below are two given positive constants):

(i) $f(x, y) = 3x^4 e^{-3y/5}$, $\qquad\qquad\qquad 0 < x < 1, \ y \geq 0$;

(ii) $f(x, y) = 8xy^3$, $\qquad\qquad\qquad\qquad 0 < x < 1, \ 0 < y < 1$;

(iii) $f(x, y) = (ab)^{-1}$, $\qquad\qquad\qquad\quad 0 < x < a, \ 0 < y < b$;

(iv) $f(x, y) = \dfrac{1}{2\pi} \exp\left[-\dfrac{1}{8}(x - 1)^2 - 2(y - 3)^2 \right], x \in \mathbb{R}, y \in \mathbb{R}$.

9. Assume that X and Y have a joint density function as

$$f(x, y) = \begin{cases} ce^{-x-y}, & \text{if } 0 < x < y, \\ 0, & \text{otherwise.} \end{cases}$$

(i) Find the value of c.

(ii) Calculate the probabilities

$$P(X > 3), \quad P(Y < 2) \quad \text{and} \quad P(X > 3, Y < 2),$$

and verify that X, Y are not independent.

10. The lifetimes (in thousands of hours) of two microorganisms, found in a contaminated area, are described by two continuous random variables X and Y with joint distribution function

$$F(x, y) = (1 - xe^{-x} - e^{-x}) \left(1 - e^{-3y^2} \right), \qquad\qquad x, y \geq 0.$$

(i) Examine whether X and Y are independent.

(ii) Calculate the probability that

(a) exactly one of the two microorganisms survives for 1000 hours;

(b) at least one of the microorganisms survives for 2000 hours.

11. Assume that X and Y have a joint continuous distribution with

$$F(x, y) = \begin{cases} (9x^8 - 8x^9)(1 - ye^{-y} - e^{-y}), & 0 \leq x \leq 1, \ y \geq 0, \\ 1 - ye^{-y} - e^{-y}, & x \geq 1, \ y \geq 0, \\ 0, & \text{otherwise.} \end{cases}$$

(i) Identify the range for the variables $X, Y, (X, Y)$.

(ii) Show that X and Y are independent.

(iii) Verify that the marginal distribution of Y is the Gamma distribution and identify the parameters of that distribution.

(iv) Calculate the probabilities

$$P(X \leq 1/2, \ Y \leq 3), \quad P(X \leq 2, \ Y \leq 3), \quad P(Y \leq 3),$$

$$P(1 < X < Y < 3), \quad P(\max\{X, Y\} \leq 1/2).$$

12. The following table gives the joint probability function $f(x, y)$ of the variables X and Y as well as the marginal probability functions $f_X(x)$ and $f_Y(y)$. Fill in the missing entries from the table if it is known that X and Y are independent random variables:

$$f(x, y) = P(X = x, Y = y)$$

x \ y	0	1	2	6	$f_X(x)$
1				0.08	0.20
3				0.12	
5				0.20	0.50
$f_Y(y)$	0.20		0.30		

Group B

13. A small aircraft has two engines. Let X and Y be the lifetimes of these engines (in hundreds of thousands of hours) and assume that their joint density function is given by

$$f(x, y) = xe^{-xy}, \qquad\qquad x \geq 0, \ y \geq 1.$$

    (i) Show that X and Y are not independent.

    (ii) Calculate the probabilities

$$P(X \geq 2), \quad P(Y \geq 2), \quad P(X \geq 2, \ Y \geq 2).$$

    (iii) If the aircraft would operate properly if at least one of the two engines functions, what is the probability that there will be no failure during the first 250 000 hours of flight?

14. The measurement errors, X and Y, of two instruments form a continuous, two-dimensional random variable whose joint density function is

$$f(x, y) = \frac{1}{4\pi} \exp\left(-\frac{x^2}{2} - \frac{y^2}{8}\right), \qquad x, y \in \mathbb{R}.$$

    (i) Verify that X and Y are independent random variables and identify the distribution for each of them.

    (ii) Calculate the probabilities

$$P(-1 \leq X \leq 1), \quad P(-1 \leq Y \leq 1), \quad P(-1 \leq X \leq 1 \text{ and } -1 \leq Y \leq 1).$$

    (iii) What is the probability that at least one of the two error measurements, X and Y, has a magnitude less than one?

    (iv) What is the probability that both error measurements are greater than one in magnitude?

15. Consider the two following joint densities for (X, Y):

    (i) $f(x, y) = \dfrac{1}{\Gamma(a)\Gamma(b)} (xy)^{a-1} y^b \, e^{-x-y}, \qquad x, y \geq 0;$

    (ii) $f(x, y) = \dfrac{1}{\Gamma(a)\Gamma(b)} x^{a-1}(y - x)^{b-1} e^{-y}, \qquad 0 < x < y,$

    for some positive constants a and b.

    (i) Verify that the marginal distributions of X and Y are the same in the two cases above.

    (ii) Show that in case (i), X and Y are independent, while in Case (ii) they are not.

    (iii) Calculate $E(X), E(X|Y = 2), E(Y), E(Y|X = 1)$ in each case.

16. The joint density function of (X, Y) is given by

$$f(x, y) = \frac{x^2 - y^2}{8e^x}, \qquad |y| < x.$$

(i) Show that the marginal distribution of X is Gamma with density

$$f_X(x) = \frac{1}{6}x^3 e^{-x}, \qquad x \geq 0.$$

(ii) Show that the marginal density of Y is given by

$$f_Y(y) = \frac{1}{4}(1 + |y|)e^{-|y|}, \qquad y \in \mathbb{R}.$$

17. For a certain type of motorcycle, there are two tire types, a and b, that can be used. For tires of type a, it is known that the number of miles which they run (in tens of thousands), before they need replacement, is a random variable X with density function

$$f_X(x) = \frac{1}{3}e^{-x/3}, \qquad x > 0.$$

Similarly, for a tire of type b the number of miles they run, before they need replacement, is another variable Y with density function

$$f_Y(y) = \frac{1}{4}e^{-y/4}, \qquad y > 0.$$

If a motorcycle is equipped with one tire of each type, and assuming that the variables X and Y are independent, what is the probability that

(i) the tire of type a needs replacement before that of type b?

(ii) the tire of type b lasts for at least 10 000 miles more than the tire of type a?

18. Two doctors work at the same surgery clinic and are scheduled to start seeing patients at 9:00 am every morning. The arrival time of each doctor at the clinic is uniformly distributed in the time interval between 8:40 am and 9:05 am. Arrival times between the two doctors are assumed independent.

(i) What is the probability that at least one of the two doctors arrives at the clinic after 9:00 am?

(ii) What is the probability that the doctor who arrives first waits for the other doctor to arrive at least for 10 minutes?

19. The daily profit of a super-market store, in thousands of dollars, is a continuous random variable X with density function

$$f_X(x) = cx^2(3 - x), \qquad 1 \leq x \leq 3,$$

for a suitable constant c.

(i) Find the value of c.

(ii) Assume that each of two stores from the same super-market chain has the density above to describe its daily profit. Let these profits for a certain day be denoted by X and Y, respectively, and suppose that X and Y are independent random variables.

   (a) Explain why the probability statement

$$P(X > Y + a) = P(Y > X + a)$$

   holds for each positive real number a.

   (b) Find the probability that the difference between the profits of the two stores on a day is at least US\$600.

3.3 PROPERTIES OF INDEPENDENT RANDOM VARIABLES

In this section, we describe several useful properties when X and Y are independent variables. The first result below states that functions of X and Y are also independent random variables.

Proposition 3.3.1 *Let X and Y be two independent random variables and assume that $g : \mathbb{R} \mapsto \mathbb{R}$ and $h : \mathbb{R} \mapsto \mathbb{R}$ are two real-valued functions. Then, the random variables $g(X)$ and $h(Y)$ are also independent.*

Proof: In view of Proposition 3.2.1, it is sufficient to show that the following holds, for any $z, w \in \mathbb{R}$:

$$P(g(X) \le z, \ h(Y) \le w) = P(g(X) \le z)P(h(Y) \le w). \tag{3.7}$$

Let us define the following sets:

$$A = \{x \in \mathbb{R} : g(x) \le z\}, \qquad B = \{y \in \mathbb{R} : h(y) \le w\}.$$

Then, we have

$$P(g(X) \le z, \ h(Y) \le w) = P(X \in A, \ Y \in B),$$

$$P(g(X) \le z) = P(X \in A),$$

$$P(h(Y) \le w) = P(Y \in B),$$

so that (3.7) is equivalent to the relation

$$P(X \in A, \ Y \in B) = P(X \in A)P(Y \in B).$$

But, this is clearly true as X and Y are independent, and this completes the proof of the proposition.   □

As a consequence of the last result, we observe for example that, if two variables X and Y are independent, then each of the following pairs consists of independent random variables:

$$\left(X^2 - 4X + 2, \ e^Y\right), \qquad (\ln X + 4, 3Y + 2), \qquad \left(X^4 - 2X^3 + 5X, \sqrt{Y^2 + 2}\right).$$

This illustrates the practical use of Proposition 3.3.1 since, even if we know the joint distribution of (X, Y), it is typically difficult to obtain the joint distribution of the pairs above, in order to check for their independence. We mention in particular that for the use of Proposition 3.3.1, the functions g and h do not have to be one-to-one or even continuous. As another example, if X and Y are independent, then so are the variables Z and W defined as

$$
Z = \begin{cases} 0, & \text{if } X < 0, \\ 1, & \text{if } 0 \le X < 1, \\ 3, & \text{if } X \ge 1, \end{cases} \qquad W = \begin{cases} -1, & \text{if } Y < 3, \\ 1, & \text{if } Y \ge 3. \end{cases}
$$

A particularly useful result concerning the expectation of a product for two independent random variables is as follows.

Proposition 3.3.2 *If X and Y are independent random variables, then*

$$
E(XY) = E(X)E(Y),
$$

provided both expectations on the right-hand side exist.

Proof: We present the proof for the continuous case (for the proof in the discrete case, simply replace integrals by sums). Let $f(x, y)$ be the joint density of (X, Y) and let $f_X(x)$, and $f_Y(y)$ be the marginal densities of X and Y, respectively. Due to the assumption of independence we have $f(x, y) = f_X(x)f_Y(y)$, and so we obtain

$$
E(XY) = \int_{-\infty}^{\infty} \int_{-\infty}^{\infty} xyf(x, y) \, dx \, dy = \int_{-\infty}^{\infty} \int_{-\infty}^{\infty} xyf_X(x)f_Y(y) \, dx \, dy
$$

$$
= \int_{-\infty}^{\infty} yf_Y(y) \left(\int_{-\infty}^{\infty} xf_X(x)dx \right) dy = \int_{-\infty}^{\infty} yf_Y(y)E(X) \, dy
$$

$$
= E(X) \int_{-\infty}^{\infty} yf_Y(y) \, dy = E(X)E(Y). \qquad \square
$$

By combining Propositions 3.3.1 and 3.3.2, we readily obtain the following corollary.

Corollary 3.3.1 *Let X and Y be independent random variables and $g : \mathbb{R} \mapsto \mathbb{R}$, $h : \mathbb{R} \mapsto \mathbb{R}$ be two functions for which the expectations $E\left[g(X)\right]$ and $E\left[h(Y)\right]$ exist (i.e. they are both finite). Then, we have*

$$
E\left[g(X)h(Y)\right] = E\left[g(X)\right] E\left[h(Y)\right].
$$

From this corollary we obtain, for example, that if X and Y are independent, then

$$
E\left(X^2 \sin Y\right) = E(X^2)E(\sin Y), \quad E\left(e^{X-1}Y^3\right) = E\left(e^{X-1}\right) E(Y^3),
$$

and so on.

We note that the converse of Proposition 3.3.2 is *not true* in general, as shown in the following counterexample (some additional examples can be found in the Exercises at the end of this section).

Example 3.3.1 Suppose the joint probability function of X and Y is as follows (here, a and b are two constants such that $0 < a, b < 1$):

$$f(x, y) = P(X = x, Y = y)$$

x \ y	-1	0	1
-1	a	b	a
0	b	0	b
1	a	b	a

(i) Show that $a + b = 1/4$, and use it to demonstrate that $E(XY) = E(X)E(Y)$.

(ii) Are the variables X, Y independent?

SOLUTION

(i) First we observe that the sum of all probabilities is equal to $4a + 4b$. Therefore, for f to be a valid joint probability function, we must have $4(a + b) = 1$, that is, $a + b = 1/4$.

Next, in order to calculate the expectations of X and Y, we need to find their (marginal) probability functions. For this, we add the row and column margins in the table given, so that we obtain the following table:

x \ y	-1	0	1	$f_X(x)$
-1	a	b	a	$2a + b$
0	b	0	b	$2b$
1	a	b	a	$2a + b$
$f_Y(y)$	$2a + b$	$2b$	$2a + b$	$4(a + b) = 1$

It is now easy to see that (since X and Y have the same marginal probability function)

$$E(X) = E(Y) = (-1) \cdot (2a + b) + 0 \cdot (2b) + 1 \cdot (2a + b) = 0,$$

while the expectation of their product, $E(XY)$, equals

$$E(XY) = \sum_{x=-1}^{1} \sum_{y=-1}^{1} xy f(x, y)$$

$$= (-1) \cdot (-1) \cdot a + (-1) \cdot 1 \cdot a + 1 \cdot (-1) \cdot a + 1 \cdot 1 \cdot a = 0.$$

Consequently, we have $E(XY) = E(X)E(Y)$, as required.

(ii) For X and Y to be independent, we must have $f(x, y) = f_X(x)f_Y(y)$ for any $(x, y) \in R_{X,Y}$. This is clearly not true here as

$$f(0, 0) = 0 \neq (2b)(2b) = f_X(0)f_Y(0).$$

So X and Y are not independent.

EXERCISES

Group A

1. Let (X, Y) be a pair of discrete random variables with joint probability function

$$f(-4, 1) = f(4, -1) = f(2, 2) = f(2, -2) = \frac{1}{4},$$

and $f(x, y) = 0$ for any other (x, y).

(i) Show that $E(XY) = E(X)E(Y)$.

(ii) Explain why X and Y are *not* independent.

2. Let X be a discrete random variable with probability function

$$f_X(-1) = f_X(0) = f_X(1) = \frac{1}{3},$$

and $f_X(x) = 0$ for any other x, and define another variable $Y = X^2$.

(i) Show that $E(XY) = E(X)E(Y)$, although X and Y are clearly dependent.

(ii) Show that $E(XY^n) = E(X)E(Y^n)$ for any positive integer n.

3. The random pair (X, Y) is uniformly distributed over the triangle of the plane having vertices at the points $(-a, a)$, $(a, 0)$ and $(0, a)$, for some $a > 0$. Show that $E(XY) = E(X)E(Y)$, but X and Y are not independent.

4. The joint density function of (X, Y) is given by

$$f(x, y) = \frac{1}{x^2 y^2}, \qquad x \geq 1 \text{ and } y \geq 1.$$

(i) Prove that X and Y are independent, and find the marginal distribution of X.

(ii) What is the conditional density of X, given $Y = y$, for $y \geq 1$?

(iii) Find the conditional expected value of $\sqrt{X}$, given $Y = y$, where $y \geq 1$.

5. In a small city, there are two weekly local newspapers. Let X and Y be the number of copies (in tens of thousands) for these two newspapers sold weekly. Suppose the joint density of X and Y is given by

$$f(x, y) = 4\left(1 - \frac{1}{x^2}\right)\left(1 - \frac{1}{y^2}\right), \qquad 1 \leq x \leq 2, \ 1 \leq y \leq 2.$$

Show that

$$E(X|Y = y) = E(Y|X = x)$$

for any x and y in the interval $(1, 2]$. How do you interpret this result intuitively?

6. Let X and Y be two independent random variables for which it is known that $E(X) = E(Y) = 3$ and $Var(X) = Var(Y) = 1$.

   (i) Calculate the expectations of the following random variables:

   (a) $W_1 = (X + Y)^2$,

   (b) $W_2 = (2X - 3Y)(3X - 2Y)$,

   (c) $W_3 = (X^2 - 2X + 1)(Y^2 - 2Y + 1)$.

   (ii) Find the values of

   (d) $E(X^2 - 5X + 4|Y = 1)$,

   (e) $E(Y^2 - 7Y + 12|X = 2)$,

   (f) $E((X - 2)(X - 3)|Y = 4)$.

Group B

7. The variables X and Y have a joint density function

$$f(x, y) = \begin{cases} \dfrac{5}{2} e^{-5x}, & \text{if } x > 0 \text{ and } 1 \leq y \leq 3, \\ 0, & \text{otherwise.} \end{cases}$$

   (i) Examine whether X and Y are independent.

   (ii) Calculate

$$P(X > 1, Y \leq 2), \quad P(X > 1 \mid 3/2 \leq Y \leq 2).$$

   (iii) Find the values of

$$E(3Y - 4X), \quad E[(X - 1)(Y - 2)], \quad E[3(X - Y)^2|Y = 2].$$

8. An extract of the joint probability function for (X, Y) is as follows:

x \ y	5	10
0	1/16	
2	1/6	1/12
4		5/12

   (i) Fill in the missing values in the table if it is known that $E(X) = 3$, and thence find $E(Y)$.

   (ii) Obtain the joint probability function of (Z, W), where

$$W = Y - 2, \quad Z = 2X - 1.$$

   Is it true that $E(WZ) = E(W)E(Z)$?

9. The joint density function of (X, Y) is given by

$$f(x, y) = f_1(x) f_2(y), \qquad x \in \mathbb{R} \text{ and } y \in \mathbb{R},$$

for two real functions of a single variable (see Proposition 3.2.4).

   (i) Verify that the conditional expectation of Y, given $X = x$, is given by

   $$E(Y|X = x) = \frac{\displaystyle\int_{-\infty}^{\infty} y f_2(y) dy}{\displaystyle\int_{-\infty}^{\infty} f_2(y) dy}.$$

   (ii) Use Part (i) to find $E(Y|X = x)$ for the case when the joint density of (X, Y) is

   $$f(x, y) = 6xy^2, \qquad 0 \le x \le 1, \ 0 \le y \le 1.$$

10. In a sequence of $n + k$ (independent) Bernoulli trials, let us define

   X: the number of successes in the $n + k$ trials,
   Y: the number of successes in the trials numbered $n + 1, n + 2, \dots, n + k$.

   Find the conditional expectation

   $$E(Y|X = x) \quad \text{for } x = 0, 1, \dots, n + k.$$

3.4 MULTIVARIATE JOINT DISTRIBUTIONS

Till now, we focused on the joint distribution of two random variables. An extension to the case when there are more than two variables can be readily made. As with the bivariate case, we start with the discrete case first.

Definition 3.4.1 Let $X_1, \dots, X_n$ be a collection of discrete random variables defined on the same sample space Ω. Let $R_{X_1, \dots, X_n}$ be the range of values for the n-tuple $(X_1, \dots, X_n)$. Then, the function f, defined for any $(x_1, \dots, x_n) \in R_{X_1, \dots, X_n}$, as

$$f(x_1, \dots, x_n) = P(X_1 = x_1, \dots, X_n = x_n)$$
$$= P\left(\{\omega \in \Omega : X_1(\omega) = x_1, \dots, X_n(\omega) = x_n\}\right),$$

is called the joint probability function of the variables $X_1, \dots, X_n$, or the probability function of the n-dimensional random variable $(X_1, \dots, X_n)$.

The set $R_{X_1, \dots, X_n}$ is called the **range of values** for the n-tuple $(X_1, \dots, X_n)$ and, as seen in the bivariate case, we have

$$R_{X_1, X_2, \dots, X_n} \subseteq R_{X_1} \times R_{X_2} \times \cdots \times R_{X_n}.$$

In the definition above, the notation $(X_1, \ldots, X_n)$ indicates that the n-tuple is ordered, and so it is not the same as $(X_1, X_3, X_2, \ldots, X_n)$, for example; in general, they will not have the same probability function (the same is true for the continuous case as well).

In Definition 3.4.1, there is no harm in considering the domain of the function f to be the entire n-dimensional space $\mathbb{R} \times \mathbb{R} \times \cdots \times \mathbb{R} = \mathbb{R}^n$, with the understanding that f vanishes outside the set $R_{X_1, \ldots, X_n}$. More specifically, we have the following as a generalization of Proposition 1.2.1.

Proposition 3.4.1 *The joint probability function* $f(x_1, \ldots, x_n)$ *of an n-dimensional random variable* $(X_1, \ldots, X_n)$ *satisfies the following conditions:*

1. $f(x_1, \ldots, x_n) = 0$ *for any* $(x_1, \ldots, x_n) \notin R_{X_1, \ldots, X_n}$;

2. $f(x_1, \ldots, x_n) \geq 0$ *for any* $(x_1, \ldots, x_n) \in R_{X_1, \ldots, X_n}$;

3. $\displaystyle\sum_{(x_1, \ldots, x_n) \in R_{X_1, \ldots, X_n}} f(x_1, \ldots, x_n) = 1.$

The three properties in the proposition are again characteristic of joint probability functions, that is, if a function $f : \mathbb{R}^n \mapsto \mathbb{R}$ satisfies these properties, then there exists an ordered n-tuple, $(X_1, \ldots, X_n)$, of random variables such that f is the probability function of $(X_1, \ldots, X_n)$.

Suppose that A is a subset of $R_{X_1, \ldots, X_n}$. Then, $P[(X_1, \ldots, X_n) \in A]$ can be found as

$$P[(X_1, \ldots, X_n) \in A] = \sum_{(x_1, \ldots, x_n) \in A} f(x_1, \ldots, x_n). \tag{3.8}$$

Let x_1 be in the range R_{X_1} of the random variable X_1. Setting $A = \{x_1\} \times R_{X_2} \times \cdots \times R_{X_n}$ above, we deduce that

$$P(X_1 = x_1) = \sum_{x_j : j \neq 1} f(x_1, x_2, \ldots, x_n), \qquad x_1 \in R_{X_1},$$

where the summation extends over all $(x_2, \ldots, x_n)$ in the set $R_{X_2} \times \cdots \times R_{X_n}$. As the above result holds for any $x_1 \in R_{X_1}$, we arrive at the probability function

$$f_{X_1}(x_1) = P(X_1 = x_1), \qquad x_1 \in R_{X_1},$$

which is again called the **marginal probability function** of X_1. More generally, for $1 \leq i \leq n$, the function

$$f_{X_i}(x_i) = P(X_i = x_i) = \sum_{x_j : j \neq i} f(x_1, \ldots, x_n), \qquad x_i \in R_{X_i}, \tag{3.9}$$

is the marginal probability function of the random variable X_i (here, the summation is over all variables other than x_i).

The above concepts are obvious extensions from the bivariate case discussed in Chapter 1. However, when we have more than two variables, we may also consider the *joint marginal probability function* for a subset of them. Rather than treating the general case, we illustrate this by considering three variables X, Y and Z with joint probability function $f(x, y, z)$. Then, the joint marginal probability function of variables Y and Z is given by

$$f_{Y,Z}(y, z) = P(Y = y, Z = z) = \sum_{x \in R_X} f(x, y, z), \qquad (y, z) \in R_{Y,Z}.$$

In other words, this is just the joint (in this case, bivariate) probability function of the pair (Y, Z). The adjective "marginal" here simply suggests that it emerges from the consideration of three variables X, Y, and Z, and then summing over all values of X.

Similarly, we have the joint marginal probability functions of (X, Y) and (X, Z) as

$$f_{X,Y}(x, y) = P(X = x, Y = y) = \sum_{z \in R_Z} f(x, y, z), \qquad (x, y) \in R_{X,Y},$$

and

$$f_{X,Z}(x, z) = P(X = x, Z = z) = \sum_{y \in R_Y} f(x, y, z), \qquad (x, z) \in R_{X,Z}.$$

Example 3.4.1 Suppose the variables X, Y, and Z have joint probability function as

$$f(x, y, z) = cxyz, \qquad x = 2, 3, \ y = 3, 4, \ z = 2, 3, 4,$$

where c is a real constant. For determining c, we use the condition that

$$\sum_{x=2}^{3} \sum_{y=3}^{4} \sum_{z=2}^{4} f(x, y, z) = 1,$$

which yields

$$c(2 \cdot 3 \cdot 2 + 2 \cdot 3 \cdot 3 + 2 \cdot 3 \cdot 4 + \cdots + 3 \cdot 4 \cdot 4) = 1$$

(note that since X, Y and Z take 2, 2 and 3 values, respectively, $R_{X,Y,Z}$ has $2 \cdot 2 \cdot 3 = 12$ elements). The sum of the 12 terms inside the brackets on the left side is 315, and so $315 \cdot c = 1$, yielding $c = 1/315$. Hence, f takes the form

$$f(x, y, z) = \frac{xyz}{315}, \qquad x = 2, 3, \ y = 3, 4, \ z = 2, 3, 4.$$

For the joint marginal probability function of (X, Y), we get, for (x, y) in the domain of (X, Y),

$$f_{X,Y}(x, y) = P(X = x, Y = y) = \sum_{z \in R_Z} f(x, y, z) = \sum_{z=2}^{4} \frac{xyz}{315}$$

$$= (2 + 3 + 4) \frac{xy}{315} = \frac{xy}{35}.$$

Similarly, we find

$$f_{X,Z}(x,z) = P(X = x, Z = z) = \sum_{y=3}^{4} \frac{xyz}{315} = \frac{xz}{45}$$

and

$$f_{Y,Z}(y,z) = P(Y = y, Z = z) = \sum_{x=2}^{3} \frac{xyz}{315} = \frac{yz}{63}.$$

If we are now interested in the marginal probability function of X alone, then from (3.9) we obtain

$$f_X(x) = \sum_{y=3}^{4} \sum_{z=2}^{4} f(x,y,z) = \frac{x}{315} \sum_{y=3}^{4} \sum_{z=2}^{4} (yz)$$

$$= \frac{x}{315} (3 \cdot 2 + 3 \cdot 3 + \cdots + 4 \cdot 4) = \frac{x}{5},$$

for $x = 2, 3$. Alternatively, from the joint probability function of X and Y that we already found, we get again

$$f_X(x) = \sum_{y=3}^{4} f_{X,Y}(x,y) = \frac{x}{5}.$$

Using either of these two ways and proceeding in a similar fashion, we find that the marginal probability function of Y (alone) as

$$f_Y(y) = \frac{y}{7}, \qquad \text{for } y = 3, 4$$

and that of Z (alone) as

$$f_Z(z) = \frac{z}{9}, \qquad \text{for } z = 2, 3, 4.$$

Observe that each of these three marginal probability functions is a probability function of a single random variable in the usual sense, so that if we add the probabilities in each of them, we get a value of one.

Finally, suppose we are interested in the probability $P(X + Y + Z = 8)$. By considering all values of X, Y and Z that sum to 8, we obtain

$$P(X + Y + Z = 8) = P[(X, Y, Z) \in \{(2,3,3),\ (2,4,2),\ (3,3,2)\}]$$

$$= \frac{2 \cdot 3 \cdot 3 + 2 \cdot 4 \cdot 2 + 3 \cdot 3 \cdot 2}{315} = \frac{52}{315} \cong 16.5\%.$$

We can now present the following analogue to Definition 3.4.1 for the continuous case.

Definition 3.4.2 Let $X_1, \ldots, X_n$ be a collection of random variables defined on the sample space Ω. Assume that there exists a nonnegative function of n-variables, $f : \mathbb{R}^n \mapsto [0, \infty)$, such that for any subset Γ of $\mathbb{R}^n$ which can be written in terms of n-dimensional rectangles by a finite or infinitely countable number of set operations (unions, intersections, and complements), we have

$$P\left[(X_1, \ldots, X_n) \in \Gamma\right] = \int \int_{\Gamma} \cdots \int f(x_1, \ldots, x_n) dx_1 \cdots dx_n. \tag{3.10}$$

Then, we say that the n-dimensional random variable $(X_1, \ldots, X_n)$ has a joint continuous distribution and the function f is called the **joint density function** of the variables $X_1, \ldots, X_n$, or the (probability) density function of the random vector $(X_1, \ldots, X_n)$.

We note again that, if the vector $(X_1, \ldots, X_n)$ has a joint continuous distribution, then each of the variables X_i for $i = 1, \ldots, n$, is continuous, and its (marginal) density is given by

$$f_{X_i}(x_i) = \underbrace{\int_{-\infty}^{\infty} \int_{-\infty}^{\infty} \cdots \int_{-\infty}^{\infty}}_{n-1 \text{ terms}} f(x_1, \ldots, x_n) dx_1 \cdots dx_{i-1} dx_{i+1} \cdots dx_n.$$

For the case when the set Γ in (3.10) can be expressed as a Cartesian product, $\Gamma = A_1 \times \cdots \times A_n$, of subsets of the real line, then the integral in (3.10) reduces to the familiar n-fold Riemann integral of the form

$$P\left(X_1 \in A_1, \ \ldots, \ X_n \in A_n\right)$$
$$= \int_{A_n} \cdots \int_{A_1} f(x_1, \ldots, x_n) dx_1 \cdots dx_n.$$

Moreover, when $A_i = \mathbb{R}$ for all i, we obtain immediately that

$$\int_{-\infty}^{\infty} \cdots \int_{-\infty}^{\infty} f(x_1, \ldots, x_n) dx_1 \cdots dx_n = 1,$$

which, as in the discrete case, along with the nonnegativity of f, is a characteristic property for a joint density function.

Next, from the joint density function f for a given set of continuous variables $X_1, \ldots, X_n$, we can obtain the joint marginal density for a subset of them by "integrating out" the remaining variables. For an illustration, let $X, Y, Z,$ and W have joint density $f(x, y, z, w)$. Then, the joint marginal density of X and Z is given by

$$f_{X,Z}(x, z) = \int_{-\infty}^{\infty} \int_{-\infty}^{\infty} f(x, y, z, w) dy \, dw$$

while the joint marginal density of (X, Y, W) is given by

$$f_{X,Y,W}(x, y, w) = \int_{-\infty}^{\infty} f(x, y, z, w) dz.$$

Example 3.4.2 Suppose the joint density function of X, Y, and Z is given by

$$f(x, y, z) = c(x + y)e^{-z}, \qquad 0 < x < 1, \ 0 < y < 1, \ z > 0.$$

The value of c can be found from the condition

$$\int_{-\infty}^{\infty} \int_{-\infty}^{\infty} \int_{-\infty}^{\infty} f(x, y, z) dx \, dy \, dz = 1,$$

which in this case gives

$$1 = \int_0^1 \int_0^1 \int_{-\infty}^{\infty} c(x + y)e^{-z} \, dz \, dy \, dx = c \int_0^1 \int_0^1 (x + y) \left[-e^{-z}\right]_{z=0}^{\infty} dy \, dx$$

$$= c \int_0^1 \int_0^1 (x + y) dy \, dx = c \int_0^1 \left[xy + \frac{y^2}{2}\right]_{y=0}^1 dx = c \int_0^1 \left(x + \frac{1}{2}\right) dx$$

$$= c \left[\frac{x^2 + x}{2}\right]_{x=0}^1 = c,$$

that is, we have $c = 1$. The marginal density of (X, Y) is then

$$f_{X,Y}(x, y) = \int_{-\infty}^{\infty} f(x, y, z) dz = \int_{-\infty}^{\infty} (x + y)e^{-z} \, dz = x + y, \quad 0 < x, y < 1.$$

For the pairs (X, Z) and (Y, Z), we similarly obtain their marginal joint densities to be

$$f_{X,Z}(x, z) = \int_{-\infty}^{\infty} f(x, y, z) dy = \int_0^1 (x + y)e^{-z} \, dy = \left(x + \frac{1}{2}\right) e^{-z}, \quad 0 < x < 1, \ z > 0,$$

and

$$f_{Y,Z}(y, z) = \int_{-\infty}^{\infty} f(x, y, z) dx = \int_0^1 (x + y)e^{-z} \, dx = \left(y + \frac{1}{2}\right) e^{-z}, \quad 0 < y < 1, \ z > 0.$$

From the above expressions, it is easy to obtain the marginal densities of X, Y, and Z. The marginal density of X is given by

$$f_X(x) = \int_{-\infty}^{\infty} f_{X,Y}(x, y) dy = \int_0^1 (x + y) dy = x + \frac{1}{2}, \quad 0 < x < 1.$$

Similarly, we find the marginal density of Y and the marginal density of Z to be

$$f_Y(y) = \int_{-\infty}^{\infty} f_{X,Y}(x, y) dx = \int_0^1 (x + y) dy = y + \frac{1}{2}, \quad 0 < y < 1,$$

and

$$f_Z(z) = \int_{-\infty}^{\infty} f_{X,Z}(x, z) dx = \int_0^1 \left(x + \frac{1}{2}\right) e^{-z} \, dx = e^{-z}, \quad z > 0.$$

Observe that the marginal densities of X and Y are the same, as the joint density $f(x, y, z)$ is symmetric in x and y. Also, the density of Z is that of a standard exponential distribution.

The **joint (cumulative) distribution function** of an n-tuple of random variables, $(X_1, \ldots, X_n)$ (both in the discrete and continuous cases) is defined as

$$F(x_1, \ldots, x_n) = P(X_1 \leq x_1, \ldots, X_n \leq x_n), \qquad x_1, \ldots, x_n \in \mathbb{R},$$

while the marginal distribution functions can be readily found from F as follows:

$$F_{X_i}(x_i) = \lim_{\substack{x_j \to \infty \ for \\ all \ j \neq i}} F(x_1, \ldots, x_n),$$

$$F_{X_i, X_j}(x_i, x_j) = \lim_{\substack{x_r \to \infty \ for \\ all \ r \neq i, j}} F(x_1, \ldots, x_n),$$

and so on. As an illustration, assume that $F(x, y, z, w)$ is the joint distribution function of (X, Y, Z, W). Then, we have

$$F_{Y, Z, W}(y, z, w) = \lim_{x \to \infty} F(x, y, z, w),$$

$$F_{X, W}(x, w) = \lim_{y \to \infty, z \to \infty} F(x, y, z, w),$$

$$F_Z(z) = \lim_{\substack{x \to \infty, y \to \infty \\ w \to \infty}} F(x, y, z, w).$$

A joint distribution function is always right-continuous and nondecreasing in each of its arguments.

When the variables $X_1, \ldots, X_n$ are discrete with joint probability function $f(x_1, \ldots, x_n)$, then we have

$$F(x_1, \ldots, x_n) = \sum_{t_1, \ldots, t_n} f(t_1, \ldots, t_n),$$

where the summation is over all $(t_1, \ldots, t_n) \in R_{X_1, \ldots, X_n}$ such that $t_i \leq x_i$, for $i = 1, \ldots, n$.

In the continuous case, if $X_1, \ldots, X_n$ have a joint density function f, then

$$F(x_1, \ldots, x_n) = \int_{-\infty}^{x_1} \cdots \int_{-\infty}^{x_n} f(t_1, \ldots, t_n) \, dt_n \, \cdots \, dt_1.$$

On the other hand, if we know F the joint density f can be found as

$$f(x_1, \ldots, x_n) = \frac{\partial F(x_1, \ldots, x_n)}{\partial x_1 \, \cdots \, \partial x_n}.$$

Example 3.4.3 Suppose the lifetimes, in thousands of hours, for three components of an electronic device are described by continuous variables X, Y, and Z having a joint distribution function

$$F(x, y, z) = 1 - e^{-x} - e^{-y} - e^{-z^2} + e^{-(x+y)} + e^{-(y+z^2)} + e^{-(x+z^2)} - e^{-(x+y+z^2)}, \qquad x, y, z > 0.$$

Then, the pairwise marginal joint distribution functions are given by

$$F_{X,Y}(x,y) = \lim_{z \to \infty} F(x,y,z) = 1 - e^{-x} - e^{-y} + e^{-(x+y)}, \quad x,y > 0,$$

$$F_{X,Z}(x,z) = \lim_{y \to \infty} F(x,y,z) = 1 - e^{-x} - e^{-z^2} + e^{-(x+z^2)}, \quad x,z > 0,$$

$$F_{Y,Z}(y,z) = \lim_{x \to \infty} F(x,y,z) = 1 - e^{-y} - e^{-z^2} + e^{-(y+z^2)}, \quad y,z > 0,$$

while the marginal distribution functions are given by

$$F_X(x) = \lim_{y \to \infty, z \to \infty} F(x,y,z) = 1 - e^{-x}, \qquad x > 0,$$

$$F_Y(y) = \lim_{x \to \infty, z \to \infty} F(x,y,z) = 1 - e^{-y}, \qquad y > 0,$$

$$F_Z(z) = \lim_{x \to \infty, y \to \infty} F(x,y,z) = 1 - e^{-z^2}, \qquad z > 0.$$

Notice that both X and Y have a standard exponential distribution (with mean equal to one). Further, upon differentiating the function $F(x,y,z)$ with respect to z, y, and x in turn, we obtain

$$\frac{\partial F(x,y,z)}{\partial z} = 2z\, e^{-z^2} - 2ze^{-(y+z^2)} - 2z\, e^{-(x+z^2)} + 2z\, e^{-(x+y+z^2)},$$

$$\frac{\partial^2 F(x,y,z)}{\partial y \partial z} = \frac{\partial}{\partial y}\left(\frac{\partial F(x,y,z)}{\partial z}\right) = 2z\, e^{-(y+z^2)} - 2z\, e^{-(x+y+z^2)},$$

and

$$\frac{\partial^3 F(x,y,z)}{\partial x \partial y \partial z} = \frac{\partial}{\partial x}\left(\frac{\partial^2 F(x,y,z)}{\partial y \partial z}\right) = 2z\, e^{-(x+y+z^2)}.$$

This shows that the joint density function of (X,Y,Z) is

$$f(x,y,z) = 2ze^{-(x+y+z^2)}, \qquad x,y,z > 0.$$

We can also readily find probabilities of various events. For example, the probability that all three components of the electronic device will work for at most 2000 hours is

$$P(X \le 2, Y \le 2, Z \le 2) = F(2,2,2) \cong 73\%,$$

while the probability that all three components will work for more than 2000 hours can be found, by an appeal to the inclusion–exclusion formula, as follows:

$$P(X > 2, Y > 2,\ Z > 2) = 1 - P(X \le 2 \text{ or } Y \le 2 \text{ or } Z \le 2)$$

$$= 1 - [P(X \le 2) + P(Y \le 2) + P(Z \le 2)] + [P(X \le 2, Y \le 2)$$

$$+ P(X \le 2, Z \le 2) + P(Y \le 2, Z \le 2)] - P(X \le 2, Y \le 2, Z \le 2)$$

$$= 1 - F_X(2) - F_Y(2) - F_Z(2) + F_{X,Y}(2,2) + F_{X,Z}(2,2) + F_{Y,Z}(2,2)$$

$$- F(2,2,2) \cong 0.0003.$$

Another issue that arises when we are dealing with more than two random variables is that we may consider a number of conditional distributions, by assuming the value(s) of one or more variables to be known. Suppose we have four continuous variables X_1, X_2, X_3, and X_4 with joint density function $f(x_1, x_2, x_3, x_4)$. Then we have,

(i) the joint density of X_1, given $X_2 = x_2, X_3 = x_3$ and $X_4 = x_4$, as

$$f_{X_1|X_2,X_3,X_4}(x_1|x_2,x_3,x_4) = \frac{f(x_1,x_2,x_3,x_4)}{f_{X_2,X_3,X_4}(x_2,x_3,x_4)};$$

(ii) the joint density of X_3, given $X_2 = x_2$ and $X_4 = x_4$, as

$$f_{X_3|X_2,X_4}(x_3|x_2,x_4) = \frac{f_{X_2,X_3,X_4}(x_2,x_3,x_4)}{f_{X_2,X_4}(x_2,x_4)};$$

(iii) the joint density of X_1, X_2, and X_3, given $X_4 = x_4$, as

$$f_{X_1,X_2,X_3|X_4}(x_1,x_2,x_3|x_4) = \frac{f(x_1,x_2,x_3,x_4)}{f_{X_4}(x_4)},$$

and so on. Note that these also hold for the case when the X_i's are discrete, with density functions being replaced by probability functions.

Following the above, we can also define conditional expectations. Thus, in Case (i) above, for example, we will have

$$E(X_1|X_2 = x_2, X_3 = x_3, X_4 = x_4) = \int_{-\infty}^{\infty} x_1 f_{X_1|X_2,X_3,X_4}(x_1|x_2,x_3,x_4)dx_1.$$

The (unconditional) expectation for a function $h(X_1, \ldots, X_n)$ of a random n-tuple is defined as

$$E\left[h(X_1, \ldots, X_n)\right] = \sum_{(x_1,\ldots,x_n) \in R_{X_1,\ldots,X_n}} h(x_1, \ldots, x_n) f(x_1, \ldots, x_n)$$

when the X_i's are discrete, while in the continuous case we have

$$E\left[h(X_1, \ldots, X_n)\right] = \int_{-\infty}^{\infty} \cdots \int_{-\infty}^{\infty} h(x_1, \ldots, x_n) f(x_1, \ldots, x_n)dx_1 \cdots dx_n.$$

The properties of expectations here are analogous to those of two-dimensional random variables (Sections 1.4 and 2.4). One particular formula of interest in applications is the linearity of expectation, in its general form when a linear combination of n random variables is considered:

$$E(a_1 X_1 + \cdots + a_n X_n) = a_1 E(X_1) + \cdots + a_n E(X_n), \tag{3.11}$$

valid for any $a_1, \ldots, a_n \in \mathbb{R}$.

Example 3.4.4 (Example 3.4.2 continued)
Consider (X, Y, Z) with joint density function

$$f(x, y, z) = c(x + y)e^{-z}, \qquad 0 < x < 1, \ 0 < y < 1, \ z > 0,$$

as in Example 3.4.2. There, we found various marginal density functions and by using them, we may find the following conditional densities:

$$f_{X,Y|Z}(x, y|z) = \frac{f(x, y, z)}{f_Z(z)} = \frac{(x + y)e^{-z}}{e^{-z}} = x + y, \qquad 0 < x < 1, \ 0 < y < 1,$$

$$f_{X,Z|Y}(x, z|y) = \frac{f(x, y, z)}{f_Y(y)} = \frac{(x + y)e^{-z}}{y + 1/2}, \qquad 0 < x < 1, \ z > 0,$$

$$f_{Z|X,Y}(z|x, y) = \frac{f(x, y, z)}{f_{X,Y}(x, y)} = \frac{(x + y)e^{-z}}{x + y} = e^{-z}, \qquad z > 0,$$

$$f_{X|Y,Z}(x|y, z) = \frac{f(x, y, z)}{f_{Y,Z}(y, z)} = \frac{(x + y)e^{-z}}{(y + 1/2)e^{-z}} = \frac{x + y}{y + 1/2}, \qquad 0 < x < 1.$$

Observe, in particular, that the conditional density of (X, Y), given $Z = z$, does not depend on z and, in fact, it is the same as the (unconditional) joint density of (X, Y) found in Example 3.4.2. This suggests that the pair (X, Y) is "independent" of Z; this concept, which generalizes the idea of independence between two variables considered in Section 3.2 is discussed in the following section.

We also observe that the conditional distribution of Z, given $X = x$ and $Y = y$, is standard exponential, which agrees with the marginal density of Z found in Example 3.4.2. Thus, we have

$$E(Z|X = x, Y = y) = 1.$$

Further, we find the conditional expectation of X, given $Y = y$ and $Z = z$, to be

$$E(X|Y = y, Z = z) = \int_{-\infty}^{\infty} x f_{X|Y,Z}(x|y, z) dx = \int_0^1 \frac{x(x + y)}{y + 1/2} \, dx = \frac{3y + 2}{6y + 3}.$$

Example 3.4.5 If a random variable X has a binomial distribution with parameters n and p, then we know that $E(X) = np$. A simpler proof can be established through (3.11) by using the fact that any binomial random variable, $X \sim b(n, p)$, can be represented as

$$X = X_1 + \cdots + X_n,$$

where each X_i has a Bernoulli distribution with parameter (success probability) p. This representation is immediate since, when X_i takes the value 1 if the ith trial results in a success (otherwise, $X_i = 0$), then X is the total number of successes in the n (identical) trials.

But, the fact that X_i is Bernoulli with parameter p implies that $E(X_i) = p$, and so from (3.11), we immediately get

$$E(X) = np.$$

A similar reasoning can be used to obtain the expectation of a random variable Y having a negative binomial distribution. Specifically, let $Y \sim Nb(r,p)$. Then, Y can be expressed as

$$Y = Y_1 + \cdots + Y_r,$$

where Y_i has the geometric distribution with parameter p, for $i = 1, 2, \ldots, r$. To see this, consider a sequence of identical Bernoulli trials with success probability p. Then define Y_1 to be the number of trials until the occurrence of first success. Similarly, define Y_2 to be the number of trials *after* the first success until the second success occurs, and so on, and finally let Y_r be the number of trials between the $(r-1)$th and the rth successes. Then, Y given above is the total number of trials until we observe r successes in this experiment, so that $Y \sim Nb(r,p)$. As the expectation of a geometric random variable is $1/p$, we then obtain readily from (3.11) that

$$E(Y) = E(Y_1 + \cdots + Y_r) = r \cdot \frac{1}{p} = \frac{r}{p},$$

and this proof is clearly much simpler than its direct derivation.

EXERCISES

Group A

1. The triple (X, Y, Z) of discrete random variables has the following joint probability function:

$$f(x, y, z) = c(xy + 2z), \qquad x = 1, 2, \ y = 1, 2, 3, \ z = 2, 3.$$

   (i) Find the value of the constant c.

   (ii) Obtain the marginal probability functions of X, Y, and Z.

   (iii) Calculate the probabilities

   $$P(X = 1, Y \geq 2, Z \geq 2), \quad P(X - Y = 1, Z \geq 2).$$

2. Find the value of c so that, in each of the following cases, $f(x, y, z)$ represents a valid joint density function of variables (X, Y, Z):

   (i) $f(x, y, z) = c(x + y)z^2$, $0 \leq x \leq 1, \ 0 \leq y \leq 1, \ 0 \leq z \leq 2$

   (ii) $f(x, y, z) = cx^2 y^3 z^4$, $0 \leq x \leq 2, \ 0 \leq y \leq 1, \ 0 \leq z \leq 1$

   (iii) $f(x, y, z) = cx^2 y(1 - z)$, $0 \leq x, y, z \leq 1$

   (iv) $f(x, y, z) = c(x + y)e^{-x-z}$, $0 \leq y \leq 1, \ x > 0, \ z > 0.$

3. Suppose X, Y, and Z have a joint continuous distribution with joint density function

$$f(x, y, z) = (2x + y)e^{-4z}, \qquad\qquad 0 < x < 1, \ 0 < y < 2, \ 0 < z < \infty.$$

(i) Find

$$P\left(X < \frac{1}{3}, Y \le \frac{2}{3}, Z > 1\right).$$

(ii) Investigate whether any of the pairs $(X, Y), (X, Z)$, and (Y, Z) consist of independent random variables.

4. The joint probability function of X, Y, Z and W is given by

$$f(x, y, z, w) = c(x + y)(2z + 2w),$$

where each of the four variables takes only the values 1 and 2 (so that there are $2^4 = 16$ combinations of values for (X, Y, Z, W)).

(i) What is the value of c?

(ii) Write down the marginal probability functions for each of the four variables.

(iii) Find

$$P(X = 2, Y + Z = 3, Z + W = 2), \ \ P(X + Y + Z + W = 6), \ \ P(X + Y \ne Z + W).$$

5. Suppose three components of an electronic device have lifetimes X, Y, and Z with the joint distribution function as given in Example 3.4.3.

(i) Calculate

$$P\left(X \le \frac{3}{2}, Y \le \frac{3}{2}\right) \text{ and } P\left(X > \frac{3}{2}, Z > \frac{3}{2}\right).$$

(ii) Assume that the device would work if at least two of the components work. Calculate the probability that the device will operate for at least 1500 hours.

6. The joint density function of (X, Y, Z, W) is given by

$$f(x, y, z, w) = \frac{2}{7}(x + 2w)e^{-y-3z}, \quad 0 \le x \le 1, \ y > 0, \ z > 0, \ 0 \le w \le 3.$$

(i) Calculate the (marginal) joint probability densities of

    (a) X and Y (b) X, Y and W

    (c) X, W and Z (d) Y and W.

(ii) Examine whether each of the following pairs consists of independent random variables:

    (a) (X, Y) (b) (Y, W) (c) (Z, W).

(iii) Obtain

$$E(X), \ E(W), \ E(X + Y), \ E(Y + 3W), \ E(X - 2W + 3Z).$$

Group B

7. Suppose X, Y, and Z have a joint continuous distribution with density function

$$f(x, y, z, w) = \begin{cases} \dfrac{c}{xyz}, & 0 < w \leq z \leq y \leq x \leq 1, \\ 0, & \text{otherwise,} \end{cases}$$

for a real constant c.

(i) Show that the joint density function of Y, Z and W is given by

$$f(y, z, w) = -\frac{\ln y}{yz}, \qquad 0 < w \leq z \leq y \leq 1.$$

(ii) Verify that the joint density function of (X, W) is

$$f_{X,W}(x, w) = \frac{1}{2x}\left(\ln \frac{x}{w}\right)^2, \qquad 0 < w \leq x \leq 1.$$

(iii) Obtain the distributions of X and of Z.

8. From an urn that contains a red and b black balls, we select n balls at random without replacement, where $1 \leq n \leq a + b$. Let X be the number of red balls selected.

(i) Check that X can be expressed as $X = X_1 + \cdots + X_n$, where X_i is an indicator random variable defined as

$$X_i = \begin{cases} 1, & \text{if the } i\text{th ball drawn is red,} \\ 0, & \text{otherwise,} \end{cases}$$

for $i = 1, 2, \ldots, n$.

(ii) Show that $E(X_i) = n/(a + b)$, for all $i = 1, 2, \ldots, a$.

(iii) Use the results of Parts (i) and (ii) to find $E(X)$ (since the distribution of X is hypergeometric, confirm that this agrees with the expectation of a random variable following this distribution).

9. Find the value of c, so that each of the following functions is a joint density function for a three-dimensional random variable (X, Y, Z):

(i) $f(x, y, z) = \begin{cases} cx^2 \, e^{-x(1+y+z)}, & x, y, z > 0, \\ 0, & \text{otherwise;} \end{cases}$

(ii) $f(x, y, z) = \begin{cases} c \, e^{-x-y-z}, & 0 < x < y < z, \\ 0, & \text{otherwise;} \end{cases}$

(iii) $f(x, y, z) = \begin{cases} \dfrac{c \ln x}{xy}, & 0 < z \leq y \leq x \leq 1, \\ 0, & \text{otherwise.} \end{cases}$

Then, in each case, find

(a) the marginal densities of X, Y, Z;

(b) the conditional density of X, given $Y = y, Z = z$, and the associated conditional expectation;

(c) the conditional distribution function of (X, Y), given $Z = z$.

10. An insurance company offers its clients two different types of life insurance coverage (a_1 and a_2) and two types of home coverage (a_3 and a_4). The proportions of customers who selected insurance coverage a_i is represented by a random variable X_i, for $i = 1, 2, 3, 4$. The joint density of X_i's is given by

$$f(x_1, x_2, x_3) = \begin{cases} cx_1x_2(1 - x_3), & 0 \le x_i \le 1, x_1 + x_2 + x_3 \le 1, \\ 0, & \text{otherwise} \end{cases}$$

(we assume that each customer has selected only one type of insurance coverage, so that $X_4 = 1 - X_1 - X_2 - X_3$).

(i) Show that $c = 144$.

(ii) Obtain the joint density function of (X_1, X_2).

(iii) What is the probability that the proportion of customers who have selected home insurance coverage is at least 60%?

(iv) Derive the marginal distributions of X_1 and of X_2.

11. The joint density function of the variables X, Y, and Z is given by

$$f(x, y, z) = 60x^2y^3z^4, \qquad 0 < x, y, z < 1.$$

(i) Find the probability $P(X \ge YZ)$.

(ii) Calculate the joint distribution function of (X, Y, Z).

(iii) Obtain the density functions of the pairs (X, Y), (X, Z), and (Y, Z).

(iv) Find the value of

$$P\left(X > \frac{1}{2}, \ Y > \frac{1}{2}, \ Z > \frac{1}{2}\right).$$

12. For which value of $c \in \mathbb{R}$ is the function

$$f(x_1, x_2, \ldots, x_n) = \begin{cases} c\, e^{-x_1}, & \text{if } x_1 > x_2 > \cdots > x_n > 0, \\ 0, & \text{otherwise,} \end{cases}$$

a valid joint density of an n-dimensional random variable?

3.5 INDEPENDENCE OF MORE THAN TWO VARIABLES

As we have seen in Volume I, the concept of independence between two events on a probability space extends in a natural way to the case where more than two events are considered. The same also applies to independence of random variables. In this regard, we have the following definition.

Definition 3.5.1 For $n \geq 2$, the random variables $X_1, \ldots, X_n$ are said to be independent if for any subsets, $A_1, \ldots, A_n$ of the real line, we have

$$P(X_1 \in A_1, \ldots, X_n \in A_n) = P(X_1 \in A_1) \cdots P(X_n \in A_n).$$

As in the bivariate case, sometimes it will be obvious from the statement of the problem that the variables we are dealing with are (or *are not*) independent. When this is not clear, we typically resort to one of the following two propositions for testing independence. Their proofs are not given here, as they are similar to the proofs of Propositions 3.2.1–3.2.3.

Proposition 3.5.1 *Let $X_1, \ldots, X_n$ be a set of random variables with joint distribution function $F(x_1, \ldots, x_n)$ and with marginal distribution functions $F_{X_1}(x_1), \ldots, F_{X_n}(x_n)$, respectively. Then, the variables $X_1, \ldots, X_n$ are independent if and only if*

$$F(x_1, \ldots, x_n) = F_{X_1}(x_1) \cdots F_{X_n}(x_n)$$

holds for any $x_1, \ldots, x_n \in \mathbb{R}$.

Proposition 3.5.2 *Let $X_1, \ldots, X_n$ be a set of random variables and assume that their joint probability function (if the X_i's are discrete) or joint density function (if the X_i's are continuous) is $f(x_1, \ldots, x_n)$, while their marginal probability or density functions (in the discrete and continuous case respectively) are $f_{X_1}(x_1), \ldots, f_{X_n}(x_n)$. Then $X_1, \ldots, X_n$ are independent if and only if the relation*

$$f(x_1, \ldots, x_n) = f_{X_1}(x_1) \cdots f_{X_n}(x_n)$$

holds for any $x_1, \ldots, x_n \in \mathbb{R}$.

A natural question that arises is: given a set of $n > 2$ variables that are independent, does it follow that any pair of them are also independent (in the sense discussed in the last two sections)? The (affirmative) answer is given in the next proposition.

Proposition 3.5.3 *If the variables $X_1, \ldots, X_n$ are independent, then any k among them (for $2 \leq k \leq n$) are also independent.*

To understand why this is true, consider X_1, X_2, and X_3, and assume them to be independent. Then, for any sets $A_1, A_2 \subseteq \mathbb{R}$, we see that

$$P(X_1 \in A_1, X_2 \in A_2)$$
$$= P(X_1 \in A_1, X_2 \in A_2, X_3 \in \mathbb{R})$$
$$= P(X_1 \in A_1)P(X_2 \in A_2)P(X_3 \in \mathbb{R}) \text{ (by the independence of } X_1, X_2 \text{ , and } X_3)$$
$$= P(X_1 \in A_1)P(X_2 \in A_2) \quad \text{(because } P(X_3 \in \mathbb{R}) = 1),$$

which shows that X_1 and X_2 are independent variables. The same argument works for the pairs (X_1, X_3) and (X_2, X_3), but also more generally for the case when n independent random variables are considered and one wants to check independence for a subset of these.

In view of Proposition 3.5.3, we can also define independence for an infinite set of random variables. Specifically, the variables $X_1, \dots, X_n, \dots$, are said to be independent if and only if any finite collection of X_i's consists of independent random variables.

We note that all the results from the preceding sections concerning a pair of random variables are also valid in the general case when more than two variables are considered. Thus, given a set of independent variables $X_1, \dots, X_n$ (either discrete or continuous), we have:

- for any real-valued functions $g_1, \dots, g_n$, the variables $Y_1 = g(X_1), \dots, Y_n = g(X_n)$, are independent;

- $E(X_1 \cdots X_n) = E(X_1) \cdots E(X_n)$;

- the conditional distribution of k variables among X_i's, for $1 \leq k \leq n - 1$, given the values of the remaining $n - k$ variables, is the same as the unconditional distribution of these k variables;

- the conditional expectation of X_i, given the value of one or more of the remaining variables, is the same as the unconditional expectation of X_i; so, we have

$$E(X_1 | X_2 = x_2, \dots, X_n = x_n) = E(X_1),$$
$$E(X_2 | X_1 = x_1, X_3 = x_3) = E(X_2),$$

and so on.

Analogous independence results hold when we consider disjoint sets of X_i's. For example, let X_1, X_2, X_3, X_4, X_5 be independent. Then:

- for any real-valued functions g_1 and g_2, the variables $Y_1 = g_1(X_1, X_3)$ and $Y_2 = g_2(X_2, X_4, X_5)$ are independent;

- for any function $g : \mathbb{R}^2 \mapsto \mathbb{R}$, we have $E(g(X_1, X_2)|X_3 = x_3, X_4 = x_4, X_5 = x_5) = E(g(X_1, X_2))$,

and so on.

Example 3.5.1 Suppose X, Y, and Z have a joint continuous distribution with density function

$$f(x, y, z) = \begin{cases} \dfrac{y^3 z}{18}, & 0 \le x \le 1,\ 0 \le y \le 2,\ 0 \le z \le 3, \\ 0, & \text{otherwise.} \end{cases}$$

The marginal density of X is then

$$f_X(x) = \int_{-\infty}^{\infty} \int_{-\infty}^{\infty} f(x, y, z)\, dy\, dz = \int_{0}^{3} z \left(\int_{0}^{2} \frac{y^3}{18}\, dy \right) dz$$

$$= \int_{0}^{3} z \left[\frac{y^4}{72} \right]_{0}^{2} dz = \int_{0}^{3} z \frac{16}{72}\, dz = \left[\frac{16 z^2}{144} \right]_{0}^{3} = 1, \qquad 0 \le x \le 1,$$

and so X has a uniform distribution over $[0, 1]$.

Similarly, we find

$$f_Y(y) = \int_{0}^{3} \int_{0}^{1} \frac{y^3 z}{18}\, dx\, dz = \frac{y^3}{4}, \qquad 0 \le y \le 2,$$

and

$$f_Z(z) = \int_{0}^{2} \int_{0}^{1} \frac{y^3 z}{18}\, dx\, dy = \frac{2z}{9}, \qquad 0 \le z \le 3.$$

It is clear that, for any $x \in [0, 1]$, $y \in [0, 2]$, and $z \in [0, 3]$,

$$f(x, y, z) = \frac{y^3 z}{18} = 1 \cdot \frac{y^3}{4} \cdot \frac{2z}{9} = f_X(x) f_Y(y) f_Z(z),$$

and so, by Proposition 3.5.2, X, Y, and Z are independent.

This means that any information about one or more among these variables does not affect the other variable(s). Thus, for example,

$$f_{X|Y,Z}(x|y, z) = f_X(x) = 1, \qquad\qquad 0 \le x \le 1,$$

$$f_{X,Z|Y}(x, z|y) = f_{X,Z}(x, z) = f_X(x) f_Z(z) = \frac{2z}{9}, \qquad 0 \le x \le 1,\ 0 \le z \le 3,$$

$$E(Y|X = 1/3,\ Z = 5/2) = E(Y) = \int_{0}^{2} y f_Y(y)\, dy = \int_{0}^{2} \frac{9 y^4}{2}\, dy = \frac{8}{5},$$

$$E(Z^2|Y = 1) = E(Z^2) = \int_{0}^{3} z^2 f_Z(z)\, dy = \int_{0}^{3} \frac{2 z^3}{9}\, dz = \frac{9}{2}.$$

Example 3.5.2 Suppose the joint distribution function of X, Y, and Z is given by

$$F(x, y, z) = \left(1 - e^{-\lambda_1 x}\right)\left(1 - e^{-\lambda_2 y}\right)\left(1 - e^{-\lambda_3 z}\right), \qquad x, y, z > 0,$$

where λ_1, λ_2 and λ_3 are three given positive parameters.

For the marginal distribution function of X, we find

$$F_X(x) = \lim_{\substack{y \to \infty \\ z \to \infty}} F(x, y, z) = \left(1 - e^{-\lambda_1 x}\right) \lim_{y \to \infty} \left(1 - e^{-\lambda_2 y}\right) \lim_{z \to \infty} \left(1 - e^{-\lambda_3 z}\right)$$

$$= 1 - e^{-\lambda_1 x}, \qquad\qquad x > 0.$$

Similarly, we find

$$F_Y(y) = \lim_{\substack{x \to \infty \\ z \to \infty}} F(x, y, z) = 1 - e^{-\lambda_2 y}, \qquad\qquad y > 0,$$

and

$$F_Z(z) = \lim_{\substack{x \to \infty \\ y \to \infty}} F(x, y, z) = 1 - e^{-\lambda_3 z}, \qquad\qquad z > 0.$$

Clearly,

$$F(x, y, z) = F_X(x) F_Y(y) F_Z(z), \quad \text{for any } x > 0, \ y > 0, \ z > 0,$$

and therefore X, Y, and Z are independent. If we wish to find $P(X > 2, \ Y < 1, \ 2 < Z < 3)$, we get

$$P(X > 2, \ Y < 1, \ 2 < Z < 3) = P(X > 2)P(Y < 1)P(2 < Z < 3)$$

$$= (1 - F_X(2))F_Y(1)\left(F_Z(3) - F_Z(2)\right)$$

$$= e^{-2\lambda_1}\left(1 - e^{-\lambda_2}\right)\left(e^{-2\lambda_3} - e^{-3\lambda_3}\right).$$

It is also evident that the marginal distributions of X, Y, and Z are exponential distributions with parameters $\lambda_1, \lambda_2, \lambda_3$, respectively. As a result, their joint density function is given by

$$f(x, y, z) = f_X(x) f_Y(y) f_Z(z) = \lambda_1 \lambda_2 \lambda_3 e^{-\lambda_1 x - \lambda_2 y - \lambda_3 z}, \quad x, y, z > 0.$$

Suppose now we want to find $P(X < Y < Z)$. For this, we cannot use the marginal densities (or distribution functions) of X, Y, and Z, because there is no way to write the required probability as a product of functions involving X, Y, and Z separately. We can instead use the expression

$$P(X < Y < Z) = \int \int_A \int f(x, y, z) dx \, dy \, dz,$$

where

$$A = \left\{(x, y, z) \in \mathbb{R}^3 \text{ with } x > 0, \ y > 0, \ z > 0 \text{ and } x < y < z\right\}.$$

In this way, we obtain

$$P(X < Y < Z) = \int_0^\infty \int_x^\infty \int_y^\infty \lambda_1 \lambda_2 \lambda_3 \, e^{-\lambda_1 x - \lambda_2 y - \lambda_3 z} \, dz \, dy \, dx$$

$$= \int_0^\infty \int_x^\infty \lambda_1 \lambda_2 \, e^{-\lambda_1 x - \lambda_2 y} \left[-e^{-\lambda_3 z}\right]_{z=y}^\infty \, dy \, dx$$

$$= \int_0^\infty \int_x^\infty \lambda_1 \lambda_2 \, e^{-\lambda_1 x - (\lambda_2 + \lambda_3) y} \, dy \, dx$$

$$= \int_0^\infty \lambda_1 \lambda_2 \, e^{-\lambda_1 x} \left[-\frac{e^{-(\lambda_2 + \lambda_3) y}}{\lambda_2 + \lambda_3} \right]_{y=x}^\infty dx$$

$$= \frac{\lambda_1 \lambda_2}{\lambda_1 + \lambda_2} e^{-(\lambda_1 + \lambda_2 + \lambda_3) x} \, dx = \frac{\lambda_1 \lambda_2}{(\lambda_2 + \lambda_3)(\lambda_1 + \lambda_2 + \lambda_3)}.$$

Observe that for $\lambda_1 = \lambda_2 = \lambda_3 = \lambda$, we get $P(X < Y < Z) = 1/6$, which is to be expected due to symmetry.

In the last two examples, establishing independence for the variables in question was rather easy. In some other cases, however, and particularly when the number of variables is not small, checking independence might be a difficult task. However, as mentioned earlier, in a variety of situations it may be clear from the description of the experiment itself that we can take independence for granted.

A typical case where this happens is when we consider **repeated experiments**. Suppose a chance experiment has sample space, Ω and X is a random variable defined on it. If we repeat this experiment n times, in a way that the outcome of each repetition does not affect the outcome of any other, and we denote by X_i the variable corresponding to the ith repetition, for $i = 1, 2, \ldots, n$, then $X_1, \ldots, X_n$ are independent random variables.

In the special case when the experiment is carried out under identical conditions in all n repetitions, it is natural to assume that the variables $X_1, \ldots, X_n$ have *the same distribution*, that is,

$$f_{X_1}(x) = \cdots = f_{X_n}(x) = f_X(x).$$

We then say that $X_1, \ldots, X_n$ constitute a **random sample** of size n from the distribution with probability function (or density function) $f_X(x)$.

Definition 3.5.2 (Random sample)   When the random variables $X_1, \ldots, X_n$ are independent and follow the same distribution, we say that they form a random sample from that distribution.

The definition applies both for discrete and continuous X_i's. If this distribution has probability function (or density function) f_X, then it is apparent that the joint probability function (respectively density function) of X_i's is

$$f(x_1, \ldots, x_n) = f_X(x_1) \cdots f_X(x_n) = \prod_{i=1}^n f_X(x_i).$$

In statistical theory, a random variable which is a function of the variables $X_1, \ldots, X_n$ of a random sample is called a *statistic*. Some common examples of interest in statistical applications include

- the sum $S = X_1 + X_2 + \cdots + X_n$;

- the sample mean, denoted by $\overline{X}$, which is simply the arithmetic average of the X_i, i.e.

$$\overline{X} = \frac{1}{n}\sum_{i=1}^{n} X_i = \frac{S}{n};$$

- the minimum and the maximum of the sample, i.e.

$$X_{(1)} = \min \left\{ X_1, \dots, X_n \right\}, \quad X_{(n)} = \max \left\{ X_1, \dots, X_n \right\};$$

- the sample **range**, $R = X_{(n)} - X_{(1)}$.

Example 3.5.3 (Series and parallel systems)
Suppose n units (e.g. water pumps) are connected in series forming a system, as shown in the graph below. This system (water network) is considered to be in operation if and only if all n units function so that water can flow from Point A to Point B on the graph. It is clear that, if $X_1, \dots, X_n$ denote the lifetimes of the n units, then the system lifetime will be

$$X_{(1)} = \min \left\{ X_1, \dots, X_n \right\}.$$

Figure 3.1 Serial connection.

Such a system is usually referred to as a *series system*, or a serial connection.

Assume, on the other hand, that the units are connected as in the graph below, so that the system operates if and only if at least one of the n units function. Then, the system lifetime in this case is

$$X_{(n)} = \max \left\{ X_1, \dots, X_n \right\},$$

where $X_1, \dots, X_n$ are again the lifetimes of the n units.

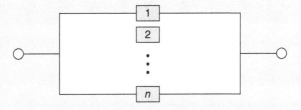

Figure 3.2 Parallel connection.

Such a system is called a *parallel system* or a parallel connection.

In both cases, a typical assumption is that the variables X_i are independent and identically distributed (i.i.d.), so that they form a random sample from a (usually, continuous) distribution.

Following upon the above discussion, it will be of interest to see how the distribution of the minimum and the maximum of a random sample is related to the distribution of X_i's. Consider first the distribution function of the maximum, $X_{(n)}$. We observe that for $X_{(n)}$ to be less than or equal to a value x, each of the X_i's must be less than or equal to x; that is,

$$P\left(X_{(n)} \leq x\right) = P(X_1 \leq x, \ldots, X_n \leq x).$$

As the X_i's are assumed to be independent, the probability on the right above is simply equal to

$$P(X_1 \leq x) \cdots P(X_n \leq x) = \prod_{i=1}^{n} P(X_i \leq x).$$

Denoting the distribution function of $X_{(n)}$ by $F_{X_{(n)}}(x)$ and $F(x)$ for the (common) distribution function of X_i, for $i = 1, \ldots, n$, we therefore have

$$F_{X_{(n)}}(x) = P\left(X_{(n)} \leq x\right) = (F(x))^n. \tag{3.12}$$

Now, let us consider the distribution of $X_{(1)}$, in which case the same argument works but for the complementary event. Specifically, the event $\{X_{(1)} > x\}$ occurs if and only if each of the X_i's is greater than x. Consequently,

$$P\left(X_{(1)} > x\right) = P(X_1 > x) \cdots P(X_n > x) = \prod_{i=1}^{n} P(X_i > x).$$

Writing $F_{X_{(1)}}(x)$ for the distribution function of $X_{(1)}$, we obtain

$$F_{X_{(1)}}(x) = P\left(X_{(1)} \leq x\right) = 1 - P\left(X_{(1)} > x\right) = 1 - \prod_{i=1}^{n} P(X_i > x)$$

$$= 1 - \prod_{i=1}^{n} \left(1 - P(X_i \leq x)\right),$$

and since each X_i has distribution function F, we arrive at

$$F_{X_{(1)}}(x) = 1 - (1 - F(x))^n.$$

If, in addition, F is a continuous distribution with density function f, differentiating both sides of the above expression, we obtain the density of the sample minimum as

$$f_{X_{(1)}}(x) = \left(F_{X_{(1)}}(x)\right)' = \left[1 - (1 - F(x))^n\right]' = nf(x)[1 - F(x)]^{n-1}.$$

Similarly, a differentiation on both sides of (3.12) yields the density of the sample maximum to be

$$f_{X_{(n)}}(x) = \left(F_{X_{(n)}}(x)\right)' = \left[(F(x))^n\right]' = nf(x)[F(x)]^{n-1}.$$

Example 3.5.4 Suppose in a serial system the unit lifetimes are exponentially distributed, with each X_i ($i = 1, \ldots, n$) having density

$$f(x) = \lambda\, e^{-\lambda x}, \qquad\qquad x > 0,$$

for some $\lambda > 0$. Then, $F(x) = 1 - e^{-\lambda x}$ and so the lifetime of the serial system has its density as

$$f_{X_{(1)}}(x) = nf(x)[1 - F(x)]^{n-1} = n\lambda\, e^{-\lambda x}\left(e^{-\lambda x}\right)^{n-1} = n\lambda\, e^{-n\lambda x}, \quad x > 0.$$

It is then clear that the distribution of the system lifetime is also exponential with parameter $n\lambda$, and so the **expected system lifetime** is simply

$$E\left[X_{(1)}\right] = \frac{1}{n\lambda}.$$

Similarly, for a parallel system we have the density function of the lifetime to be

$$f_{X_{(n)}}(x) = nf(x)[F(x)]^{n-1} = n\lambda\, e^{-\lambda x}\left(1 - e^{-\lambda x}\right)^{n-1}, \qquad x > 0.$$

EXERCISES

Group A

1. The joint density function of X, Y, and Z is given by

$$f(x, y, z) = \begin{cases} 30e^{-2x-5y-3z}, & x, y, z > 0, \\ 0, & \text{otherwise.} \end{cases}$$

   (i) Show that X, Y, and Z are independent.

   (ii) Find

   $$P(X > 1,\ Y \le 2,\ Z > 3), \quad P(X > 1,\ Y \le 2 | 1 \le Z \le 3), \quad P(Y \le 2 | X + Z > 3).$$

   (iii) Find the value of

   $$E(3X + 4Y | Z = 3/4), \quad E(Z - Y | 1 < X < 4), \quad E(2X^2 | Y = 1, Z > 2).$$

2. The joint density function of X, Y, Z, and W is given by

$$f(x, y, z, w) = \begin{cases} c(x + 2y + 3z + 4w), & 0 < x, y, z, w \le 1, \\ 0, & \text{otherwise.} \end{cases}$$

   (i) Find the value of c.

   (ii) Verify that the variables X, Y, Z and W are not independent.

   (iii) Calculate

   $$P\left(X < \frac{1}{2} \,\Big|\, Y > \frac{1}{2},\ Z > \frac{1}{3},\ W < \frac{1}{4}\right).$$

3. Write down the joint probability function or density function of a random sample $X_1, \ldots, X_n$ from each of the following distributions:

   (a) binomial with parameters n and p;

   (b) Poisson with parameter λ;

(c) geometric with parameter p;

(d) exponential with parameter λ;

(e) standard normal;

(f) uniform over the interval $(0, 1)$.

4. We ask a computer to generate a sample of size 3 from the uniform distribution over the interval $(0, 1)$. What is the probability that the first value generated will be greater than the product of the other two values? What is the expected value of the product of the three values?

5. Nick tosses four dice. Find the probability that

   (i) all dice outcomes are smaller than 5;

   (ii) the largest outcome is 5;

   (iii) two outcomes are equal to 5 and the other two are smaller than 5.

6. The variables X_1, X_2, and X_3 have a joint distribution with density

$$f(x_1, x_2, x_3) = \lambda_1 \, \lambda_2 \, \lambda_3 \, \exp\left[-\lambda_1 x_1 - \lambda_2 x_2 - \lambda_3 x_3\right], \quad x_1 > 0, \; x_2 > 0, \; x_3 > 0,$$

where $\lambda_i > 0 \; (i = 1, 2, 3)$ are the parameters of that distribution.

   (i) Examine whether the variables X_1, X_2, and X_3 are independent.

   (ii) Calculate

$$P(X_1 > 1, \; X_3 > 2), \qquad P(X_1 > 1, \; X_2 < 2, \; X_3 > 3).$$

   (iii) Find the distribution of $X_{(1)} = \min\{X_1, X_2, X_3\}$. Deduce that

$$E\left(X_{(1)}\right) = \frac{1}{\lambda_1 + \lambda_2 + \lambda_3}, \qquad Var\left(X_{(1)}\right) = \frac{1}{(\lambda_1 + \lambda_2 + \lambda_3)^2}.$$

7. Let X_1, X_2, X_3, and X_4 be a random sample of size 4 from the exponential distribution with parameter $1/\lambda$. Calculate

$$P\left(X_{(1)} \geq \lambda\right), \quad P\left(\lambda \leq X_{(4)} \leq 2\lambda\right).$$

Group B

8. A system has n units serially connected. The joint distribution function of the unit lifetimes $X_1, \ldots, X_n$ is given by

$$F(x_1, \ldots, x_n) = \left(1 - e^{-3x_1^2}\right) \cdots \left(1 - e^{-3x_n^2}\right), \quad x_1, \ldots, x_n > 0.$$

   (i) Verify that the n units work independent of one another and that their lifetimes have the same (marginal) distribution.

   (ii) Derive the common density function of X_i for $i = 1, \ldots, n$.

(iii) Obtain the density function of the system lifetime.

(iv) What is the expected lifetime of the system?

9. A computer produces n values at random from the uniform distribution over the interval $(0, 1)$, and these values are assumed independent. Let Y be the smallest value in this sample and Z be the largest value.

   (i) Show that the density functions of Y and Z are given, respectively, by

$$f_Y(y) = n(1 - y)^{n-1}, \qquad\qquad 0 \le y \le 1,$$
$$f_Z(z) = nz^{n-1}, \qquad\qquad 0 \le z \le 1.$$

   (ii) Using the result in Part (i), calculate the expected values of Y and Z, and show that

$$\lim_{n \to \infty} nE(Y) = \lim_{n \to \infty} E(Z) = 1.$$

10. Let $X_1, \dots, X_n$ be a random sample from the uniform distribution over the interval $(0, 1)$, and let T_n be defined as

$$T_n = nX_{(1)} = n \min \{X_1, \dots, X_n\}.$$

Prove that

$$\lim_{n \to \infty} P(T_n \le x) = 1 - e^{-x}, \qquad x > 0,$$

that is, asymptotically (as $n \to \infty$), T_n has a standard exponential distribution (with mean 1).

11. Let $X_1, \dots, X_n$ be a set of n (≥ 2) independent random variables, and assume that $A_1, \dots, A_n$ are subsets of the real line. Prove that the events

$$\{X_1 \in A_1\}, \ \{X_2 \in A_2\}, \ \dots, \ \{X_n \in A_n\}$$

are (completely) independent.

12. Let $X_1, \dots, X_n$ be a random sample from a continuous distribution with distribution function $F(x)$, and assume that $F(x) = 0$ for $x < 0$.

   (i) Show that

$$E\left(X_{(n)}\right) = \int_0^\infty \left[1 - (F(x))^n\right] dx.$$

   (ii) Apply the result in Part (i) to the case when the distribution of X_i is uniform over the interval $(0, 2)$.

3.6 DISTRIBUTION OF AN ORDERED SAMPLE

In the latter part of Section 3.5, we discussed the distributions of sample minimum and sample maximum. We used the notation $X_{(1)}$ for the minimum and $X_{(n)}$ for the maximum. This can clearly be generalized to cover the case of second or third smallest values in a sample, and so on. We generally write $X_{(k)}$ for the kth smallest value of a sample $X_1, \ldots, X_n$, which is called **k th order statistic.** As order statistics have found key applications in a number of diverse areas, and the methods used are frequently different from other parts of statistical theory, the area of order statistics has grown as an independent branch of statistical theory and substantial theoretical developments have been made over the years. Interested readers may refer to the encyclopedia handbooks by Balakrishnan and Rao (1998a,b) for details on all these developments. In this section, we merely present some of the main results and give some examples which serve a twofold purpose: first, they illustrate the theoretical results, and secondly they demonstrate some applications of order statistics.

To begin with, for a sample $X_1, \ldots, X_n$, it is obvious that (with probability one)

$$X_{(1)} \leq X_{(2)} \leq \cdots \leq X_{(n)}.$$

The n-dimensional random variable $(X_{(1)}, \ldots, X_{(n)})$ is called an **ordered sample**, since its elements are in increasing (more precisely, nondecreasing) order.

Example 3.6.1 In the 100-m final of Olympic games, *eight* athletes compete for the gold, silver, and bronze medals. If we put the participating runners in a random order, e.g. numbered by the lane in which they run in the final, and denote by $X_1, \ldots, X_8$ the times of these eight runners in the race, then

- the variable $X_{(1)} = \min\{X_1, \ldots, X_8\}$ represents the time of the winner of the race;

- the variable $X_{(n)} = \max\{X_1, \ldots, X_8\}$ gives the time of the athlete finishing last;

- the variables $X_{(2)}$ and $X_{(3)}$ correspond to the times of the athletes who win silver and bronze medals, respectively;

- the variable $X_{(2)} - X_{(1)}$ gives the time separating the first two athletes in the final.

Example 3.6.2 (Mean and range charts)
The manufacturer of a milk chocolate bar wishes to produce chocolate bars each weighing 1.55 ounces. Then, the process is set so as to produce chocolate bars weighing 1.55 ounces, but will have a natural variability. From a large production from the past, the standard deviation of weights of bars has been estimated to be $\sigma = 0.015$ ounce. Now, as part of the on-line quality monitoring process, a random sample of $n = 25$ bars are selected. Then, with the median of the selected sample being $X_{(13)}$, a *median chart* would declare the production process to be out-of-control if $|X_{(13)} - 1.55| > k\sigma$, where k could be determined so as to have a prespecified false alarm rate (i.e. probability of declaring the production process to be out-of-control when, in fact, it is in-control). Note that for the determination of k, it is common to assume a specific distribution for the weight of chocolate bars such as normal; then, all we need to do is to use the distribution of the sample median (see Proposition 3.6.1). Instead of using the sample median, we can use the sample mean $\overline{X}$ and in this case we will get the $\overline{X}$-chart. But, the median chart may be preferable as it will be robust to outliers in the sample.

While the above median chart would monitor the (center) weight of the chocolate bars, it may also be of interest to monitor the variability in the weight of bars. Keeping in mind that the bars of less weight and of more weight than the targeted weight of 1.55 ounces are both undesirable, a *range chart* would use the sample range $R = X_{(25)} - X_{(1)}$ and declare the production process to be out-of-control (with more variability) if $R > \ell\sigma$, where ℓ could be determined so as to have a prespecified false alarm rate. The distribution of the sample range R can be used, along with the assumed distribution for the weight of chocolate bars, for the determination of ℓ.

Example 3.6.3 (A k-out-of-n system)
An airplane with four engines can operate safely if at least two of its engines work. The plane is about to take off for a long journey. Let X_i be the time until the ith engine presents an operational problem, starting from the take-off time. Then, the time at which the airplane faces a problem in its operation is the random variable $X_{(3)}$, while the probability that there is no problem during a 20-hour flight is $P(X_{(3)} > 20)$. In this case, the question in mathematical terms can be formulated as follows: given the distribution of X_i's (which is assumed to be the same, for $i = 1, 2, 3, 4$), how do we determine the distribution of $X_{(3)}$?

More generally, when we have a system consisting of n components and the system works if at least k of these components work, we discuss a **k-out-of-n system**. In this case, when the lifetimes of components $X_1, \dots, X_n$ are independent and identically distributed, the system lifetime corresponds to the ordered $(n - k + 1)$th observation, i.e. $X_{(n-k+1)}$.

Even though ordered samples are applicable for both discrete and continuous random variables, in practice they are more useful in the continuous case. In the discrete case, one has to account for possible **ties**, i.e. sample values which are equal. For the remainder of this chapter, we restrict our attention to the case when a random sample $X_1, \dots, X_n$ is drawn from a continuous distribution with density function $f(x)$ and distribution function $F(x)$; however, the first part of the following proposition concerning the distribution function of an order statistic is also valid for discrete case (the proof applies without change).

In Section 3.5, we obtained the distributions of the sample minimum and maximum, $X_{(1)}$ and $X_{(n)}$, respectively. The following proposition is a generalization of this result.

Proposition 3.6.1 (*Distribution of order statistics*) *Let* $X_{(1)} \leq \cdots \leq X_{(n)}$ *be the ordered sample corresponding to a random sample of size* n, $X_1, \dots, X_n$, *from a continuous distribution with density* $f(x)$ *and distribution function* $F(x)$. *Then, the distribution function of the kth order statistic* $X_{(k)}$, *for* $1 \leq k \leq n$, *is*

$$F_{X_{(k)}}(x) = \sum_{r=k}^{n} \binom{n}{r} [F(x)]^r [1 - F(x)]^{n-r}, \qquad x \in \mathbb{R}, \qquad (3.13)$$

while the density of $X_{(k)}$ *is*

$$f_{X_{(k)}}(x) = \frac{n!}{(k-1)!(n-k)!} f(x)[F(x)]^{k-1}[1 - F(x)]^{n-k}, \qquad x \in \mathbb{R}.$$

Proof: The quantity $F_{X_{(k)}}(x)$ corresponds to the probability of the event $\{X_{(k)} \leq x\}$. This event occurs if and only if at least k among $X_1, \ldots, X_n$ take values in the interval $(-\infty, x]$, while the remaining variables take a value greater than x. We define the auxiliary random variables

$$Y_i = \begin{cases} 1, & \text{if } X_i \leq x, \\ 0, & \text{if } X_i > x, \end{cases}$$

for $i = 1, \ldots, n$. Then, Y_i's are independent binary (or, in other words, Bernoulli) random variables with

$$p = P(Y_i = 1) = P(X_i \leq x) = F(x), \quad q = 1 - p = P(Y_i = 0) = 1 - F(x)$$

for all i. So, their sum, $Y = Y_1 + \cdots + Y_n$, has the binomial distribution, $b(n, p)$, with probability function

$$P(Y = r) = \binom{n}{r} p^r q^{n-r} = \binom{n}{r} [F(x)]^r [1 - F(x)]^{n-r}, \quad r = 0, 1, \ldots, n. \tag{3.14}$$

The link between X_i's and Y is provided by the observation that

$$\{X_{(k)} \leq x\} \quad \text{if and only if} \quad \{Y \geq k\}$$

(try to explain why this is true!). As the two events above are equivalent, they have the same probability and, consequently, the distribution function of $X_{(k)}$ is given by

$$F_{X_{(k)}}(x) = P\left(X_{(k)} \leq x\right) = P(Y \geq k) = \sum_{r=k}^{n} P(Y = r)$$

and the result of the first part of the proposition now follows from (3.14).

For the density function of $X_{(k)}$, we differentiate the distribution function to obtain

$$f_{X_{(k)}}(x) = \sum_{r=k}^{n} \binom{n}{r} r[F(x)]^{r-1}[1 - F(x)]^{n-r} F'(x)$$

$$- \sum_{r=k}^{n-1} \binom{n}{r} (n - r)[F(x)]^r [1 - F(x)]^{n-r-1} F'(x)$$

$$= nf(x) \left\{ \sum_{r=k}^{n} \binom{n-1}{r-1} [F(x)]^{r-1}[1 - F(x)]^{n-r} \right.$$

$$\left. - \sum_{r=k}^{n-1} \binom{n-1}{r} [F(x)]^r [1 - F(x)]^{n-r-1} \right\}.$$

A careful inspection of the two sums above shows that, when we take the difference between these sums, all terms cancel out except the first term in the first sum (every other term of the first sum is equal to a term of the second sum). Thus, we obtain

$$f_{X_{(k)}}(x) = nf(x) \binom{n-1}{k-1} [F(x)]^{k-1}[1 - F(x)]^{n-k}$$

$$= \frac{n!}{(k-1)!(n-k)!} f(x)[F(x)]^{k-1}[1 - F(x)]^{n-k}, \quad x \in \mathbb{R},$$

as required. Note that in the proof above, we have used the well-known combinatorial formulas

$$r\binom{n}{r} = n\binom{n-1}{r-1}, \quad (n-r)\binom{n}{r} = n\binom{n-1}{r}.$$ □

For the case $k = n$, the expressions of the distribution function and the density function for the maximum are exactly the same as derived in Section 3.5. For the minimum ($k = 1$), this is less obvious and it is left as an exercise (see Exercise 2 of this section).

Example 3.6.4 In Example 3.6.1, suppose the times of the *eight* finalists in the 100 m race are independent and each of them has a uniform distribution in the interval $(9.5, 10.5)$. Calculate the expected value for the time of the runner who wins

(i) the gold medal;

(ii) the silver medal.

SOLUTION Here, we want the expected values of $X_{(1)}$ and $X_{(2)}$. For this, we need the density and distribution functions of X_i, which are given by

$$f(x) = \begin{cases} 1, & 9.5 \le x \le 10.5, \\ 0, & \text{otherwise,} \end{cases} \qquad F(x) = \begin{cases} 0, & x < 9.5, \\ x - 9.5, & 9.5 \le x \le 10.5, \\ 1, & x > 10.5. \end{cases}$$

(i) The density function of the winning time in the race is given by

$$f_{X_{(1)}}(x) = 8f(x)[1 - F(x)]^7$$

$$= 8 \cdot 1 \cdot [1 - (x - 9.5)]^7 = 8(10.5 - x)^7, \quad 9.5 \le x \le 10.5;$$

of course, the density vanishes outside the interval $(9.5, 10.5)$. Thus, the expected time for the winner is

$$E\left(X_{(1)}\right) = \int_{9.5}^{10.5} x f_{X_{(1)}}(x)dx = \int_{9.5}^{10.5} 8x(10.5 - x)^7 dx.$$

Making the substitution $y = 10.5 - x$, this yields

$$E\left(X_{(1)}\right) = \int_0^1 8y^7(10.5 - y)dy = \int_0^1 (84y^7 - 8y^8)dy$$

$$= \left[\frac{84y^8}{8} - \frac{8y^9}{9}\right]_0^1 = \frac{84}{8} - \frac{8}{9} \cong 9.61 \quad \text{(seconds)},$$

a time certainly worthy of an Olympic champion!

(ii) We now have to find the density of $X_{(2)}$. From Proposition 3.6.1, we have

$$f_{X_{(2)}}(x) = \frac{8!}{(2-1)!(8-2)!} f(x)[F(x)]^{2-1}[1 - F(x)]^{8-2}$$

$$= 56(x - 9.5)(10.5 - x)^6.$$

Working in a similar way as before, we obtain

$$E\left(X_{(2)}\right) = \int_0^1 56y^6(10.5 - y)(1 - y)dy = \frac{175}{18} \cong 9.72 \text{ (seconds)}.$$

In practical applications, we may also be interested in random variables that are related to more than one order statistic associated with a random sample $X_1, \ldots, X_n$. In such cases, we need to know the joint distribution of order statistics, which are provided in the next two propositions.

Proposition 3.6.2 *Let $X_{(1)} \leq \cdots \leq X_{(n)}$ be the ordered sample associated with the random sample $X_1, \ldots, X_n$ of size n, arising from a continuous distribution with density $f(x)$ and distribution function $F(x)$. Then, for any i, j with $1 \leq i < j \leq n$, the joint density function of $X_{(i)}$ and $X_{(j)}$ is given by*

$$f_{X_{(i)}, X_{(j)}}(x, y) = \begin{cases} \dfrac{n!}{(i-1)!(j-i-1)!(n-j)!} f(x)f(y) \\ \quad \times [F(x)]^{i-1} \left[F(y) - F(x)\right]^{j-i-1} \left[1 - F(y)\right]^{n-j}, & \text{if } x < y, \\ 0, & \text{otherwise.} \end{cases}$$

Proof: The proof requires some auxiliary results, hence it is deferred to Chapter 6 (see Exercise 14, Section 6.2). □

Proposition 3.6.3 *Let $X_1, \ldots, X_n$ be a random sample from a continuous distribution with density $f(x)$, and let $X_{(1)} \leq \cdots \leq X_{(n)}$ be the ordered sample associated with it. Then the joint density function of the n-dimensional random variable $(X_{(1)}, \ldots, X_{(n)})$ is given by*

$$f_{X_{(1)}, \ldots, X_{(n)}}(x_1, \ldots, x_n) = \begin{cases} n! f(x_1) \cdots f(x_n), & \text{if } x_1 < \cdots < x_n, \\ 0, & \text{otherwise.} \end{cases}$$

Proof: For the joint distribution function of order statistics, considering all possible arrangements for the relative order of the variables $X_1, \ldots, X_n$, we obtain

$$F_{X_{(1)}, \ldots, X_{(n)}}(x_1, x_2, \ldots, x_n) = P\left(X_{(1)} \leq x_1, \ldots, X_{(n)} \leq x_n\right)$$

$$= \sum_i P\left(X_{i_1} \leq x_1, \ldots, X_{i_n} \leq x_n\right),$$

where the summation is over all permutations $(i_1, \ldots, i_n)$ of the set $\{1, 2, \ldots, n\}$. The number of such permutations is $n!$ and, by symmetry, each term in the summation on the RHS has the same probability, which is $F_{X_1, \ldots, X_n}(x_1, \ldots, x_n)$. Thus, the value of the sum is simply

$$n! F(x_1) \cdots F(x_n),$$

using the independence of X_i's. The result for the joint density then follows by differentiation (it is clear that this density vanishes at all points $(x_1, \ldots, x_n)$ except those for which we have $x_1 < \cdots < x_n$). □

Example 3.6.5 Let $X_1, \ldots, X_n$ be a random sample from the uniform distribution over the interval $(0, 1)$ with density and distribution functions as

$$f(x) = \begin{cases} 1, & 0 \le x \le 1, \\ 0, & \text{otherwise,} \end{cases} \quad \text{and} \quad F(x) = \begin{cases} 0, & x < 0, \\ x, & 0 \le x \le 1, \\ 1, & x > 1. \end{cases}$$

Then, we have the joint density of $X_{(1)}$ and $X_{(n)}$ as

$$f_{X_{(1)}, X_{(n)}}(x, y) = \frac{n!}{0!(n-2)!0!} f(x) f(y) [F(x)]^0 [F(y) - F(x)]^{n-2} [1 - F(y)]^0$$

$$= n(n-1)(y-x)^{n-2}$$

if $x < y$, and zero otherwise.

For the marginal densities of $X_{(1)}$ and $X_{(n)}$, we obtain, either by integrating out one variable in the above joint density, or directly from the results of the last section, that

$$f_{X_{(1)}}(x) = \begin{cases} n(1-x)^{n-1}, & 0 \le x \le 1, \\ 0, & \text{otherwise,} \end{cases}$$

and

$$f_{X_{(n)}}(y) = \begin{cases} n y^{n-1}, & 0 \le y \le 1, \\ 0, & \text{otherwise.} \end{cases}$$

We do observe that $f_{X_{(1)}, X_{(n)}}(x, y) \neq f_{X_{(1)}}(x) f_{X_{(n)}}(y)$, revealing that the random variables $X_{(1)}$ and $X_{(n)}$ are dependent (although the original variables $X_1, \ldots, X_n$ are all independent). Intuitively, this is obvious since, if $X_{(1)}$ and $X_{(n)}$ were independent, then knowing the value of $X_{(1)}$, say, would give no information about the distribution of $X_{(n)}$. But, this is clearly not true; if we know that $X_{(1)} = x$, then $X_{(n)}$ cannot be smaller than x.

Example 3.6.6 Suppose the lifetimes for a certain type of batteries are distributed as exponential with mean lifetime of 60 hours, i.e. $\lambda = 1/60$. Further, suppose we take two batteries as a sample and place them on a life-test simultaneously. Let X_1 and X_2 denote their lifetimes, and $X_{(1)}$ and $X_{(2)}$ denote their ordered lifetimes. Evidently, $X_{(1)}$ is the lifetime of the weaker battery and $X_{(2)}$ is the lifetime of the stronger between the two batteries. Let us compute the probability that one of the batteries fails in the first 30 hours while the other one lasts for at least 60 hours after the failure of the first. Then, from Proposition 3.6.3, we have this to be

$$P\left(X_{(1)} \le 30, X_{(2)} > 60 + X_{(1)}\right) = \frac{2}{60} \int_0^{30} e^{-\frac{1}{60}x_1} \left\{ \int_{60+x_1}^{\infty} \frac{1}{60} e^{-\frac{1}{60}x_2} \, dx_2 \right\} dx_1$$

$$= \frac{2}{60} \int_0^{30} e^{-\frac{1}{60}x_1} e^{-1-\frac{1}{60}x_1} \, dx_1$$

$$= e^{-1} \int_0^{30} \frac{1}{30} e^{-\frac{1}{30}x_1} \, dx_1$$

$$= e^{-1} \left(1 - e^{-1}\right) \cong 23\%.$$

By proceeding in a similar way, we can find more generally that

$$P\left(X_{(1)} \le a, X_{(2)} - X_{(1)} > b\right) = P\left(X_{(1)} \le a, X_{(2)} > b + X_{(1)}\right)$$

$$= \left(1 - e^{-\frac{a}{30}}\right) e^{-\frac{b}{60}}, \quad \text{for } a, b > 0.$$

Interestingly, from the above expression, by letting $b \downarrow 0$ and $a \uparrow \infty$, respectively, we observe that

$$P\left(X_{(1)} \le a\right) = 1 - e^{-\frac{a}{30}}, \quad \text{for } a > 0,$$

and

$$P\left(X_{(2)} - X_{(1)} > b\right) = e^{-\frac{b}{60}}, \quad \text{for } b > 0.$$

Thus, we find $X_{(1)}$ is distributed as exponential with mean lifetime of 30 hours (i.e. $\lambda = 1/30$) and $X_{(2)} - X_{(1)}$ is distributed as exponential with mean lifetime of 60 hours (i.e. $\lambda = 1/60$). Furthermore, we observe these two variables to be independent.

EXERCISES

Group A

1. Consider a k-out-of-n system, as described in Example 3.6.3. What system corresponds to the case $k = 1$? Which system corresponds to the case $k = n$?

2. Verify that for $k = 1$, the results for the distribution and density function for the minimum of a random sample from Proposition 3.6.1 agree with the corresponding results obtained in Section 3.5 before Example 3.5.4.

3. Let X_1, X_2, X_3, and X_4 be a random sample from a continuous distribution with density

$$f(x) = cx^9, \qquad\qquad 0 < x < 4,$$

for some constant c. Obtain the density function and the expected value of the third order statistic from this sample.

4. The time (in hours) to service a car at a garage follows a continuous distribution with density

$$f(x) = 4x^2 e^{-2x}, \qquad\qquad x > 0.$$

Assume that at the start of a working day, three cars arrive for service and the work begins immediately by three different technicians, one for each car. Calculate the density function and the distribution function of the time until two cars have completed service. Thence, find the expected value of this time.

5. Using Proposition 3.6.1 and the fact that

$$F_{X_{(k)}}(x) = \int_{-\infty}^{x} f_{X_{(k)}}(t)dt,$$

prove the identity

$$\sum_{r=k}^{n} \binom{n}{r} p^r (1-p)^{n-r} = \frac{n!}{(k-1)!(n-k)!} \int_{0}^{p} u^{k-1}(1-u)^{n-k} \, du,$$

for $1 \leq k \leq n$ and $0 < p \leq 1$. Which distribution is related to the left-hand side of this identity and which one to the right-hand side?

6. A consumer buys five light bulbs of the same type in order to use them for lighting an area. The lifetimes of the bulbs (in tens of thousands of hours) X_1, X_2, X_3, X_4, X_5 are assumed to be independent random variables, each having exponential distribution with mean $\lambda = 1$. The consumer will buy new bulbs at the time where only one of the five light bulbs will be working.

   (i) Find the expected time until the new purchase of light bulbs is made.

   (ii) Calculate the expected time between the failure of the first light bulb and the failure of the second bulb.

7. An automatic mechanism produces sounds whose intensity has density function

$$f(x) = 2x, \qquad\qquad 0 < x < 1.$$

If the mechanism produces four sounds in turn, find the probability that the highest intensity observed is at least twice as much as the lowest one.

8. For the four-engine aircraft of Example 3.6.3, assume that the lifetimes X_1, X_2, X_3, X_4 of the four engines (which are assumed independent and identically distributed) have an exponential distribution with a mean of 500 hours.

   (i) Obtain the density of $X_{(3)}$, the failure time of the aircraft.

   (ii) Find the probability that the aircraft operates without any problem during a

       (a) 10-hour flight;

       (b) 24-hour flight.

9. For a random sample of size 4 from the uniform distribution over the interval $(0, 1)$, calculate the joint density of the second and third smallest observation.

10. In Exercise 6 of Section 3.5, we considered the variables X_1, X_2, X_3 having joint density

$$f(x_1, x_2, x_3) = \lambda_1 \lambda_2 \lambda_3 \exp\left[-\lambda_1 x_1 - \lambda_2 x_2 - \lambda_2 x_2\right], \quad x_1 > 0, \ x_2 > 0, \ x_3 > 0,$$

where $\lambda_i > 0$ for $i = 1, 2, 3$.

   (i) Obtain the joint density of $X_{(1)}$ and $X_{(3)}$.

   (ii) Calculate the expectation $E\left(X_{(1)} X_{(3)}\right)$.

   (iii) Examine whether any of the following relations is true:

$$f_{X_{(1)}, X_{(3)}}(x, y) = f_{X_{(1)}}(x) f_{X_{(3)}}(y) \quad \text{for any } x, y \in \mathbb{R}$$

and

$$E\left(X_{(1)} X_{(3)}\right) = E\left(X_{(1)}\right) E\left(X_{(3)}\right).$$

What do you conclude about the two variables?

11. Peter needs to pass three exams to finish his degree. The courses he has decided to take are: Social Psychology (SP), Psychology in Education (PE), and Organizational Psychology (OP). Based on his previous results, it is assumed that Peter's exam mark in a course, on a scale 0–100, follows the uniform distribution over the interval (70,100). In order to obtain a first class in his degree, Peter needs to pass at least two courses with a mark of 85 or higher and to get no less than 75 in any subject.

   (i) Express the event that he gets a first class degree in terms of the order statistics for his marks and hence calculate the probability of that event.

   (ii) Calculate the probability of the event that Peter receives a mark of 75 or higher in each of the three subjects but he is not awarded a first class degree.

   (iii) What is the probability that he gets a mark of 85 or higher in OP, given that he achieves a mark of 75 or higher in all three subjects?

   (iv) What is the probability that he gets a mark of 85 or higher in SP, given that his maximum mark is less than 95?

Group B

12. Let $X_1, X_2, \ldots$ be a sequence of independent continuous random variables. Then, *upper record values* are defined as follows. The first upper record value is trivially $R_1 \equiv X_1$. Next, the second upper record value is $R_2 = X_k$ if $X_2, \ldots, X_{k-1}$ are all less than X_1 and X_k is larger than X_1.

   (i) Derive the joint density function of the first two upper record values, R_1 and R_2.

   (ii) Deduce the marginal density function of R_2.

(iii) Let us similarly define the *lower record values* as follows. The first lower record value is trivially $R'_1 \equiv X_1$. Next, the second lower record value is $R'_2 = X_\ell$ if $X_2, \ldots, X_{\ell-1}$ are all larger than X_1 and X_ℓ is smaller than X_1. Then, derive the joint density function of the first two lower record values R'_1 and R'_2, and use it to derive the marginal density function of R'_2.

13. Let $X_1, \ldots, X_n$ be a random sample of size $n = 2k + 1$, for some positive integer k. The "middle" observation $X_{(k+1)}$ is called the **sample median**. Assuming that the distribution of X_i's is uniform over $(0, 1)$, show that $X_{(k+1)}$ has a Beta distribution with parameters $\alpha = k + 1$ and $\beta = k + 1$, with density function

$$f_{X_{(k+1)}}(x) = \frac{(2k + 1)!}{k!k!} x^k (1 - x)^k, \quad 0 < x < 1.$$

14. Let $X_1, \ldots, X_n$ be a random sample from a distribution with density $f(x)$ and distribution function $F(x)$. Let $R = X_{(n)} - X_{(1)}$ be the sample range.

(i) Write down the joint density of the variables $X_{(1)}$ and $X_{(n)}$.

(ii) By a suitable integration of the joint density found in Part (i), show that

$$P(R \le r) = n \int_{-\infty}^{\infty} [F(x) - F(x - r)]^{n-1} f(x) dx, \qquad r > 0.$$

(iii) Give an expression, in terms of the functions $f(x)$ and $F(x)$, for the density function of R.

Application: Five persons are at a bus stop, each waiting for a different bus. The times, X_i (in minutes), for $i = 1, 2, 3, 4, 5$, until the five buses arrive are independent random variables, each having a uniform distribution over $(0, 20)$. Find the density function, and the expected value, for the time duration between the arrival of the first bus and the arrival of the last one.

15. During the automatic production of metal bars in a factory, a machine makes three spots on every bar of length 4 cm produced. The distance, from one particular edge of the bar, until the point where a spot is made, has distribution function

$$F(x) = \frac{x^2}{16}, \qquad 0 \le x \le 4.$$

Let A, B, and C be the three points where the spots are made on the bar, with A being the point closest to the edge, denoted by O, and C being the one farthest to the edge. Then, calculate the probabilities for each of the following events:

(i) both segments AB and BC have a length greater than 1 cm;

(ii) both segments AB and BC have a length less than 1 cm;

(iii) the length of the segment OC is at least three times larger than the length of OA and at least twice as much as the length of the segment OB;

(iv) the segment BC is longer than the segment AB.

3.7   BASIC CONCEPTS AND FORMULAS

Independent random variables	$P(X \in A, \ Y \in B) = P(X \in A)P(Y \in B)$ for any A, B
Sufficient and necessary conditions for X and Y to be independent	• $F(x, y) = F_X(x)F_Y(y)$ for any $x, y \in \mathbb{R}$ • $f(x, y) = f_X(x)f_Y(y)$ for any $x, y \in \mathbb{R}$
Properties of two independent random variables X and Y	If X, Y are independent, then • the random variables $g(X)$ and $h(Y)$ are independent for any real functions $g(\cdot)$ and $h(\cdot)$ • $E(XY) = E(X)E(Y)$ • $E\left[g(X)h(Y)\right] = E\left[g(X)\right]E\left[h(Y)\right]$ for any real functions $g(\cdot)$, and $h(\cdot)$ • $f_{X|Y}(x|y) = f_X(x), f_{Y|X}(y|x) = f_Y(y)$ for any x, y • $F_{X|Y}(x|y) = F_X(x), F_{Y|X}(y|x) = F_Y(y)$ for any x, y • $E(X|Y = y) = E(X)$ for any y, $E(Y|X = x) = E(Y)$ for any x, and, more generally, $E\left[g(X)|Y = y\right] = E[g(X)]$ for any y, $E\left[h(Y)|X = x\right] = E[h(Y)]$ for any x
Properties of a joint probability function of an n-dimensional discrete random variable	• $f(x_1, \dots, x_n) = 0$ if $(x_1, \dots, x_n) \notin R_{X_1, \dots, X_n}$ • $f(x_1, \dots, x_n) \geq 0$ for any $(x_1, \dots, x_n) \in R_{X_1, \dots, X_n}$ • $\displaystyle\sum_{(x_1, \dots, x_n) \in R_{X_1, \dots, X_n}} f(x_1, \dots, x_n) = 1$
Marginal probability functions	$f_{X_i}(x_i) = P(X_i = x_i) = \displaystyle\sum_{x_j, j \neq i} f(x_1, \dots, x_n)$ for $1 \leq i \leq n$ $f_{X_i, X_j}(x_i, x_j) = P(X_i = x_i, X_j = x_j) = \displaystyle\sum_{x_k, k \neq i, j} f(x_1, \dots, x_n)$ for any $i \neq j$, and so on
Properties of a joint density function of an n-dimensional continuous random variable	• $f(x_1, x_2, \dots, x_n) \geq 0$ for any $x_i \in \mathbb{R}, i = 1, 2, \dots, n$ • $\displaystyle\int_{-\infty}^{\infty} \cdots \int_{-\infty}^{\infty} f(x_1, \dots, x_n)dx_1 \cdots dx_n = 1$
Marginal density functions	$f_{X_1}(x_1) = \underbrace{\displaystyle\int_{-\infty}^{\infty} \cdots \int_{-\infty}^{\infty}}_{n-1 \text{ integrals}} f(x_1, \dots, x_n)dx_2 \cdots dx_n$ $f_{X_1, X_2}(x_1, x_2) = \underbrace{\displaystyle\int_{-\infty}^{\infty} \cdots \int_{-\infty}^{\infty}}_{n-2 \text{ integrals}} f(x_1, \dots, x_n)dx_3 \cdots dx_n$ and so on

Joint distribution function of an n-dimensional random variable	$F(x_1, \ldots, x_n) = P(X_1 \leq x_1, \ldots, X_n \leq x_n)$
Relationship between a distribution function and the associated density function for an n-dimensional continuous random variable	$F(x_1, \ldots, x_n) = \int_{-\infty}^{x_1} \cdots \int_{-\infty}^{x_n} f(t_1, \ldots, t_n) dt_n \cdots dt_1$ $f(x_1, \ldots, x_n) = \dfrac{\partial^n F(x_1, \ldots, x_n)}{\partial x_1 \cdots \partial x_n}$
Expectation for a function of n random variables $X_1, \ldots, X_n$	$E\left[h(X_1, \ldots, X_n)\right] = \displaystyle\sum_{(x_1, \ldots, x_n) \in R_{X_1, \ldots, X_n}} h(x_1, \ldots, x_n) f(x_1, \ldots, x_n)$ (discrete case) $E\left[h(X_1, \ldots, X_n)\right] = \int_{-\infty}^{\infty} \cdots \int_{-\infty}^{\infty} h(x_1, \ldots, x_n) f(x_1, \ldots, x_n)$ $dx_1 \cdots dx_n$ (continuous case)
Expectation of a linear combination	$E(a_1 X_1 + \cdots + a_n X_n) = a_1 E(X_1) + \cdots + a_n E(X_n)$
Independent random variables $X_1, \ldots, X_n$	$P(X_1 \in A_1, \ldots, X_n \in A_n) = \displaystyle\prod_{i=1}^{n} P(X_i \in A_i)$ for any $A_1, \ldots, A_n$
Sufficient and necessary conditions for $X_1, \ldots, X_n$ to be independent	• $F(x_1, \ldots, x_n) = F_{X_1}(x_1) \cdots F_{X_n}(x_n)$ for any $x_i \in \mathbb{R}$ • $f(x_1, \ldots, x_n) = f_{X_1}(x_1) \cdots f_{X_n}(x_n)$ for any $x_i \in \mathbb{R}$
Random sample $X_1, \ldots, X_n$ of size $n \geq 2$	A set of n independent variables having the same distribution
Statistic	Any function of the variables $X_1, \ldots, X_n$ which form a random sample
Some important statistics based on a random sample $X_1, \ldots, X_n$	$S = X_1 + \cdots + X_n = \displaystyle\sum_{i=1}^{n} X_i$ $\overline{X} = \dfrac{1}{n} \displaystyle\sum_{i=1}^{n} X_i = \dfrac{S}{n}$ $X_{(1)} = \min\{X_1, \ldots, X_n\}, \quad X_{(n)} = \max\{X_1, \ldots, X_n\}$ $R = X_{(n)} - X_{(1)}$
Distribution of the minimum and maximum of a random sample	$F_{X_{(1)}}(x) = 1 - [1 - F(x)]^n$ $f_{X_{(1)}}(x) = nf(x)[1 - F(x)]^{n-1}$ $F_{X_{(n)}}(x) = [F(x)]^n$ $f_{X_{(n)}}(x) = nf(x)[F(x)]^{n-1}$

Distribution of the kth order statistic	$$F_{X_{(k)}}(x) = \sum_{r=k}^{n} \binom{n}{r} [F(x)]^r [1 - F(x)]^{n-r}$$ $$f_{X_{(k)}}(x) = \frac{n!}{(k-1)!(n-k)!} f(x)[F(x)]^{k-1}[1 - F(x)]^{n-k}$$
Joint density function of two ordered statistics, $X_{(i)}$ and $X_{(j)}$ $(i < j)$	$$f_{X_{(i)},X_{(j)}}(x,y) = \begin{cases} \dfrac{n!}{(i-1)!(j-i-1)!(n-j)!}f(x)f(y) \\ \quad \times [F(x)]^{i-1} \big[F(y) - F(x)\big]^{j-i-1} \\ \quad \times \big[1 - F(y)\big]^{n-j}, & \text{if } x < y \\ 0, & \text{otherwise} \end{cases}$$
Joint density function of the ordered sample $X_{(1)}, \ldots, X_{(n)}$	$$f_{X_{(1)},\ldots,X_{(n)}}(x_1,\ldots,x_n) = n! f(x_1)\cdots f(x_n)$$ for $x_1 < x_2 < \cdots < x_n$ (and zero otherwise)

3.8 COMPUTATIONAL EXERCISES

Once we are acquainted with handling double integrals in Mathematica, performing an n-fold integration should present no difficulty. For instance, we need only the following two lines to calculate the constant c for the joint density function in Example 3.4.2. In the first line, we define the function and then evaluate the threefold integral over the entire range of values for the density function (equating this to one, we get $c = 1$, as in the example).

```
In[1]:= f[x_, y_, z_]:= (x + y)*Exp[-z]
a:=Integrate[f[x, y, z], {x, 0, 1}, {y, 0, 1}, {z, 0, Infinity}]
Print["c=", 1/a]

Out[1]:= c=1
```

It is also very easy to integrate out one or two variables in order to obtain marginal (joint) densities of the variables $X, Y,$ and Z. For example, we find the marginal joint density of (X, Y) and the marginal density of X as follows:

```
In[2]:= fX[x_]:= Integrate[f[x, y, z], {y, 0, 1}, {z, 0, Infinity}]
fXY[x_, y_]:= Integrate[f[x, y, z], {z, 0, Infinity}]
Print["Marginal density of X is ", fX[x]]
Print["Joint marginal density of X, Y is ", fXY[x, y]]

Out[2]:= Marginal density of X is 1/2+x
Joint marginal density of X, Y is x+y
```

Using similar ideas, do the following.

1. Find the value of c, so that each of the following functions is a valid joint density function of (X, Y, Z):

   (i) $f(x, y, z) = \begin{cases} cx^2(1 + y + 2z) \exp\left[-x^2(3 + y + z)\right], & x, y, z > 0, \\ 0, & \text{otherwise;} \end{cases}$

   (ii) $f(x, y, z) = \begin{cases} cx^3 \ln(y + 1)e^{-x-3z}, & 0 < x < 2,\ 0 < y < 3,\ 0 < z < \infty, \\ 0, & \text{otherwise;} \end{cases}$

   (iii) $f(x, y, z) = \begin{cases} \dfrac{c(x^2 + z)e^{-4x}}{y^2 z^4}, & x > 0,\ y > 1,\ z > 2, \\ 0, & \text{otherwise.} \end{cases}$

   Then, in each case calculate

   (a) the marginal densities of X, Y, Z;

   (b) the joint marginal density of (Y, Z);

   (c) the conditional density of X, given $Y = y$ and $Z = z$, and the associated conditional expectation;

   (d) the conditional distribution of (X, Y), given $Z = z$.

2. Consider a bivariate random variable (X, Y) with probability function

   $$f(x, y) = \frac{2x + 3y}{236}, \qquad x = 1, 2, 4,\ y = 2, 4, 6, 8.$$

   The following sequence of commands in Mathematica enables us to examine whether X and Y are independent.

```
In[4]:= f[x_, y_]:= (2 x + 3 y)/236
n = 3; m = 4;
xval = {1, 2, 4};
yval = {2, 4, 6, 8};
Print["Joint Probability Function of X and Y "]
t = Table[f[xval[[i]], yval[[j]]], {i, 1, n}, {j, 1, m}];
Print[MatrixForm[t]]
(*Calculation of the Marginal Distribution of X*)
Do[Xmarg[i] = Sum[t[[i, j]], {j, 1, m}], {i, 1, n}]
(*Calculation of the Marginal Distribution of Y*)
Do[Ymarg[j] = Sum[t[[i, j]], {i, 1, n}], {j, 1, m}]
Print["Product of the Marginal Probability Functions"]
Do[Do[xymult[i, j] = Xmarg[i]*Ymarg[j], {i, 1, n}], {j, 1, m}]
Print[MatrixForm[Table[xymult[i, j], {i, 1, n}, {j, 1, m}]]]
(*Check to see if the variables are independent*)
count = 0;
Do[Do[count = count + If[xymult[i, j] == t[[i, j]], 1, 0],
{i, 1,n}], {j, 1, m}]
```

```
If[count == n*m, Print["X and Y ARE INDEPENDENT"],
Print["X and Y ARE NOT INDEPENDENT"]]


Joint Probability Function of X and Y
(2/59 7/118 5/59  13/118
5/118 4/59  11/118 7/59
7/118 5/59  13/118 8/59)


Product of the Marginal Probability Functions
(136/3481 425/6962 289/3481 731/6962
152/3481  475/6962 323/3481 817/6962
184/3481  575/6962 391/3481 989/6962)
X and Y ARE NOT INDEPENDENT
```

In a similar manner, investigate whether X and Y are independent in each of the following cases:

(i) $f(x,y) = \dfrac{x^2 y^3}{295590}$, $x = 1, 2, 3, 4, \ y = 5, 6, 8, 10, 20$;

(ii) $f(x,y) = \dfrac{661500 x^2}{143963 y^3}$, $x = 1, 2, 3, \ y = 5, 6, 7$;

(iii) $f(x,y) = \dfrac{x^2 + y^3}{5883}$, $x = 1, 2, 4, 8, 16, \ y = 3, 6, 9$;

(iv) $f(x,y) = \dfrac{|x - y|}{64}$, $x = 1, 2, 4, \ y = 3, 6, 9, 12$.

3. In the following Mathematica program, we examine graphically whether the variables X and Y with joint distribution function

$$F(x,y) = 1 - e^{-2x} - e^{-y^2} + e^{-2x-y^2}, \qquad x, y \geq 0,$$

are independent. Specifically, we do the following:

(i) find the marginal distribution functions $F_X(x)$ and $F_Y(y)$;

(ii) make a three-dimensional plot of the function

$$g(x,y) = \frac{F(x,y)}{F_X(x) F_Y(y)}.$$

```
In[4]:= F[x_, y_]:= 1 - Exp[-2 x] - Exp[-y^2] + Exp[-2 x - y^2]
Print["Joint Distribution Function of X,Y: ", F[x, y]]
Plot3D[F[x, y], {x, 0, 2}, {y, 0, 2}]
FX[x_] = Limit[F[x, y], y -> Infinity];
Print["Marginal Distribution Function of X :", FX[x]]
FY[y_] = Limit[F[x, y], x -> Infinity];
Print["Marginal Distribution Function of Y :", FY[y]]
g[x_, y_]:= F[x, y]/(FX[x]*FY[y]);
Print["3-D Diagram of the Function g(x,y):"]
Plot3D[g[x, y], {x, 0, 2}, {y, 0, 2}]
Print["g(x,y)=F(x,y)/(FX(x)*FY(y))= ", g[x, y]]
```

Out[6]:=Joint Distribution Function
of X,Y: 1-E^(-2x)-E^-y^2+E^(-2x-y^2)

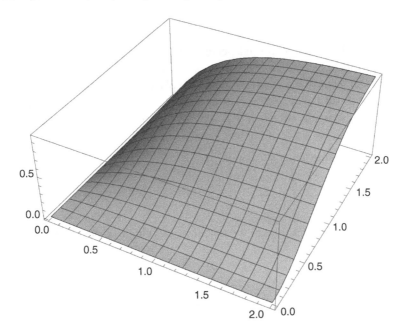

Marginal Distribution Function of X :1-Exp[-2 x]
Marginal Distribution Function of Y :1-Exp[-y^2]
3-D Diagram of the Function g(x,y):

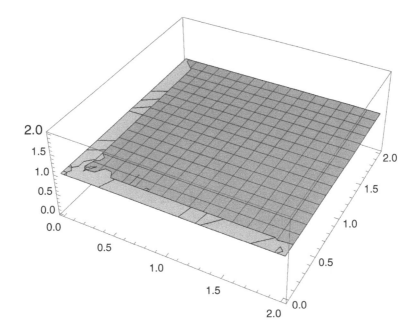

```
g(x,y)=F(x,y)/(FX(x)*FY(y))
= (1-Exp[-2 x]-Exp[-y^2]+Exp[-2 x-y^2])/((1-Exp[-2 x])
(1-Exp[-y^2]))
```

The variables X and Y are independent if and only if the function $g(x,y)$ defined above is identically equal to 1, or equivalently, the 3D plot of $g(x,y)$ produces a horizontal plane at a unit height above the plane xOy. In some cases, Mathematica may not do all the simplifications necessary to show that g is a constant function (you can also try the command *FullSimplify*); in such cases, the plot helps to identify independence for a pair of random variables.

Consider now the joint distribution

$$F(x,y) = 1 - e^{-x} - e^{-y} + e^{-x-y-0.5xy}, \qquad x,y \geq 0.$$

The corresponding 3D plot of the function $g(x,y)$ in this case is as follows:

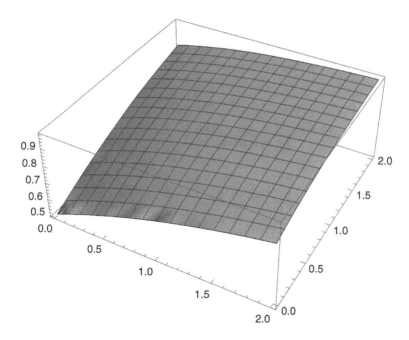

It is clear from the above figure that X and Y are *not independent*.

In a similar manner, investigate (with the aid of a graph) for each of the following three cases whether the associated jointly continuous variables X and Y are independent:

(i) $f(x,y) = \begin{cases} \dfrac{1}{54}x^2y^3, & 0 \leq x \leq 2,\ 0 \leq y \leq 3, \\ 0, & \text{otherwise;} \end{cases}$

(ii) $f(x,y) = \begin{cases} \dfrac{2}{97}(x^2 + y^3), & 0 \leq x \leq 2,\ 0 \leq y \leq 3, \\ 0, & \text{otherwise;} \end{cases}$

(iii) $f(x, y, z) = \begin{cases} \dfrac{4}{5}(x + y + xy), & 0 \le x \le 1,\ 0 \le y \le 1, \\ 0, & \text{otherwise.} \end{cases}$

4. The joint density function of (X, Y, Z, W) is given by

$$f(x, y, z, w) = \frac{c(x + 2y + 3z^2)}{(y + 2)^4} e^{-2x-4z-6w}, \qquad x, y, z, w > 0.$$

(i) Show that $c = 9216/23$.

(ii) Calculate the (marginal) joint probability densities of

        (a) X and Y (b) X, Y and W,
        (c) Y, Z and W (d) Y and W.

(iii) Examine whether each of the following random vectors consists of independent random variables:

        (a) (X, Y, Z, W) (b) (Y, Z, W) (c) (X, Y, W) (d) (X, Y).

(iv) Calculate

$$E(4X + Z^2), \ E(5X \mid Y = 1), \ E(X^2 Y Z^3 W^4).$$

5. In Example 3.6.4, we obtained the expected time of the gold and silver medalists in the 100 m Olympic final. Under the assumptions of that example (see also Example 3.6.1 earlier), calculate the expected time for each of the other 6 runners of the race, i.e. the one who finishes 3rd, 4th, and so on, until the last runner. What do you observe?

6. For this and the following exercise you should first read Section 3.11.1, the last section of this chapter. There we discuss the situation where we have many production lines in a manufacturing process and we want each of them to be represented in the sample under inspection.

(i) Assuming that there are n different production lines, find the smallest number of products k that should be sampled so that the expected number of *different lines* represented in the sample (rounded to the nearest integer) is n. Do this for

        (a) $n = 10$ (b) $n = 40$ (c) $n = 80$.

(ii) For each of the three values of n above, draw a diagram with the points (k, y), for $1 \le k \le 5n$, where k is the number of sampled items and $y = E(X)$ is the expected number of different lines present in the sample. Hence, in particular, confirm graphically the result found in Part (i).

7. In Section 3.11.1, the exact value for the expected number of products that should be sampled in order to have all lines present in the sample is found to be

$$E(Y) = n \sum_{i=1}^{n} \frac{1}{i}.$$

An approximation which has been suggested for this is $E(Y) \cong a_n$, where

$$a_n = n \cdot \ln n + \gamma n + \frac{1}{2}.$$

Here $\gamma \cong 0.5772$ is known as the Euler constant (this is available in Mathematica through *EulerGamma*).

Check the quality of this approximation by calculating its relative percentage error, viz.,

$$\frac{a_n - E(Y)}{E(Y)} \times 100$$

for $n = 5(5)200$ (this notation simply means all multiples of 5 up to 200) and find for which values of n the approximation performs the best; note that in the last formula, $E(Y)$ also depends on n although this is suppressed in the notation.

8. The command

   RandomReal[dist,n]

   produces a random sample of size n from the distribution *dist*. For instance,

   RandomReal[ExponentialDistribution[1],10]

   produces a random sample of size 10 from the exponential distribution with parameter $\lambda = 1$, while the command

   RandomReal[NormalDistribution[2,4],30]

   produces a random sample of size 30 from the normal distribution with parameters $\mu = 2$, $\sigma^2 = 4^2$.

   (i) Produce a random sample of size 20 from the uniform distribution over $[0, 1]$. Calculate the sample mean and sample variance, and compare them to the mean and variance of the uniform distribution. Repeat this for sample sizes $n = 30, 50, 100, 300, 1000$. What do you observe?

   (ii) Generate a random sample of size $m = 20$ from the binomial distribution with parameters $n = 10$ and $p = 0.2$. Compare the sample mean and variance to those of the binomial distribution $b(10, 0.2)$. Repeat this for the following values of the parameters m and n; note here that m is the size of the sample you have to generate, while n is one of the parameters of the binomial distribution:

   (a) $m = 100$, $n = 10$,

   (b) $m = 500$, $n = 10$,

   (c) $m = 20$, $n = 30$,

   (d) $m = 100$, $n = 30$,

   (e) $m = 500$, $n = 30$,

   (f) $m = 20$, $n = 80$,

   (g) $m = 100$, $n = 80$,

   (h) $m = 500$, $n = 80$.

   What can you conclude?

3.9 SELF-ASSESSMENT EXERCISES

3.9.1 True–False Questions

1. When we roll a green die and a red die, let X be the outcome on the green die and Y be the sum of the two outcomes. Then, X and Y are not independent variables.

2. If the variables X and Y are such that $E(XY) = E(X)E(Y)$, then X and Y are independent.

3. If X and Y are independent random variables, then

$$E(X|Y = 1) = E(X).$$

4. If X and Y are independent random variables, then

$$P(X \leq 3|1 \leq Y \leq 2) = P(X \leq 3).$$

5. Suppose that the joint probability function of (X, Y) is given by

$$f(x, y) = \frac{x + y}{45}, \quad x = 1, 2, 3, \ y = 2, 3, 4.$$

   Then, X and Y are independent.

6. For two independent random variables X and Y, we have

$$E\left(X^4 e^{Y+2}\right) = E(X^4)E\left(e^{Y+2}\right).$$

7. Assuming that X, Y are independent, we have

$$E(X^2(Y + 1)^4) = [E(X)]^2[E(Y) + 1]^4.$$

8. The joint density function of (X, Y) is

$$f(x, y) = \begin{cases} c \ e^{-(x+y)}x^3 y^2, & x \geq 0, \ y \geq 0, \\ 0, & \text{otherwise,} \end{cases}$$

   for a suitable constant c. Then, X and Y are independent random variables.

9. If the joint density of (X, Y) is

$$f(x, y) = 5x \ e^{-x-5y}, \quad x, y \geq 0$$

   then, for any $x, y \geq 0$, we have

$$f_{X|Y}(x|y) = f_{Y|X}(y|x).$$

10. Assume that X and Y are independent and have the same distribution. Provided the expectation of X is finite, there exists a real number c such that

$$E(X|Y = y) = E(Y|X = x) = c$$

for all x, y such that $f_X(x) \neq 0, f_Y(y) \neq 0$.

11. Let $X_1, \ldots, X_n$ be a random sample from the exponential distribution with parameter λ. Then, the distribution of $X_{(n)}$ is also exponential.

12. Let $X_1, \ldots, X_n$ be a random sample from the geometric distribution with parameter p. Then, the distribution of $X_{(1)}$ is also geometric.

13. Suppose X_1, X_2, X_3, X_4 is a random sample from the uniform distribution over the interval $(0, 3)$. Then, the distribution function of $X_{(3)}$ is given by

$$F_{X_{(3)}}(x) = P(X_{(3)} \leq x) = \begin{cases} 4\left(\dfrac{x}{3}\right)^3 \cdot \left(\dfrac{3-x}{3}\right) + \left(\dfrac{x}{3}\right)^4, & 0 < x < 3, \\ 0, & \text{otherwise.} \end{cases}$$

Questions 14–19 below refer to the discrete variables X, Y, Z having joint probability function

$$f(x, y, z) = \frac{27}{4^{x+y+z}}, \qquad x, y, z \in \{1, 2, \ldots\}.$$

14. The variables X, Y, Z are independent.

15. All three variables X, Y, and Z have the same expectation.

16. The conditional probability $P(X > x, Z \leq z|Y = y)$ does not depend on the value of y.

17. The expected value of the product XY is $E(XY) = 16$.

18. It holds that
$$P(X = 3, Y = 3) = \frac{9}{4096}.$$

19. Let $W = \min\{X, Y, Z\}$. Then,

$$P(W \geq 4) = \left(\frac{3}{4}\right)^9.$$

3.9.2 Multiple Choice Questions

1. Assuming that X, Y are independent, the conditional expectation

$$E(X^2|3Y + 1 = 2)$$

is equal to

(a) 2   (b) 4   (c) $(3Y + 1)^2$ (d) $E\left[(3Y + 1)^2\right]$ (e) $E(X^2)$

2. If the joint density of (X, Y) is

$$f(x, y) = 2x \, e^{-x-2y}, \qquad\qquad x, y \geq 0$$

then $P(Y \geq 3 | X = 2)$ is equal to

(a) $3e^{-3}$ (b) e^{-3} (c) $6e^{-6}$ (d) e^{-6} (e) e^{-12}

3. Let X be a discrete random variable with probability function

$$f_X(-3) = f_X(0) = f_X(3) = \frac{1}{3},$$

and $f_X(x) = 0$ for any other x. Further, let $Y = X^3$. Then, it follows that

(a) $E(XY) = [E(X)]^4$

(b) $f_X(x) = f_Y(x)$ for all real x

(c) the variables X and Y are independent

(d) $E(XY) = E(X)E(Y)$, but X and Y are not independent

(e) $f_{Y|X}(0|0) = 1$

4. Suppose the joint probability function of (X, Y) is given by

$$f(x, y) = \frac{x + y}{36}, \quad x = 1, 3, 5, \ y = 2, 4.$$

Then:

(a) X and Y are independent

(b) $f_X(1) = f_Y(2)$

(c) $f_{X|Y}(x|y) = f_X(x)$ for any $(x, y) \in R_{X,Y}$

(d) $E(X + Y) = E(X) + E(Y)$, but X and Y are not independent

(e) $P(Y \leq 3 | X = 1) = P(Y = 2)$

5. The table below gives the joint probability function of two discrete variables X and Y. For which values of the constants a, b, and c in the table X and Y are independent?

x \ y	5	10
1	0.28	a
2	b	0.09
3	0.21	c

(a) $a = 0.12, \ b = 0.18, \ c = 0.12$ (b) $a = 0.21, \ b = 0.09, \ c = 0.12$

(c) $a = 0.12, \ b = 0.21, \ c = 0.09$ (d) $a = 0.09, \ b = 0.21, \ c = 0.12$

(e) $a = 0.09, \ b = 0.12, \ c = 0.21$

6. Let X, Y be two random variables, each taking values in the set of positive integers, and assume that their joint distribution function is given by

$$P(X \leq x, \ Y \leq y) = (1 - s^x)(1 - q^y), \qquad x, y = 1, 2, \ldots$$

for some known constants $s, q \in (0, 1)$. Then it follows that

(a) $E(X) = E(Y)$

(b) $P(X > Y) = 1/2$

(c) the marginal distribution of X is the Bernoulli distribution

(d) the marginal distributions of both X and Y are geometric distributions

(e) $P(X > x|Y = 1) = q^x$

7. Let X and Y be two independent random variables for which it is known that

$$E(X) = E(Y) = 2, \quad Var(X) = 1, \quad Var(Y) = 3.$$

Then, the value of

$$E[(X - 1)(X + 1)|Y = 3]$$

is

(a) 0   (b) 4   (c) 5   (d) 8   (e) 9

8. Let $X_1, \ldots, X_{10}$ be a random sample of size 10 from the Bernoulli distribution with $E(X_i) = p$, for $i = 1, 2, \ldots, 10$. Then, $P(X_{(3)} = 1)$ equals

(a) $\binom{10}{8} p^8 (1 - p)^2$ (b) $\binom{10}{7} p^7 (1 - p)^3$ (c) $\binom{10}{3} p^3 (1 - p)^7$

(d) $\binom{10}{9} p^9 (1 - p) + p^{10}$ (e) $\binom{10}{8} p^8 (1 - p)^2 + \binom{10}{9} p^9 (1 - p) + p^{10}$

9. Let $X_1, \ldots, X_n$ be a random sample from the exponential distribution with parameter λ. Then, the distribution of $X_{(1)}$ is

(a) exponential with parameter $n\lambda$

(b) exponential with parameter λ

(c) exponential with parameter λ/n

(d) Gamma with parameters n and λ

(e) Gamma with parameters n and $1/\lambda$

Questions 10–14 below refer to the jointly continuous variables (X, Y, Z), having joint density function

$$g(x, y, z) = \begin{cases} \dfrac{3(y - x)(z + 1)e^{-z}}{8}, & 0 \leq x \leq y \leq 2, \ z \geq 0, \\ 0, & \text{otherwise.} \end{cases}$$

10. The marginal density of Y, for $0 \leq y \leq 2$, is

(a) $f_Y(y) = y/2$ (b) $f_Y(y) = (y-x)/2$ (c) $f_Y(y) = 3y^2/8$

(d) $f_Y(y) = y^2/8$ (e) $1/4$

11. The joint marginal density of (X, Z), for $0 \leq x \leq 2$ and $z \geq 0$, is

(a) $\dfrac{3(x-2)(z+1)e^{-z}}{16}$ (b) $\dfrac{3(x-2)(z+1)e^{-z}}{8}$ (c) $\dfrac{3(x-2)^2(z+1)e^{-z}}{16}$

(d) $\dfrac{3(x-1)(z+1)e^{-z}}{4}$ (e) $\dfrac{3(x-1)^2(z+1)e^{-z}}{8}$

12. The expectation $E(X)$ equals

$\qquad$ (a) $\dfrac{1}{2}$ (b) 1 (c) $\dfrac{1}{4}$ (d) $\dfrac{3}{2}$ (e) $\dfrac{3}{4}$

13. The value of $E(Z|X = x, Y = y)$ is (for $0 \leq x \leq y \leq 2$)

$\qquad$ (a) $2(y-x)$ (b) 1 (c) $\dfrac{3(y-x)}{2}$ (d) $\dfrac{3}{2}$ (e) 4

14. The probability $P(X \leq 1/2,\ Y > 1|Z = 1)$ equals

$\qquad$ (a) $\dfrac{1}{4}$ (b) $\dfrac{15}{32}$ (c) $\dfrac{45}{64}$ (d) $\dfrac{15}{16}$ (e) $\dfrac{7}{8}$

15. Let X_1, X_2, and X_3 be a random sample from the uniform distribution over the interval $(2, 4)$. Then, $P(X_{(1)} \leq 3,\ X_{(2)} > 3)$ equals

$\qquad$ (a) $\dfrac{1}{8}$ (b) $\dfrac{1}{4}$ (c) $\dfrac{3}{8}$ (d) $\dfrac{1}{2}$ (e) $\dfrac{5}{8}$

16. Let X_1, X_2, X_3, and X_4 be a random sample from the uniform distribution over the interval $(0, 6)$. Then, $P(X_{(1)} \leq 2,\ X_{(3)} \geq 4)$ equals

$\qquad$ (a) $\dfrac{4}{27}$ (b) $\dfrac{2}{9}$ (c) $\dfrac{1}{3}$ (d) $\dfrac{4}{9}$ (e) $\dfrac{22}{81}$

3.10 REVIEW PROBLEMS

1. Let $X_1, \ldots, X_n$ be a set of independent random variables with a common probability function

$$P(X_i = x) = p^x(1-p)^{1-x}, \quad x = 0, 1,$$

for $i = 1, \ldots, n$ (for some $0 < p < 1$).

(i) Calculate the expectations for each of the following variables:

(a) $X_1 \cdots X_n$;

(b) $1 - (1 - X_1) \cdots (1 - X_n)$;

(c) $X_1 \cdots X_n \left(\dfrac{1}{X_1} + \cdots + \dfrac{1}{X_n} \right)$.

(ii) For $n = 4$, find the expected values of the following random variables:

(a) $1 - (1 - X_1X_2)(1 - X_2X_3)(1 - X_3X_4)$;

(b) $1 - (1 - X_1)^2(1 - X_2)^2(1 - X_3)^2(1 - X_4)^2$.

(Hint: For Part (ii), use the fact that $X_i^2 = X_i$, for $i = 1, 2, 3, 4$.)

2. X and Y are two independent random variables, each having a geometric distribution with probability function $f_X = f_Y = f$ given by

$$f(x) = \frac{1}{5} \cdot \left(\frac{4}{5}\right)^x, \qquad x = 1, 2, 3, \ldots.$$

(i) Write down the joint probability function of (X, Y).

(ii) Find $E(X \mid Y = 2)$, $E(Y|X = 5)$.

(iii) Calculate

$$P(X \geq 2, Y \geq 2), \quad P(X \geq 2, Y \leq 3), \quad P(X + Y \leq 4), \quad P(X^2 + Y^2 > 10).$$

3. In a poker hand, each player receives 5 cards (out of 52 from an ordinary deck of cards). Let X be the number of aces and Y be the number of queens that a player receives in a hand. Are X and Y independent?

4. Laura tells her friend Eva that she will be at the University library at some point between 1 pm and 2 pm on a particular day. Eva decides to go to the library to meet her. Suppose the times Laura and Eva arrive at the library, denoted by X and Y, respectively, are uniformly distributed over the interval $(1, 2)$, and that they are independent.

(i) Write down the joint density function of X and Y.

(ii) Find the probability that

(a) both students arrive at the library between 1:30 pm and 1:40 pm.

(b) one student arrives at the library between 1pm and 1:10 pm and the other arrives between 1:50 pm and 2 pm. What do you observe?

(iii) If the student who arrives first waits for 15 minutes and then leaves, what is the probability that they meet?

(iv) Calculate

$$E(X|\text{Eva arrives at 1:30 pm}), \quad E(Y|\text{Laura arrives at 1:45 pm}).$$

(v) What is the expected waiting time of the student who arrives first at the library?

(vi) What is the expected waiting time of the student who arrives first at the library, given that her friend arrives at 1:45 pm?

5. For any event E on a sample space, we define the *indicator function* of E (which is, in fact, a random variable on that space) as follows:

$$I_E = \begin{cases} 1, & \text{if } E \text{ occurs,} \\ 0, & \text{otherwise.} \end{cases}$$

Verify that two events A and B are independent if and only if the associated indicator functions I_A and I_B are independent random variables.

6. At a production line in a factory, screws are produced and then put into metal objects of three types, A, B, and C. Objects of type A, B, and C have one, two, and three screws, respectively. The probability that a screw is defective is 0.5%, and the events that different screws are defective or not are independent. It is known that 30% of metal objects are of type A, 20% are of type B, and the remaining 50% are of type C.

We select one metal object at random and let X be the number of defective screws on it and Y be the number of nondefective screws.

 (i) Find the joint probability function of X and Y.

 (ii) Are X and Y independent?

 (iii) Calculate $E(X)$, $E(Y)$, and $E(XY)$.

 (iv) Find the conditional probability function of X, given $Y = 2$, and then obtain $E(X|Y = 2)$.

7. The random variables X and Y have joint density function

$$f(x, y) = \frac{4}{x^2 y^3}, \qquad x \geq 2, \ y \geq 1.$$

 (i) Find the marginal densities of X and Y, and verify that the two variables are independent.

 (ii) Show that, while the conditional expectations $E(Y|X = x)$ exist for any $x \geq 2$, the conditional expectations $E(X|Y = y)$ do not exist (i.e. they are not finite) for any $y \geq 1$.

8. The random variables X and Y have a joint density given by

$$f(x, y) = \begin{cases} n(n-1)(y-x)^{n-2}, & 0 \leq x \leq y \leq 1, \\ 0, & \text{otherwise,} \end{cases}$$

where $n > 1$ is a given positive integer.

 (i) Find the expected value of Y.

 (ii) Calculate the conditional density of Y, given $X = x$.

 (iii) Find $E(Y|X = x)$.

 (iv) Are the variables X and Y independent for some value(s) of n?

9. Let X, Y, and Z be three binary random variables, and let their joint probability function be
$$f(1, 0, 0) = f(0, 1, 0) = f(0, 0, 1) = f(1, 1, 1)$$

(f is zero for all other values).

 (i) Write down all marginal probability functions with respect to one or two variables among X, Y, and Z.

(ii) Show that all three pairs (X,Y), (X,Z), and (Y,Z) consist of independent random variables (so that (X,Y,Z) are pairwise independent).

(iii) Verify that X, Y, and Z are *not* independent random variables.

10. The intensity X of earthquakes in a particular area, measured on the Richter scale, has density function

$$f(x) = \begin{cases} 2x/25, & 0 < x < 5, \\ 0, & \text{otherwise.} \end{cases}$$

Suppose in a year, n earthquakes occur in that area and their sizes are independent random variables.

(i) Find the distribution function and the expected value of the largest earthquake during that year.

(ii) Find the distribution function and the expected value of the smallest earthquake during that year.

(iii) What is the probability that the range of the intensity of earthquakes (that is, the difference between the maximum and minimum values) does not exceed 2 points on the Richter scale?

(iv) For $n = 6$, what is the probability that there are at least 4 earthquakes of size 4.3 degrees or higher?

(Hint: For Part (iii), you may use the result of Part (ii) Exercise 14 of Section 3.6.)

11. The annual precipitation in a city, measured in inches, has distribution function

$$F(x) = 1 - \exp\left[-\left(\frac{x}{5}\right)^{1/2}\right], \qquad x > 0.$$

Suppose the levels of precipitation in different years are independent random variables.

(i) Calculate the probability that during the next three years, there will be exactly two years with precipitation over 9 in.

(ii) Let X_1, X_2, and X_3 be the precipitation levels during the following three years. Find

$$P(X_{(1)} \geq 9), \quad P(X_{(2)} \leq 12), \quad P(X_{(2)} \geq 12 | X_{(1)} \geq 9).$$

12. (**Order statistics from a discrete distribution**) The weekly number of accidents on a motorway has the Poisson distribution with parameter $\lambda = 3$ accidents. Let X_1, X_2, and X_3 be the number of accidents in a 3-week period and assume that these are independent random variables.

(i) Write down an expression for the probability functions of $X_{(1)}$ and $X_{(3)}$, the sample minimum and maximum, in terms of the (cumulative) distribution function F of the Poisson distribution.

(ii) Using either Part (i), or Eq. (3.13) which is also valid for discrete variables, find the probabilities

$$P(X_{(1)} = 1), \quad P(X_{(2)} \leq 2), \quad P(X_{(3)} \geq 3).$$

(iii) Calculate

$$P(X_{(1)} + X_{(2)} = 2), \quad P(X_{(2)} + X_{(3)} = 3).$$

13. A real number X is selected randomly from the interval $(0, 1)$. For $X = x$, we select two more numbers, represented by the random variables W and Z, uniformly from the interval $(0, x)$.

   (i) Find the marginal density functions of W and Z.

   (ii) Calculate the joint marginal density function of (W, Z). Are the variables W and Z independent?

   (iii) Obtain the joint marginal density function of (X, Z). Are the variables X and Z independent?

   (iv) Show that the following holds:

   $$f_{W,Z|X}(w, z|x) = f_{W|X}(w|x) f_{Z|X}(z|x)$$

   for any $x \in (0, 1)$ and any $0 < w < x$, $0 < z < x$.

   Can you provide a intuitive interpretation for this result?

14. (**Order statistics from independent but not identically distributed random variables**)

   Three friends make an appointment to meet at 3 pm at Times Square in New York. Let X_1, X_2, and X_3 be the times, in minutes after 2:45 pm, that each arrives at the square, and suppose $X_1 \sim U(0, 30)$, $X_2 \sim U(5, 20)$, and $X \sim U(10, 30)$; here, $U(a, b)$ stands for the uniform distribution over the interval (a, b).

   (i) Find the probability that they all arrive before 3 pm.

   (ii) What is the probability that no one arrives by 3 pm?

   (iii) Obtain the distribution function of the random variable $X_{(2)}$. Thence, find the probability that the second person to arrive at the square does so between 3 pm and 3:05 pm.

15. Suppose the joint probability density of X, Y, and Z can be expressed in the form

   $$f(x, y, z) = g(x)g(y)g(z), \qquad x, y, z \in \mathbb{R},$$

   where g is the density function of a (one-dimensional) random variable. Show that the following are true:

   (i) $P(X < Y < Z) = \dfrac{1}{6}$;

   (ii) $P(X < Y \text{ or } Y < Z) = \dfrac{5}{6}$.

   Generalize the first result when more than three variables are considered (instead of 3).

16. Let (X, Y) have a bivariate continuous distribution with joint density function of the form

   $$f(x, y) = \phi(x^2 + y^2), \qquad x, y \in \mathbb{R},$$

where $\phi(\cdot)$ is a suitable real function of a single variable (the distribution of the pair (X, Y) is then known as a **spherically symmetric distribution**).

(i) Prove that, if X and Y are independent random variables, each having the normal distribution with mean 0 and variance $\sigma^2 > 0$, then (X, Y) has a spherically symmetric distribution.

(ii) Assume that X and Y are independent and that the distribution of (X, Y) is spherically symmetric with a joint density of the form given above.
(a) Show that

$$\frac{2\phi'(x^2 + y^2)}{\phi(x^2 + y^2)} = \frac{f_X'(x)}{xf_X(x)} = \frac{f_Y'(y)}{yf_Y(y)}$$

for any $x, y \in \mathbb{R}$ with $x, y \neq 0$ and $f_X(x) \neq 0, f_Y(y) \neq 0$; here, f_X and f_Y are the marginal densities of X and Y, respectively.

(b) Deduce that, for some real r,

$$\frac{f_X'(x)}{xf_X(x)} = \frac{f_Y'(y)}{yf_Y(y)} = r, \qquad \text{for all } x, y \in \mathbb{R}.$$

Hence, show that both X and Y follow a normal distribution with zero mean and with the same variance.

17. The random variables X and Y have a joint continuous distribution function given by

$$F(x, y) = \begin{cases} 1 - e^{-x} - e^{-y} + e^{-x-y-axy}, & \text{if } x, y \geq 0, \\ 0, & \text{otherwise,} \end{cases}$$

where a is a constant such that $0 \leq a \leq 1$.

Derive the (marginal) distribution functions of X and Y. Hence, show that X and Y are independent if and only if $a = 0$.

18. Let X, Y be two binary random variables with range $R_X = R_Y = \{0, 1\}$. For their joint probability function $f(x, y) = P(X = x, Y = y)$, we are given that

$$f(0, 0) = f(0, 1) = f(1, 1) = a.$$

Prove that X and Y are independent if and only if $a = 1/4$.

3.11 APPLICATIONS

3.11.1 Acceptance Sampling

Acceptance sampling is a common quality control technique used in the industry. It is used to decide whether a lot of manufactured material is to be accepted or rejected. The origins of the method date to the early 1940s, when it was applied by the U.S. military for testing bullets during World War II.

In acceptance sampling, a representative sample of a product is collected and it is tested for conformity to the manufacturer's standards. The quality of the sample is considered to reflect the quality level of the entire manufacturing process. Let us assume that the products under test are assembled in n different and independent production lines. If the production rate of each line is similar to the others one may assume that, when the sampling in the quality control unit is random, the probability of selecting a product from any of the n lines is the same, i.e. $1/n$.

A crucial issue in acceptance sampling is determining the size of the sample to be used, so that it is representative for the entire process. Under the framework described above it seems reasonable to require our sample to contain products from all production lines. Therefore, a pertinent question to be answered is the following: what is the expected number of products that should be sampled so that at least one product has been included from each line?

Manifestly, we are interested in $E(Y)$ where Y is the number of products examined until all production lines have been included in the sample. In order to handle the expectation of Y, we decompose Y as a sum of simpler random variables, and this can be done as follows.

Let Y_1 be the number of items inspected until we have included at least one production line in the sample (then, obviously $Y_1 = 1$ with probability one). Next, let Y_2 be the number of items that will be inspected after the first one, until an item from a *different production line* is found. Then, the distribution of Y_2 is geometric with parameter

$$p_2 = \frac{n-1}{n}.$$

In a similar way, once we have items from exactly two production lines in the sample, let Y_3 be the number of items until we include an item from a line that is different from the two which are present already. Then, $Y_3 \sim G(p_3)$ with

$$p_3 = \frac{n-2}{n}.$$

Continuing in this way, if we have included items from $n-1$ distinct lines, let Y_n be the number of items that will be examined, from this point onwards, so that an item from the missing production line will be examined. Then, $Y_n \sim G(p_n)$ with

$$p_n = \frac{1}{n}.$$

It is then clear that $Y = Y_1 + \cdots + Y_n$, and so from (3.11) we get

$$E(Y) = E(Y_1 + \cdots + Y_n) = E(Y_1) + \cdots + E(Y_n).$$

As Y_i's are geometrically distributed for $i = 2, 3, \ldots, N$, we have $E(Y_i) = 1/p_i$ with $p_i = (n-i+1)/n$ for $i = 1, 2, \ldots, n$, and the expected number of items that need to be inspected so that at least one from each production line is included in the sample equals

$$E(Y) = 1 + \frac{1}{p_2} + \cdots + \frac{1}{p_{n-1}} + \frac{1}{p_n} = \frac{n}{n} + \frac{n}{n-1} + \cdots + \frac{n}{2} + \frac{n}{1}$$

$$= n \sum_{i=1}^{n} \frac{1}{i}.$$

It is of interest to mention that the representation $Y = Y_1 + \cdots + Y_n$ can also be used to calculate the variance of Y. More specifically, since $Y_1, Y_2, \ldots, Y_n$ are independent, we get that

$$Var(Y) = \sum_{i=1}^{n} Var(Y_i)$$

and making use of the variance for the geometric distribution, i.e. $Var(Y_i) = (1 - p_i)/p_i^2$, we obtain that

$$Var(Y) = n \sum_{i=2}^{n} \frac{i-1}{(n-i+1)^2} = n \sum_{i=1}^{n-1} \frac{i}{(n-i)^2}$$

$$= n \left(\frac{1}{(n-1)^2} + \frac{2}{(n-2)^2} + \cdots + \frac{n-2}{2^2} + \frac{n-1}{1^2} \right).$$

The following table presents the values of $E(Y)$ and $Var(Y)$ for several choices of n (the values in the table are rounded to the closest integer). It is clear that as n increases, both the expectation and the variance of Y increase rapidly.

n	$E(Y)$	$V(Y)$
2	3	2
3	6	7
4	8	14
5	11	25
6	15	39
7	18	56
8	22	76
9	25	99
10	29	126
11	33	155
12	37	188
13	41	224
14	46	263
15	50	306
16	54	352
17	58	400
18	63	453
19	67	508
20	72	567
25	95	908
30	120	1331
35	145	1835
40	171	2421
45	198	3089
50	225	3838

The results for the expectation and the variance are also shown graphically in the figure below.

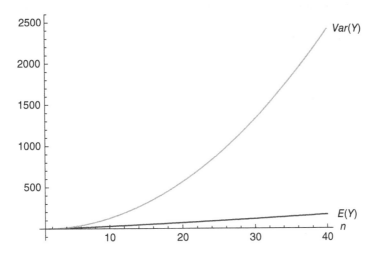

Another interesting issue under the acceptance sampling framework is the following. It goes without saying that the large (expected) sample size results in high costs for the manufacturer if we keep sampling until we observe at least one item from each line. Due to this, the quality control unit may decide to sample k products; then, a quantity of interest would be the expected number of different lines that will be represented in the sample. In order to study this quantity, let for $i = 1, 2, \ldots, n$,

$$X_i = \begin{cases} 1, & \text{if at least one item of the } i\text{th production line is found,} \\ 0, & \text{otherwise.} \end{cases}$$

Then, the number of different lines, X, present in the sample can be expressed as

$$X = X_1 + \cdots + X_n,$$

and so

$$E(X) = E(X_1 + \cdots + X_n).$$

But for each i, we have

$$E(X_i) = 1 \cdot P(X_1 = 0) + 0 \cdot P(X_i = 0) = P(X_i = 1) = 1 - P(X_i = 0) = 1 - \left(\frac{n-1}{n}\right)^k$$

(the event $\{X_i = 0\}$ means that none of the k items in the sample is from line i). Hence, the required expectation is

$$E(X) = n\left[1 - \left(\frac{n-1}{n}\right)^k\right]. \tag{3.15}$$

For $n = 10$ and $k = 15$, for example, we get from (3.15) that $E(X) \cong 7.94$, which means that in a sample of 15 items we expect to have items from about eight different production lines.

The following table presents the values of $E(X)$ for $n = 5$ and $n = 10$ (upper part), $n = 20$ and $n = 50$ (lower part) and for different values of k. It is clear that $E(X)$ is an increasing function of n and k, a fact which can be verified theoretically by noting that

(a) $((n - 1)/n)^k$ is decreasing in k for fixed n, and

(b) $E(X)$ can be expressed as

$$E(X) = n \left[1 + \left(1 - \frac{1}{n} \right) + \left(1 - \frac{1}{n} \right)^2 + \cdots + \left(1 - \frac{1}{n} \right)^{k-1} \right]$$

and each term of the sum is increasing in n for fixed k.

$n = 5$		$n = 10$	
k	$E(X)$	k	$E(X)$
1	1	1	1
2	1.80	2	1.90
3	2.44	3	2.71
4	2.95	4	3.44
5	3.36	5	4.10
6	3.69	7	5.22
7	3.95	10	6.51
8	4.16	15	7.94
9	4.33	20	8.78
10	4.46	25	9.28
11	4.57	30	9.58

$n = 20$		$n = 50$	
k	$E(X)$	k	$E(X)$
2	1.95	2	1.98
3	2.85	3	2.94
4	3.71	5	4.80
5	4.52	10	9.15
7	6.03	15	13.07
10	8.03	20	16.62
15	10.73	25	19.83
20	12.83	30	22.73
25	14.45	40	27.71
30	15.71	50	31.79
40	17.43	70	37.84
50	18.46	100	43.37
60	19.08	150	47.59
70	19.45	200	49.12

A practitioner may use these tables to determine the sample size k so that a specific (mean) number of production lines be represented in the sample. For example, if $n = 10$ and we wish at least half of the production lines to be present in the sample we should work with samples of size $k = 7$.

The problem we have considered in this section appears also, in various guises, in a number of other situations. For example, a famous problem in probability theory is the *coupon collector's problem* (where a person collects different coupons found in a commercial product), which deals essentially with the same questions as those we discussed above.

KEY TERMS

acceptance sampling

independent random variables

kth order statistic

k-out-of-n system

median (or sample median)

median chart

multivariate joint distribution

ordered statistics

ordered sample

parallel system

random sample

random vector

range (or sample range)

sample mean

serial system

statistic

CHAPTER 4

TRANSFORMATIONS OF VARIABLES

George E.P. Box (Gravesend, Kent 1919 – Madison, Wisconsin, USA 2013)

George Edward Pelham Box was born on 18 October 1919, in Gravesend, Kent, England, and died on 28 March 2013, in Madison, Wisconsin, USA. He received his PhD from the University of London, England, in 1953 under the supervision of E.S. Pearson. In 1960, he moved to the University of Wisconsin-Madison to create the Department of Statistics and stayed there till the end of his career. He is well known for his pioneering contributions to time-series analysis, design of experiments, quality control, and Bayesian inference. He also made profound contributions to distribution theory and introduced Box–Muller transformation and Box–Cox transformation. Among the many honors he received are the Shewhart Medal from the American Society for Quality Control, the Wilks Memorial Award from the American Statistical Association, and the Guy Medal in Gold from the Royal Statistical Society.

4.1   INTRODUCTION

In the first part of this chapter, we discuss the following problem: suppose (X, Y) has a known joint distribution. Can we then obtain the joint distribution of two other variables U and V, each of which is a function of X and Y? The problem is then generalized to the case when n variables are considered. For the case of two variables X and Y, we apply the results to derive the distribution of a variable which arises from X and Y by the usual arithmetic operations (addition, subtraction, multiplication, and division).

Making use of the tools developed for the previous needs, we will derive the density functions of three basic distributions that are widely used in statistics: the chi-squared, t, and F distributions.

4.2   JOINT DISTRIBUTION FOR FUNCTIONS OF VARIABLES

It is often the case that, given one or more variables from a known distribution, we want to find the distribution of some other variable(s) which are explicit functions of the original variables. The simplest case is clearly the scalar-to-scalar transformation; more precisely, given the distribution of a variable X, what is the distribution of $Y = h(X)$ for a given function h? This problem has been treated in Section 6.2 of Volume I. In this chapter, we focus our attention on the case when we are given the joint distribution of two variables, X and Y, and our interest lies in the distribution of $g(X, Y)$ for a given function g.

In the first two chapters, when we considered the joint distribution of two variables X and Y in the discrete and continuous cases, respectively, we obtained results for facilitating the calculation of the *expectation* of a function, $g(X, Y)$, of these variables. We mentioned there that the analysis does not require knowledge of the distribution of $g(X, Y)$, and typically this is a much harder exercise than finding the expectation $E[g(X, Y)]$.

Here, we concentrate on the case when X and Y have a **joint continuous distribution** and, for this case, we treat a more general problem: let the joint density of (X, Y) be $f(x, y)$, and consider two variables U and V defined in terms of X and Y as follows:

$$U = g(X, Y), \quad V = h(X, Y),$$

for two given functions g and h.

Our first result in this section shows how we can obtain an expression for the joint density of U and V under some fairly general conditions.

Proposition 4.2.1 *Let X and Y have a joint continuous distribution with density $f(x, y)$, and assume that $g(x, y)$ and $h(x, y)$ are two real functions. If the following conditions are satisfied:*

(a) for any u and v, the system of two equations on two unknowns

$$g(x, y) = u,$$

$$h(x, y) = v,$$

*has a **unique** solution for x and y, denoted by*

$$x = g^*(u, v), \quad y = h^*(u, v);$$

(b) *the functions $g^*(u, v)$ and $h^*(u, v)$ have continuous partial derivatives with respect to u and v and the determinant*

$$J(u, v) = \det \begin{pmatrix} \dfrac{\partial g^*(u, v)}{\partial u} & \dfrac{\partial g^*(u, v)}{\partial v} \\ \dfrac{\partial h^*(u, v)}{\partial u} & \dfrac{\partial h^*(u, v)}{\partial v} \end{pmatrix}$$

$$= \frac{\partial g^*(u, v)}{\partial u} \cdot \frac{\partial h^*(u, v)}{\partial v} - \frac{\partial g^*(u, v)}{\partial v} \cdot \frac{\partial h^*(u, v)}{\partial u}$$

is such that $J = J(u, v) \neq 0$ for all u and v,

then the joint density function of U and V is given by

$$f_{U,V}(u, v) = f\left(g^*(u, v), h^*(u, v)\right) |J|.$$

Proof: The proof involves calculus rather than probability considerations and arguments. The result is, in fact, a consequence of change of variables in a double integral, that can be found in most calculus textbooks and is therefore omitted here. ☐

The quantity J in Proposition 4.2.1 is known as the **Jacobian** (after the nineteenth century German mathematician Karl Jacobi) of the transformation $x = g^*(u, v)$ and $y = h^*(u, v)$. Note that it is implicit in the proposition that (x, y) belongs to the range of values for (X, Y), while the pair (u, v) belongs to the range of (U, V).

Example 4.2.1 (From Cartesian to polar coordinates)
The position of a particle on the plane can be described by its Cartesian coordinates (X, Y), which are assumed to form a two-dimensional random variable with joint density function $f(x, y)$. If O denotes the origin, then we may be interested in expressing the location of the particle in polar coordinates, (R, Θ), where R is the distance of the particle position A from the origin and Θ is the angle formed by the segment OA and the x-axis in the figure below.

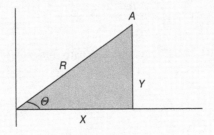

In probabilistic terms, and as an application of Proposition 4.2.1, we wish to derive the joint density function of the random variables

$$R = \sqrt{X^2 + Y^2} = g(X, Y),$$

$$\Theta = \arctan \frac{y}{x} = h(X, Y).$$

Solving the system of two equations

$$\left.\begin{array}{l} g(x, y) = r, \\ h(x, y) = \theta, \end{array}\right\} \quad \Leftrightarrow \quad \left.\begin{array}{l} \sqrt{x^2 + y^2} = r, \\ \arctan \frac{y}{x} = \theta, \end{array}\right\} \quad \Leftrightarrow \quad \left.\begin{array}{l} x^2 + y^2 = r^2, \\ \tan \theta = \frac{y}{x}, \end{array}\right.$$

with respect to x and y, we obtain

$$x = r \cos \theta = g^*(r, \theta),$$

$$y = r \sin \theta = h^*(r, \theta).$$

The corresponding Jacobian is

$$J(r, \theta) = \det \begin{pmatrix} \dfrac{\partial g^*(u, v)}{\partial r} & \dfrac{\partial g^*(u, v)}{\partial \theta} \\ \dfrac{\partial h^*(u, v)}{\partial r} & \dfrac{\partial h^*(u, v)}{\partial \theta} \end{pmatrix} = \det \begin{pmatrix} \cos \theta & -r \sin \theta \\ \sin \theta & r \cos \theta \end{pmatrix} = r.$$

Then, we have the joint density function of the pair (R, Θ) as

$$f_{R,\Theta}(r, \theta) = f\left(g^*(r, \theta), h^*(r, \theta)\right) |J| = f(r \cos \theta, r \sin \theta)|r| = rf(r \cos \theta, r \sin \theta),$$

$$\text{for } r \geq 0 \text{ and } 0 < \theta < 2\pi.$$

Let us now assume that X and Y are independent and both follow the standard normal distribution, $N(0, 1)$. Then, we have

$$f(x, y) = f_X(x) f_Y(y) = \frac{1}{\sqrt{2\pi}} e^{-x^2/2} \cdot \frac{1}{\sqrt{2\pi}} e^{-y^2/2}, \qquad x, y \in \mathbb{R}.$$

Upon applying the above result to this density, we get

$$f_{R,\Theta}(r, \theta) = \frac{r}{2\pi} \exp\left\{-\frac{r^2(\cos^2\theta + \sin^2\theta)}{2}\right\} = \frac{r}{2\pi} \exp\left(-\frac{r^2}{2}\right),$$

for any $r \geq 0$ and $\theta \in (0, 2\pi)$. The marginal density function of Θ can now be readily obtained as

$$f_\Theta(\theta) = \int_{-\infty}^{\infty} f_{R,\Theta}(r, \theta) \, dr = \int_0^{\infty} \frac{r}{2\pi} e^{-r^2/2} dr = \frac{1}{2\pi}, \quad 0 < \theta < 2\pi,$$

which is the density of the uniform distribution over $(0, 2\pi)$.

Similarly, the marginal density of R is found to be

$$f_R(r) = \int_{-\infty}^{\infty} f_{R,\Theta}(r,\theta)\, d\theta = \int_0^{2\pi} \frac{r}{2\pi} e^{-r^2/2} d\theta = re^{-r^2/2},\ r \geq 0,$$

which is called the standard *Rayleigh* distribution. From the above results, it is apparent that

$$f_{R,\Theta}(r,\theta) = f_R(r)f_\Theta(\theta)$$

for any $r > 0$ and $0 < \theta < 2\pi$, which means that the variables R and Θ, representing the polar coordinates of the particle, are independent.

Note that Proposition 4.2.1 covers the case when we know the joint distribution of X and Y and we wish to find the joint distribution of *two* other variables. What if we want to find the distribution of a *single variable*, $U = g(X, Y)$ (for example, the sum $X + Y$, or the product XY)? The general result for this case, which follows from Proposition 4.2.1, is given in Section 4.3 (see (4.1)). Although the precise results will be presented there, we describe below the main steps involved in accomplishing this. More explicitly, given the joint density $f(x, y)$ of two variables X and Y, the procedure for deriving the density of $U = g(X, Y)$ is as follows:

- We consider the pair of random variables

$$U = g(X, Y), \quad V = X,$$

  or the pair

$$U = g(X, Y), \quad V = Y.$$

- We then use Proposition 4.2.1 to find the joint density $f_{U,V}(u, v)$ of (U, V).

- Finally, we obtain the marginal density of U by integrating out v from the obtained joint density.

Example 4.2.2 Let X and Y be two independent random variables, each following an exponential distribution with parameter λ. Find the distribution of the variable $U = X/(X + Y)$.

SOLUTION   First, let us consider the pair of variables

$$U = \frac{X}{X + Y} = g(X, Y), \quad V = X = h(X, Y).$$

Then, we get

$$\left. \begin{array}{l} g(x, y) = u, \\ h(x, y) = v, \end{array} \right\} \Leftrightarrow \left. \begin{array}{l} \dfrac{x}{x + y} = u, \\ x = v, \end{array} \right\} \Leftrightarrow \begin{array}{l} x = v = g^*(u, v), \\ y = \dfrac{v(1 - u)}{u} = h^*(u, v), \end{array}$$

with the associated Jacobian as

$$J(u,v) = \det \begin{pmatrix} \dfrac{\partial g^*(u,v)}{\partial u} & \dfrac{\partial g^*(u,v)}{\partial v} \\ \dfrac{\partial h^*(u,v)}{\partial u} & \dfrac{\partial h^*(u,v)}{\partial v} \end{pmatrix} = \det \begin{pmatrix} 0 & 1 \\ -\dfrac{v}{u^2} & \dfrac{1-u}{u} \end{pmatrix} = \dfrac{v}{u^2}.$$

As X and Y are exponential, their range is the positive half-line, and that means that the range of values for U is the open interval $(0,1)$. Thus, the range of values for u and v above is $0 < u < 1$ and $v > 0$. Further, from the independence of X and Y we have the joint density of (X,Y) as

$$f(x,y) = f_X(x)f_Y(y) = \lambda^2 e^{-\lambda(x+y)}$$

for all $x, y > 0$. Then, from Proposition 4.2.1, we obtain the joint density of U and V as

$$f_{U,V}(u,v) = f\left(v, \frac{v(1-u)}{u}\right)|J| = \lambda^2 e^{-\lambda v/u} \cdot \frac{v}{u^2} = \frac{\lambda^2 v e^{-\lambda v/u}}{u^2}.$$

Now, integrating out v, we obtain the required (marginal) density of U to be

$$f_U(u) = \int_{-\infty}^{\infty} f_{U,V}(u,v)\,dv = \frac{\lambda}{u}\int_0^{\infty} v\left(\frac{\lambda}{u}\right)e^{-(\lambda/u)v}\,dv = 1.$$

This follows because the last integral is the expectation of an exponential random variable with parameter λ/u, and so the integral equals $(\lambda/u)^{-1}$. Keeping in mind that the range of values for U is the interval $(0,1)$, we thus observe that the variable $U = X/(X+Y)$ has the uniform distribution over the unit interval.

It is not difficult to extend the result of Proposition 4.2.1 to the case when more than two variables are involved. Specifically, let $X_1, \ldots, X_n$ be a collection of random variables (not necessarily independent), for which their joint density, $f(x_1, x_2, \ldots, x_n)$ is given. Suppose that we are interested in the joint density of the variables

$$U_1 = g_1(X_1, \ldots, X_n),$$
$$U_2 = g_2(X_1, \ldots, X_n),$$
$$\cdots$$
$$U_n = g_n(X_1, \ldots, X_n),$$

for some (known) functions $g_1, \ldots, g_n$.

Further, suppose the following system of n equations

$$u_1 = g_1(x_1, \ldots, x_n),$$
$$u_2 = g_2(x_1, \ldots, x_n),$$
$$\cdots$$
$$u_n = g_n(x_1, \ldots, x_n),$$

with n unknowns, $x_1, \ldots, x_n$, has a unique solution, that can be expressed as

$$x_1 = g_1^*(u_1, \ldots, u_n),$$

$$x_2 = g_2^*(u_1, \ldots, u_n),$$

$$\cdots$$

$$x_n = g_n^*(u_1, \ldots, u_n).$$

The Jacobian associated with the above transformation now takes the form

$$J = \det \begin{pmatrix} \dfrac{\partial g_1^*}{\partial u_1} & \dfrac{\partial g_1^*}{\partial u_2} & \cdots & \dfrac{\partial g_1^*}{\partial u_n} \\[2ex] \dfrac{\partial g_2^*}{\partial u_1} & \dfrac{\partial g_2^*}{\partial u_2} & \cdots & \dfrac{\partial g_2^*}{\partial u_n} \\[2ex] \vdots & \vdots & \vdots & \vdots \\[2ex] \dfrac{\partial g_n^*}{\partial u_1} & \dfrac{\partial g_n^*}{\partial u_2} & \cdots & \dfrac{\partial g_n^*}{\partial u_n} \end{pmatrix}.$$

Then, provided that $J \neq 0$ for all $u_1, \ldots, u_n$, the joint density function of the variables $U_1, \ldots, U_n$ is given by

$$f_{U_1, \ldots, U_n}(u_1, \ldots, u_n) = f\left(g_1^*(u_1, \ldots, u_n), \ldots, g_n^*(u_1, \ldots, u_n)\right)|J|.$$

Example 4.2.3 Let X, Y, and Z be three variables with joint density $f(x, y, z)$ and U, V, and W be three variables defined as

$$U = X + Y + Z, \quad V = X + 2Y, \quad W = 2Y + Z.$$

Find the joint density function of (U, V, W).

SOLUTION We first need to solve the system of three equations

$$u = x + y + z, \quad v = x + 2y, \quad w = 2y + z.$$

Adding the last two equations and then subtracting the first, we obtain $y = (v + w - u)/3$. Upon substituting for y, we find the unique solution of the system to be

$$x = \frac{2u + v - 2w}{3}, \qquad y = \frac{-u + v + w}{3}, \qquad z = \frac{2u - 2v + w}{3}.$$

The Jacobian is

$$J = J(u, v, w) = \det \left[\frac{1}{3} \begin{pmatrix} 2 & 1 & -2 \\ -1 & 1 & 1 \\ 2 & -2 & 1 \end{pmatrix} \right] = \frac{1}{3}.$$

Then, the joint density of $U, V,$ and W is given by

$$f_{U,V,W}(u, v, w) = \frac{1}{3} f\left(\frac{2u + v - 2w}{3}, \frac{-u + v + w}{3}, \frac{2u - 2v + w}{3}\right).$$

If we assume in addition that the variables $X, Y,$ and Z are independent and all have exponential distribution with parameter 1, we have $f(x, y, z) = e^{-(x+y+z)}$ for $x, y, z > 0$. As the joint density function f is zero if one of its arguments is negative, we then obtain the joint density of $U, V,$ and W to be

$$f_{U,V,W}(u, v, w) = \begin{cases} e^{-u}, & \text{if } w \le u + v/2, \ u \le v + w, \ v \le u + w/2, \\ 0, & \text{otherwise.} \end{cases}$$

EXERCISES

Group A

1. Let X and Y be two independent random variables, each having a uniform distribution over $(0, 1)$ and let the variables U and V be defined as

$$U = -2 \ln X, \quad V = -2 \ln Y.$$

   (i) Find the joint density of U and V.

   (ii) Show that both U and V have a marginal distribution with mean 2.

   (iii) Verify that U and V are independent.

   How could you answer all three questions above without an appeal to Proposition 4.2.1?

2. Let X and Y be two random variables with joint density function $f(x, y)$, and U and V be two new variables defined as

$$U = \sqrt{X^2 + Y^2}, \qquad V = X.$$

   Find the joint density function of (U, V). Thence, obtain the marginal density function of U. What do you observe in relation to the result of Example 4.2.1?

3. Let X and Y be two exponential random variables with parameters λ and μ, respectively, so that

$$f_X(x) = \lambda e^{-\lambda x}, \quad f_Y(y) = \mu e^{-\mu y}, \qquad x \ge 0, \ y \ge 0.$$

   Assuming that X and Y are independent, obtain the joint density of the variables

$$U = 3X, \quad V = 5Y + 1,$$

   and thence confirm that U and V are independent.

4. Let X and Y be uniformly distributed over the interval $(0, 1)$, and assume that they are independent. Find the joint density function of U and V where

$$U = X + Y, \qquad V = 5X - 2Y.$$

5. Assume that X, Y, and Z are independent random variables, each having a uniform distribution over $(0, 1)$. Find the joint density function of the variables

$$U = X^2, \qquad V = YZ,$$

and then show that $P(U < V) = 4/9$. Verify this result by a direct computation of $P(X^2 \leq YZ)$ through an integral expression.

6. Let X and Y be two continuous random variables that are independent and with joint density function $f(x, y)$. Verify that the joint density of the variables

$$U = X + Y, \quad V = X - Y$$

is given by

$$f_{U,V}(u, v) = \frac{1}{2} f \left(\frac{u + v}{2}, \frac{u - v}{2} \right).$$

Application: If, in addition, both X and Y have a standard normal distribution, prove that the variables U and V

(i) are independent;

(ii) follow a normal distribution, and identify the parameters μ and σ^2 of the two distributions.

7. Assume that X and Y are independent random variables and each of them has an exponential distribution with parameter $\lambda = 1$.

(i) Prove that the variables $U = X + Y$ and $V = X/(X + Y)$ are independent and confirm the result of Example 4.2.2 that the distribution of V is uniform over $(0, 1)$.

(ii) It follows from Part (i) (or by Example 4.2.2) that

$$E(V) = 1/2.$$

Phil, who attends a probability course at an University, heard about this result from his lecturer and thought that this was obvious since

$$E(V) = E \left(\frac{X}{X + Y} \right) = \frac{E(X)}{E(X + Y)} = \frac{1}{2}.$$

Is Phil's line of argument correct?

Group B

8. Assume that X and Y are independent and identically distributed with common density function

$$f(x) = \begin{cases} \dfrac{5}{x^2}, & x \geq 5, \\[2mm] 0, & \text{otherwise.} \end{cases}$$

Obtain the joint density function of the variables $U = X/Y$ and $V = XY$.

9. Let X and Y be independent random variables having an exponential distribution with parameter λ. Define two new variables Z and W as

$$Z = X + Y, \quad W = X/Y.$$

(i) Examine whether Z and W are independent.

(ii) Find the marginal density of W (the marginal density of Z will be given later in Example 4.3.1; see also the discussion following that example).

10. Let X_1 and X_2 be a random sample of size 2 from the exponential distribution with parameter $\lambda > 0$.

(i) Obtain the joint density function of the random variables

$$X_{(1)} = \min\{X_1, X_2\} \quad \text{and} \quad X_{(2)} = \max\{X_1, X_2\}.$$

(ii) Find the joint density of the variables $X = X_{(1)}$ and $V = X_{(2)} - X_{(1)}$.

(iii) Show that X and V are independent.

4.3 DISTRIBUTIONS OF SUM, DIFFERENCE, PRODUCT AND QUOTIENT

So far, we have seen many examples involving sum of two or more random variables. In particular, it has been seen that the expectation for the sum of two or more random variables is equal to the sum of their individual expectations. Yet, we have not presented a general method for finding the *distribution* of the sum, or indeed of a random variable which arises by any usual arithmetic operations such as subtraction, multiplication, division, etc., between two variables. This is discussed in this section.

Given two random variables X and Y, we discussed in Section 4.2 the problem of finding the joint distribution between *two variables* that are functions of X and Y. The problem of finding the distribution of a single variable that is a function of X and Y seems conceptually simpler, but a standard way to approach it is through the previous method. Let us now suppose that X and Y are continuous variables and we want to obtain the density function of the variable $U = g(X, Y)$ for a given real function g. We assume that the equation $u = g(x, y)$ has a unique solution with respect to x (for any fixed y), that is, there exists a function g^* such that

$$u = g(x, y) \Leftrightarrow x = g^*(u, y).$$

By introducing the auxiliary variable $V = Y = h(X, Y)$, we have the system of equations

$$\left.\begin{array}{l} g(x, y) = u, \\ h(x, y) = v, \end{array}\right\} \Leftrightarrow \left.\begin{array}{l} x = g^*(u, y), \\ y = v, \end{array}\right\} \Leftrightarrow \begin{array}{l} x = g^*(u, y), \\ y = v = h^*(u, v). \end{array}$$

The associated Jacobian is

$$J(u, v) = \det \begin{vmatrix} \dfrac{\partial g^*(u, v)}{\partial u} & \dfrac{\partial g^*(u, v)}{\partial v} \\ \dfrac{\partial h^*(u, v)}{\partial u} & \dfrac{\partial h^*(u, v)}{\partial v} \end{vmatrix} = \det \begin{pmatrix} \dfrac{\partial g^*(u, v)}{\partial u} & \dfrac{\partial g^*(u, v)}{\partial v} \\ 0 & 1 \end{pmatrix}$$

$$= \dfrac{\partial g^*(u, v)}{\partial u}.$$

Therefore, in view of the results of Section 4.2, provided that $J \neq 0$, the joint density function of U and V is given by

$$f_{U,V}(u, v) = f\left(g^*(u, v), h^*(u, v)\right) \left| \frac{\partial g^*(u, v)}{\partial u} \right| = f\left(g^*(u, v), v\right) \left| \frac{\partial g^*(u, v)}{\partial u} \right|.$$

Integrating this with respect to v, we obtain the following expression for the density function of the variable $U = g(X, Y)$, in terms of the joint density function f of the pair (X, Y):

$$f_U(u) = \int_{-\infty}^{\infty} f\left(g^*(u, v), v\right) \left| \frac{\partial g^*(u, v)}{\partial u} \right| dv. \tag{4.1}$$

This is a general result, valid for any function g for which the associated Jacobian has $J \neq 0$. Note also that we have not assumed anywhere that X and Y are independent variables. In the special case when they are indeed independent, we have $f(x, y) = f_X(x) f_Y(y)$ in which case (4.1) becomes

$$f_U(u) = \int_{-\infty}^{\infty} f_X\left(g^*(u, v)\right) f_Y(v) \left| \frac{\partial g^*(u, v)}{\partial u} \right| dv,$$

or (which is the same since we put $v = y$ above),

$$f_U(u) = \int_{-\infty}^{\infty} f_X\left(g^*(u, y)\right) f_Y(y) \left| \frac{\partial g^*(u, y)}{\partial u} \right| dy. \tag{4.2}$$

In the rest of this section, we use (4.2) to cases when the function g represents one of the four usual arithmetic operations. Consequently, we are assuming in the rest of this section that the variables X and Y are **independent**.

Proposition 4.3.1 *For two continuous, independent random variables X and Y with density functions $f_X(x)$ and $f_Y(y)$, respectively, the density function of their sum, $U = X + Y$, is given by*

$$f_U(u) = \int_{-\infty}^{\infty} f_X(u - y) f_Y(y) dy = \int_{-\infty}^{\infty} f_X(x) f_Y(u - x) dx. \tag{4.3}$$

Proof: By setting $g(x, y) = x + y$, we have

$$g(x, y) = u \Leftrightarrow x + y = u \Leftrightarrow x = u - y = g^*(u, y),$$

whence

$$\frac{\partial g^*(u, y)}{\partial u} = 1.$$

Substituting this into (4.2), we obtain

$$f_U(u) = \int_{-\infty}^{\infty} f_X(u - y) f_Y(y) \cdot 1 \, dy = \int_{-\infty}^{\infty} f_X(u - y) f_Y(y) dy.$$

The second expression in (4.3) is obtained readily by a change of variable $u - y = x$.   □

The function f_U in (4.3), which can be found by either of the integrals there, is called the **convolution** of the functions f_X and f_Y.

In the important special case when X and Y take on only nonnegative values, and so f_X and f_Y vanish on the negative half-axis, (4.3) reduces to

$$f_U(u) = \int_0^u f_X(u - y) f_Y(y) dy = \int_0^u f_X(x) f_Y(u - x) dx. \tag{4.4}$$

Up until now, we have focused on the case when the two variables X and Y are continuous. However, the need for the distribution of sums of random variables is equally important in the discrete case. The result for this case, which parallels Proposition, 4.3.1, is presented next. Though the results of the two propositions look quite similar, the proof below is completely different from that of Proposition 4.3.1.

Proposition 4.3.2 *Let X and Y be two discrete random variables that are independent and have probability functions $f_X(x)$ and $f_Y(y)$ and ranges R_X and R_Y, respectively. Then, the probability function of their sum, $U = X + Y$, is given by*

$$f_U(u) = \sum_{y \in R_Y : u - y \in R_X} f_X(u - y) f_Y(y) = \sum_{x \in R_X : u - x \in R_Y} f_X(x) f_Y(u - x).$$

Proof: For the joint probability function of Y and $U = X + Y$, we have

$$f_{Y,U}(y, u) = P(Y = y, U = u) = P(Y = y, X + Y = u) = P(Y = y, X = u - y).$$

Using the fact that X and Y are independent, we have

$$f_{Y,U}(y, u) = P(X = u - y) P(Y = y) = f_X(u - y) f_Y(y).$$

The first equality in the result of the proposition now follows by adding the above probabilities $f_{Y,U}(y, u)$ over all $y \in R_Y$, and realizing that this sum equals $f_U(u) = P(U = u)$. The second expression then follows by setting $u - y = x$, and observing that any $x \in R_X$ can be written in the form $u - y$, where u and y are in the ranges of U and Y, respectively.   □

In analogy with the continuous case, the function f_U in the last proposition which gives the probability function for the sum $X + Y$ of two independent random variables, is called the **convolution** of the probability functions f_X, f_Y.

Proposition 4.3.2 might be somewhat awkward to use in practice as it involves an infinite summation, and especially so when X and Y take both positive and negative values. However, and in analogy to the continuous case, if both X and Y are nonnegative with probability one, and assuming for simplicity that their range of values is the set of nonnegative integers (as it is the case in many applications), then the result of Proposition 4.3.2 becomes

$$f_U(u) = \sum_{y=0}^{u} f_X(u - y)f_Y(y) = \sum_{x=0}^{u} f_X(x)f_Y(u - x), \qquad u = 0, 1, 2, \dots \qquad (4.5)$$

Example 4.3.1 Suppose that X and Y are independent random variables, with X having a Gamma distribution with parameters n and λ and Y having a Gamma distribution with parameters m and λ, and with densities

$$f_X(x) = \frac{\lambda^n x^{n-1} e^{-\lambda x}}{\Gamma(n)}, \qquad f_Y(y) = \frac{\lambda^m y^{m-1} e^{-\lambda y}}{\Gamma(m)},$$

for $x \geq 0$, $y \geq 0$. In order to find the density of their sum, $U = X + Y$, we make use of (4.4) (since X and Y are nonnegative variables) to get

$$f_U(u) = \frac{\lambda^{n+m} e^{-\lambda u}}{\Gamma(n)\Gamma(m)} \int_0^u x^{n-1}(u - x)^{m-1} \, dx.$$

By a change of variable, $t = x/u$, so that $0 \leq t \leq 1$, we obtain

$$f_U(u) = \frac{\lambda^{n+m} e^{-\lambda u}}{\Gamma(n)\Gamma(m)} \cdot u^{n+m-1} \int_0^1 t^{n-1}(1 - t)^{m-1} \, dt.$$

As the integral does not involve u, we can write $f_U(u)$ in the form

$$f_U(u) = c \, e^{-\lambda u} u^{n+m-1}, \qquad u > 0,$$

where the constant c equals

$$c = \frac{\lambda^{n+m}}{\Gamma(n)\Gamma(m)} \int_0^1 t^{n-1}(1 - t)^{m-1} \, dt.$$

The value of c can be determined by the condition

$$1 = \int_{-\infty}^{\infty} f_U(u) du = c \int_0^{\infty} u^{n+m-1} e^{-\lambda u} \, du,$$

which, upon substituting $\lambda u = t$ and using the definition of the Gamma function, gives

$$1 = \frac{c}{\lambda^{n+m}} \int_0^\infty t^{n+m-1} e^{-t} \, dt = \frac{c\Gamma(n+m)}{\lambda^{n+m}}.$$

This shows

$$c = \frac{\lambda^{n+m}}{\Gamma(n+m)},$$

so that we finally obtain the density of $U = X + Y$ as

$$f_U(u) = \frac{\lambda^{n+m}}{\Gamma(n+m)} u^{n+m-1} e^{-\lambda u}, \qquad\qquad u > 0.$$

But, this is just the density of a Gamma distribution with parameters $n + m$ and λ. Note that n and m need not be integers.

It is worth noting that, by equating the two expressions obtained above for the constant c, we arrive at the identity

$$\int_0^1 t^{n-1}(1-t)^{m-1} \, dt = \frac{\Gamma(n)\Gamma(m)}{\Gamma(n+m)}.$$

The quantity on the left-hand side is called the **Beta function**, and is usually denoted by $B(n, m)$.

Example 4.3.1 can be extended to the case when we have the sum of more than two random variables. More specifically, the following result holds:

Let $X_1, \ldots, X_n$ be, for $n \geq 2$, a collection of independent random variables, so that for $i = 1, 2, \ldots, n$, X_i has the Gamma distribution with parameters a_i and λ. Then, their sum follows a Gamma distribution with parameters $a_1 + a_2 + \cdots + a_n$ and λ.

In other words,

$$X_i \sim Ga(a_i, \lambda) \quad \text{and} \quad X_i \text{ independent for } i = 1, 2, \ldots, n$$

imply that

$$\sum_{i=1}^n X_i \sim Ga\left(\sum_{i=1}^n a_i, \lambda\right).$$

As an important special case of this, if we take $a_i = 1$ for $i = 1, 2, \ldots, n$, i.e. each X_i has an exponential distribution with parameter λ, then their sum, $X_1 + X_2 + \cdots + X_n$ follows a Gamma distribution with parameters n and λ, that is,

$$X_i \sim \mathcal{E}(\lambda) \text{ for } i = 1, 2, \ldots, n \text{ and } X_i \text{ independent} \Rightarrow \sum_{i=1}^n X_i \sim Ga(n, \lambda).$$

We have just seen that the sum of Gamma random variables has also a Gamma distribution (provided that the second parameter, λ, is the same for all variables). A similar result also holds for the Poisson distribution as discussed in the following example.

Example 4.3.2 Let X and Y be two independent, discrete random variables, with X following a Poisson distribution with parameter λ_1 and Y following a Poisson distribution with parameter λ_2. We then use Proposition 4.3.2 (in particular, the expression in (4.5) since X and Y are nonnegative) to obtain the probability function of their sum, $U = X + Y$.

First, the probability functions of X and Y are

$$f_X(x) = e^{-\lambda_1} \cdot \frac{\lambda_1^x}{x!}, \qquad f_Y(y) = e^{-\lambda_2} \cdot \frac{\lambda_2^y}{y!},$$

for $x, y \in \{0, 1, 2, \dots\}$. From (4.5), we then have

$$f_U(u) = \frac{1}{u!} e^{-(\lambda_1 + \lambda_2)} \sum_{x=0}^{u} \binom{u}{x} \lambda_1^x \lambda_2^{u-x}, \qquad u = 0, 1, 2, \dots,$$

and now by using the binomial expansion, this reduces to

$$f_U(u) = e^{-(\lambda_1 + \lambda_2)} \cdot \frac{(\lambda_1 + \lambda_2)^u}{u!}, \qquad u = 0, 1, 2, \dots$$

It is then apparent that the distribution of $U = X + Y$ is Poisson with parameter $\lambda_1 + \lambda_2$.

The above result can be extended to the case when more than two random variables are considered, as follows.

For $n \geq 2$, assume that $X_1, X_2, \dots, X_n$ are independent Poisson random variables with parameters $\lambda_1, \lambda_2, \dots, \lambda_n$, respectively. Then, the distribution of their sum, $X_1 + X_2 + \cdots + X_n$, is Poisson with parameter $\lambda_1 + \lambda_2 + \cdots + \lambda_n$.

In other words,

$$X_i \sim \mathcal{P}(\lambda_i) \text{ for } i = 1, 2, \dots, n \text{ and } X_i \text{ independent } \Rightarrow \sum_{i=1}^{n} X_i \sim \mathcal{P}\left(\sum_{i=1}^{n} \lambda_i \right).$$

The difference of variables, instead of the sum, can be handled with minor changes. In fact, using the function $g(x, y) = x - y$ in (4.2), it is easy to obtain the following result (the arguments are quite similar to those in the proof of Proposition 4.3.1, and so the proof is not presented here and left as an exercise to the reader).

Proposition 4.3.3 *Let X and Y be two independent continuous random variables with densities $f_X(x)$ and $f_Y(y)$, respectively. Then, the density function of their difference, $U = X - Y$, is given by*

$$f_U(u) = \int_{-\infty}^{\infty} f_X(x) f_Y(x - u) dx = \int_{-\infty}^{\infty} f_X(y + u) f_Y(y) dy. \qquad (4.6)$$

As an illustration of the last result, we consider the difference between two exponential random variables in the following example.

Example 4.3.3 *(Laplace distribution)*

Let X and Y be independent exponential random variables with the same parameter λ, so that

$$f_X(x) = \lambda e^{-\lambda x}, \quad x \geq 0, \quad \text{and} \quad f_Y(y) = \lambda e^{-\lambda y}, \quad y \geq 0.$$

Now, we need to insert $f_X(x)$ and $f_Y(x - u)$ in Proposition 4.3.3, but the arguments of both these functions must be nonnegative, i.e. we must have $x \geq 0$ and $x - u \geq 0$, or $x \geq 0$ and $x \geq u$. Thus, we have

- for $u < 0$, the limits of integration are given by the condition $x \geq 0$ (see the graph below),

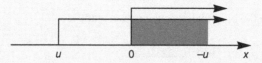

- for $u \geq 0$, the limits of integration are given by the condition $x \geq u$, as shown in the next graph.

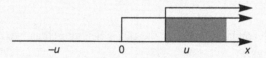

We thus obtain the density of $U = X - Y$ as

$$f_U(u) = \begin{cases} \lambda^2 e^{-\lambda u} \displaystyle\int_0^\infty e^{-2\lambda x} dx, & \text{if } u < 0 \\ \lambda^2 e^{-\lambda u} \displaystyle\int_u^\infty e^{-2\lambda x} dx, & \text{if } u \geq 0 \end{cases} = \begin{cases} \dfrac{1}{2} \lambda e^{\lambda u}, & \text{if } u < 0, \\ \dfrac{1}{2} \lambda e^{-\lambda u}, & \text{if } u \geq 0. \end{cases}$$

The above expression can be written in a concise form as

$$f_U(u) = \frac{1}{2} \lambda e^{-\lambda |u|}, \qquad u \in \mathbb{R}.$$

This function has been plotted in Figure 4.1. The distribution with the above density is known as the **Laplace distribution** or the **double exponential distribution**, and has found key applications in the computer analysis of speech recognition.

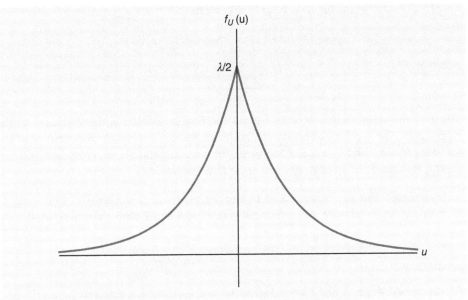

Figure 4.1 Density function of the Laplace distribution.

Next, there are many practical situations when one is interested in the distribution of the product between two random variables. For instance,

- suppose the length X and width Y of a rectangle, representing the surface of an apartment (or a plot, a floor, a wall, etc.), are random variables (typically, due to measurement errors); then the area of the rectangle is XY;

- let I be the intensity of the electric current, passing through a resistor and R be the value of the electrical resistance. In many cases, I and R are not known with certainty, and so they can be regarded as random variables. In this case their product, $V = IR$, which represents the voltage at the end of the electric circuit, is also a random variable.

The following proposition gives a general result for the density of the product of two random variables, in terms of their densities.

Proposition 4.3.4 *Let X and Y be two independent, continuous random variables with densities $f_X(x)$ and $f_Y(y)$, respectively. Then, the density function of the product $U = XY$ is given by*

$$f_U(u) = \int_{-\infty}^{\infty} \frac{1}{|y|} f_X(u/y) f_Y(y)\, dy = \int_{-\infty}^{\infty} \frac{1}{|x|} f_X(x) f_Y(u/x)\, dx. \qquad (4.7)$$

Proof: Considering the function $g(x, y) = xy$, we have

$$g(x, y) = u \Leftrightarrow xy = u \Leftrightarrow x = u/y = g^*(u, y),$$

which gives

$$\frac{\partial g^*(u, y)}{\partial u} = \frac{1}{y}.$$

When this is substituted into (4.2), we get the first expression in the statement of the proposition. The second expression can be found by considering the joint density of the variables $U = XY$ and $V = X$ and following the same steps as above, or alternatively, by making the change of variable $u/y = x$ in the first integral in (4.7).   □

In the case when the variables X and Y are not independent, the density of the variable $U = XY$ can be found from the joint density $f(x, y)$ of X and Y through the formulas

$$f_U(u) = \int_{-\infty}^{\infty} \frac{1}{|y|} f(u/y, y) \, dy = \int_{-\infty}^{\infty} \frac{1}{|x|} f(x, u/x) \, dx.$$

Example 4.3.4 Let X and Y be independent random variables, each having a uniform distribution over $[0, 1]$. Suppose we want to obtain the density of their product, $U = XY$. Note that the range of U is also the unit interval. The densities f_X and f_Y of X and Y, respectively, are $f_X(x) = f_Y(y) = 1$ for any $x, y, \in [0, 1]$, and they are zero outside this interval. Thus, for any value u inside the unit interval, using (4.7) we obtain

$$f_U(u) = \int_{-\infty}^{\infty} \frac{1}{|y|} f_X(u/y) f_Y(y) \, dy = \int_u^1 \frac{1}{y} \, dy$$

(if y is outside the range $[u, 1]$, then either y or u/y is outside the unit interval, so that the value of the integrand is zero). This yields immediately the density of U as

$$f_U(u) = \begin{cases} -\ln u, & 0 < u \le 1, \\ 0, & \text{otherwise.} \end{cases}$$

As a result we see, for example, that the probability the product $U = XY$ is greater than half equals

$$P\left(U > \frac{1}{2}\right) = \int_{1/2}^1 (-\ln y) \, dy = \frac{1 - \ln 2}{2} \cong 0.1534.$$

Finally, we obtain the density function for the ratio of two variables. Here again, the result is presented for the case of continuous variables.

Proposition 4.3.5 *For two continuous random variables X and Y with density functions $f_X(x)$ and $f_Y(y)$, respectively, the density function of their ratio, $U = X/Y$, is given by*

$$f_U(u) = \int_{-\infty}^{\infty} |y| f_X(uy) f_Y(y) \, dy.$$

Proof: The proof is analogous to that of Proposition 4.3.4. It is therefore omitted and left as an exercise to the reader.   □

In this case, there is also a more general result concerning the density of $U = X/Y$ when X and Y are not necessarily independent. The formula is then

$$f_U(u) = \int_{-\infty}^{\infty} |y| f(uy, y) \, dy,$$

where $f(x, y)$ is the joint density of X and Y. Clearly, for independent variables, this would coincide with the result of Proposition 4.3.5.

Example 4.3.5 (The Cauchy distribution)

Suppose X and Y are standard normal random variables. Let us assume that X and Y are independent. Then, we obtain the density function of $U = X/Y$ as

$$f_U(u) = \int_{-\infty}^{\infty} |y| \phi(uy)\phi(y) \, dy, \qquad u \in \mathbb{R},$$

where

$$\phi(x) = \frac{1}{\sqrt{2\pi}} e^{-x^2/2}, \qquad x \in \mathbb{R}$$

is the density function of the standard normal distribution, $N(0, 1)$. Then, we obtain

$$f_U(u) = \frac{1}{2\pi} \int_{-\infty}^{\infty} |y| \exp\left\{-\frac{1}{2}(u^2 + 1)y^2\right\} dy.$$

Upon noting that the integrand is an even function of y, we see that

$$f_U(u) = \frac{1}{\pi} \int_{0}^{\infty} y \exp\left\{-\frac{1}{2}(u^2 + 1)y^2\right\} dy.$$

By setting $t = (u^2 + 1)y^2/2$, we find

$$y^2 = \frac{2}{u^2 + 1} t, \quad y \, dy = \frac{1}{u^2 + 1} dt;$$

we then obtain

$$f_U(u) = \frac{1}{\pi} \int_{0}^{\infty} \frac{1}{u^2 + 1} e^{-t} \, dt = \frac{1}{\pi(u^2 + 1)} \int_{0}^{\infty} e^{-t} \, dt = \frac{1}{\pi(u^2 + 1)}.$$

The distribution with density function

$$f_U(u) = \frac{1}{\pi(u^2 + 1)}, \qquad u \in \mathbb{R}$$

is known as the **Cauchy distribution**. We have already seen this distribution in Exercise 3 of Section 2.8, where it was also mentioned that the expectation for a random variable having this distribution is undefined (explain why!).

Example 4.3.6 Let $X_1, X_2, \ldots, X_n$ be a random sample from the uniform distribution over the interval $(0, 1)$. Suppose we are interested in the distribution of the ratio of smallest to largest observation, namely, $U = X_{(1)}/X_{(n)}$. Using Proposition 3.6.2, we find the joint density of $X_{(1)}$ and $X_{(n)}$ to be

$$f_{X_{(1)}, X_{(n)}}(x, y) = n(n-1)(y-x)^{n-2}, \qquad 0 < x < y < 1,$$

see also Example 3.6.5. As the variables $X_{(1)}$ and $X_{(n)}$ are *not independent*, we use the formula given after Proposition 4.3.5 to obtain

$$f_U(u) = \int_{-\infty}^{\infty} |y| f(uy, y) \, dy = \int_0^1 y n(n-1)(y - uy)^{n-2} \, dy$$

$$= n(n-1)(1-u)^{n-2} \int_0^1 y^{n-1} \, dy = (n-1)(1-u)^{n-2}, \qquad 0 < u < 1.$$

EXERCISES

Group A

1. Let X and Y be two continuous random variables with joint density function $f(x, y)$. Consider $U = X + Y$ and $V = X - Y$. The result of Exercise 6 from Section 4.2 states that the joint density of U and V is given by

$$f_{U,V}(u, v) = \frac{1}{2} f\left(\frac{u+v}{2}, \frac{u-v}{2}\right).$$

   (i) Using this, obtain expressions for the density functions of the variables $U = X + Y$ and $V = X - Y$.

   (ii) Assuming in addition that X and Y are independent, verify that the results of Part (i) reduce to those given in Propositions 4.3.1 and 4.3.3.

2. A businessman has invested in two stocks, A and B. His return (profit or loss) from stock A is a random variable which has the uniform distribution over the interval $(-1, 4)$, while the return from stock B is uniformly distributed over $(-2, 3)$. Find the probability that the businessman's return from these two stocks combined is

   (i) positive,

   (ii) greater than 2 but less than 5,

   (iii) between -2 and $+3$.

3. Let X and Y have joint probability function

$$f(x, y) = c(x^2 + y^2), \qquad x = 1, 2, 3, \ y = 1, 2, 3.$$

   After finding the value of c and the marginal distributions of X and Y, calculate the probabilities

$$P(X + Y = 3), \quad P(X + Y \le 3), \quad P(X + Y \ge 5).$$

4. Suppose X has a uniform distribution over the interval $(1, 2)$ and Y has a uniform distribution over $(2, 4)$. If X and Y are independent, obtain the density functions of the variables $U = XY$ and $V = X/Y$.

5. Let X and Y have a joint density function $f(x, y)$. Extend the arguments used in the proofs of Propositions 4.3.1 and 4.3.4 to find the density function of the variable

$$W = aX + bY$$

for some given positive constants a and b.

   Application: A company sells two products, A and B. The amounts of each product sold daily, in tens of kilograms, are represented by random variables, X and Y respectively, for A and B. The joint density function of X and Y is given by

$$f(x, y) = \begin{cases} \dfrac{120x^2 + 40xy^3(1 - y)}{41}, & 0 < x < 1,\ 0 < y < 1, \\ 0, & \text{otherwise.} \end{cases}$$

   If the company makes a profit of US$1.3 for each kilogram of Product A sold and a profit of US$2.5 for each kilogram of Product B sold, obtain the density function and the expected value of the company's daily profit.

6. In an electronic game, the player aims at a rectangular target whose dimensions, X and Y, are continuous random variables with densities

$$f_X(x) = \frac{x}{2}, \qquad 0 \le x \le 2,$$

$$f_Y(y) = \frac{y^3}{64}, \qquad 0 < y \le 4.$$

   Find the density function of the area of the target, and thence calculate the expected target area.

7. In Example 4.3.5, we observed that the Cauchy distribution arises as the distribution of the ratio of two independent standard normal random variables. Another way to obtain this distribution is as follows: let W be a uniform random variable over $(-\pi, \pi)$. Then, $Y = \tan W$ has a Cauchy distribution. Verify this result!

8. The lifetime for a certain type of laptop batteries, measured in years under regular use, can be described by a variable exponentially distributed with mean 5. A company selling these laptops, operating with a single battery of the above type, offers with each sale an extra battery as a replacement when the first one fails.

   What is the probability that, within the first eight years of regular use, the buyer will not have to buy a new battery? (Hint: Consider the distribution of the sum $X + Y$, where X and Y are independent exponential random variables and use the result of Example 4.3.1.)

9. Mark and Laura, who work for the same company, travel a lot as part of their job requirements. Let X and Y be the number of return journeys by airplane that Mark and Laura make during a month. Suppose the joint distribution of X and Y is as follows:

y \ x	1	2	3
0	0.13	0.05	0.04
1	0.17	0.18	0.06
2	0.07	0.14	0.16

(i) Calculate the probability function

$$P(X + Y = i), \qquad \text{for } i = 1, 2, 3, 4, 5$$

of the random variable $X + Y$, the total number of journeys made by both Mark and Laura in a given month. Do this in two ways:

(a) directly from the table above, as we did in Example 1.4.1;

(b) by finding the marginal distributions of X and Y and then using one of the two expressions in (4.3),

and check that the results from these two methods agree.

(ii) Do the same for the probability function of the random variable $Y - X$, the random variable which denotes the number of return journeys that Laura makes more than Mark in a given month.

10. Let X and Y be two independent random variables with the distribution of X being binomial with parameters n_1 and p, and the distribution of Y being binomial with parameters n_2 and p. Show that the distribution of $X + Y$ is binomial with parameters $n_1 + n_2$ and p. Generalize this result to the case of more than two variables.

Application: During the inspection of items coming out of a production line, a batch of n items is taken and the number of defectives found in the batch is recorded. Suppose the overall proportion of defective items during the production process is p, for a given $p \in (0, 1)$.

(i) Write down the distribution of the total number of defective items found after $m \geq 2$ batches are examined.

(ii) If $p = 0.01$ and $n = 15$, what is the probability that in five batches that will be examined, no more than two defectives will be found?

(Hint: You may use the following combinatorial identity

$$\sum_{x=0}^{k} \binom{n_1}{x} \binom{n_2}{k-x} = \binom{n_1 + n_2}{k},$$

which is known as the Cauchy formula.)

11. In Example 4.3.2, we observed that the sum of two independent random variables X and Y following Poisson distribution with parameters λ_1 and λ_2, respectively, is also Poisson with parameter $\lambda_1 + \lambda_2$. Use this result to prove the following: if X and Y are independent Poisson random variables with parameters λ_1 and λ_2, then the *conditional distribution* of X, given $X + Y = n$, is binomial with parameters n and $\lambda_1/(\lambda_1 + \lambda_2)$.

    Application: The numbers of traffic accidents in two motorways, A and B, during a week follow the Poisson distribution with parameters $\lambda_1 = 4$ and $\lambda_2 = 7$, respectively. If we know that in a particular week, there were nine accidents in these two motorways combined, find the probability that at most three of them occurred in motorway A.

Group B

12. Consider the bivariate discrete random variable from Example 1.3.3 with joint probability function

$$f(x, y) = \frac{6}{3^x 4^y}, \qquad x = 1, 2, \ldots \quad \text{and} \quad y = 1, 2, \ldots$$

    Use the marginal probability functions of X and Y found there to calculate the probability function of the variable $U = X + Y$.

13. Let X_1 and X_2 be two independent random variables, each having a normal distribution. We are given that $E(X_1) = \mu_1$, $E(X_2) = \mu_2$, $Var(X_1) = \sigma_1^2$ and $Var(X_2) = \sigma_2^2$. Then, show that

    (i) the variable $U = X_1 + X_2$ has normal distribution $N(\mu, \sigma^2)$ with $\mu = \mu_1 + \mu_2$ and $\sigma^2 = \sigma_1^2 + \sigma_2^2$;

    (ii) the variable $V = X_1 - X_2$ has normal distribution $N(\mu', \sigma^2)$ with $\mu' = \mu_1 - \mu_2$ and $\sigma^2 = \sigma_1^2 + \sigma_2^2$.

    How do you interpret intuitively the fact that the two variances are the same?

14. Let X and Y be two independent random variables with the distribution of X being Beta with parameters α and β and the distribution of Y being Beta with parameters $\alpha + \beta$ and γ. Show that the variable $U = XY$ has a Beta distribution with parameters α and $\beta + \gamma$.

4.4 χ^2, t AND F DISTRIBUTIONS

In this section, we focus on three continuous (univariate) distributions, all of which are highly useful in statistics. All three distributions emerge as distributions of random variables that can be expressed in terms of other variables through the usual arithmetic operations that we have seen so far.

First, we present the so-called χ^2 (pronounced **chi-squared**) **distribution**.[1]

---

[1] Traditionally, the term chi-squared in statistics is associated with the name of Karl Pearson, but this is due to a statistical goodness-of-fit test, known as the chi-squared test, rather than the distribution that we study here. The first use of this distribution appears to be by the German geodecist F.R. Helmert in 1876 in connection with the distribution of the sample variance, but this passed unnoticed in the English-speaking community of statisticians for some time. The British statistician R.A. Fisher was the first to study the chi-squared as a family of distributions relating to the normal; he also introduced the term degrees of freedom.

Definition 4.4.1 Let $Z_1, Z_2, \ldots, Z_n$ be independent random variables, each having a standard normal distribution, $N(0, 1)$. Then, the sum of squares

$$X = Z_1^2 + Z_2^2 + \cdots + Z_n^2$$

is a continuous random variable which is said to follow a chi-squared distribution with n degrees of freedom, and we denote it by $X \sim \chi_n^2$.

Our primary concern below is to obtain the density function of X in the above definition. For this purpose, we calculate first the distribution of $X_i = Z_i^2$, for $i = 1, 2, \ldots, n$.

The cumulative distribution function of X_i is given by

$$F_{X_i}(x) = P(X_i \leq x) = P\left(Z_i^2 \leq x\right) = P\left(-\sqrt{x} \leq Z_i \leq \sqrt{x}\right),$$

which gives

$$F_{X_i}(x) = \Phi\left(\sqrt{x}\right) - \Phi\left(-\sqrt{x}\right) = 2\Phi\left(\sqrt{x}\right) - 1, \qquad x \geq 0;$$

here Φ is the cumulative distribution function of a standard normal random variable. Differentiating the above expression, we obtain the density function of X_i as

$$f_{X_i}(x) = F'_{X_i}(x) = 2\Phi'\left(\sqrt{x}\right)\left(\sqrt{x}\right)' = 2\phi\left(\sqrt{x}\right)\frac{1}{2\sqrt{x}}$$

$$= \phi\left(\sqrt{x}\right)\frac{1}{\sqrt{x}}, \qquad x > 0.$$

Replacing now the quantity $\phi\left(\sqrt{x}\right)$ by using the standard normal density

$$\phi(x) = \frac{1}{\sqrt{2\pi}}e^{-x^2/2}, \qquad x \in \mathbb{R}$$

we obtain the density of X_i for $i = 1, 2, \ldots, n$, to be

$$f_{X_i}(x) = \frac{1}{\sqrt{2\pi x}}e^{-x/2}, \qquad x > 0.$$

It may not appear immediate, but this is in fact a special case of one of the continuous distributions we have seen in Volume I, but also in several instances so far in this book. Specifically, writing the above density function in the form (using the fact that $\Gamma(1/2) = \sqrt{\pi}$)

$$f_{X_i}(x) = \frac{(1/2)^{1/2}}{\Gamma(1/2)}x^{1/2-1}e^{-x/2}, \qquad x > 0,$$

we see that the distribution of the variable X_i is the Gamma distribution with parameters $a = 1/2$ and $\lambda = 1/2$. As the variables X_i are also independent, as functions of the independent variables Z_i, by using the result about the sum of independent Gamma variables in Example 4.3.1, we can arrive at the density function of χ^2_n distribution to be

$$f_X(x) = \frac{1}{2^{n/2}\Gamma(n/2)}x^{n/2-1}e^{-x/2}, \qquad\qquad x > 0. \qquad (4.8)$$

From this, we see that the χ^2_n distribution is nothing but a Gamma distribution with parameters $a = n/2$ and $\lambda = 1/2$.

The mean of a random variable X with χ^2_n distribution is n, while the variance is $2n$. Both these properties follow immediately from the corresponding general expressions for the mean and variance of the Gamma distribution. Figure 4.2 presents plots of the density of χ^2_n distribution for various values of n. Observe that as n increases, the shape of the density becomes "more symmetric", and in fact for $n \geq 30$, the shape of the χ^2_n density is close to that of a normal density function. This observation is made clearer in Figure 4.3

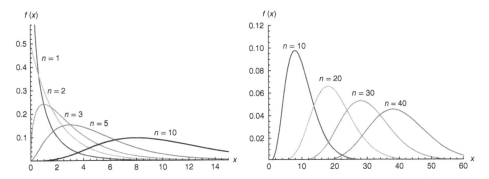

Figure 4.2   Density function of the χ^2_n distribution for various values of n.

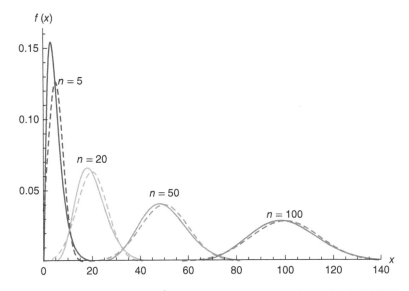

Figure 4.3   Density function of the χ^2_n distribution for various values of n (solid line) and the approximating normal density (dashed line).

where we show, in the same plot, the densities of the χ_n^2 distribution for $n = 5, 20, 50$, and 100 along with the corresponding normal densities of $N(n, 2n)$ distribution.

For the cumulative distribution function of χ_n^2, there is in general no simple analytic expression, but simple expressions are possible in some special cases (see, for instance Exercise 2 at the end of this section). As the χ_n^2 distribution plays a key role in statistics, most statistical textbooks have, usually in the form of an appendix, tables of this distribution. Specifically, such a table would give the (upper) α-quantiles of χ_n^2 distribution, for various values of α and n; namely, the points $\chi_n^2(\alpha)$ for which

$$P(X > \chi_n^2(\alpha)) = \alpha.$$

Here, $0 < \alpha < 1$ and X is a random variable with $X \sim \chi_n^2$. Such a table is also presented in the Appendix of this book.

For large values of n (as $n \to \infty$, but for practical purposes, it is sufficient to assume that $n \geq 30$), the chi-squared distribution can be approximated adequately by the $N(n, 2n)$ distribution, as illustrated in Figure 4.3. For this reason, the tables of χ_n^2 distribution do not go beyond $n = 30$ since for larger values of n, we can rely on the normal approximation.

Next, we discuss another important distribution known as the **t distribution**. This was derived originally by the English statistician, William S. Gosset, when he was working at the Guinness brewery in Dublin in 1908, and published a paper under the pseudonym[2] *Student*. Hence, the distribution is also known as Student's t distribution, or simply the Student distribution.

> **Definition 4.4.2** Let Z be a random variable having the standard normal distribution and S_n be another variable, independent of Z, having χ_n^2 distribution. Then, the random variable T_n defined by
>
> $$T_n = \frac{Z}{\sqrt{\dfrac{S_n}{n}}}$$
>
> is said to follow the t distribution, or the Student distribution, with n degrees of freedom. This is denoted by $T_n \sim t_n$.

To find the density function of T_n, we first observe that the random variable

$$Y = \sqrt{\frac{S_n}{n}},$$

appearing in the denominator of the definition of T_n, has its density function as

$$f_Y(y) = 2ny f_{S_n}(ny^2) = 2ny f_{\chi_n^2}(ny^2), \qquad\qquad y > 0 \qquad (4.9)$$

---

[2]The most likely reason that Gosset did not publish the paper under his real name seems to be due to the fact that another Guinness employee had earlier published a paper containing trade secrets of the brewery, and so Guinness prohibited its employees from publishing any papers regardless of their content.

(see Exercise 6), where we use $f_{\chi_n^2}$ to denote the density of the chi-squared distribution with n degrees of freedom. So, using (4.8), we see that (4.9) gives

$$f_Y(y) = \frac{n^{n/2}}{2^{n/2-1}\Gamma(n/2)} y^{n-1} \exp\left(-\frac{ny^2}{2}\right), \qquad y > 0.$$

As $T_n = Z/Y$ and Z has the standard normal density,

$$f_Z(z) = \frac{1}{\sqrt{2\pi}} \exp\left(-\frac{z^2}{2}\right), \qquad z \in \mathbb{R},$$

using Proposition 4.3.5, we get immediately

$$f_{T_n}(u) = \int_0^\infty |y| \frac{1}{\sqrt{2\pi}} \exp\left(-\frac{u^2 y^2}{2}\right) \frac{n^{n/2}}{2^{n/2-1}\Gamma(n/2)} y^{n-1} \exp\left(-\frac{ny^2}{2}\right) dy$$

$$= \frac{n^{n/2}}{\sqrt{2\pi}2^{n/2-1}\Gamma(n/2)} \int_0^\infty y^n \exp\left(-\frac{(u^2+n)}{2}y^2\right) dy.$$

Observe that the variable Y takes only nonnegative values, and so the range of integration above is the interval $[0, \infty)$, although u can clearly take on any real value.

Next, making the change of variable $\tau = (u^2 + n)y^2/2$, we get

$$f_{T_n}(u) = \frac{n^{n/2}}{\sqrt{2\pi}\,2^{n/2-1}\Gamma(n/2)} \cdot \frac{2^{(n-1)/2}}{(u^2+n)^{(n+1)/2}} \int_0^\infty \tau^{(n+1)/2-1} e^{-\tau}\, d\tau.$$

The integral on the right is now easily calculated because it is the value of the Γ function at the point $(n + 1)/2$. Therefore, using x rather than u for the argument of the density of t_n distribution, we finally arrive at this density as

$$f_{T_n}(x) = \frac{\Gamma\left(\dfrac{n+1}{2}\right)}{\sqrt{\pi n}\,\Gamma\left(\dfrac{n}{2}\right)} \frac{1}{\left(1+\dfrac{x^2}{n}\right)^{(n+1)/2}}, \qquad x \in \mathbb{R}.$$

It is apparent that this satisfies

$$f_{T_n}(-x) = f_{T_n}(x), \qquad \text{for any } x \in \mathbb{R},$$

which means that the density function of the Student distribution is symmetric around the y-axis. This is shown graphically in Figure 4.4, which plots the density of t_n distribution for various values of n; for comparison, we have also included the standard normal density (dashed line on the graph).

The symmetry of the t distribution implies in particular that the mean of a random variable T_n following the t_n distribution is zero (for $n > 1$). Further, it can be shown that the variance of T_n, for $n > 2$, is $n/(n - 2)$.

As with the case of χ_n^2 distribution (and, of course, that of the normal distribution), there is no simple form for the cumulative distribution function of T_n. In the Appendix, we have

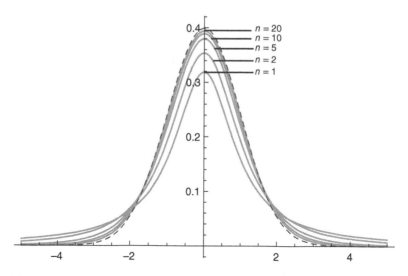

Figure 4.4 Density function of Student's t distribution with n degrees of freedom (solid lines), for various values of n, and standard normal density (dashed line).

presented a table of upper α-quantiles of t_n distribution for $\alpha = 0.1, 0.05, 0.025, 0.01, 0.005$ and $n = 1, 2, \ldots, 30$. Note that the upper α-quantile, denoted here by $t_n(\alpha)$, is such that

$$P(T_n > t_n(\alpha)) = \alpha.$$

There are a number of differences between the two distributions we discussed so far in this section, viz. the χ_n^2 and t_n distributions. The most important ones are:

- the t_n distribution is, for all values of n, symmetric, while the chi-squared distribution is always skewed to the right;

- a random variable with t_n distribution can take on any value on the real line, while a random variable with χ_n^2 distribution (as it is defined as a sum of squares) takes only nonnegative values;

- a random variable with t_n distribution has (for any $n > 1$) a zero mean, while the mean of a variable having χ_n^2 distribution is n, and so the mean increases with the number n of degrees of freedom.

Finally, we focus our attention on the **F distribution** or the **Snedecor distribution**, named after the American statistician George W. Snedecor (1881–1974).

Definition 4.4.3 Let S_n and S_k be two independent random variables having χ_n^2 and χ_k^2 distributions, respectively. Then, we say that the random variable W defined by

$$W = \frac{\dfrac{S_n}{n}}{\dfrac{S_k}{k}} = \frac{k}{n} \cdot \frac{S_n}{S_k}$$

has F distribution (or Snedecor distribution) with n and k degrees of freedom. This distribution is denoted by $F_{n,k}$ and, for W as above, we write $W \sim F_{n,k}$.

For obtaining the density function of F distribution, we write the variable W in the definition above in the form

$$W = \frac{k}{n} U,$$

where $U = S_n/S_k$ is the ratio of two independent χ^2 random variables. Using the density function of χ^2 distribution in (4.8) and Proposition 4.3.5, we find the density of U as

$$f_U(u) = \int_{-\infty}^{\infty} |y| f_{\chi_n^2}(uy) f_{\chi_k^2}(y) \, dy$$

$$= \frac{u^{n/2-1}}{2^{(n+k)/2} \Gamma\left(\frac{n}{2}\right) \Gamma\left(\frac{k}{2}\right)} \int_0^{\infty} y^{(n+k)/2-1} \exp\left(-\frac{u+1}{2} y\right) \, dy,$$

which reduces to

$$f_U(u) = \frac{\Gamma\left(\frac{n+k}{2}\right)}{\Gamma\left(\frac{n}{2}\right) \Gamma\left(\frac{k}{2}\right)} \cdot \frac{u^{n/2-1}}{(1+u)^{(n+k)/2}}, \qquad u > 0.$$

From this, it is easy to find the desired density function of $W = kU/n$ as

$$f_{n,k}(x) = \frac{n}{k} \cdot f_U\left(\frac{n}{k} x\right),$$

which yields

$$f_{n,k}(x) = \frac{\Gamma\left(\frac{n+k}{2}\right)}{\Gamma\left(\frac{n}{2}\right) \Gamma\left(\frac{k}{2}\right)} \cdot \left(\frac{n}{k}\right)^{n/2} \cdot \frac{x^{n/2-1}}{\left(1 + \frac{nx}{k}\right)^{(n+k)/2}}$$

$$= \frac{\Gamma\left(\frac{n+k}{2}\right) n^{n/2} k^{k/2}}{\Gamma\left(\frac{n}{2}\right) \Gamma\left(\frac{k}{2}\right)} \cdot \frac{x^{n/2-1}}{(k + nx)^{(n+k)/2}}, \qquad x > 0.$$

Figure 4.5 presents the plots of the density for $F_{n,k}$ distribution, for various values of n and k. The cumulative distribution function for a random variable which has the $F_{n,k}$ distribution is again not available in an explicit form in general. In the Appendix, we have presented a table of points $F_{n,k}(\alpha)$ such that

$$P(W > F_{n,k}(\alpha)) = \alpha,$$

when $W \sim F_{n,k}$, for $\alpha = 0.01$ and 0.05 and various values of n and k.

Like χ^2 distribution, F distribution is on the nonnegative half-axis and is skewed to the right. When both n and k are large (say, larger than 30), the density of $F_{n,k}$ distribution approaches that of the normal distribution (Figure 4.6).

Example 4.4.1 Suppose that a system consists of n particles, each having a mass m. The velocities of these particles along a certain axis are described by random variables $Z_1, Z_2, \ldots, Z_n$, which are independent and standard normally distributed. The total

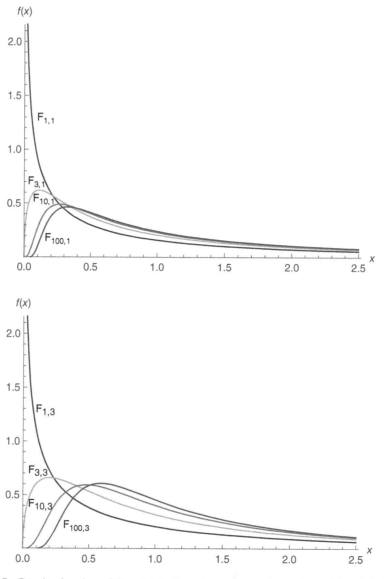

Figure 4.5   Density function of $F_{n,k}$ distribution, for various values of n and $k = 1$ (up), $k = 3$ (down).

kinetic energy of the system is then

$$K = \sum_{i=1}^{n} \frac{1}{2} m Z_i^2 = \frac{1}{2} m \sum_{i=1}^{n} Z_i^2 = \frac{1}{2} m S_n,$$

where S_n is a random variable with the chi-squared distribution with n degrees of freedom. Using the facts that $E(S_n) = n$ and $Var(S_n) = 2n$, we find the expected kinetic

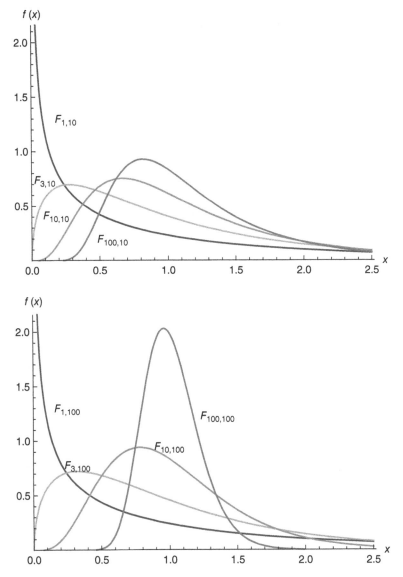

Figure 4.6 Density function of $F_{n,k}$ distribution, for various values of n and $k = 10$ (up), $k = 100$ (down).

energy of the system to be

$$E(K) = E\left(\frac{1}{2}mS_n\right) = \frac{1}{2}mE(S_n) = \frac{mn}{2},$$

while the variance of K is

$$Var(K) = Var\left(\frac{1}{2}mS_n\right) = \frac{1}{4}m^2 Var(S_n) = \frac{m^2 n}{2}.$$

Application: Suppose we have $n = 60$ particles each with $m = 4$ (mass units). Then, $E(K) = 120$, $Var(K) = 480$, and the exact distribution of

$$S_{60} = Z_1^2 + Z_2^2 + \cdots + Z_{60}^2$$

is χ_{60}^2. As the number of degrees of freedom is large (greater than 30), we can approximate the distribution of S_{60} by the normal distribution. Thus, the distribution of

$$K = \frac{1}{2} \cdot 4 \cdot S_{60} = 2\, S_{60}$$

is also approximately normal, with mean $\mu = 120$ and variance $\sigma^2 = 480$. If we wish to find the probability $P(K > 140)$, using the normal approximation, we find it to be

$$P(K > 140) = 1 - P(K \leq 140) = 1 - P\left(\frac{K - 120}{\sqrt{480}} \leq \frac{140 - 120}{\sqrt{480}} \right)$$

$$= 1 - \Phi\left(\frac{20}{\sqrt{480}} \right) \cong 1 - \Phi(0.9129) = 0.1807.$$

By comparison, the exact probability using χ_{60}^2 distribution (this cannot be found from the table in the Appendix B, but using a software, such as Mathematica) is 0.1770.

Example 4.4.2 We assume that, in addition to the system with n particles from the last example, we have another independent (i.e. not interacting with the first one) system consisting of n' particles, each with a mass m'. Let the velocities of these new particles also follow the standard normal distribution. We are now interested in the probability that the kinetic energy of the new system, denoted by K', is at least three times as much as that of the first system. For finding this probability, we first observe that K' can be expressed as

$$K' = \frac{1}{2} m' S_{n'},$$

where $S_{n'} \sim \chi_{n'}^2$. Then, we find the required probability to be

$$p = P(K' \geq 3K) = P\left(\frac{1}{2} m' S_n' \geq 3 \cdot \frac{1}{2} m S_n \right) = P\left(\frac{S_n}{S_{n'}} \leq \frac{m'}{3m} \right)$$

$$= P\left(\frac{n' S_n}{n S_{n'}} \leq \frac{n' m'}{3 n m} \right).$$

As S_n and $S_{n'}$ are independent chi-squared random variables with n and n' degrees of freedom respectively, the quantity

$$W = \frac{n' S_n}{n S_{n'}} = \frac{S_n / n}{S_{n'} / n'}$$

has an $F_{n,n'}$ distribution, and so the probability of interest

$$p = P\left(W \leq \frac{n'm'}{3nm}\right)$$

can be found readily from the cumulative distribution function of $F_{n,n'}$ distribution.

Application: For $n = 60$, $m = 4$, $n' = 40$, and $m' = 24$, we have $W \sim F_{60,40}$. So, the desired probability is

$$p = P\left(W \leq \frac{40 \cdot 24}{3 \cdot 60 \cdot 4}\right) = P(W \leq 1.33) = 0.8297$$

(this value is again found by the use of software such as Mathematica).

EXERCISES

Group A

1. If X has a chi-squared distribution with n degrees of freedom, show that the density function of the variable $Y = \sqrt{X}$ is given by

$$f_Y(y) = \frac{1}{2^{r/2-1}\Gamma\left(\dfrac{n}{2}\right)} y^{n-1} \exp\left(-\frac{y^2}{2}\right), \qquad y > 0$$

(the distribution with this density is called the *chi* distribution with n degrees of freedom). Then, show that

$$E(Y^r) = \frac{2^{r/2}\Gamma\left(\dfrac{n+r}{2}\right)}{\Gamma\left(\dfrac{n}{2}\right)}, \qquad r = 1, 2, \ldots,$$

and from this, obtain expressions for the mean and variance of Y.

2. Calculate the cumulative distribution function of χ_n^2 distribution for $n = 2$ and $n = 4$.

3. Let X be a random variable that is uniformly distributed over $(0, 1)$. Show that the variable $Y = -2\ln X$ has a χ_2^2 distribution.

   (Hint: The simplest way to do this is by using the result of Exercise 2 for $n = 2$.)

4. Let $X_1, X_2, \ldots, X_n$ be a random sample from the normal distribution with

$$E(X_i) = \mu, \quad V(X_i) = \sigma^2, \qquad \text{for } i = 1, 2, \ldots, n.$$

   Find the density function of the random variable[3]

$$\frac{1}{n}\sum_{i=1}^{n}(X_i - \mu)^2.$$

---

[3]We mentioned earlier that, historically, the χ^2 distribution arose in the study of the variance of a sample. Readers acquainted with the basics of statistics will identify this quantity here as the variance of a *population* of size n, whose units have values $X_1, X_2, \ldots, X_n$ and arithmetic mean μ. Note, however, that when all values X_i are known, this quantity is no longer a random variable, but is simply a measure for the dispersion of the values around their mean μ.

(Hint: First, find the distribution of the variable

$$Y = \sum_{i=1}^{n} \left(\frac{X_i - \mu}{\sigma} \right)^2$$

and then find the distribution of $(\sigma^2/n)Y$.)

5. Let X and Y be two independent random variables having chi-squared distribution with n and k degrees of freedom, respectively. Find the distribution of $X + Y$ in two ways:

   (i) from Definition 4.4.1, and

   (ii) using Proposition 4.3.1.

6. Let $S_n \sim \chi_n^2$, as in Definition 4.4.2. Explain why the random variable $Y = \sqrt{S_n/n}$ has density

$$f_Y(y) = 2nyf_{S_n}(ny^2), \qquad\qquad y > 0,$$

   as stated in (4.9).

7. Assume that X is a continuous variable having $F_{n,k}$ distribution.

   (i) Obtain the distribution of the variable $1/X$.

   (ii) For $n = k$, deduce that X and $1/X$ have the same distribution.

   (Hint: For Part (i), use Definition 4.4.3.)

8. Suppose $0 < \alpha < 1$. Exploiting the symmetry of t_n distribution around the y-axis, verify the validity of the following statements:

   (i) $t_n(1 - \alpha) = -t_n(\alpha)$.

   (ii) $P\left[-t_n(\alpha/2) \le T_n \le t_n(\alpha/2)\right] = 1 - \alpha$.

9. Let T_n be a random variable having the t distribution with n degrees of freedom. Show that the variable T_n^2 has an F distribution and identify the degrees of freedom for that distribution.

10. Assume that X has a Gamma distribution with parameters $n/2$ and $\lambda > 0$, where n is a positive integer. Prove that the variable $Y = 2\lambda X$ has a χ_n^2 distribution.

Group B

11. Let $X_1, X_2, \ldots, X_{60}$ be a random sample from the normal distribution $N(15, 5^2)$, and $Y_1, Y_2, \ldots, Y_{120}$ be a random sample from the normal distribution $N(20, 5^2)$. Let us further assume that these two samples are independent. Then, calculate each of the following probabilities:

    (i) $P\left(\sum_{i=1}^{60} (X_i - 15)^2 < 1500 \right)$;

(ii) $P\left(\sum_{i=1}^{60} (X_i - 15)^2 < \sum_{i=1}^{120} (Y_i - 20)^2 \right);$

(iii) $P\left(\sum_{i=1}^{60} [(X_i - 15)^2 + (Y_i - 20)^2] < 3000 \right);$

(iv) $P\left(\sum_{i=1}^{60} (X_i + Y_i - 35)(X_i - Y_i + 5) > 0 \right).$

It is given that $F_{60,60}(0.50) = 1$ and $F_{60,120}(0.001) = 2$.

(Hint: Use the results of Exercises 4 and 5 as well as the normal approximation to χ_{60}^2 and χ_{120}^2 distributions.)

12. Let X and Y be two independent random variables having Gamma distribution with parameters (α, λ) and (β, λ), respectively.

(i) Derive the joint density function of the variables $U = X + Y$ and $V = X/Y$.

(ii) Are the variables U and V independent?

(iii) Find the marginal density function of V, and then show that the variable $\beta V / \alpha$ has $F_{2\alpha, 2\beta}$ distribution, provided that 2α and 2β are positive integers.

13. Let W be a random variable having the F distribution with n and k degrees of freedom, that is, $W \sim F_{n,k}$.

(i) Verify that

$$Y = \frac{nW}{k + nW} = \frac{\frac{n}{k} W}{1 + \frac{n}{k} W}$$

has a density function of the form

$$f_Y(y) = \frac{\Gamma\left(\frac{n+k}{2}\right)}{\Gamma\left(\frac{n}{2}\right) \Gamma\left(\frac{k}{2}\right)} y^{n/2-1}(1-y)^{k/2-1}, \qquad 0 < y < 1,$$

so that it follows a Beta distribution with parameters $\alpha = n/2$ and $\beta = k/2$.

(ii) Find the mean and variance of $(k + nW)^{-1}$.

14. Suppose that T_n has a t distribution with n degrees of freedom.

(i) Show that the random variable

$$Y = \frac{n}{n + T_n^2} = \frac{1}{1 + \frac{T_n^2}{n}}$$

has a Beta distribution with parameters $\alpha = n/2$ and $\beta = 1/2$.

(ii) Find the mean and variance of the random variable $\frac{1}{T_n^2 + n}$.

4.5 BASIC CONCEPTS AND FORMULAS

Joint density function of continuous random variables $U = g(X, Y)$, and $V = h(X, Y)$	$f_{U,V}(u, v) = f\left(g^*(u, v), h^*(u, v)\right)\lvert J\rvert,$ where $g^*(u, v)$ and $h^*(u, v)$ is the unique solution (for x, y) of the system $$\left\{\begin{array}{l} g(x, y) = u, \\ h(x, y) = v, \end{array}\right\}$$ and $J \neq 0$ is the Jacobian of the transformation
Distribution of the sum $U = X + Y$ of two independent random variables X and Y	$f_U(u) = \int_{-\infty}^{\infty} f_X(u - y)f_Y(y)\, dy = \int_{-\infty}^{\infty} f_X(x)f_Y(u - x)\, dx$ (continuous case) $f_U(u) = \sum_{y \in R_Y} f_X(u - y)f_Y(y) = \sum_{x \in R_X} f_X(x)f_Y(u - x)$ (discrete case)
Distribution of the difference $U = X - Y$ of two independent continuous random variables X and Y	$f_U(u) = \int_{-\infty}^{\infty} f_X(x)f_Y(x - u)\, dx = \int_{-\infty}^{\infty} f_X(y + u)f_Y(y)\, dy$
Distribution of the product $U = XY$ of two independent continuous random variables X and Y	$\begin{aligned} f_U(u) &= \int_{-\infty}^{\infty} \frac{1}{\lvert y\rvert} f_X(u/y)f_Y(y)\, dy \\ &= \int_{-\infty}^{\infty} \frac{1}{\lvert x\rvert} f_X(x)f_Y(u/x)\, dx \end{aligned}$
Distribution of the quotient $U = X/Y$ of two independent continuous random variables X and Y	$f_U(u) = \int_{-\infty}^{\infty} \lvert y\rvert f_X(uy)f_Y(y)\, dy$
χ^2 distribution with n degrees of freedom (χ_n^2 distribution)	It is the distribution of $X = Z_1^2 + Z_2^2 + \cdots + Z_n^2$ where, for $i = 1, 2, \ldots, n$, $Z_i \sim N(0, 1)$. The density is $$f_X(x) = \frac{1}{2^{n/2}\Gamma(n/2)} x^{n/2 - 1} e^{-x/2}, \qquad x > 0.$$ For large n, χ_n^2 distribution can be approximated by $N(n, 2n)$ distribution

Student's t distribution with n degrees of freedom (t_n distribution)	It is the distribution of $T_n = Z/\sqrt{S_n/n}$, where $Z \sim N(0,1)$ and $S_n \sim \chi_n^2$ independently of Z. The density is $$f_{T_n}(x) = \frac{\Gamma\left(\dfrac{n+1}{2}\right)}{\sqrt{\pi n}\,\Gamma\left(\dfrac{n}{2}\right)} \frac{1}{\left(1 + \dfrac{x^2}{n}\right)^{(n+1)/2}}, \quad x \in \mathbb{R}.$$ For large n, t_n distribution can be approximated by the $N\left(0, \dfrac{n}{n-2}\right)$ distribution
Snedecor distribution (or F distribution) with n and k degrees of freedom ($F_{n,k}$ distribution)	It is the distribution of $$W = \frac{\dfrac{S_n}{n}}{\dfrac{S_k}{k}} = \frac{k}{n} \cdot \frac{S_n}{S_k},$$ where $S_n \sim \chi_n^2, S_k \sim \chi_k^2$, and S_n, S_k are independent. The density is $$\begin{aligned} f_{n,k}(x) &= \frac{\Gamma\left(\dfrac{n+k}{2}\right)}{\Gamma\left(\dfrac{n}{2}\right)\Gamma\left(\dfrac{k}{2}\right)} \cdot \left(\frac{n}{k}\right)^{n/2} \cdot \frac{x^{n/2-1}}{\left(1 + \dfrac{nx}{k}\right)^{(n+k)/2}} \\ &= \frac{\Gamma\left(\dfrac{n+k}{2}\right)n^{n/2}k^{k/2}}{\Gamma\left(\dfrac{n}{2}\right)\Gamma\left(\dfrac{k}{2}\right)} \cdot \frac{x^{n/2-1}}{(k+nx)^{(n+k)/2}}, \quad x > 0 \end{aligned}$$

4.6 COMPUTATIONAL EXERCISES

Mathematica has built-in functions for all distributions we have discussed in this chapter. Table 4.1 presents the notation used in Mathematica for the most common continuous distributions we have met here.

The following sequence of commands can been used to produce a graph similar to Figure 4.4.

```
In[1]:= f1[x_]:= PDF[StudentTDistribution[1], x]
f2[x_]:= PDF[StudentTDistribution[2], x]
f5[x_]:= PDF[StudentTDistribution[5], x]
f10[x_]:= PDF[StudentTDistribution[10], x]
f20[x_]:= PDF[StudentTDistribution[20], x]
f[x_]:= PDF[NormalDistribution[0, 1], x]
Plot[{f1[x], f2[x], f5[x], f10[x], f20[x], f[x]}, {x, -5, 5},
PlotStyle -> {Black, Orange, Red, Brown, Green,
Directive[Blue, Thickness[.0024], Dashed]}]
```

Mathematica function	Distribution
LaplaceDistribution$[0, 1/\lambda]$	Laplace distribution (see Example 4.3.3)
CauchyDistribution[ ]	Cauchy distribution (see Example 4.3.5)
ChiSquareDistribution$[n]$	χ^2 distribution with n degrees of freedom
ChiDistribution$[n]$	χ distribution with n degrees of freedom (see Exercise 1, Section 4.4)
StudentTDistribution$[n]$	t distribution with n degrees of freedom
FRatioDistribution$[n, m]$	F (or Snedecor) distribution with n and m degrees of freedom

Table 4.1   Mathematica functions for continuous distributions.

1. (i) Draw, on the same graph, the density function of the χ_n^2 distribution along with the $N(n, 2n)$ density for $5, 10, 20, 50, 100,$ *and* 200. What do you observe?

   (ii) Do the same for the *cumulative distribution functions* of the distributions in Part (i).

2. (i) Draw, on the same graph, the density function of the $F_{n,n}$ distribution along with the density function of the normal distribution with the same mean and variance as the $F_{n,n}$ distribution, for $5, 10, 20, 50, 100,$ *and* 200. What do you observe?

   (ii) Do the same for the *cumulative distribution functions* of the distributions in Part (i).

3. For a random sample, the **sample median**, M, is defined as the middle observation of the sample values, when these values are arranged in an increasing order. When the sample size n is an odd integer, the median is equal to the $(n + 1)/2$th order statistic, i.e. $X_{((n+1)/2)}$, while if n is an even integer, the median is taken as

$$M = \frac{X_{(n/2)} + X_{(n/2+1)}}{2}.$$

   Now, consider a sample $X_1, X_2, \dots, X_6$ of size $n = 6$ from a continuous distribution with density function f and distribution function F.

   (i) Using the results about the joint distribution of order statistics from the last chapter, along with the results of Section 4.2, derive the density function of the sample median in terms of f and F. Thence calculate the mean and variance of M.

   (ii) As an application, calculate the distribution of the sample median when the variables X_i

      (a) follow the exponential distribution with mean 1,

      (b) follow the uniform distribution over the interval $(0, 1)$.

4. A supermarket has five cashiers. The service time at each cashier has an exponential distribution with a mean value of one minute. Assume that five customers, one at

each cashier, begin to be served at the same time, and let R be the time between the completion of the first customer's service and the time when all five customers have been served. In other words, if X_1, X_2, X_3, X_4, X_5 are the service times of the customers, then $R = X_{(5)} - X_{(1)}$.

(i) Calculate the density function of R, and thence find its mean and variance.

(ii) Find the density function, mean and variance of R by assuming that service times, instead of the exponential distribution, follow

    a. a uniform distribution over the interval $(0, 2)$,

    b. a Gamma distribution with density $f(x) = xe^{-x}$

(times here are measured in minutes). Note that in all three cases, the mean service time is one minute. Which distribution corresponds to the largest expected value of R? Which distribution corresponds to the smallest one? (You may wish to compare your results to those in Exercise 14 of Section 3.6.)

5. In Example 3.4.5, we observed that a negative binomial random variable Y possesses a representation as a sum of independent geometrically distributed random variables. Specifically, if $Y \sim Nb(r, p)$, then Y can be expressed as

$$Y = Y_1 + Y_2 + \cdots + Y_r,$$

where Y_i has geometric distribution with parameter p, for $i = 1, 2, \ldots, r$.

It is then easy to see that if X and Y are two independent random variables with $X \sim Nb(r, p)$ and $Y \sim Nb(m, p)$, then

$$X + Y \sim Nb(r + m, p).$$

(i) Verify this result directly, by obtaining the probability function of $X + Y$, using those of X and Y along with the result of Proposition 4.3.2.

(ii) Use induction to obtain a general result about the sum of k independent negative binomial random variables, similar to the ones given in Section 4.3 for Gamma and Poisson distributions (the corresponding result for the binomial distribution can be found in Exercise 10 of Section 4.3).

For χ^2, t, and F distributions we have mentioned earlier that there is no "simple" explicit analytic expression except in some special cases. Although finding the distribution functions of these distributions can be hard when working it out by hand, it should present no difficulty when a computer algebra software, such as Mathematica, is employed. For instance, the following command calculates the distribution function of t distribution with $n = 4$ degrees of freedom.

```
In[1]:= F4[x_]:= CDF[StudentTDistribution[4], x]
```

Another way to obtain the same result is to integrate the associated density function.

```
In[2]:= f4dens[x_]:= PDF[StudentTDistribution[4], x]
F4b[x_]:= Integrate[f4dens[y], {y, -Infinity, x}]
FullSimplify[F4b[x]]
Out[4]:= (6 x + x^3 + (4 + x^2)^(3/2))/(2 (4 + x^2)^(3/2))
```

This yields the distribution function of T_4 to be

$$F_{T_4}(x) = \frac{\left(4+x^2\right)^{3/2} + x^3 + 6x}{2\left(4+x^2\right)^{3/2}}, \qquad x \in \mathbb{R}.$$

Once we have obtained the distribution function, we can now solve (numerically, and not analytically) the equation

$$P(T_n > t_n(\alpha)) = \alpha$$

to find the upper-α quantile of t_4 distribution. For example, for $\alpha = 0.05$, we get

```
In[6]:= Print[Quantile[StudentTDistribution[4], 0.95]]
2.13185
```

which agrees with the value 2.132 found in the t-table in the Appendix (keep in mind that all values in that table are rounded to three decimal places). Note also that the command *Quantile* in Mathematica calculates the *lower quantile* for a distribution, that is why we have used $1 - \alpha = 0.95$ in the above syntax.

6. Working as above, find an expression for the distribution function of t_n distribution when the number of degrees of freedom is $n = 6, 8, 10, 15, 20$. Notice, in particular, the increasing complexity of these functions as n becomes larger. For each of the above values of n, find the upper α-quantiles of the associated distribution, for all values of α given in the t-table in the Appendix, and check that your results agree with those in the table.

7. In Exercise 2 of Section 4.4, you were asked to find analytically the distribution function corresponding to χ_n^2 distribution for $n = 2$ and $n = 4$. Because in both cases $n/2$ is an integer, and bearing in mind the connection between χ_n^2 distribution and Gamma distribution with parameters $n/2$ and $1/2$, an integration by parts does the trick. When n becomes large, say $n = 30$, the result becomes more complicated although it should be noted that for the calculation of the distribution function, no further skills than integrating by parts are again needed.

   (i) Use Mathematica to find analytically the distribution function of the χ_n^2 distribution for $n = 5, 10, 15, 20, 25, 30$.

   (ii) For the above values of n, and for each value of α found in the χ^2-table in the Appendix, calculate the upper-α quantile of the associated distribution and verify that your results agree with the values presented in that table.

8. Let X be a random variable with $X \sim \chi_n^2$.

   (i) Obtain directly, using Mathematica and the density of X, the results

   $$E(X) = n, \qquad Var(X) = 2n$$

   about the mean and variance of X stated in Section 4.4.

(ii) In Section 4.4, it was mentioned that χ_n^2 distribution is "skewed to the right". Quantitatively, this can be shown (and be made more precise) by calculating the coefficient of skewness for X, given by

$$\gamma_1 = \frac{E[(X - E(X))^3]}{(Var(X))^{3/2}}.$$

When $X \sim \chi_n^2$, obtain an expression for γ_1 and show in particular that, as $n \to \infty$, γ_1 tends to zero, which agrees with the fact that for large n, χ_n^2 can be approximated by the normal distribution.

9. In this, and the following two exercises, we intend to assess the quality of the normal approximation to the three distributions discussed in Section 4.4.

Suppose a random variable X follows χ_n^2 distribution.

(i) Calculate the exact value of the probability

$$G_{1,n}(x) = P(X \le x)$$

as well as the approximation, $G_{2,n}(x)$, of this probability based on the normal distribution with mean n and variance $2n$, when $n = 5, 10, 30, 50$ and for $x \in (0, 5)$.

(ii) Find the relative percentage error

$$E_n(x) = \frac{G_{2,n}(x) - G_{1,n}(x)}{G_{1,n}(x)}$$

of this approximation, and plot this function, for x in the interval $(0, 5)$. For which values of n (and x) does the approximation perform the best?

10. Suppose the variable T_n has t_n distribution. Calculate the probability

$$H_{1,n}(x) = P(T_n \le x)$$

for $n = 5, 10, 30, 50$ and for $x \in (0, 3)$. (Note that for $x > 3$, this probability is quite close to one.) Find also the approximation for the probability above based on normal distribution, and plot the relative percentage error, for $x \in (0, 3)$ and for each value of n above. What do you conclude?

11. As in Exercises 9 and 10, assess the normal approximation to $F_{n,k}$ distribution. For this purpose, consider the following pairs of values for n, k:

(i) $n = 10$, $k = 10$,

(ii) $n = 10$, $k = 40$,

(iii) $n = 40$, $k = 10$,

(iv) $n = 40$, $k = 40$,

and plot the relative percentage error, as a function of x in the range $x \in (0, 5)$.

4.7 SELF-ASSESSMENT EXERCISES

4.7.1 True–False Questions

1. If X is a random variable having the uniform distribution over $(0, 1)$, then the variable $Y = 3X - 1$ follows the uniform distribution over the interval $(-1, 2)$.

2. Let X, Y, and Z be three random variables having a joint continuous distribution. Let
$$U = X + Y, \quad V = Y + Z, \quad W = Z + X.$$
Then, the Jacobian of this transformation is $1/2$.

3. Suppose X and Y represent the number of accidents that occur weekly in two motorways. Then, assuming that X and Y are independent, and that they have Poisson distribution with parameters λ_1 and λ_2, respectively, the total number of accidents has probability function
$$P(X + Y = r) = e^{-(\lambda_1 + \lambda_2)} \frac{(\lambda_1 \lambda_2)^r}{\lambda_1! \lambda_2!}, \qquad \text{for } r = 0, 1, 2, \ldots$$

4. Let X and Y be independent random variables from the uniform distribution over $(0, 1)$. Then, the density function of $W = X - Y$ is given by
$$f_W(w) = 2(1 - w), \qquad 0 < w < 1.$$

5. Let X and Y be independent random variables with X having an exponential distribution with mean λ_1 and Y having an exponential distribution with mean λ_2, where $\lambda_1 \neq \lambda_2$. Then, the distribution of the sum $X + Y$ is Gamma.

6. Let X_1, X_2, and X_3 be a random sample from the uniform distribution over $(0, 1)$. Then, for $0 < y < 1$, we have
$$P(X_1 + X_2 + X_3 \leq y) = \frac{y^3}{6}.$$

7. Suppose that X has an exponential distribution with parameter λ. Then, the random variables
$$U = X - Y, \quad V = X/Y$$
are independent.

8. Assuming that X and Y are two independent Poisson random variables with parameters λ_1 and λ_2, respectively, the distribution of $X - Y$ is also Poisson with parameter $\lambda_1 - \lambda_2$.

9. Let X and Y be two independent exponential random variables with the same parameter λ, and define
$$U = 2X - 3, \quad V = 2Y - 3.$$
Then, the variable $U + V$ follows a Gamma distribution.

10. If X and Y are independent, each having standard normal distribution, then the variable $U = X/Y$ follows a Cauchy distribution.

11. If X follows χ_n^2 distribution and Y has χ_m^2 distribution, then the distribution of $U = X + Y$ is χ_{n+m}^2 (here n and m are positive integers).

12. Suppose X and Y are independent, and it is known that $X \sim \chi_{10}^2$ and $Y \sim \chi_7^2$. Then, the distribution of $X - Y$ is χ_3^2.

13. Assume that X has χ_{10}^2 distribution. Then, the distribution of $Y = 3X$ is χ_{30}^2.

14. Suppose $X \sim N(1, 2^2)$. Then, $Y = (X/2)^2$ has χ_1^2 distribution.

15. Assume that a variable X has chi-squared distribution with 7 degrees of freedom. Then,
$$P(X \leq 2.167) = 0.95.$$

16. If a random variable X has a chi-squared distribution, then the distribution of X is symmetric.

17. Let S_n and S_k be two independent random variables having χ_n^2 and χ_k^2 distributions, respectively. Then, the distribution of S_n/S_k is $F_{n,k}$.

18. Let $t_n(\alpha)$ be the upper α-quantile of a t distribution with n degrees of freedom. Then, we have
$$t_n(\alpha) = 1 - t_n(1 - \alpha).$$

19. Both the χ^2 and F distributions are skewed to the right.

4.7.2 Multiple Choice Questions

1. Let X and Y be two independent random variables having exponential distribution with the same parameter $\lambda = 1$. Then, the probability $P(X + Y > w)$, for a given $w > 0$, is equal to

   (a) e^{-w} (b) $2e^{-w}$ (c) we^{-w} (d) $(1 + w)e^{-w}$ (e) $(1 + 2w)e^{-2w}$

2. Let X and Y be continuous random variables with a joint density $f(x, y)$. Then, the density function of the pair (U, V), where
$$U = 2X + Y, \quad V = 3Y + 1,$$

   is given by

   (a) $f_{U,V}(u, v) = f(2u + v, 3v + 1)$

   (b) $f_{U,V}(u, v) = \dfrac{1}{3} f\left(\dfrac{3u + v - 1}{3}, \dfrac{v - 1}{3}\right)$

   (c) $f_{U,V}(u, v) = \dfrac{1}{3} f\left(\dfrac{3u - 2v + 2}{3}, \dfrac{v - 1}{3}\right)$

   (d) $f_{U,V}(u, v) = \dfrac{1}{6} f\left(\dfrac{3u - v + 1}{6}, \dfrac{v - 1}{3}\right)$

   (e) $f_{U,V}(u, v) = \dfrac{4}{3} f\left(\dfrac{3u - 2v + 2}{3}, \dfrac{3u - 4v - 4}{3}\right)$

3. The radius of a circle is a random variable having the uniform distribution over the interval $(1, 2)$. Let V be the area of the circle. Then, the density function of V is given by

   (a) $f_V(v) = \dfrac{1}{\sqrt{\pi v^2}}$, $\pi < v < 4\pi$

   (b) $f_V(v) = \dfrac{1}{2\sqrt{\pi v}}$, $\pi < v < 4\pi$

   (c) $f_V(v) = \dfrac{2}{\pi v}$, $\pi < v < 4\pi$

   (d) $f_V(v) = \dfrac{\sqrt{2}}{\sqrt{\pi v}}$, $\pi < v < 4\pi$

   (e) $f_V(v) = \dfrac{2}{\pi v^2}$, $\pi < v < 4\pi$

4. Assume that X and Y are independent random variables with X having uniform distribution over $(0, 2)$ and Y having uniform distribution over $(0, 1)$. Then, the density function of the variable $U = X + Y$ is given by

   (a) $f(x) = \begin{cases} x/2, & 0 \le x \le 2 \\ 1/2, & 2 < x \le 3 \end{cases}$

   (b) $f(x) = \begin{cases} x/2, & 0 \le x \le 1 \\ (3-x)/2, & 1 < x \le 3 \end{cases}$

   (c) $f(x) = \begin{cases} x/2, & 0 \le x \le 1 \\ 1/2, & 1 < x \le 2 \\ (3-x)/2, & 2 < x \le 3 \end{cases}$

   (d) $f(x) = \begin{cases} x, & 0 \le x \le 1 \\ 1, & 1 < x \le 2 \\ 2-x, & 2 < x \le 3 \end{cases}$

   (e) $f(x) = \begin{cases} x/2, & 0 \le x \le 2 \\ (3-x)/2, & 2 < x \le 3 \end{cases}$

5. Let X be a random variable with $P(X = i) = 1/4$, for $i \in \{1, 2, 3, 4\}$, and Y be another variable, independent of X, such that $P(Y = j) = 1/3$, for $j \in \{1, 2, 3\}$. Then, $P(X + Y = 5)$ is

   (a) $\dfrac{1}{4}$ (b) $\dfrac{1}{3}$ (c) $\dfrac{3}{8}$ (d) $\dfrac{5}{12}$ (e) $\dfrac{1}{2}$

6. Assume that X and Y are independent Poisson random variables with $E(X) = 3$ and $E(Y) = 8$. Then, $P(X + Y = 6)$ is equal to

   (a) $e^{-11} \dfrac{11^6}{6!}$ (b) $e^{-6} \dfrac{6^{11}}{6!}$ (c) $e^{-11} \dfrac{24^6}{(6!)^2}$

   (d) $e^{-6} \dfrac{11^6}{24!}$ (e) $e^{-11} \dfrac{11^6}{24!}$

7. John plays an electronic game, shooting against a rectangular target whose dimensions, X and Y, are independent continuous random variables with densities

$$f_X(x) = 2x, \qquad 0 \le x \le 1,$$

$$f_Y(y) = \frac{y^2}{9}, \qquad 0 < y \le 3.$$

The density function for the area U of this target is, for $0 \le u \le 3$, given by

(a) $\dfrac{4u^3}{81}$ (b) $\dfrac{2u(3-u)}{27}$ (c) $\dfrac{2u(3-u)}{9}$

(d) $\dfrac{4u^2(3-u)}{27}$ (e) $\dfrac{5u^4}{243}$

8. Let X and Y be independent uniform random variables over the interval $(0, 1)$. Then, $P(X\,Y > 1/5)$ is equal to

(a) $\dfrac{1}{5}$ (b) $\dfrac{4-\ln 5}{5}$ (c) $\dfrac{\ln 5 - 1}{5}$ (d) $\dfrac{1+\ln 5}{5}$ (e) $\dfrac{24}{25}$

9. Let X, Y, and Z be independent exponential random variables with the same parameter $\lambda = 2$. Then, the density function of their sum $U = X + Y + Z$ is, for $u \ge 0$, given by

(a) $f_U(u) = 6e^{-6u}$ (b) $f_U(u) = 4ue^{-3u}$ (c) $f_U(u) = 4u^2e^{-2u}$

(d) $f_U(u) = \dfrac{2u^2e^{-2u}}{3}$ (e) $f_U(u) = 9ue^{-3u}$

10. The χ^2 distribution is a special case of

(a) normal distribution

(b) uniform distribution

(c) exponential distribution

(d) Gamma distribution

(e) both normal and Gamma distributions

11. Let X be a random variable having the $N(1, 1)$ distribution and Y be another variable, independent of X, having χ_5^2 distribution. Then:

(a) $P(X \le Y) = 1$

(b) $P(X^2 \le Y) = 1$

(c) $X + Y \sim \chi_6^2$

(d) $X^2 + Y \sim \chi_6^2$

(e) none of the above

12. Assume that X has a t distribution with 17 degrees of freedom, and it is known that $P(X \le x) = 0.05$. Then, the value of x is

(a) 1.333 (b) 1.74 (c) 2.11 (d) -2.11 (e) -1.74

13. Assume that X has a t distribution with 5 degrees of freedom, while it is known that $P(|X| \leq a) = 0.95$. Then, the value of a is

    (a) 1.476   (b) 2.015   (c) 2.571   (d) 4.032

    (e) none of the above

14. It is known that T_n has a t distribution with n degrees of freedom and that $Var(T_n) = 5/4$. The value of n is

    (a) 5   (b) 8   (c) 10   (d) 12   (e) 15

15. Let X be a random variable having $F_{3,5}$ distribution and Y be another variable, independent of X, having $F_{4,8}$ distribution. Then:

    (a) $P(X + Y \geq 0) = 1$

    (b) both X and Y have a symmetric distribution

    (c) the variable $X + Y$ has an $F_{7,13}$ distribution

    (d) the variable $X - Y$ also has an F distribution

    (e) $P(X > 5.41) = 0.01$

16. Let X be a random variable having χ_5^2 distribution and Y be another variable, independent of X, having χ_{10}^2 distribution. Then, which of the following statements is **not true**:

    (a) $E(3X - 2Y) = -5$

    (b) $P\left(\dfrac{2X}{Y} > 5.64\right) = 0.01$

    (c) X^2 and Y^2 are independent

    (d) $2X + 1$ and $4X - Y$ are independent

    (e) $P(X + Y > 30.578) = 0.01$

4.8 REVIEW PROBLEMS

Problems 2–4 below are on Laplace distribution (see Example 4.3.3 for the definition of this distribution).

1. Let X be a random variable having uniform distribution over the interval $(0, 2\pi)$ while Y is another variable, independent of X, having the exponential distribution with mean 1. Prove that the variables

$$U = \sqrt{2Y} \cos X, \qquad V = \sqrt{2Y} \sin X$$

   are independent, and that they both follow the standard normal distribution.

2. Assume that X has a Laplace distribution with density

$$f_X(x) = 3e^{-6|x|}, \qquad\qquad x \in \mathbb{R}.$$

(i) Find the mean and variance of X.

(ii) Calculate the conditional distribution function

$$G(x) = P(X \le x|X > 0), \qquad\qquad x \in \mathbb{R}$$

(clearly, $G(x) = 0$ for $x \le 0$).

(iii) Obtain the conditional expectation $E(X|X > 0)$.

3. Let X be a random variable having standard normal distribution and Y be another variable, independent of X, having its density function as

$$f_Y(y) = ye^{-y^2/2}, \qquad\qquad y \ge 0.$$

Establish that the distribution of $U = XY$ is Laplace.

4. Let X and Y be two independent random variables uniformly distributed over the interval $(0, 1)$.

(i) Find the density function and the distribution function of $U = X/Y$.

(ii) Show that the distribution of $\ln(X/Y)$ is Laplace.

5. Assume that X_1, X_2 and X_3 are independent random variables, each having exponential distribution with parameter λ.

(i) Write down the joint density of the order statistics $X_{(1)}, X_{(2)}, X_{(3)}$ based on these three variables. Then, use it to show that the joint density function of

$$U_1 = X_{(1)}, \quad U_2 = X_{(2)} - X_{(1)}, \quad U_3 = X_{(3)} - X_{(2)}$$

is given by

$$f(u_1, u_2, u_3) = 6\lambda^3 \exp\left[-\lambda(3u_1 + 2u_2 + u_3)\right], \qquad u_1, u_2, u_3 > 0.$$

(ii) Examine whether the variables U_i for $i = 1, 2, 3$, are independent. Does any of them follow an exponential distribution?

(iii) Verify that $V_1 = 3U_1$, $V_2 = 2U_2$ and $V_3 = U_3$ are independent and have the same distribution.

Application: For the lighting of a room, we use three light bulbs. We assume that three new light bulbs are installed in the room and that each of them has a lifetime (in thousands of hours) which follows an exponential distribution with parameter $\lambda = 2$. The light bulbs are replaced only after all three of them have failed. Calculate the probability

(i) that the time elapsed between the first and the second bulb failure is at least 500 hours;

(ii) that the time elapsed between the second and the third bulb failure is at least 750 hours.

6. We know that the sum $X_1 + X_2 + \cdots + X_n$ of n independent variables, having the geometric distribution with the same parameter p, has a Negative binomial distribution with parameters n and p. For $n = 2$, prove this result by using Proposition 4.3.2.

7. The random variables X and Y are independent and have a discrete uniform distribution over the range $\{1, 2, \ldots, n\}$ for some $n \geq 2$, so that

$$f_X(i) = f_Y(i) = \frac{1}{n}, \qquad \text{for } i = 1, 2, \ldots, n.$$

Prove that the probability function of their sum $U = X + Y$ is given by

$$f_U(u) = \begin{cases} \dfrac{u - 1}{n^2}, & 2 \leq u \leq n + 1, \\[2mm] \dfrac{2n - u + 1}{n^2}, & n + 2 \leq u \leq 2n. \end{cases}$$

8. Suppose X and Y are independent, having a uniform distribution over the interval $(0, 2)$. Calculate the probabilities

$$P(X - Y \leq 1 | X = 2), \quad P(X - Y \leq 1 | X \leq 2),$$

$$P(X - Y \leq 1 | X + Y \geq 2), \quad P(X + Y \geq 2 | X - Y \leq 1).$$

9. Let $X_1, X_2, \ldots, X_n$ be a random sample from the uniform distribution over the interval $(0, 1)$, and let $X_{(i)}$ and $X_{(j)}$ denote the ith and jth order statistics, respectively, for $1 \leq i < j \leq n$.

   (i) Use Proposition 3.6.2 to obtain the joint density of $X_{(i)}$ and $X_{(j)}$.

   (ii) Show that the difference $X_{(j)} - X_{(i)}$ has a Beta distribution with parameters $\alpha = j - i$ and $\beta = n - j + i + 1$.

   (iii) Show that the ratio $X_{(i)}/X_{(j)}$ also has a Beta distribution with parameters $\alpha' = i$ and $\beta' = j - i$.

10. Assume that X_1, X_2, and X_3 are independent random variables, each having standard normal distribution. Show that the variables

$$U_1 = \frac{1}{\sqrt{2}}(X_1 - X_2),$$

$$U_2 = \frac{1}{\sqrt{6}}(X_1 + X_2 - 2X_3),$$

$$U_3 = \frac{1}{\sqrt{3}}(X_1 + X_2 + X_3)$$

are also independent standard normal random variables.

11. Diana goes to her local bank and joins a queue at a cashier waiting to be served. She estimates that the time, in minutes, until all customers in the queue in front of her are served is exponential with parameter α, while the time that she spends being

served is exponential with parameter β. Prove that the total time X she spends at the bank has density function

$$f_X(x) = \frac{\alpha\beta}{\beta - \alpha} \left(e^{-\alpha x} - e^{-\beta x}\right), \qquad\qquad x \geq 0,$$

provided $\alpha \neq \beta$. What happens if $\alpha = \beta$?

12. Let X be a random variable representing the outcome of a random selection from the interval $(0, 1)$, and define another variable Y as

$$Y = \begin{cases} X, & \text{if } X < 1/2, \\ 1 - X, & \text{otherwise.} \end{cases}$$

(i) Show that Y is uniformly distributed over the interval $(0, 1/2)$.

(ii) Let $W = 1 - Y$. Write down the joint density function of (Y, W) and the marginal density of W.

(iii) Check that

$$\frac{E(W)}{E(Y)} = 3.$$

(iv) Show that the distribution function of the ratio W/Y is given by

$$P\left(\frac{W}{Y} \leq t\right) = \begin{cases} 0, & t \leq 1, \\ \dfrac{t - 1}{t + 1}, & t > 1. \end{cases}$$

(v) Verify that the random variable W/Y has infinite expectation (compare the result with that in Part (iii)).

13. Let X and Y be independent random variables following exponential distribution with the same parameter λ, and let $U = X + Y$. Find the joint density function of X and U.

14. Let $X, Y,$ and Z be independent variables that are exponentially distributed with parameter 1. Find the density function and the distribution function of

(i) $X + Y/X$;

(ii) $X + Y/Z$.

15. Let $X_1, X_2, \ldots, X_{n+k}$ be independent random variables following $N(0, \sigma^2)$ distribution. Find the distribution for each of the variables below:

(i) $X_1^2 + X_2^2 + \cdots + X_n^2$;

(ii) $X_{n+1}^2 + X_{n+2}^2 + \cdots + X_{n+k}^2$;

(iii) $\dfrac{k\left(X_1^2 + X_2^2 + \cdots + X_n^2\right)}{n\left(X_{n+1}^2 + X_{n+2}^2 + \cdots + X_{n+k}^2\right)}$;

(iv) $\dfrac{\sqrt{k}\, X_1}{\sqrt{X_{n+1}^2 + X_{n+2}^2 + \cdots + X_{n+k}^2}}$;

(v) $\dfrac{n\, X_1^2}{X_1^2 + X_2^2 + \cdots + X_n^2}$.

4.9 APPLICATIONS

4.9.1 Random Number Generators Coverage – Planning Under Random Event Occurrences

In this section, we shall first describe two practical problems which, despite the fact that they are different in nature, share the same underlying stochastic model. Then we shall present the technical details of the model and establish some useful formulas for the quantities of interest. Finally, we shall provide the answers to the questions raised in the original practical problems.

> **Problem 1.** A random number generator is a device (usually a computer program) that can generate sequences of random numbers (i.e. series of numbers that cannot be predicted) within a predetermined scope. Perhaps the most common "device" that generates random numbers, in the range $1, 2, \ldots, 6$, is a die. However, the most important case of random numbers are the ones that belong to the unit interval $(0, 1)$; when no specific reference is made, the term random number generation refers to this case. Random number generation has numerous applications in statistical sampling, cryptography, computer simulation, experimental design, etc. Most computer programming languages include routines (activated by a function named RANDOM, RND, RAN, etc.) that provide sequences of random numbers. The output of a random number generator is a sample $X_1, X_2, \ldots, X_n$ of n independent random variables following the Uniform distribution on the internal $(0, 1)$. A reasonable question, when dealing with a sample of random numbers, is the efficacy of our sample in terms of coverage; this may be accessed by the range of the sample $R = max(X_1, X_2, \ldots, X_n) - min(X_1, X_2, \ldots, X_n)$, in which case it makes sense to seek answers to the next three questions:
>
> (a) For given n, what is the probability of achieving a range that exceeds a threshold r_0 i.e. $R > r_0$?
>
> (b) How large should n be selected so that $R > r_0$ with probability (at least) $1 - a$?
>
> (c) How large should n be selected so that $E(R) > r_0$?
>
> **Problem 2.** Let us assume that the times of occurrence of a random event (e.g. arrival of a person at a meeting, nondestructive shocks affecting the operation of an engineering system) are described by the random variables $X_1, X_2, \ldots, X_n$. In this case, the range R describes the time–distance between the first and last occurrence of the event, a quantity which is quite useful for planning. For example, if $X_1, X_2, \ldots, X_n$ are the arrival times (in minutes) of the persons that have booked for a guided tour, two questions of interest might be
>
> (a) how large should the group size be so that no one has to wait more than t_0 minutes before the tour commences?
>
> (b) how large should the group size be so that the mean waiting time of any person participating in the tour does not exceed t_0?
>
> Analogous questions could be raised for the nondestructive shock model, in terms of the total time the system is under stress when n shocks occur.
>
> Let us next build the stochastic model that will provide the tools to answer the questions stated above. Assume that $X_1, X_2, \ldots, X_n$ are independent continuous random

variables from a distribution with density function $f(x)$ and cumulative distribution function $F(x)$. Denote by $X_{(1)} = min(X_1, X_2, \dots, X_n)$ and $X_{(n)} = max(X_1, X_2, \dots, X_n)$ the smallest and the largest observation among $X_1, X_2, \dots X_n$ and by $R = X_{(n)} - X_{(1)}$ the range of the sample $X_1, X_2, \dots, X_n$. Our aim is to derive a formula for the distribution of R and $E(R)$.

Since R is the difference of two random variables, a direct application of formula (4.1) for $g(x, y) = y - x$ (note that $u = g(x, y) \Leftrightarrow x = y - u = g^*(u, y)$ and $\partial g^*(u, y)/\partial u = -1$) yields the following for the density function of R,

$$f_R(r) = \int_{-\infty}^{\infty} f_{X_{(1)}, X_{(n)}}(v - r, v) \cdot |-1| dv, \tag{4.10}$$

where $f_{X_{(1)}, X_{(n)}}(s, t)$ denotes the joint density of $X_{(1)}$ and $X_{(n)}$. Employing Proposition 3.6.2 for $i = 1, j = n$ we get

$$f_{X_{(1)}, X_{(n)}}(s, t) = \frac{n!}{0!(n-2)!0!} f(s)f(t)[F(s)]^0[F(t) - F(s)]^{n-2}[1 - F(t)]^0$$

$$= n(n-1)f(s)f(t)[F(t) - F(s)]^{n-2}, \quad s < t. \tag{4.11}$$

To provide the answers to the questions set in Problem 1 above, we now assume that the distribution of $X_1, X_2, \dots, X_n$ is uniform over the interval $(0, 1)$; in this case, (4.11) reduces to (see also Example 3.6.5)

$$f_{X_{(1)}, X_{(n)}}(s, t) = n(n-1)(t-s)^{n-2}, \quad 0 < s < t < 1.$$

Therefore, in view of (4.10), we deduce that

$$f_R(r) = \int_r^1 n(n-1)r^{n-2}dv = n(n-1)(1-r)r^{n-2}, \quad \text{for } 0 < r < 1.$$

Further, the cumulative distribution function of R is given by

$$F_R(r) = \int_0^r f_R(u)du = n(n-1)\int_0^r (1-u)u^{n-2}du$$

$$= nr^{n-1} - (n-1)r^n, \quad 0 < r < 1$$

while the expectation $E(R)$ is

$$E(R) = \int_0^1 rf_R(r)dr = n(n-1)\int_0^1 (1-r)r^{n-1} dr = \frac{n-1}{n+1}. \tag{4.12}$$

We are now ready to deal with questions (a)–(c) of Problem 1.

(a) The probability of interest equals

$$P(R > r_0) = 1 - F(r_0) = 1 - nr_0^{n-1} + (n-1)r_0^n. \tag{4.13}$$

As an illustration, the probability that the n numbers created by our random number generator cover at least half of the interval $(0, 1)$ is given by

$$P\left(R > \frac{1}{2}\right) = 1 - \frac{n+1}{2^n},$$

while the probability of achieving a 90% coverage is

$$P\left(R > \frac{9}{10}\right) = 1 - \frac{(n+9)9^{n-1}}{10^n}.$$

Both these probabilities are increasing functions of n and approach the value one as n becomes large (theoretically, as $n \to \infty$); see Table 4.2 and Figures 4.7 and 4.8.

(b) We are now seeking the value of n so that the following condition holds true:

$$P(R > r_0) \geq 1 - \alpha.$$

Recalling (4.13) we get

$$1 - nr_0^n + (n-1)r_0^{n+1} \geq 1 - \alpha,$$

or equivalently,

$$nr_0^n - (n-1)r_0^{n+1} \leq \alpha.$$

n	r_0	$P(R > r_0)$
5	0.5	0.8125
5	0.75	0.3672
5	0.9	0.0815
10	0.5	0.9893
10	0.75	0.7560
10	0.9	0.2639
15	0.5	0.9995
15	0.75	0.9198
15	0.9	0.4510
20	0.5	1.0000
20	0.75	0.9757
−20	0.9	0.6083
50	0.5	1.0000
50	0.75	1.0000
50	0.9	0.9662
100	0.5	1.0000
100	0.75	1.0000
100	0.9	0.9997

Table 4.2   The probabilities $P(R > r_0)$ for different values of n and r_0.

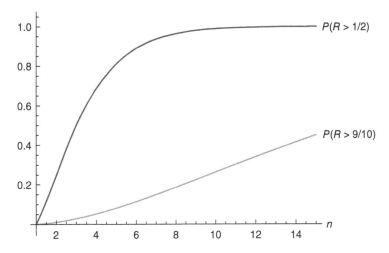

Figure 4.7 The probabilities $P(R > 1/2)$ and $P(R > 9/10)$ for $1 \le n \le 15$.

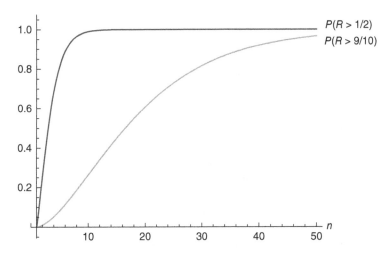

Figure 4.8 The probabilities $P(R > 1/2)$ and $P(R > 9/10)$ for $1 \le n \le 50$.

Solving the last inequality for n can only be achieved through numerical methods. Alternatively, one may produce, for a given r_0, a table similar to that for part (a) and identify the minimum value of n that leads to a coverage probability that is at least $1 - \alpha$. For example, if we wish to have a 99% probability for 90% coverage, we should use a sample of size $n \ge 64$ (the interested reader may verify this!).

(c) Using (4.12) we get

$$\frac{n-1}{n+1} \ge r_0 \Leftrightarrow n \ge \frac{1}{1-r_0} - 1.$$

Table 4.3 compares the values of n for achieving a $100r_0\%$ coverage (for $r_0 = 0.5, 0.7, 0.9, 0.95, 0.99$) with probabilities 90%, 95% and 99% (case (b) with $\alpha = 0.1, 0.05, 0.01$) and in mean (case (c)).

	Case (b)			Case (c)
r_0	$\alpha = 0.1$	$\alpha = 0.05$	$\alpha = 0.01$	$E(R) \geq r_0$
0.50	7	8	11	1
0.70	12	14	20	3
0.90	38	46	64	9
0.95	77	93	129	19
0.99	388	472	662	100

Table 4.3   The minimum value of n required to achieve a $100r_0$ coverage for different values of r_0, for case (b) with $\alpha = 0.1, 0.05, 0.01$ and for case (c)

Let us next discuss in brief Problem 2. Suppose that the tour is scheduled to begin at 10:00 a.m. Then, one may assume that the people who have booked for that time will be expected to arrive at a random point in time between, say, 09:50 a.m. and 10:10 a.m. Therefore, using 09:50am as our starting point in time, the arrival times $X_1, X_2, \ldots, X_n$ could be considered as uniform random variables over the interval $(0, \theta)$, with $\theta = 20$ minutes. It is not difficult to verify that, in this case, we get (in a similar manner as before),

$$f_{X_{(1)}, X_{(n)}}(s, t) = \frac{n(n-1)}{\theta^n}(t - s)^{n-2}, \qquad 0 < s < t < \theta,$$

$$f_R(r) = \frac{n(n-1)}{\theta^n}(\theta - r)r^{n-2}, \quad \text{for } 0 < r < \theta,$$

$$F_R(r) = n\left(\frac{r}{\theta}\right)^{n-1} - (n-1)\left(\frac{r}{\theta}\right)^n \qquad 0 < r < \theta$$

and, finally, the expectation of the range is

$$E(R) = \frac{n-1}{n+1} \cdot \theta.$$

Question (b) is now answered by solving the following inequality

$$\frac{n-1}{n+1} \cdot \theta \geq t_0$$

for $\theta = 20$ and t_0 as desired. In addition, if we wish to determine the group size n so that, with probability $1 - \alpha$ (at least), no one has to wait for more than t_0 minutes, the answer comes by solving for n the inequality

$$n\left(\frac{t_0}{\theta}\right)^{n-1} - (n-1)\left(\frac{t_0}{\theta}\right)^n \leq \alpha.$$

The first inequality leads to the explicit solution

$$n \geq \frac{\theta + t_0}{\theta - t_0},$$

while the second one can be solved numerically (or by generating extensive tables for several n).

KEY TERMS

Cauchy distribution
chi-squared (χ^2) distribution
convolution
degrees of freedom
F (or Snedecor) distribution
Jacobian (of a transformation)
Laplace distribution
quantile (upper α–quantile) of a distribution
t distribution (Student distribution)
transformation of random variables

CHAPTER 5

COVARIANCE AND CORRELATION

George U. Yule (Morham, Scotland 1871 – Cambridge, England 1951)

George Udny Yule was born on 18 February 1871, in Morham, Scotland, and died on 26 June 1951, at the age of 80, in Cambridge, England. At the age of 16, he entered University College London to study Engineering. Then, in 1893, he returned to University College to work as a demonstrator for Karl Pearson, whose interest in the field of Statistics he followed. During 1897–1899, he wrote many important papers on correlation and regression, while three decades later he published a key paper dealing with spurious correlation. In 1912, he moved to Cambridge University and stayed there till the end of his career. In 1911, he published an influential textbook *Introduction to the Theory of Statistics*, which was highly successful and went through fourteen editions. Through his work on the theory of evolution in 1925, he introduced a stochastic process which is now called *Yule process*, and the resulting distribution is also called *Yule distribution*.

Introduction to Probability: Multivariate Models and Applications, Volume 2, First Edition.
N. Balakrishnan, Markos V. Koutras, and Konstadinos G. Politis.
© 2022 John Wiley & Sons, Inc. Published 2022 by John Wiley & Sons, Inc.

5.1   INTRODUCTION

In Chapter 3, we discussed the concept of independence between two or more random variables. When two variables are independent, then the value that one of them takes on does not have any effect on the values that the other variable may take on. Although this statement seems to deserve no further elaboration, when two variables are *not* independent, we may wish to have an insight about how (and how much) the two variables depend on each other. There are cases, for instance, when the value that one variable takes on completely determines the value of the other. For example, suppose X is an arbitrary random variable and that Y and W are two other variables defined as

$$Y = 3X - 2, \qquad W = e^X + 1.$$

If we observe the value of X in a realization of the experiment, we then know exactly the values of Y and W for this realization. Moreover, before conducting the experiment, once we know the probability distribution of X, we can derive from it the distributions of Y and W. This is, of course, a rather simple case of dependence between two variables.

In contrast, consider the situation discussed in Example 1.2.1 wherein two balls are selected at random, and without replacement, from a box containing three balls numbered 1–3. Now, let

   X: the number on the first ball drawn,
   Y: the largest number on the two balls selected.

It seems clear from what we have discussed thus far that X and Y are not independent. Suppose, in particular, that we select a ball from the box and it has the number 2 on it. This immediately implies that Y can be either 2 or 3, but the value $Y = 1$ is certainly not possible. So, our knowledge about the possible values for Y has changed, but not to the extent that we know with certainty what will be the value of Y for this experiment. Conversely, suppose the experiment is conducted by someone else and this person tells us that, after selecting the two balls, the largest number on them was $Y = 3$. Can we then be certain about the value of X on the first ball selected? The answer is no, but $X = 3$ is now more likely than it was before we received this information. In fact, without any knowledge of Y, the probability that $X = 3$ is 1/3, since all three choices for the first ball are equally likely. When we know that $Y = 3$, however, this probability increases to 1/2, as one can easily verify.

In this example, the relationship between X and Y is not "perfect", but it is useful to know the value of one of them, because this would enable us to make more precise probability statements about the other. In this chapter, we discuss cases like this in a general framework, which includes both discrete and continuous random variables, by introducing two important concepts for measuring the dependence between two variables: covariance and correlation.

5.2   COVARIANCE

Suppose we have two random variables X and Y with corresponding expectations $\mu_X = E(X)$ and $\mu_Y = E(Y)$. We know that the variance $\sigma_X^2 = Var(X)$ measures the dispersion of the values that X takes on around its mean, μ_X, and similarly σ_Y^2 expresses how much the values of Y depart from the expected value, μ_Y.

For two constants $a, b \in \mathbb{R}$, let us consider the random variable $Z = aX + bY$, which is a linear combination of X and Y. We then have

$$E(Z) = E(aX + bY) = aE(X) + bE(Y) = a\mu_X + b\mu_Y.$$

Using this, we then obtain the variance of Z as

$$Var(Z) = E\left[(Z - E(Z))^2\right] = E\left[(aX + bY - a\mu_X - b\mu_Y)^2\right]$$
$$= E\left\{\left[a(X - \mu_X) + b(Y - \mu_Y)\right]^2\right\}.$$

Expanding the square term and then using the linearity of expectation (which has already been used above), we obtain

$$Var(Z) = E\left[a^2(X - \mu_X)^2 + b^2(Y - \mu_Y)^2 + 2ab(X - \mu_X)(Y - \mu_Y)\right]$$
$$= a^2 E\left[(X - \mu_X)^2\right] + b^2 E\left[(Y - \mu_Y)^2\right] + 2ab E\left[(X - \mu_X)(Y - \mu_Y)\right]$$
$$= a^2 Var(X) + b^2 Var(Y) + 2ab E\left[(X - \mu_X)(Y - \mu_Y)\right]$$
$$= a^2 \sigma_X^2 + b^2 \sigma_Y^2 + 2ab E\left[(X - \mu_X)(Y - \mu_Y)\right].$$

We observe that among the three summands on the right-hand side above, only the last one depends on both X and Y. In particular, the sign of this quantity depends on both the sign of the product ab and the sign of $E\left[(X - \mu_X)(Y - \mu_Y)\right]$. This last quantity has special importance and is defined next.

Definition 5.2.1 Let X and Y be two random variables with expectations μ_X and μ_Y, respectively. Then, the quantity

$$Cov(X, Y) = E\left[(X - \mu_X)(Y - \mu_Y)\right]$$

is called the **covariance** between X and Y. Another symbol we shall use for the covariance, apart from $Cov(X, Y)$, is σ_{XY}.

From the definition, it is evident that

$$Cov(X, X) = Var(X), \qquad Cov(Y, Y) = Var(Y),$$

and also that the covariance as an operation between two variables is commutative, i.e.

$$Cov(X, Y) = Cov(Y, X).$$

Thus, in view of Definition 5.2.1, we can rewrite the expression we found above for the variance of the linear combination $Z = aX + bY$ as

$$Var(aX + bY) = a^2 Var(X) + b^2 Var(Y) + 2ab \, Cov(X, Y).$$

For $b = 0$ and $a = 0$, we recover the known relations $Var(aX) = a^2 Var(X)$ and $Var(bY) = b^2 Var(Y)$, respectively. Another consequence of the above relation corresponds to cases

$a = b = 1$ and $a = 1, b = -1$, when we can express the variance of the sum or the difference between two variables using their covariance as follows.

Proposition 5.2.1 *For any random variables X and Y, we have*

$$Var(X + Y) = Var(X) + Var(Y) + 2\ Cov(X, Y),$$

$$Var(X - Y) = Var(X) + Var(Y) - 2\ Cov(X, Y).$$

From the proposition, we readily see that the variance of the sum between two variables is larger (smaller) than the sum of their individual variances, according to the covariance being positive (negative) between the variables. So, it is natural to wonder whether the covariance between two variables can be zero and as to when this could happen. A sufficient, though not necessary, condition for this is detailed next.

Proposition 5.2.2 *For any two variables X and Y, we have*

$$Cov(X, Y) = E(XY) - E(X)E(Y). \tag{5.1}$$

In particular, when the variables X and Y (either both discrete or both continuous) are independent, then $Cov(X, Y) = 0$.

Proof: For the first part of the proposition, we use once again the linearity of expectation. This, with Definition 5.2.1, gives

$$Cov(X, Y) = E(XY - X\mu_Y - Y\mu_X + \mu_X\mu_Y)$$

$$= E(XY) - E(X)\mu_Y - E(Y)\mu_X + \mu_X\mu_Y$$

$$= E(XY) - 2E(X)E(Y) + E(X)E(Y) = E(XY) - E(X)E(Y),$$

as required. The second part follows immediately since it is known that, for independent X and Y, both in the discrete and in the continuous case it holds that $E(XY) = E(X)E(Y)$. □

Example 5.2.1 Table 1.2 presents the joint probability function of the variables X and Y, mentioned already in Section 5.1. In this case, it is easy to see from the table that the (marginal) probability functions, f_X and f_Y, of X and Y are given by

$$f_X(1) = f_X(2) = f_X(3) = \frac{1}{3}$$

and

$$f_Y(2) = \frac{1}{3}, \quad f_Y(3) = \frac{2}{3},$$

while the probability function of their product, $U = XY$, is given in the following table:

i	$f_U(i) = P(U = i)$
2	1/6
3	1/6
4	1/6
6	1/6
9	1/3

We can now find the expectations of X, Y and $U = XY$ as follows:

$$E(X) = 1 \cdot \frac{1}{3} + 2 \cdot \frac{1}{3} + 3 \cdot \frac{1}{3} = 2,$$

$$E(Y) = 2 \cdot \frac{1}{3} + 3 \cdot \frac{2}{3} = \frac{8}{3}$$

and

$$E(U) = \sum_{i \in R_U} iP(U = i) = (2 + 3 + 4 + 6) \cdot \frac{1}{6} + 9 \cdot \frac{2}{6} = \frac{11}{2}.$$

Thence, we see that the covariance between X and Y is

$$Cov(X, Y) = E(XY) - E(X)E(Y) = \frac{11}{2} - 2 \cdot \frac{8}{3} = \frac{1}{6}.$$

At this stage, it may not be clear what this value actually "means" for the relation between the two variables. This will become clear after we introduce the concept of correlation in Section 5.3. It needs to be kept in mind, however, that the covariance here is positive (see the discussion after Example 5.2.3).

Example 5.2.2 Suppose the discrete variables X and Y have joint probability function

$$f(x, y) = \begin{cases} \dfrac{x + y}{45}, & \text{for } x = 1, 2, 3 \text{ and } y = 2, 3, 4, \\ \\ 0, & \text{otherwise.} \end{cases}$$

For the marginal probability functions of X and Y, we find

$$f_X(x) = \sum_{y=2}^{4} f(x, y) = \frac{x + 2}{45} + \frac{x + 3}{45} + \frac{x + 4}{45} = \frac{3x + 9}{45} = \frac{x + 3}{15}, \quad x = 1, 2, 3,$$

and

$$f_Y(y) = \sum_{x=1}^{3} f(x, y) = \frac{1 + y}{45} + \frac{2 + y}{45} + \frac{3 + y}{45} = \frac{6 + 3y}{45} = \frac{y + 2}{15}, \quad y = 2, 3, 4,$$

while the corresponding expectations are

$$E(X) = 1 \cdot \frac{4}{15} + 2 \cdot \frac{5}{15} + 3 \cdot \frac{6}{15} = \frac{32}{15}$$

and

$$E(Y) = 2 \cdot \frac{4}{15} + 3 \cdot \frac{5}{15} + 4 \cdot \frac{6}{15} = \frac{47}{15}.$$

Finally, for the expectation of their product, XY, we find

$$E(XY) = \sum_{x=1}^{3}\sum_{y=2}^{4} xyf(x,y) = 1 \cdot 2 \cdot \frac{3}{45} + 1 \cdot 3 \cdot \frac{4}{45} + \cdots + 3 \cdot 4 \cdot \frac{7}{45} = \frac{20}{3}.$$

Consequently, we obtain the covariance between X and Y to be

$$Cov(X,Y) = E(XY) - E(X)E(Y) = \frac{20}{3} - \frac{32}{15} \cdot \frac{47}{15} = -\frac{4}{225}.$$

Example 5.2.3 The last two examples involved discrete variables. However, both Definition 5.2.1 and Propositions 5.2.1 and 5.2.2 do not distinguish between discrete and continuous variables (we only need to assume that X and Y are of the same type, otherwise we do not have a way to calculate the expectation $E(XY)$).

Let (X, Y) now have a joint continuous distribution with density

$$f(x,y) = \begin{cases} \dfrac{1}{4}(1 - x^3y + xy^3), & -1 \le x \le 1, -1 \le y \le 1, \\ 0, & \text{otherwise.} \end{cases}$$

The marginal density functions of X and Y are then

$$f_X(x) = \int_{-1}^{1} f(x,y)\, dy = \left[\frac{1}{4}\left(y - \frac{x^3y^2}{2} + \frac{xy^4}{4} \right) \right]_{y=-1}^{y=1} = \frac{1}{2}, \qquad -1 \le x \le 1,$$

and

$$f_Y(y) = \int_{-1}^{1} f(x,y)\, dx = \frac{1}{2}, \qquad -1 \le y \le 1.$$

Thus, we observe that both X and Y have uniform distribution on $(-1, 1)$. It then follows that

$$E(X) = E(Y) = 0,$$

while the expectation $E(XY)$ is found to be

$$E(XY) = \int_{-1}^{1}\int_{-1}^{1} xyf(x,y)\, dx\, dy = \int_{-1}^{1}\int_{-1}^{1} \frac{xy}{4}(1 - x^3y + x^3y)\, dx\, dy = 0,$$

after some straightforward calculations. Therefore, we find

$$Cov(X, Y) = E(XY) - E(X)E(Y) = 0.$$

We also observe that

$$f(x, y) \neq f_X(x)f_Y(y),$$

and so X and Y are not independent, and yet their covariance is zero. This shows that the converse of Proposition 5.2.2 is not true, that is two variables may not be independent and still have a zero covariance.

It has become clear from the above examples that the covariance between two random variables can be positive, zero, or negative. Trying to get a better understanding about what each of these means, let us assume first that $Cov(X, Y) > 0$ for two variables X and Y. Then, the expected value of the quantity $(X - \mu_X)(Y - \mu_Y)$ is positive, which suggests that the quantities $X - \mu_X$ and $Y - \mu_Y$ tend to have the same sign, so that when X takes values larger than its expected value, Y does the same, and when X takes values smaller than its expected value, Y tends to do the same. The positive sign of their covariance therefore means that X and Y move in the same direction "on average", i.e. large (small) values of X are associated with large (small) values of Y. Similarly, when $Cov(X, Y) < 0$, the quantities $X - \mu_X$ and $Y - \mu_Y$ have opposite signs "on average", with large (small) values of X being associated with small (large) values of Y. For this reason, when $Cov(X, Y) > 0$ we say that X and Y are **positively correlated**, while if $Cov(X, Y) < 0$, we say that X and Y are **negatively correlated**. If $Cov(X, Y) = 0$, we say that X and Y are **uncorrelated**.

The concept of positive/negative/zero correlation detailed above is in accordance with the use of the term in daily language. For example, we expect the mark a student gets in a *Probability* course at a University to be positively correlated with the mark the same student gets in a *Statistics* course. On the other hand, it is reasonable to expect that a driver's ability to react to an emergency is negatively correlated with the alcohol level in his/her blood. Finally, for the case $Cov(X, Y) = 0$, we should first note that if X and Y are independent, then they are uncorrelated; in fact, for independent random variables, we have $E(XY) = E(X)E(Y)$, and so

$$Cov(X, Y) = E(XY) - E(X)E(Y) = 0.$$

The converse of this, as we have seen in Example 5.2.3 above is not true, that is, two variables that are uncorrelated are not necessarily independent. This is stated formally in the following proposition.

Proposition 5.2.3 *If two variables X and Y are independent, then $Cov(X, Y) = 0$, so that they are uncorrelated. The converse is not always true.*

Further, from Proposition 5.2.1 we see that for uncorrelated variables, we have

$$Var(X + Y) = Var(X) + Var(Y) \quad \text{and} \quad Var(X - Y) = Var(X) + Var(Y).$$

Clearly, these two relations hold in the particular case when X and Y are independent. The notion of independence extends to more than two random variables, as we have seen in Chapter 3. The first of the two formulas above can be generalized to the case when more than two variables are involved. Specifically, if $X_1, \dots, X_n$ are independent random variables, then

$$Var(X_1 + \cdots + X_n) = Var(X_1) + \cdots + Var(X_n) = \sum_{i=1}^{n} Var(X_i).$$

The proof of this follows from a more general result given later in Proposition 5.2.5.

Example 5.2.4 Let $X_1, \dots, X_n$ be a random sample from a distribution with finite variance σ^2. Let the sample mean be $\overline{X}$, given by

$$\overline{X} = \frac{X_1 + \cdots + X_n}{n}.$$

Calculate the variance of the sample mean.

SOLUTION   Since $X_1, \dots, X_n$ have been assumed to constitute a random sample, it follows that the variables X_i are independent, for $i = 1, 2, \dots, n$. Therefore, we have

$$Var\left(\overline{X}\right) = Var\left(\frac{X_1 + \cdots + X_n}{n}\right) = \frac{1}{n^2} Var(X_1 + X_2 + \cdots + X_n)$$

$$= \frac{1}{n^2}\left[Var(X_1) + Var(X_2) + \cdots + Var(X_n)\right] = \frac{1}{n^2} \cdot n\sigma^2 = \frac{\sigma^2}{n}.$$

This result has very important consequences in statistics, where a common technique is to use the sample mean as an "estimate" for the mean of the population where the data come from. Since the variance of the sample mean decreases as n grows, we observe formally the intuitively appealing fact that the sample mean becomes more precise (i.e. it has smaller variance) as the sample size becomes larger.

Note that the finiteness of the variance is essential here; if the X_i's have an infinite variance, then the variance of the sample mean will also be infinite.

The following proposition lists some important properties for the covariance of two variables.

Proposition 5.2.4 *(i) For any two random variables X and Y and real numbers a, b, c, and d, we have*

$$Cov(aX + b, cY + d) = ac\, Cov(X, Y).$$

(ii) For any random variables X, Y, and Z, we have

$$Cov(X + Y, Z) = Cov(X, Z) + Cov(Y, Z)$$

and

$$Cov(X, Y + Z) = Cov(X, Y) + Cov(X, Z).$$

Proof:

(i) Setting $\mu_X = E(X)$ and $\mu_Y = E(Y)$, it is clear that

$$E(aX + b) = a\mu_X + b, \quad E(cY + d) = c\mu_Y + d,$$

thus we get

$$\begin{aligned}
Cov(aX + b, cY + d) &= E\left\{\left[(aX + b) - (a\mu_X + b)\right]\left[(cY + d) - (c\mu_Y + d)\right]\right\} \\
&= E\left[a(X - \mu_X)c(Y - \mu_Y)\right] = acE\left[(X - \mu_X)(Y - \mu_Y)\right] \\
&= ac\ Cov(X, Y).
\end{aligned}$$

(ii) For the first formula, we use Proposition 5.2.2 and known properties of expectation to find

$$\begin{aligned}
Cov(X + Y, Z) &= E\left[(X + Y)Z\right] - E(X + Y)E(Z) \\
&= E(XZ + YZ) - \left[E(X) + E(Y)\right]E(Z) \\
&= E(XZ) + E(YZ) - E(X)E(Z) - E(Y)E(Z) \\
&= \left[E(XZ) - E(X)E(Z)\right] + \left[E(YZ) - E(Y)E(Z)\right] \\
&= Cov(X, Z) + Cov(Y, Z),
\end{aligned}$$

as required. The second formula follows similarly. $\square$

Next, we present a generalization of Proposition 5.2.4.

Proposition 5.2.5 *Let $X_1, \dots, X_n$ and $Y_1, \dots, Y_k$ be two sets of random variables, and let $a_1, \dots, a_n$ and $b_1, \dots, b_k$ be real constants.*

(i) *We have*

$$Cov\left(\sum_{i=1}^{n} a_i X_i, \sum_{j=1}^{k} b_j Y_j\right) = \sum_{i=1}^{n}\sum_{j=1}^{k} a_i b_j Cov(X_i, Y_j).$$

(ii) *The variance of the linear combination $\sum_{i=1}^{n} a_i X_i$ is given by*

$$\begin{aligned}
Var\left(\sum_{i=1}^{n} a_i X_i\right) &= \sum_{i=1}^{n} a_i^2 Var(X_i) + \sum_{j=1}^{n}\sum_{i \neq j} a_i a_j Cov(X_i, X_j) \\
&= \sum_{i=1}^{n} a_i^2 Var(X_i) + 2\sum_{j=2}^{n}\sum_{i=1}^{j-1} a_i a_j Cov(X_i, X_j).
\end{aligned}$$

(iii) *Assuming in addition that the variables $X_1, \ldots, X_n$ are pairwise uncorrelated, that is,*

$$Cov(X_i, X_j) = 0 \quad for \ i \neq j,$$

we have

$$Var\left(\sum_{i=1}^{n} X_i\right) = \sum_{i=1}^{n} Var(X_i).$$

Proof:

(i) For ease of notation, let us put $X = \sum_{i=1}^{n} a_i X_i$. Then, using Proposition 5.2.4, we find

$$Cov\left(X, \sum_{j=1}^{k} b_j Y_j\right) = \sum_{j=1}^{k} Cov(X, b_j Y_j) = \sum_{j=1}^{k} b_j Cov(X, Y_j).$$

Similarly, we obtain, for any $j \in \{1, \ldots, k\}$, that

$$Cov(X, Y_j) = Cov\left(\sum_{i=1}^{n} a_i X_i, Y_j\right) = \sum_{i=1}^{n} Cov(a_i X_i, Y_j) = a_i \sum_{i=1}^{n} Cov(X_i, Y_j),$$

and the required result now follows from the above expressions (note that we can interchange the order of summation).

(ii) This follows from the identity

$$Var\left(\sum_{i=1}^{n} a_i X_i\right) = Cov\left(\sum_{i=1}^{n} a_i X_i, \sum_{j=1}^{n} a_j X_j\right),$$

which gives

$$Var\left(\sum_{i=1}^{n} a_i X_i\right) = \sum_{i=1}^{n}\sum_{j=1}^{n} a_i a_j Cov\left(X_i, X_j\right)$$

$$= \sum_{i=1}^{n} a_i^2 Cov(X_i, X_i) + \sum\sum_{i \neq j} a_i a_j Cov(X_i, X_j)$$

$$= \sum_{i=1}^{n} a_i^2 Var(X_i) + 2\sum_{j=2}^{n}\sum_{i=1}^{j-1} a_i a_j Cov(X_i, X_j).$$

(iii) This part follows immediately upon setting $a_i = 1$, for $i = 1, \ldots, n$ in Part (ii). □

Example 5.2.5   We roll a die n times. Find the covariance between the number of aces and the number of sixes observed in these n rolls.

SOLUTION For $i = 1, \ldots, n$, let us introduce the random variables

$$X_i = \begin{cases} 1, & \text{if an ace shows up at the } i\text{th roll,} \\ 0, & \text{otherwise,} \end{cases}$$

and

$$Y_i = \begin{cases} 1, & \text{if a six shows up at the } i\text{th roll,} \\ 0, & \text{otherwise.} \end{cases}$$

Then, we are interested in $Cov(X, Y)$, where

$$X = \sum_{i=1}^{n} X_i, \quad Y = \sum_{j=1}^{n} Y_j.$$

It is clear that, for $i \neq j$, the random variables X_i and Y_j are independent (as they are associated with different rolls) and so $Cov(X_i, Y_j) = 0$. Now, by using Proposition 5.2.5, we obtain

$$Cov(X, Y) = Cov\left(\sum_{i=1}^{n} X_i, \sum_{j=1}^{n} Y_j\right) = \sum_{i=1}^{n}\sum_{j=1}^{n} Cov(X_i, Y_j)$$

$$= \sum_{i=1}^{n} Cov(X_i, Y_i).$$

Next, we note that the product $X_i Y_i$ cannot take any value other than zero, since X_i and Y_i cannot simultaneously be equal to one. Thus, $P(X_i Y_i = 0) = 1$ which implies that $E(X_i Y_i) = 0$ and this, in turn, yields

$$Cov(X_i, Y_i) = E(X_i Y_i) - E(X_i)E(Y_i) = 0 - \frac{1}{6} \cdot \frac{1}{6} = -\frac{1}{36}.$$

Substituting this in the previous expression, we finally deduce that

$$Cov(X, Y) = -\frac{n}{36}.$$

The negative sign in the covariance between X and Y is rather easy to interpret intuitively. For a fixed number of throws, n, the larger the number of aces, the smaller the number of sixes (and vice versa).

Exercises
Group A

1. For any random variables X and Y, show that

$$Var(X + Y) + Var(X - Y) = 2Var(X) + 2Var(Y).$$

Also, can you explain why the following is incorrect?

$$Var(X + Y) + Var(X - Y) = Var\left[(X + Y) + (X - Y)\right] = Var(2X) = 4Var(X).$$

2. Suppose the joint density function of X and Y is

$$f(x, y) = 28e^{-4x}e^{-7y}, \qquad\qquad x, y > 0.$$

Show that X and Y are uncorrelated.

(Hint: Show that X and Y are independent, and the result then follows.)

3. Let X be a random variable having the uniform distribution over $(-1, 1)$, and define another variable $Y = X^2$.

(i) Verify that

$$E(X^r) = \frac{1 - (-1)^{r+1}}{2(r + 1)}, \qquad\qquad r = 1, 2, \ldots$$

(ii) Show that X and Y are uncorrelated, but not independent.

4. Let X and Y be two discrete random variables, each of which takes four values only, and their joint probability function is given by

$$P(X = a_i, Y = b_i) = \frac{1}{4}, \qquad\qquad \text{for } i = 1, 2, 3, 4,$$

where a_1, a_2, a_3, a_4 and b_1, b_2, b_3, b_4 are real numbers ($a_i \neq a_j$ for $i \neq j$ and $b_i \neq b_j$ for $i \neq j$) such that

$$\sum_{i=1}^{4} a_i = \sum_{i=1}^{4} b_i = \sum_{i=1}^{4} a_i b_i = 0.$$

Prove that X and Y are uncorrelated, but not independent.

5. Let X and Y be two variables, either discrete or continuous, which are independent having the same distribution, and define two new variables

$$U = X + Y, \qquad\qquad V = X - Y.$$

Show that $Cov(U, V) = 0$ despite the obvious fact that U and V are not independent.

Application: Let X and Y be the outcomes in two successive rolls of a die, and define U and V to be their sum and difference.

Obtain the conditional probability function of U, given V, that is, $f_{U|V}(u|v)$, for $v = -5, -2, 0, 2, 5$, and check whether this is the same as the unconditional probability function of U. Explain what this means about the independence, or the lack of it, between U and V.

6. The joint probability function of the discrete pair (X, Y) is as follows:

$$f(x, y) = P(X = x, Y = y)$$

x \ y	-1	0	1
-1	1/20	1/5	1/20
0	1/5	0	1/5
1	1/20	1/5	1/20

Find the covariance between X and Y. Are X and Y independent?

7. Calculate the covariance of the discrete random pair (X, Y), for each of the following joint probability functions:

   (i) $f(x, y) = \dfrac{2x + y}{18}$, $x = 1, 2$ and $y = 1, 2$;

   (ii) $f(x, y) = \dfrac{3x + 2y}{84}$, $x = 1, 2, 3, 4$ and $y = 1, 2$;

   (iii) $f(x, y) = \dfrac{xy}{18}$, $x = 1, 2, 3$ and $y = 1, 2$;

   (iv) $f(x, y) = \dfrac{1}{15}$, $x = 1, 2, 3, 4, 5$ and $y = 1, 2, \dots, x$.

8. Let I_A and I_B be the indicator functions of two events A and B; that is,

$$I_A = \begin{cases} 1, & \text{if } A \text{ occurs,} \\ 0, & \text{otherwise,} \end{cases}$$

$$I_B = \begin{cases} 1, & \text{if } B \text{ occurs,} \\ 0, & \text{otherwise.} \end{cases}$$

   (i) Prove that the covariance between the variables I_A and I_B is given by

$$Cov(I_A, I_B) = P(B)\left[P(A|B) - P(A)\right].$$

   Thence, show that $P(A|B) = P(A)$ if and only if the variables I_A and I_B are uncorrelated.

   (ii) Suppose $Cov(I_A, I_B) > 0$. How would you interpret intuitively the result of Part (i)?

9. Let X, Y, and Z be independent random variables with $E(Y) = 3$ and $E(Y^2) = 10$. Calculate the covariance between the variables $U = 3X + Y$ and $V = Y + 2Z$.

10. Let X and Y be variables with $Var(X) = \sigma_X^2$ and $Var(Y) = \sigma_Y^2$. If $\alpha \sigma_X = \beta \sigma_Y$, for some constants $\alpha, \beta \in \mathbb{R}$, show that the variables $\alpha X + \beta Y$ and $\alpha X - \beta Y$ are uncorrelated.

11. Let X and Y be random variables that are independent and have the same distribution with mean μ and a finite variance σ^2. Let $U = (X + Y)/2$.

   (i) Show that
$$Cov(X - U, U) = 0.$$

   (ii) Calculate the covariance between $X - U$ and $Y - U$.

12. Let $X_1, \dots, X_n$ be a set of random variables such that $Var(X_i) = \sigma^2$, for $i = 1, \dots, n$, while
$$Cov(X_i, X_j) = \kappa, \qquad \text{for any } i, j \text{ with } i \neq j.$$

   Show that the variance of $\overline{X} = \frac{1}{n}\sum_{i=1}^{n} X_i$ is given by
$$Var\left(\overline{X}\right) = \frac{\sigma^2 + (n-1)\kappa}{n}.$$

   This shows that, as the covariance between X_i's increases (or decreases), the variance of the mean increases (or decreases) accordingly.

13. For two independent random variables X and Y, show that

$$Cov(X, XY) = E(Y)Var(X).$$

14. Suppose U has a uniform distribution over the interval $(0, 2\pi)$. Prove that the variables $X = \sin U$ and $Y = \cos U$ are uncorrelated.

Group B

15. The time, in hours, that Bill needs to travel to work in the morning is a random variable X with density function

$$f(x) = \begin{cases} \dfrac{32x}{15}, & \dfrac{1}{4} < x < 1, \\ 0, & \text{otherwise,} \end{cases}$$

while the time he needs to return home from work in the afternoon is another variable Y, independent of X, having density function

$$f(y) = \begin{cases} \dfrac{12\sqrt{y}}{7}, & \dfrac{1}{4} < y < 1, \\ 0, & \text{otherwise.} \end{cases}$$

Find the mean and the variance of the total time that Bill spends traveling during a day.

16. A computer selects n real numbers at random from the interval $(0, 5)$. Find the mean and variance for the sum of the numbers selected.

17. Suppose (X, Y) have a joint continuous distribution with a joint density function

$$f(x, y) = \begin{cases} 1, & |x| < y < 1, \\ 0, & \text{otherwise.} \end{cases}$$

Show that X and Y are uncorrelated, but they are not independent.

18. Calculate the covariance between the random variables X and Y for each of the following joint density functions:

(i)
$$f(x, y) = \begin{cases} 24e^{-4x-6y}, & x, y > 0, \\ 0, & \text{otherwise;} \end{cases}$$

(ii)
$$f(x, y) = \begin{cases} e^{-y}, & x > 0, y > x, \\ 0, & \text{otherwise;} \end{cases}$$

(iii)
$$f(x, y) = \begin{cases} \dfrac{e^{-y+1}}{e-2}, & 0 < x < y < 1, \\ 0, & \text{otherwise;} \end{cases}$$

(iv)
$$f(x,y) = \begin{cases} \dfrac{1}{2x}, & 0 < y < x < 2, \\ 0, & \text{otherwise}; \end{cases}$$

(v)
$$f(x,y) = \begin{cases} x+y, & 0 < x < 1, 0 < y < 1, \\ 0, & \text{otherwise}. \end{cases}$$

19. Nick rolls a die twice. Let X be the outcome of the first roll and Y be the largest of the two outcomes. Find the covariance between X and Y.

20. Consider the pair (X, Y) with joint density function

$$f(x,y) = \begin{cases} 8xy, & 0 < x < y < 1, \\ 0, & \text{otherwise}. \end{cases}$$

(i) Obtain the expectation and variance of the sum $X + Y$.

(ii) Calculate the expectation and variance for each of X and Y, and verify that $Var(X + Y) \neq Var(X) + Var(Y)$.

(iii) Are X and Y correlated or not? If yes, are they positively or negatively correlated?

21. Suppose the variables X_1 and X_2 take only two values, so that $R_{X_1} = R_{X_2} = \{0, 1\}$. Show that, in this case, X_1 and X_2 are independent if and only if they are uncorrelated.

22. A domestic appliance uses an electronic device whose lifetime is represented by a continuous variable Y. The lifetime of the appliance is another continuous variable X. If the device fails, then the appliance breaks down, but not vice versa. Suppose the joint density function of X and Y, when time is measured in years, is given by

$$f(x,y) = \begin{cases} ce^{-y/7}, & 0 < x < y < \infty, \\ 0, & \text{otherwise}. \end{cases}$$

(i) Show that $c = 1/49$.

(ii) Verify that the expected excess time the device works after the appliance has failed is seven years.

(iii) Demonstrate that X and Y are positively correlated (show in particular that $Cov(X, Y) = 49$). How do you interpret this result intuitively?

(iv) Calculate the variance of the excess lifetime of the device compared with that of the appliance.

23. In a sequence of n independent Bernoulli trials with the same success probability p, we define the variables

$$X_i = \begin{cases} 1, & \text{if the } i\text{th trial results in a success}, \\ 0, & \text{otherwise}, \end{cases}$$

for $i = 1, \ldots, n$.

(i) Show that $E(X_i) = E(X_i^2) = p$ and

$$Var(X_i) = p(1 - p),$$

for any $i = 1, \ldots, n$.

(ii) Use the fact that X_i's are uncorrelated to calculate the variance of

$$X = \sum_{i=1}^{n} X_i,$$

which corresponds to the total number of successes in n trials.

[This is an alternate way of deriving the variance of the binomial distribution.]

5.3 CORRELATION COEFFICIENT

When we have two random variables X and Y with nonzero covariance, then we are certain that X and Y are not independent. As we have seen in Section 5.2, in such a case X and Y are positively or negatively correlated, depending on the sign of their covariance.

For two variables that are either positively or negatively correlated, we would like to have a measure for the degree of their dependence. The covariance, discussed in Section 5.2, cannot serve this purpose as it depends on the units of measurement for the two variables. For example, suppose we wish to study the relationship between the height (in meters) X of the oldest child in a family and the height of their father, Y. Assuming we have all the available information, we calculate the value of their covariance $Cov(X, Y)$. Suppose now someone else, using the same information, wishes to study the same relationship, but uses centimeters, rather than meters, as the unit of measurement for height. Then, the height of the oldest child would be represented by $X_1 = 100X$, while that of the father by $Y_1 = 100Y$. From Part (i) of Proposition 5.2.4(i), we then have

$$Cov(X_1, Y_1) = Cov(100X, 100Y) = 10^4 Cov(X, Y).$$

We thus obtain two totally different measures of the relationship between the two quantities.

For this reason, it is clear that if we want to find a reliable index for the strength of the relationship between X and Y, we must remove the effect that the units of measurement have on this index. In the univariate case, a quantity which arises from a variable X and is unit-free is the standardized variable

$$X^* = \frac{X - \mu_X}{\sigma_X},$$

where $\mu_X = E(X)$, and $\sigma_X^2 = Var(X)$. Assume now that we have another variable Y, with mean μ_Y and variance σ_Y^2, and with the associated standardized variable being

$$Y^* = \frac{Y - \mu_Y}{\sigma_Y}.$$

Then, for the covariance between X^* and Y^*, we find

$$Cov(X^*, Y^*) = Cov\left(\frac{1}{\sigma_X} \cdot X + \left(-\frac{\mu_X}{\sigma_X}\right), \frac{1}{\sigma_Y} \cdot Y + \left(-\frac{\mu_Y}{\sigma_Y}\right)\right)$$

$$= \frac{1}{\sigma_X} \cdot \frac{1}{\sigma_Y} \cdot Cov(X, Y) \text{ (using Part (i) of Proposition 5.2.4)}$$

$$= \frac{Cov(X, Y)}{\sqrt{Var(X)}\sqrt{Var(Y)}},$$

which leads to the following definition.

Definition 5.3.1 Let X and Y be two random variables with $Cov(X, Y) = \sigma_{XY}$, $Var(X) = \sigma_X^2 > 0$ and $Var(Y) = \sigma_Y^2 > 0$. Then, the quantity

$$\rho = \rho_{X,Y} = \frac{Cov(X, Y)}{\sqrt{Var(X)}\sqrt{Var(Y)}} = \frac{\sigma_{XY}}{\sigma_X \sigma_Y}$$

is called the **correlation coefficient** between X and Y.

The quantity $\rho_{X,Y}$ contains all the information that is contained in $Cov(X, Y)$ about the way X and Y vary simultaneously; it has also the same sign:

$$\rho_{X,Y} > 0 \Leftrightarrow Cov(X, Y) > 0, \qquad \rho_{X,Y} < 0 \Leftrightarrow Cov(X, Y) < 0$$

and

$$\rho_{X,Y} = 0 \Leftrightarrow Cov(X, Y) = 0, \tag{5.2}$$

so that $\rho_{X,Y} = 0$ means that X and Y are uncorrelated. However, in contrast with the covariance, the correlation coefficient is invariant under linear transformations of the variables, as stated in Proposition 5.3.1.

Example 5.3.1 We mentioned in Section 5.2 that if X and Y are independent, then $Cov(X, Y) = 0$, but the converse is not true. In view of (5.2), the same is also true for the correlation coefficient, namely, that independent variables have a zero correlation coefficient, but not (always) vice versa. Suppose X takes the values, $\pm 1, \pm 2$, each with probability $1/4$, and $Y = X^2$. Then, the joint probability function of (X, Y) is

$$f(1, 1) = f(-1, 1) = f(2, 4) = f(-2, 4) = \frac{1}{4},$$

while $f(x, y) = 0$ for any other pair (x, y). It is easy to check that X has mean zero. For the expectation of the product XY, we find

$$E(XY) = \sum_{(x,y) \in R_{X,Y}} xyf(x, y) = (1 \cdot 1 + (-1) \cdot 1 + 2 \cdot 4 + (-2) \cdot 4) \cdot \frac{1}{4} = 0.$$

Thus, we have

$$Cov(X, Y) = E(XY) - E(X)E(Y) = 0 - 0 = 0,$$

i.e. X and Y are uncorrelated, and, in view of (5.2), their correlation coefficient is also zero. Keep in mind that the value of Y is completely determined by that of X (there is a deterministic relationship between the two variables). This example illustrates the fact that it is *wrong* to interpret the value of the correlation coefficient as a measure of how closely dependent, or related, the two variables are. In fact, we may only view $\rho_{X,Y}$ as a measure of the strength of the **linear relationship** between X and Y; for $Y = X^2$ as in the present example, such a relationship is absent.

Proposition 5.3.1 *Let $a, b, c, d \in \mathbb{R}$ such that $ac > 0$. Then, the correlation coefficient of the variables $aX + b$ and $cY + d$ is the same as the correlation coefficient between X and Y, that is,*

$$\rho_{aX+b,cY+d} = \rho_{X,Y}.$$

Proof: Using Definition 5.3.1 and the identities

$$Cov(aX + b, cY + d) = ac\ Cov(X, Y)$$

and

$$Var(aX + b) = a^2 Var(X), \quad Var(cY + d) = c^2 Var(Y),$$

we find

$$\rho_{aX+b,cY+d} = \frac{Cov(aX + b, cY + d)}{\sqrt{Var(aX + b)} \cdot \sqrt{Var(cY + d)}} = \frac{ac\ Cov(X, Y)}{\sqrt{a^2 Var(X)} \cdot \sqrt{c^2 Var(Y)}}$$

$$= \frac{ac}{|ac|} \cdot \frac{Cov(X, Y)}{\sqrt{Var(X)} \cdot \sqrt{Var(Y)}} = \rho_{X,Y},$$

since it is assumed that a and c have the same sign. $\qquad\qquad\qquad\qquad\qquad\square$

Another key difference between the correlation coefficient and the covariance between two random variables is that, while the latter can take any real value, the correlation coefficient can only take values between -1 and 1, and the endpoints ± 1 occur when one of the variables is a linear transformation of the other. These properties are stated in the following proposition.

Proposition 5.3.2 *Let X and Y be two random variables with correlation coefficient $\rho_{X,Y}$. Then, we have the following:*

(i) *$-1 \le \rho_{X,Y} \le 1$.*

(ii) *If $\rho_{X,Y} = 1$, then there exist $a, b \in \mathbb{R}$, with $a > 0$, such that*

$$Y = aX + b,$$

and vice versa.

(iii) *If $\rho_{X,Y} = -1$, then there exist $a, b \in \mathbb{R}$, with $a < 0$, such that*

$$Y = aX + b,$$

and vice versa.

Proof: Let us set $\sigma_X^2 = Var(X)$ and $\sigma_Y^2 = Var(Y)$. Then, from Proposition 5.2.1 we have

$$Var\left(\frac{X}{\sigma_X} \pm \frac{Y}{\sigma_Y}\right) = Var\left(\frac{X}{\sigma_X}\right) + Var\left(\frac{Y}{\sigma_Y}\right) \pm 2Cov\left(\frac{X}{\sigma_X}, \frac{Y}{\sigma_Y}\right)$$

$$= \frac{1}{\sigma_X^2}Var(X) + \frac{1}{\sigma_Y^2}Var(Y) \pm 2 \cdot \frac{1}{\sigma_X\sigma_Y}Cov(X, Y)$$

$$= 2 \pm 2\rho_{X,Y}.$$

Thus, we have shown that

$$Var\left(\frac{X}{\sigma_X} + \frac{Y}{\sigma_Y}\right) = 2(1 + \rho_{X,Y}), \qquad Var\left(\frac{X}{\sigma_X} - \frac{Y}{\sigma_Y}\right) = 2(1 - \rho_{X,Y}). \qquad (5.3)$$

(i) As

$$Var\left(\frac{X}{\sigma_X} + \frac{Y}{\sigma_Y}\right) \geq 0, \qquad Var\left(\frac{X}{\sigma_X} - \frac{Y}{\sigma_Y}\right) \geq 0,$$

we see from (5.3) that

$$1 + \rho_{X,Y} \geq 0, \qquad 1 - \rho_{X,Y} \geq 0,$$

yielding $-1 \leq \rho_{X,Y} \leq 1$, as required.

(ii) Assume that $\rho_{X,Y} = 1$. Then, from (5.3) again, we get

$$Var\left(\frac{X}{\sigma_X} - \frac{Y}{\sigma_Y}\right) = 0.$$

But, if a random variable has variance zero, then the variable takes only one value (with probability one), i.e. there is a real constant c such that

$$\frac{X}{\sigma_X} - \frac{Y}{\sigma_Y} = c,$$

and a simple rearrangement of the equality yields

$$Y = \frac{\sigma_Y}{\sigma_X}X - c\sigma_Y = aX + b,$$

where $a = \sigma_Y/\sigma_X > 0$ and $b = -c\,\sigma_Y \in \mathbb{R}$.

Conversely, assume that $Y = aX + b$, for some $a > 0, b \in \mathbb{R}$. Then, by using Proposition 5.3.1, we get

$$\rho_{X,Y} = \rho_{X,aX+b} = \rho_{X,X} = \frac{Cov(X, X)}{\sqrt{Var(X)} \cdot \sqrt{Var(X)}} = 1.$$

(iii) This is proved in a manner similar to Part (ii). □

Examples 5.3.2 and 5.3.3 illustrate how we calculate the correlation coefficient in the discrete and continuous cases respectively.

Example 5.3.2 Let (X, Y) be a discrete pair of random variables with joint probability function

$$f(x, y) = \begin{cases} \dfrac{x^2 + y^2}{43}, & x = 1, 2, 3 \text{ and } y = 1, 2, \\ 0, & \text{otherwise.} \end{cases}$$

In order to find the correlation coefficient between X and Y, we need to find first the covariance between X and Y and the variances of X and Y. All these quantities are easily derived from the marginal probability functions of X and Y which are obtained from $f(x, y)$ above. In particular, we have

$$f_X(x) = \sum_{y=1}^{2} f(x, y) = \frac{x^2 + 1^2}{43} + \frac{x^2 + 2^2}{43} = \frac{2x^2 + 5}{43}, \qquad x = 1, 2, 3,$$

from which we find

$$f_X(1) = \frac{7}{43}, \quad f_X(2) = \frac{13}{43}, \quad f_X(3) = \frac{23}{43}.$$

Similarly, the marginal probability function of Y is obtained to be

$$f_Y(y) = \sum_{x=1}^{3} f(x, y) = \frac{1^2 + y^2}{43} + \frac{2^2 + y^2}{43} + \frac{3^2 + y^2}{43} = \frac{14 + 3y^2}{43}, \quad y = 1, 2,$$

so that

$$f_Y(1) = \frac{17}{43}, \qquad f_Y(2) = \frac{26}{43}.$$

Thus, we get

$$E(X) = \sum_{x=1}^{3} x f_X(x) = 1 \cdot \frac{7}{43} + 2 \cdot \frac{13}{43} + 3 \cdot \frac{23}{43} = \frac{102}{43}$$

and

$$E(Y) = \sum_{y=1}^{2} y f_Y(y) = 1 \cdot \frac{17}{43} + 2 \cdot \frac{26}{43} = \frac{69}{43}.$$

Similarly, we have

$$E(X^2) = \sum_{x=1}^{3} x^2 f_X(x) = 1^2 \cdot \frac{7}{43} + 2^2 \cdot \frac{13}{43} + 3^2 \cdot \frac{23}{43} = \frac{266}{43}$$

and

$$E(Y^2) = \sum_{y=1}^{2} y^2 f_Y(y) = 1^2 \cdot \frac{17}{43} + 2^2 \cdot \frac{26}{43} = \frac{121}{43},$$

from which we get

$$Var(X) = E(X^2) - [E(X)]^2 = \frac{266}{43} - \left(\frac{102}{43}\right)^2 = \frac{1034}{1849} \cong 0.5592,$$

and

$$Var(Y) = E(Y^2) - [E(Y)]^2 = \frac{121}{43} - \left(\frac{69}{43}\right)^2 = \frac{442}{1849} \cong 0.2390.$$

Finally, we need the covariance of X and Y, for which we first find

$$E(XY) = \sum_{x=1}^{3}\sum_{y=1}^{2} xy\frac{x^2+y^2}{43} = 1\cdot 1\cdot\frac{1^2+1^2}{43} + 1\cdot 2\cdot\frac{1^2+2^2}{43} + \cdots + 3\cdot 3\cdot\frac{3^2+2^2}{43}$$

$$= \frac{162}{43},$$

so that

$$Cov(X,Y) = E(XY) - E(X)E(Y) = \frac{162}{43} - \frac{102}{43}\cdot\frac{69}{43} = -\frac{72}{1849} \cong -0.0390.$$

Upon substituting these values in the formula for the correlation coefficient, we find

$$\rho_{X,Y} = \frac{Cov(X,Y)}{\sqrt{Var(X)}\sqrt{Var(Y)}} \cong \frac{-0.0390}{\sqrt{0.5592}\sqrt{0.2390}} = -0.1065.$$

The value of the correlation coefficient here implies that X and Y have a (rather weak) negative correlation (10.65%).

Example 5.3.3 Let X and Y be two continuous variables with joint density function

$$f(x,y) = \begin{cases} a^2 e^{-ay}, & 0 < x < y < \infty, \\ 0, & \text{otherwise}, \end{cases}$$

where $a > 0$ is a given constant. For $r = 1, 2$, we find

$$E(X^r) = \int_0^\infty \int_0^y x^r a^2 e^{-ay}\, dx\, dy = \int_0^\infty \frac{a^2 y^{r+1}}{r+1} e^{-ay}\, dy = \frac{\Gamma(r+1)}{a^r} = \frac{r!}{a^r}$$

and

$$E(Y^r) = \int_0^\infty \int_0^y y^r a^2 e^{-ay}\, dx\, dy = \int_0^\infty a^2 y^{r+1} e^{-ay}\, dy = \frac{\Gamma(r+2)}{a^r} = \frac{(r+1)!}{a^r}.$$

Moreover, we find

$$E(XY) = \int_0^\infty \int_0^y xy a^2 e^{-ay}\, dx\, dy = \int_0^\infty \frac{a^2}{2} y^3 e^{-ay}\, dy = \frac{3}{a^2}.$$

Consequently,

$$E(X) = \frac{1}{a}, \qquad E(X^2) = \frac{2}{a^2}, \qquad Var(X) = \frac{2}{a^2} - \frac{1}{a^2} = \frac{1}{a^2},$$

$$E(Y) = \frac{2}{a}, \qquad E(Y^2) = \frac{6}{a^2}, \qquad Var(Y) = \frac{6}{a^2} - \frac{4}{a^2} = \frac{2}{a^2},$$

and

$$Cov(X, Y) = E(XY) - E(X)E(Y) = \frac{3}{a^2} - \left(\frac{1}{a}\right) \cdot \left(\frac{2}{a}\right) = \frac{1}{a^2}.$$

The correlation coefficient between X and Y is thus found to be

$$\rho_{X,Y} = \frac{Cov(X, Y)}{\sigma_X \sigma_Y} = \frac{\frac{1}{a^2}}{\sqrt{\frac{1}{a^2} \cdot \frac{2}{a^2}}} = \frac{1}{\sqrt{2}} \cong 0.707.$$

We observe that, while X and Y have a strong positive correlation, their covariance ranges, depending on the value of a, from very small to very large values; for instance, when $a = 1000$, we have $Cov(X, Y) = 10^{-6}$, while for $a = 0.001$, we have $Cov(X, Y) = 10^6$. The correlation coefficient, however, does not depend on a.

Exercises
Group A

1. Calculate the correlation coefficient for each pair of discrete random variables in Exercise 7 of Section 5.2.

2. Let X, Y, and Z be three random variables that are pairwise independent and have variances $3, 5$, and 10, respectively. Find the correlation coefficient between $X + Y$ and $Y + Z$.

3. Let X, Y, and Z be three random variables that are (completely) independent and follow exponential distribution with parameter λ. Show that the correlation coefficient between $U = X + Y$ and $V = Y + Z$ is $1/2$ irrespective of the value of λ. Does the covariance between U and V depend on λ?

4. Let X be a random variable having standard normal distribution, that is, $X \sim N(0, 1)$. Show that the variables X and X^2 are uncorrelated, so that their correlation coefficient is zero (although clearly, they are not independent). More generally, explain why for any positive integer m, the variables X and X^{2m} are uncorrelated.

5. Assume that X is a Bernoulli random variable with success probability $p, 0 < p < 1$, and $Y = 2X^2 + 1$. Calculate $\rho_{X,Y}$.

6. Suppose X and Y are uncorrelated variables. Then, show that the correlation coefficient between $U = X + Y$ and $V = X - Y$ is given by

$$\rho_{U,V} = \frac{Var(X) - Var(Y)}{Var(X) + Var(Y)}.$$

7. For the random variables X and Y, suppose it is known that

$$\sigma_X = 3, \quad \sigma_Y = 5, \quad \rho_{X,Y} = \frac{1}{2}.$$

Find the variance of $W = 5X - 2Y + 12$.

8. A medical center has two admission points. Let X and Y be the number of people waiting for service at the first and second admission points respectively, at 10.30 p.m. of a given day. Using data from the last six months, the following table has been determined for the joint probability function of X and Y:

x \ y	0	1	2
0	0.10	0.05	0.03
1	0.18	0.10	0.06
2	0.16	0.23	0.09

Find the covariance and the correlation coefficient between X and Y.

9. An urn contains two red, three white, and four black balls. We select two balls at random without replacement. Let X be the number of red balls and Y be the number of white balls among those selected.

   (i) Write the joint probability function of X and Y in a tabular form.

   (ii) Obtain the marginal probability functions of X and Y, and thence find their expectations.

   (iii) Calculate $Cov(X, Y)$ and $\rho_{X,Y}$.

10. Suppose X has a binomial distribution with parameters $n = 2$ and $p = 1/2$. Let

$$Y = 3X + 1, \quad Z = X^2.$$

Calculate the correlation coefficients $\rho_{X,Y}$, $\rho_{X,Z}$ and $\rho_{Y,Z}$.

11. Let X_1, X_2, and X_3 be a random sample from the exponential distribution with parameter λ. Define

$$U = X_1, \quad V = X_1 + X_2, \quad W = X_1 + X_2 + X_3.$$

Calculate the covariance and the correlation coefficient between

   (i) U and V,

   (ii) U and W,

   (iii) V and W.

(Hint: Use the fact that the sum of independent exponential random variables has a Gamma distribution.)

12. What can we conclude about the random variables X and Y if it is known that

$$Var(X) = Var(Y) = Cov(X, Y) \neq 0?$$

(Hint: Calculate the correlation coefficient between X and Y.)

13. Let X be a random variable having a t distribution with n degrees of freedom, for some $n > 2$, while Y is another variable, having the χ^2 distribution with n degrees of freedom. If it is given that

$$E(XY) = 5, \quad \text{and} \quad \rho_{X,Y} = \frac{10}{17},$$

what is the value of n?

Group B

14. For any two random variables X and Y, show the following inequality, known as the **variance–covariance inequality**:

$$[Cov(X, Y)]^2 \leq Var(X)Var(Y). \tag{5.4}$$

The proof of this is immediate if one uses that $1 \leq \rho_{X,Y} \leq 1$. Another way is as follows:

(i) For the special case when $E(X) = E(Y) = 0$, derive the Cauchy–Schwarz inequality

$$[E(XY)]^2 \leq E(X^2)E(Y^2).$$

(ii) Suppose instead $E(X) = \mu_X$ and $E(Y) = \mu_Y$. Apply the result in Part (i) to the random variables $X - \mu_X$ and $Y - \mu_Y$ to prove (5.4).

(iii) Using the above, show that there do not exist two random variables X and Y such that

$$\sigma_X = 4, \quad \sigma_Y = 5, \quad Cov(X, Y) = 21.$$

15. Calculate the correlation coefficient between the continuous random variables described in Exercise 18 of Section 5.2.

16. A pharmaceutical company produces a multi-vitamin product in boxes of 3 g. The product contains vitamins A, B, and C. Let X be the weight (in grams) of vitamin A in a box and Y be the weight of vitamin B (so that the weight of vitamin C is $3 - X - Y$). Assume that the joint density function of X and Y is given by

$$f(x, y) = \begin{cases} cxy, & 0 \leq x \leq 3, 0 \leq y \leq 3, 0 \leq x + y \leq 3, \\ 0, & \text{otherwise}, \end{cases}$$

for a suitable constant c.

(i) What is the value of c?

(ii) Find the marginal density function, the mean and variance of X and Y.

(iii) Let Z denote the weight of vitamin C in a box. For each of the pairs $(X, Y), (X, Z), (Y, Z)$, obtain the covariance and correlation coefficient.

17. In a certain population, 10% of the families have no children, 20% have one child, 40% have two children and 30% have three children. We assume that the two sexes are equally likely, and let X and Y denote the number of boys and girls in a family chosen at random from the above population.

    (i) Give the joint probability function of (X, Y) in the form of a two-way table.

    (ii) Verify that X and Y are not independent.

    (iii) Calculate the covariance between X and Y.

    (iv) Find $\rho_{X,Y}$.

    (Hint: Let Z be the number of children in the family. Observe that

    $$f(x, y) = P(X = x, Y = y) = P(X = x, Z = x + y)$$
    $$= P(X = x | Z = x + y)P(Z = x + y),$$

    so that

    $$f(0, 0) = P(X = 0 | Z = 0)P(Z = 0) = 1 \cdot (0.10) = 0.10$$

    $$f(1, 2) = P(X = 1 | Z = 3)P(Z = 3) = \frac{3}{8} \cdot (0.30) = 0.1125$$

    and so on.)

5.4 CONDITIONAL EXPECTATION AND VARIANCE

Let (X, Y) be a pair of random variables with a joint continuous distribution, joint density function $f(x, y)$, and marginal density functions $f_X(x)$ and $f_Y(y)$, respectively. For a given y, such that $f_Y(y) \neq 0$, the conditional density function of X, given $Y = y$, is known to be

$$f_{X|Y}(x|y) = \frac{f(x, y)}{f_Y(y)}, \qquad x \in \mathbb{R},$$

while the associated conditional expectation is given by

$$E(X|Y = y) = \int_{-\infty}^{\infty} x f_{X|Y}(x|y) \, dx = \int_{-\infty}^{\infty} x \frac{f(x, y)}{f_Y(y)} \, dx = \frac{1}{f_Y(y)} \int_{-\infty}^{\infty} x f(x, y) \, dx. \qquad (5.5)$$

In general, if $g(X)$ is a function of X then the conditional expectation of $Z = g(X)$, given $Y = y$, is provided by

$$E\left(g(X)|Y = y\right) = \frac{1}{f_Y(y)} \int_{-\infty}^{\infty} g(x)f(x, y) \, dx.$$

Note that the quantity $E(X|Y = y)$, or more generally $E\left(g(X)|Y = y\right)$, involves y only, and so it is a function of a single variable. For convenience, we introduce the notation

$$m(y) = E(X|Y = y) = \frac{1}{f_Y(y)} \int_{-\infty}^{\infty} x f(x, y) \, dx. \qquad (5.6)$$

The random variable $m(Y)$, which arises by replacing y in (5.6) by the random variable Y, is called the **conditional expectation of X, given Y,** and is denoted by $E(X|Y)$. Similarly, for any $x \in R_X$ such that $f_X(x) \neq 0$, we can define the conditional expectation of Y, given $X = x$, as

$$s(x) = E(Y|X = x) = \frac{1}{f_X(x)} \int_{-\infty}^{\infty} yf(x, y) \, dy,$$

and when x is replaced by the random variable X, we obtain the **conditional expectation of Y, given X,** denoted by $E(Y|X)$. The above definitions extend easily to the case when the random variables X and Y are discrete. Specifically, if their joint probability function is $f(x, y)$, then

$$m(y) = E(X|Y = y) = \frac{1}{f_Y(y)} \sum_{x \in R_X} xf(x, y)$$

and

$$s(x) = E(Y|X = x) = \frac{1}{f_X(x)} \sum_{y \in R_Y} yf(x, y).$$

Now, since the quantity $E(X|Y)$ represents a random variable (which, in fact, is a function of the variable Y), we can consider its expectation which is given by

$$E[m(Y)] = E[E(X|Y)].$$

Sometimes, the notation that is used for the quantity on the right-hand side is $E_Y[E_X(X|Y)]$ in order to make it clear that the expectation inside the square brackets is taken with respect to X, while the one outside is expectation with respect to Y.

Although this repeated use of the expectation operator might appear complicated, we have the following useful result which asserts that the stated quantity is just the (unconditional) expectation of X.

Proposition 5.4.1 *Let X and Y be any two random variables. Then, we have*

$$E[E(X|Y)] = E(X), \qquad E[E(Y|X)] = E(Y),$$

provided the above expectations exist.

Proof: We present the proof for the continuous case. Using the notation introduced above and substituting $m(y)$ from (5.6), we have

$$E[E(X|Y)] = E[m(Y)] = \int_{-\infty}^{\infty} m(y)f_Y(y) \, dy$$

$$= \int_{-\infty}^{\infty} \left(\frac{1}{f_Y(y)} \int_{-\infty}^{\infty} xf(x, y) \, dx \right) f_Y(y) \, dy$$

$$= \int_{-\infty}^{\infty} \int_{-\infty}^{\infty} xf(x, y) \, dx \, dy = E(X).$$

The second equation can be proved similarly. Further, the proof for the case when X and Y are discrete variables is obtained by a similar argument, simply by replacing the integrals by sums.   □

We note that Proposition 5.4.1 is valid for *any* pair of variables X and Y, that is, even in the case when one of the variables is discrete and the other is continuous, a case not treated in this book.

The equation $E[E(X|Y)] = E(X)$ can be interpreted as follows. The expected value of X can be found as a weighted average of the conditional expectations of X, given $Y = y$, over all possible values of the variable Y. Moreover, in the case when X and Y are independent, then $f_{X|Y}(x|y) = f_X(x)$, so that

$$m(y) = E(X|Y = y) = \int_{-\infty}^{\infty} x f_{X|Y}(x|y) \, dx = \int_{-\infty}^{\infty} x f_X(x) \, dx = E(X),$$

which does not depend on y. The result holds in the discrete case as well. Similarly, we have $s(x) = E(Y)$, since independence is a commutative relation.

Example 5.4.1 Suppose X and Y have a joint density function

$$f(x, y) = \begin{cases} \dfrac{2(x + 2y)}{3}, & 0 < x < 1, \ 0 < y < 1, \\ 0, & \text{otherwise.} \end{cases}$$

For the marginal density functions of X and Y, we have

$$f_X(x) = \int_{-\infty}^{\infty} f(x, y) \, dy = \int_0^1 \frac{2(x + 2y)}{3} \, dy = \frac{2(x + 1)}{3}, \qquad 0 < x < 1,$$

and

$$f_Y(y) = \int_{-\infty}^{\infty} f(x, y) \, dx = \int_0^1 \frac{2(x + 2y)}{3} \, dx = \frac{4y + 1}{3}, \qquad 0 < y < 1.$$

Then, the corresponding conditional densities are

$$f_{X|Y}(x|y) = \frac{f(x, y)}{f_Y(y)} = \frac{2x + 4y}{1 + 4y}, \qquad 0 < x < 1,$$

and

$$f_{Y|X}(y|x) = \frac{f(x, y)}{f_X(x)} = \frac{x + 2y}{x + 1}, \qquad 0 < y < 1,$$

from which we find the conditional expectations as follows. For a given $y \in (0, 1)$, we have

$$m(y) = E[X|Y = y] = \int_{-\infty}^{\infty} x f_{X|Y}(x|y) \, dx = \int_0^1 \frac{2x(x + 2y)}{4y + 1} \, dx = \frac{2(3y + 1)}{3(4y + 1)},$$

while for a given $x \in (0, 1)$, we have

$$s(x) = E[Y|X = x] = \int_{-\infty}^{\infty} y f_{Y|X}(y|x) \, dy = \int_0^1 \frac{y(x + 2y)}{x + 1} \, dy = \frac{3x + 4}{6(x + 1)}.$$

Replacing y and x in the above two equations by Y and X, respectively, we find the conditional expectations (random variables) of X and Y to be

$$E(X|Y) = \frac{2(3Y + 1)}{3(4Y + 1)}, \qquad E(Y|X) = \frac{3X + 4}{6(X + 1)}.$$

Suppose now we want to calculate the expectation of $m(Y) = E(X|Y)$. Then, we could either use the formula

$$E[m(Y)] = \int_{-\infty}^{\infty} m(y) f_Y(y) \, dy = \int_0^1 \frac{2(3y + 1)}{3(4y + 1)} \cdot \frac{4y + 1}{3} \, dy = \frac{5}{9},$$

or employ Proposition 5.4.1 stating that this is simply $E(X)$; that is,

$$E[m(Y)] = E[E(X|Y)] = E(X) = \int_{-\infty}^{\infty} x f_X(x) \, dx = \int_0^1 \frac{2x(x + 1)}{3} \, dx = \frac{5}{9}$$

and the two results obviously agree.

In a similar fashion, for the expectation of $s(X) = E(Y|X)$, we find

$$E[s(X)] = \int_{-\infty}^{\infty} s(x) f_X(x) \, dx = \int_0^1 \frac{3x + 4}{6(x + 1)} \cdot \frac{2(x + 1)}{3} \, dy = \frac{11}{18},$$

or, alternatively,

$$E[s(X)] = E[E(Y|X)] = E(Y) = \int_{-\infty}^{\infty} y f_Y(y) \, dy = \int_0^1 \frac{y(4y + 1)}{3} \, dy = \frac{11}{18}.$$

Example 5.4.2 Let $N(t)$ be the number of customers entering a shop in the time interval $[0, t]$ and $M(t)$ be the number of those who buy a certain product in the same time interval. We further assume that $N(t)$ is a linear function of t, namely $N(t) = \lambda t$, for some $\lambda > 0$, and that each customer buys the product with probability p $(0 < p < 1)$. What is the expected number of customers who buy the product in the interval $[0, t]$?

SOLUTION In cases like this, it is easier to work out the conditional expectation of $M(t)$, for a given value of $N(t)$, and then employ Proposition 5.4.1 instead of finding the unconditional expectation directly. Thus, if we assume $N(t) = n$, then the distribution of $M(t)$ is binomial with parameters n and p. Consequently,

$$E[M(t)|N(t) = n] = np.$$

As this holds for any value of $N(t)$, replacing now n by $N(t)$ above, we obtain

$$E[M(t)|N(t)] = N(t)p,$$

and now Proposition 5.4.1 yields

$$E[M(t)] = E[E[M(t)|N(t)]] = E[N(t)p] = pE[N(t)] = p\lambda t,$$

which is the desired result.

Example 5.4.3 Suppose we want to calculate the expected number of trials needed until two consecutive successes appear in a sequence of Bernoulli trials. Let X denote the number of trials until two successes appear consecutively and Y stand for the number of trials until the first failure. We then have

$$E(X) = E[E(X|Y)] = \sum_{i=1}^{\infty} E(X|Y = i)P(Y = i)$$

$$= \sum_{i=1}^{2} E(X|Y = i)P(Y = i) + \sum_{i=3}^{\infty} E(X|Y = i)P(Y = i).$$

Now, if p and q denote the success and failure probabilities, respectively, in a trial, since

$$E(X|Y = i) = \begin{cases} i + E(X), & \text{for } i = 1, 2, \\ 2, & \text{for } i \geq 3, \end{cases}$$

and

$$P(Y = i) = qp^{i-1}, \qquad i = 1, 2, \ldots,$$

we readily obtain

$$E(X) = (1 + E(X))q + (2 + E(X))qp + 2qp^2(1 + p + p^2 + \cdots).$$

This leads to

$$E(X) = \left(q + 2qp + 2qp^2 \frac{1}{1 - p} \right) + qE(X) + qpE(X).$$

Upon solving this, we find

$$E(X) = \frac{q + 2pq + 2p^2}{1 - q - pq} = \frac{1 + p}{p^2}.$$

As demonstrated by the above examples, the main use of Proposition 5.4.1 is in cases when, for two variables X and Y, it is easier to find a conditional expectation, say $E(X|Y)$, rather than calculate $E(X)$ directly. A similar approach can be taken when we want the probability of a certain event, A, associated with a variable X. There are many cases in which it is difficult to find $P(A)$ directly, but it may be easier to calculate the probability of A, given that X has taken a specific value, $X = x$. For a case like this, let us introduce the auxiliary variable

$$Y = \begin{cases} 1, & \text{if } A \text{ occurs,} \\ 0, & \text{otherwise.} \end{cases}$$

We then have

$$E(Y) = 1 \cdot P(Y = 1) + 0 \cdot P(Y = 0) = P(Y = 1) = P(A).$$

However, putting $s(X) = E(Y|X)$, Proposition 5.4.1 yields

$$E(Y) = E[E(Y|X)] = E[s(X)] = \int_{-\infty}^{\infty} s(x) f_X(x) \, dx = \int_{-\infty}^{\infty} E(Y|X = x) f_X(x) \, dx.$$

Moreover, since

$$E(Y|X = x) = 1 \cdot P(Y = 1|X = x) + 0 \cdot P(Y = 1|X = x) = P(Y = 1|X = x)$$

$$= P(A|X = x),$$

we finally arrive at the expression

$$P(A) = \int_{-\infty}^{\infty} P(A|X = x) f_X(x) \, dx.$$

A similar argument works for the discrete case as well, and we thus arrive at the following result.

Proposition 5.4.2 *Let A be an event and X be a random variable associated with a chance experiment.*

(i) *If X is discrete with probability function $f_X(x)$, then*

$$P(A) = \sum_{x \in R_X} P(A|X = x) f_X(x).$$

(ii) *If X is continuous with density function $f_X(x)$, then*

$$P(A) = \int_{-\infty}^{\infty} P(A|X = x) f_X(x) \, dx.$$

Note that the formula in Part (i) is in fact the law of total probability applied to the partition $\{\{x\}, x \in R_X\}$.

Example 5.4.4 John and Cindy work as cashiers at a shop. The time they need to serve a customer has exponential distribution with parameters λ_1 and λ_2, respectively. If two customers start to be served at the same time, one at each cashier, what is the probability that the customer served by Cindy leaves the shop first?

SOLUTION Let X and Y be the times that John and Cindy, respectively, take to serve their customers. The density functions of X and Y are given by

$$f_X(x) = \lambda_1 e^{-\lambda_1 x}, \qquad x > 0,$$

and

$$f_Y(y) = \lambda_2 e^{-\lambda_2 y}, \qquad y > 0$$

and X and Y are assumed independent. We require the probability $P(X > Y)$. Due to Proposition 5.4.2, this can be expressed as

$$P(X > Y) = \int_{-\infty}^{\infty} P(X > Y | Y = y) f_Y(y) \, dy.$$

But, by using the independence of X and Y, we have

$$P(X > Y | Y = y) = P(X > y | Y = y) = P(X > y) = 1 - F_X(y) = e^{-\lambda_1 y},$$

using which we immediately obtain

$$P(X > Y) = \int_0^{\infty} e^{-\lambda_1 y} \lambda_2 e^{-\lambda_2 y} \, dy = \lambda_2 \int_0^{\infty} e^{-(\lambda_1 + \lambda_2)y} \, dy = \frac{\lambda_2}{\lambda_1 + \lambda_2}.$$

Example 5.4.5 Let X and Y be two independent, continuous random variables with density functions f_X and f_Y, and distribution functions F_X and F_Y, respectively. Find the distribution function of $X + Y$.

SOLUTION   Let x be a real number. Applying the result in Part (ii) of Proposition 5.4.2 to the event $\{X + Y \leq x\}$ and conditioning on the value of Y, we obtain

$$P(X + Y \leq x) = \int_{-\infty}^{\infty} P(X + Y \leq x | Y = y) f_Y(y) \, dy$$

$$= \int_{-\infty}^{\infty} P(X \leq x - y) f_Y(y) \, dy.$$

If F_{X+Y} denotes the distribution function of $X + Y$, the above equation can be expressed as

$$F_{X+Y}(x) = \int_{-\infty}^{\infty} F_X(x - y) f_Y(y) \, dy.$$

This gives the required distribution function of $X + Y$. Differentiating both sides of this equation with respect to x, and interchanging the order of integration and differentiation, we arrive at the density of $X + Y$ (of course, this will be the same with the result we have found in Section 4.3).

Proposition 5.4.1 can be generalized for functions of random variables. Specifically,

$$E[E[g(X)|Y]] = E[g(X)], \qquad E[E[h(Y)|X]] = E[h(Y)]$$

hold for any functions $g(X)$ and $h(Y)$ of the random variables X and Y, respectively. Further, the familiar properties of expectation remain valid for conditional expectations. For example, we have

$$E(aX + b|Y = y) = aE(X|Y = y) + b,$$

$$E[g(X) + h(X)|Y = y] = E[g(X)|Y = y] + E[h(X)|Y = y].$$

Suppose now we consider the random variable

$$g(X) = (X - m(y))^2,$$

where $m(y) = E(X|Y = y)$. The conditional expectation of this, given that $Y = y$, is

$$\delta(y) = E[g(X)|Y = y] = E[(X - E(X|Y = y))^2|Y = y].$$

This is called the conditional variance of X given $Y = y$. We use $Var(X|Y = y)$ to denote this conditional variance. Replacing y by the random variable Y, we write $\delta(Y)$ for the random variable representing the **conditional variance of X, given Y**, and denote it by $Var(X|Y)$ or $\sigma^2_{X|Y}$. For this quantity, we have the following result.

Proposition 5.4.3 *For any random variables X and Y, we have*

$$Var(X) = E[Var(X|Y)] + Var[E(X|Y)].$$

Proof: In analogy with the relation

$$Var(X) = E(X^2) - [E(X)]^2,$$

we have

$$Var(X|Y) = E(X^2|Y) - [E(X|Y)]^2.$$

Upon taking expectations on both sides, we obtain

$$E[Var(X|Y)] = E[E(X^2|Y)] - E\left\{[E(X|Y)]^2\right\} = E(X^2) - E\left\{[E(X|Y)]^2\right\}. \qquad (5.7)$$

Next, for the variance of the random variable $m(Y)$, we can write

$$Var(m(Y)) = E[m(Y)^2] - [E(m(Y)]^2,$$

which implies that

$$Var(E(X|Y)) = E\left\{[E(X|Y)]^2\right\} - \{E[E(X|Y)]\}^2 = E\left\{[E(X|Y)]^2\right\} - [E(X)]^2.$$

Adding this with (5.7) we obtain the required result. □

Example 5.4.6 The number of paper jams in a photocopying machine, when it works for t hours, is a random variable whose mean and variance are both equal to $t/3$. If the photocopier works daily for a random amount of time, which ranges uniformly between 10 and 12 hours, find the expected value and the variance of paper jams during a working day.

SOLUTION Let $N(t)$ be the number of paper jams when the photocopier works for t hours, and T be the random variable representing the number of hours the photocopier works during a day. Then,

$$E[N(t)] = Var[N(t)] = \frac{t}{3}, \qquad E(T) = 11, \qquad Var(T) = \frac{(12-10)^2}{12} = \frac{1}{3},$$

using the mean and variance of the uniform distribution, since $T \sim U[10, 12]$.

We are now interested in the quantities $E[N(T)]$ and $Var[N(T)]$. First, we have $E[N(T)] = E[E[N(T)|T]]$ and since $E[N(T)|T = t] = E[N(t)] = t/3$, we obtain $E[N(T)|T] = T/3$ so that, by taking expectations on both sides,

$$E[E[N(T)|T]] = E\left(\frac{T}{3}\right) = \frac{1}{3}E(T) = \frac{11}{3},$$

which, in view of Proposition 5.4.1, is indeed $E[N(T)]$. For the variance of $N(T)$, we note that

$$Var[N(T)] = E[Var(N(T)|T)] + Var[E(N(T)|T)].$$

If we know that T takes a specific value $T = t$, then

$$Var(N(T)|T = t) = Var(N(t)) = \frac{t}{3}.$$

From this, we finally obtain

$$Var[N(T)] = E[Var(N(T)|T)] + Var[E(N(T)|T)]$$

$$= E\left(\frac{T}{3}\right) + Var\left(\frac{T}{3}\right) = \frac{1}{3}E(T) + \frac{1}{9}Var(T)$$

$$= \frac{11}{3} + \frac{1}{27} = \frac{100}{27}.$$

We conclude this section by considering the expectation and variance for a *random sum*, i.e. a sum of a random number of random variables. To be specific, let $X_1, X_2, \dots$, be a sequence of independent and identically distributed random variables, which can be either continuous or discrete, and N be a discrete variable independent of X_i's. Then, let us consider the sum

$$S_N = \sum_{i=1}^{N} X_i = X_1 + \cdots + X_N \tag{5.8}$$

(note that when $N = 0$, S is defined to be zero).

Given $N = n$, S_N is simply a sum of random variables in the ordinary sense. Thus,

$$E(S_N|N = n) = E\left(\sum_{i=1}^{n} X_i\right) = n\mu$$

and, using the independence of X_i's,

$$Var(S_N|N = n) = Var\left(\sum_{i=1}^{n} X_i\right) = n\sigma^2,$$

where $\mu = E(X_i)$ and $\sigma^2 = Var(X_i)$. We consequently obtain

$$E(S_N|N) = N\mu, \qquad Var(S_N|N) = N\sigma^2,$$

and then using the results of Propositions 5.4.1 and 5.4.3, we find

$$E(S_N) = E[E(S_N|N)] = E(N\mu) = E(N)\mu,$$

$$Var(S_N) = Var[E(S_N|N)] + E[Var(S_N|N)] = Var(N\mu) + E(N\sigma^2)$$

$$= \mu^2 Var(N) + E(N)\sigma^2.$$

These are formally stated in the following proposition.

Proposition 5.4.4 *Let* $X_1, X_2, \ldots$ *be a sequence of independent and identically distributed random variables with mean* μ *and variance* σ^2, *and* N *be a discrete variable independent of* X_i's. *Then, for the mean and variance of the random sum* S_N *in (5.8), we have*

$$E(S_N) = E(N)\mu, \qquad Var(S_N) = E(N)\sigma^2 + \mu^2 Var(N).$$

Exercises
Group A

1. The joint probability density of X and Y is given by

    $$f(x, y) = \begin{cases} 4xy, & 0 < x < 1 \text{ and } 0 < y < 1, \\ 0, & \text{otherwise.} \end{cases}$$

    (i) Verify that X and Y are independent random variables.

    (ii) Calculate $E(X|Y)$, $E(Y|X)$, $Var(X|Y)$ and $Var(Y|X)$.

2. The conditional density function of the random variable Y, given $X = x$, is given by

    $$f_{Y|X}(y|x) = \frac{1}{1 - x}, \qquad x < y < 1.$$

    If it is known that $E(X) = 1/3$, calculate the expected value of Y.

    (Hint: First, find the conditional expectation $E(Y|X)$ and then use Proposition 5.4.1.)

3. If X and Y are two independent exponential variables with the same parameter λ, then we know that $X - Y$ has the Laplace distribution. Provide an alternative proof of this result using Proposition 5.4.2.

4. Let (X, Y) be a random pair with joint density function

$$f(x, y) = \begin{cases} \dfrac{3}{2}(x^2 + y^2), & 0 < x < 1 \text{ and } 0 < y < 1, \\[2mm] 0, & \text{otherwise.} \end{cases}$$

   (i) Find the marginal densities of X and Y.

   (ii) Write down explicit expressions for

$$E(X|Y), \ E(X^2|Y), \ Var(X|Y), \ E(Y|X), \ E(Y^2|X), \ Var(Y|X).$$

   (iii) Calculate $E[Var(X|Y)]$ and $Var[E(X|Y)]$. Check that their sum is equal to $Var(X)$.

5. For two independent continuous random variables X and Y, show that

$$P(X < Y) = \int_{-\infty}^{\infty} F_X(y) f_Y(y) \, dy,$$

   where F_X is the cumulative distribution function of X and f_Y is the density function of Y.

6. For two variables X and Y associated with the same experiment, it is known that $E(X|Y = y) = 8y - 22$ and $E(Y|X = x) = 2x - 1$. Calculate the quantities $E(X)$ and $E(Y)$.

7. Let X and Y be two variables for which it is known that

   • X has the uniform distribution over the interval $(10, 20)$,

   • the conditional distribution of Y, given $X = x$, is exponential with parameter $\lambda = 1/x$.

   Find the value of the unconditional expectation $E(Y)$.

8. Let (X, Y) be a two-dimensional random variable and let $g(X, Y)$ be a function of (X, Y). Show that

$$E[E[g(X, Y)|X]] = E[E[g(X, Y)|Y]] = E[g(X, Y)].$$

9. The amount of a claim arriving at an insurance company for a certain portfolio, in thousands of dollars, is a random variable X with density function

$$f(x) = \frac{375}{(x + 5)^4}, \qquad x > 0.$$

   The number of claims that arrive for this portfolio per month is a Poisson random variable with parameter $\lambda = 75$. Calculate the expected value and the variance of the total amount of claims that the company receives during a year.

   (Hint: Use Proposition 5.4.4.)

10. The number of children born by a female animal follows the geometric distribution and it is known that the expected number of children born by the animal is 4. The weight of each newborn animal, in pounds, is assumed to follow a Uniform distribution over the interval $[3, 5]$. Find the expected value and the variance of the total weight of all newborn babies.

11. The number of cars that arrive at a gas station every hour is a random variable having Poisson distribution, with parameter $\lambda = 15$ cars. If the time, in hours, that the gas station is open during a day is uniformly distributed over the interval $(11.5, 12.5)$, calculate the mean and variance of the number of cars served during a day.

    (Hint: Proceed as in Example 5.4.6.)

Group B

12. Let X denote the number of customers arriving at a bank cashier during a 15-minute time interval, and Y be the number, among those, who withdraw money from their accounts. Suppose the distribution of X is Poisson with parameter $\lambda = 9$ and that the conditional expectation of Y, given $X = x$, is $E(Y|X = x) = x/3$, while the associated variance is

$$Var(Y|X = x) = \frac{x+4}{8}.$$

    (i) Calculate the mean and variance of Y.

    (ii) Find the covariance between X and Y.

    (Hint: For Part (ii), use the fact $E[XY] = E[E(XY|X)] = E[XE(Y|X)].$)

13. A two-dimensional random variable (X, Y) is uniformly distributed over its range, $R_{X,Y}$, which is defined in a set of Cartesian coordinates by the triangle with vertices at the points $(0, 0)$, $(1, 0)$, and $(1, 3)$.

    (i) Obtain the joint density, $f(x, y)$, of (X, Y) and the marginal densities $f_X(x)$ and $f_Y(y)$.

    (ii) Calculate the quantities $E(X), E(Y), Var(X)$, and $Var(Y)$.

    (iii) Find the values of the conditional expectations $E(X|Y)$, $E(Y|X)$ and the conditional variances $Var(X|Y)$ and $Var(Y|X)$.

14. At the quality control section of a factory, the following procedure is used: a sample of n items is selected and checked one after the other. If a defective item is found, it is replaced in the sample by a non-defective item, and any non-defective item found is retained in the sample. Assume that a given chosen sample contains r defective items, and the remaining $n - r$ are non-defective. Let X_i be the number of non-defective items included in the sample after the first i items have been inspected.

    (i) Verify that

$$E(X_i|X_{i-1}) = 1 + \left(1 - \frac{1}{n}\right)X_{i-1}, \qquad i \geq 2.$$

(ii) Obtain a recursive relationship which links the expected values of X_i and X_{i-1}.

(iii) Show that $E(X_i)$ is given by

$$E(X_i) = n - r\left(1 - \frac{1}{n}\right)^i.$$

(iv) Calculate the probability that the ith item inspected is found to be non-defective.

(Hint: For Part (iv), consider the event

A_i: the ith item inspected is non-defective

and, by conditioning on X_{i-1}, show that $P(A_i) = E(X_{i-1})/n).$)

15. The weekly production of a manufacturing company A is described by a random variable X, which has uniform distribution on $[a, b]$. The product demand from the shops to which the manufacturer supplies is another variable Y, uniformly distributed over $[a, c]$, with $c < b$. Each unit of this product sold by the manufacturer yields a profit of α dollars, while when the demand exceeds the production ($Y > X$), the excess demand is bought by another manufacturer, B, so that the profit of A per unit is $\beta(< \alpha)$ dollars.

(i) Show that the weekly profit, P, made by manufacturer A is given by

$$P = \begin{cases} \alpha Y, & \text{if } Y < X, \\ (\alpha - \beta)X + \beta Y, & \text{otherwise.} \end{cases}$$

(ii) Calculate the conditional expectation $E(P|X = x)$.

(iii) Find the average weekly profit, $E(P)$.

(iv) Determine numerical answers to Parts (i)–(iii) when $a = 1000$, $b = 3000$, $c = 2400$, $\alpha = 3.5$, and $\beta = 1$.

5.5 REGRESSION CURVES

As we have seen in Section 5.4, the conditional expectation $E(Y|X = x)$ is a function of x, denoted by $s(x)$. The curve on the plane defined by the equation

$$y = s(x)$$

(that is, the set of points $(x, s(x))$, for $x \in R_X$) is called a **regression curve** of Y on X. Similarly, the curve defined by

$$x = m(y) = E(X|Y = y)$$

is called a regression curve of X on Y; see Figure 5.1.

In the case when X and Y are independent, we know that

$$s(x) = E(Y|X = x) = E(Y) \qquad \text{for any } x,$$
$$m(y) = E(X|Y = y) = E(X) \qquad \text{for any } y,$$

so that both $s(x)$ and $m(y)$ are constants; the regression curves in this case are straight lines parallel to the x- and y-axes, respectively.

$s(x) = E(Y|X = x)$ $\qquad\qquad\qquad\qquad\qquad m(y) = E(X|Y = y)$

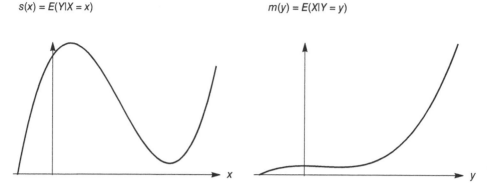

Figure 5.1 Regression curves of Y on X and X on Y.

A situation wherein the problem of finding regression curves arises naturally is the following: suppose we observe two variables X and Y that are not independent, and the value of X is known. If we want to *predict* the value of Y, then we might use a function of the observed value of X, say $g(X)$, and we expect this to be "close" to the future value of Y. Of course, the idea of "closeness" needs to be made precise on the basis of a certain criterion we need to impose. A common choice is the minimization of the quantity

$$E[(Y - g(X))^2],$$

known as **mean squared deviation**. The following result shows that the best choice of g using this minimization criterion is the function $g(x) = s(x) = E(Y|X = x)$.

Proposition 5.5.1 *Let* (X, Y) *be a pair of variables and* $s(x) = E(Y|X = x)$ *be the conditional expectation of* Y, *given* $X = x$. *Then, for any function* g, *we have*

$$E[(Y - g(X))^2] \ge E[(Y - s(X))^2].$$

Proof: We have successively

$$E[(Y - g(X))^2]$$
$$= E\left\{[(Y - s(X)) + (s(X) - g(X))]^2\right\}$$
$$= E[(Y - s(X))^2 + (s(X) - g(X))^2 + 2(Y - s(X))(s(X) - g(X))]$$
$$= E[(Y - s(X))^2] + E[(s(X) - g(X))^2] + 2E[(Y - s(X))(s(X) - g(X))]. \qquad (5.9)$$

However,

$$E[(Y - s(X))(s(X) - g(X))] = E\left[E[(Y - s(X))(s(X) - g(X))|X]\right].$$

Note now that, given X, the quantity $s(X) - g(X)$ is a constant, so that we have

$$E[(Y - s(X))(s(X) - g(X))|X] = (s(X) - g(X))E[Y - s(X)|X]$$
$$= (s(X) - g(X))\left(E(Y|X) - E[s(X)|X]\right)$$
$$= (s(X) - g(X))(s(X) - s(X)) = 0.$$

Consequently, Eq. (5.9), in view of the above equation, yields

$$E[(Y - g(X))^2] = E[(Y - s(X))^2] + E[(s(X) - g(X))^2] \geq E[(Y - s(X))^2],$$

thus establishing the required result.   □

Example 5.5.1 Let (X, Y) be a random pair having the uniform distribution over the half-circle A as shown below. In this case, it is evident that the joint density of the pair (X, Y) is given by

$$f(x, y) = \begin{cases} \dfrac{2}{\pi}, & \text{if } x^2 + y^2 \leq 1, \\[2ex] 0, & \text{otherwise.} \end{cases}$$

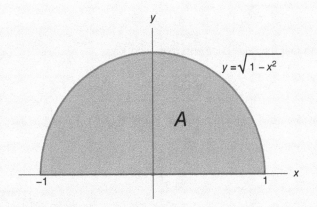

Then, the marginal density of X is given by

$$f_X(x) = \int_0^{\sqrt{1-x^2}} \frac{2}{\pi} \, dy = \frac{2}{\pi} \sqrt{1 - x^2}, \qquad -1 \leq x \leq 1,$$

and the marginal density of Y is

$$f_Y(y) = \int_{-\sqrt{1-y^2}}^{\sqrt{1-y^2}} \frac{2}{\pi} \, dx = \frac{4}{\pi} \sqrt{1 - y^2}, \qquad 0 \leq y \leq 1.$$

The corresponding conditional densities are then given by

$$f_{X|Y}(x|y) = \frac{f(x, y)}{f_Y(y)} = \frac{1}{2\sqrt{1 - y^2}}, \qquad -\sqrt{1 - y^2} \leq x \leq \sqrt{1 - y^2},$$

and

$$f_{Y|X}(y|x) = \frac{f(x, y)}{f_X(x)} = \frac{1}{\sqrt{1 - x^2}}, \qquad 0 \leq y \leq \sqrt{1 - x^2}.$$

The conditional expectations are determined as

$$m(y) = E(X|Y = y) = \int_{-\sqrt{1-y^2}}^{\sqrt{1-y^2}} x \cdot \frac{1}{2\sqrt{1 - y^2}} \, dx$$

$$= \frac{1}{2\sqrt{1 - y^2}} \left[\frac{x^2}{2} \right]_{-\sqrt{1-y^2}}^{\sqrt{1-y^2}} = 0, \qquad 0 \le y \le 1,$$

and

$$s(y) = E(Y|X = x) = \int_{0}^{\sqrt{1-x^2}} y \, \frac{1}{\sqrt{1 - x^2}} \, dy = \frac{1}{\sqrt{1 - x^2}} \left[\frac{y^2}{2} \right]_{0}^{\sqrt{1-x^2}}$$

$$= \frac{1}{2} \sqrt{1 - x^2}, \qquad -1 \le x \le 1.$$

It then follows that the regression curve of X on Y is

$$y = s(x) = \frac{1}{2} \sqrt{1 - x^2}, \qquad -1 \le x \le 1,$$

which is plotted below, while the regression curve of X on Y is the line $x = 0$.

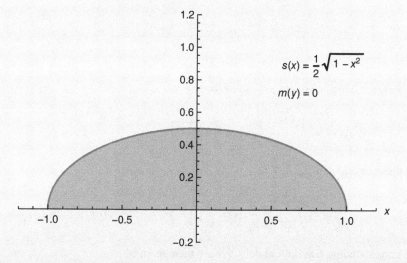

Therefore, if for a given value x of the variable X, we are interested in predicting Y, our best guess (based on the criterion of the minimum mean squared deviation) will be the value

$$y = \frac{1}{2} \sqrt{1 - x^2}.$$

Example 5.5.2 (*Linear regression curves*)

Suppose the pair (X, Y) has the uniform distribution over the triangle shown below, so that the joint density of (X, Y) is

$$f(x, y) = \begin{cases} \dfrac{1}{4}, & \text{if } 0 < y < 2x < 4, \\ \\ 0, & \text{otherwise.} \end{cases}$$

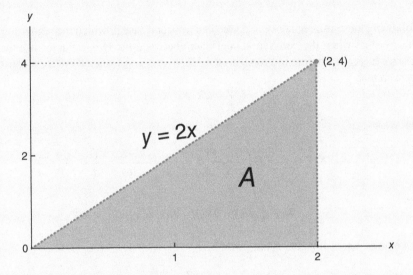

The marginal densities of X and Y are

$$f_X(x) = \int_{-\infty}^{\infty} f(x, y)\, dy = \int_0^{2x} \frac{1}{4}\, dy = \frac{x}{2}, \qquad 0 < x < 2,$$

and

$$f_Y(y) = \int_{-\infty}^{\infty} f(x, y)\, dx = \int_{y/2}^{2} \frac{1}{4}\, dx = \frac{4 - y}{8}, \qquad 0 < y < 4,$$

from which we obtain the conditional densities to be

$$f_{X|Y}(x|y) = \frac{f(x, y)}{f_Y(y)} = \frac{2}{4 - y}, \qquad \frac{y}{2} < x < 2,$$

$$f_{Y|X}(y|x) = \frac{f(x, y)}{f_x(x)} = \frac{1}{2x}, \qquad 0 < y < 2x.$$

Then, the conditional expectations are

$$m(y) = E(X|Y = y) = \int_{y/2}^{2} x\, \frac{2}{4 - y}\, dx = \frac{y + 4}{4}, \qquad 0 < y < 4,$$

and

$$s(y) = E(Y|X = x) = \int_0^{2x} y\, \frac{1}{2x}\, dy = x, \qquad 0 < x < 2.$$

We observe that in this case, the regression curve of X on Y as well as the regression curve of Y on X are both straight lines; we specifically have

$$x = m(y) = \frac{1}{4}y + 1, \qquad\qquad 0 < y < 4$$

and

$$y = s(x) = 1 \cdot x + 0 = x, \qquad\qquad 0 < x < 2.$$

Following the comment above, we note that a special case in which the regression curve is linear occurs when the correlation coefficient between the two variables is either 1 or -1. As we have seen before in Proposition 5.3.2, if $\rho_{X,Y} = \pm 1$, then there exist constants a and b such that

$$Y = aX + b,$$

so that we have

$$m(y) = E(X|Y = y) = E\left(\frac{Y - b}{a} \middle| Y = y\right) = \frac{y - b}{a} = \frac{1}{a} \cdot y + \left(-\frac{b}{a}\right)$$

and

$$s(x) = E(Y|X = x) = E(aX + b|X = x) = ax + b.$$

Thus, the two regression curves are given by

$$x = m(y) \Leftrightarrow x = \frac{y - b}{a} \Leftrightarrow y = ax + b,$$

$$y = s(x) \Leftrightarrow y = ax + b$$

(in fact, the two lines coincide).

In general, it is not necessary that $\rho_{X,Y} = \pm 1$ for the regression curves to be linear. For instance, in Example 5.5.2 we can verify that $\rho_{X,Y} = 0.16$, while it is readily seen that both regression curves are linear.

The following result gives expressions for the coefficients of the regression lines in cases when we actually know that the regression curve of one variable on the other is a straight line.

Proposition 5.5.2 *Let* (X, Y) *be a pair of random variables with*

$$E(X) = \mu_X, \ E(Y) = \mu_Y, \ Var(X) = \sigma_X^2, \ Var(Y) = \sigma_Y^2, \ \rho_{X,Y} = \rho,$$

and assume that the regression line $y = s(x) = E(Y|X = x)$ *is a linear function.*

(i) *If* $s(x) = ax + b$, *then* a *and* b *are given by*

$$a = \rho \frac{\sigma_Y}{\sigma_X}, \ b = \mu_Y - a\mu_X = \mu_Y - \rho \frac{\sigma_Y}{\sigma_X}\mu_X,$$

so that

$$E(Y|X = x) = \mu_Y + \rho \frac{\sigma_Y}{\sigma_X}(x - \mu_X).$$

(ii) *If the conditional variance Var(Y|X = x) is a constant (i.e. it does not depend on x), then*

$$Var(Y|X = x) = (1 - \rho^2)\sigma_Y^2.$$

Proof: We shall give the proof only for the continuous case.

(i) Since

$$s(x) = E(Y|X = x) = \int_{-\infty}^{\infty} y f_{Y|X}(y|x) \, dy = \frac{\int_{-\infty}^{\infty} y f(x, y) \, dy}{f_X(x)},$$

the assumption

$$s(x) = ax + b$$

leads to the relation

$$\int_{-\infty}^{\infty} y f(x, y) \, dy = a x f_X(x) + b f_X(x). \tag{5.10}$$

Integrating both sides with respect to x over $(-\infty, \infty)$, we then deduce that

$$\int_{-\infty}^{\infty} \int_{-\infty}^{\infty} y f(x, y) \, dy \, dx = a \int_{-\infty}^{\infty} x f_X(x) \, dx + b \int_{-\infty}^{\infty} f_X(x) \, dx,$$

which yields

$$\mu_Y = a\mu_X + b. \tag{5.11}$$

Multiplying both sides of (5.10) by x and integrating, with respect to x, over the entire real line, we obtain

$$\int_{-\infty}^{\infty} \int_{-\infty}^{\infty} xy f(x, y) \, dy \, dx = a \int_{-\infty}^{\infty} x^2 f_X(x) \, dx + b \int_{-\infty}^{\infty} x f_X(x) \, dx,$$

which yields

$$E(XY) = aE(X^2) + b\mu_X.$$

As this last equation and (5.11) hold simultaneously, solving these two equations for a and b, we derive that

$$a = \frac{E(XY) - \mu_X\mu_Y}{E(X^2) - \mu_X^2} = \frac{Cov(X, Y)}{\sigma_X^2} = \frac{Cov(X, Y)}{\sigma_X\sigma_Y} \cdot \frac{\sigma_Y}{\sigma_X} = \rho \frac{\sigma_Y}{\sigma_X},$$

$$b = \mu_Y - a\mu_X = \mu_Y - \rho \frac{\sigma_Y}{\sigma_X} \cdot \mu_X.$$

(ii) Upon considering

$$Var(Y|X = x) = E\left[(Y - s(X))^2|X = x\right] = \int_{-\infty}^{\infty} (y - s(x))^2 f_{Y|X}(y|x)\, dy,$$

multiplying both sides by $f_X(x)$ and integrating over $(-\infty, \infty)$, we obtain (using the fact that $Var(Y|X = x)$ is a constant function of x)

$$Var(Y|X = x) \int_{-\infty}^{\infty} f_X(x)\, dx = \int_{-\infty}^{\infty}\int_{-\infty}^{\infty} (y - s(x))^2 f_{Y|X}(y|x) f_X(x)\, dy\, dx.$$

This can also be written as

$$Var(Y|X = x) = \int_{-\infty}^{\infty}\int_{-\infty}^{\infty} (y - s(x))^2 f(x, y)\, dy\, dx,$$

where $f(x, y) = f_{Y|X}(y|x) f_X(x)$ is the joint density function of X and Y. Hence,

$$Var(Y|X = x) = E\left[(Y - s(X))^2\right]$$

and using the fact that $s(X) = \mu_Y + \rho(\sigma_Y/\sigma_X)(X - \mu_X)$, we get

$$Var(Y|X = x) = E\left\{\left[(Y - \mu_Y) - \rho\frac{\sigma_Y}{\sigma_X}(X - \mu_X)\right]^2\right\}$$

$$= E\left[(Y - \mu_Y)^2 + \rho^2\frac{\sigma_Y^2}{\sigma_X^2}(X - \mu_X)^2 - 2\rho\frac{\sigma_Y}{\sigma_X}(Y - \mu_Y)(X - \mu_X)\right]$$

$$= \sigma_Y^2 + \rho^2\frac{\sigma_Y^2}{\sigma_X^2}\,\sigma_X^2 - 2\rho\frac{\sigma_Y}{\sigma_X}Cov(X, Y).$$

The result now follows if we use that $Cov(X, Y) = \rho \cdot \sigma_X \cdot \sigma_Y$. □

The final question we are concerned with in this chapter is the following. Suppose the calculation of the regression curve $y = s(x) = E(Y|X = x)$ is difficult or leads to a cumbersome expression so that use of $s(X)$ to predict Y poses difficulties. Can we *approximate* Y via a linear combination of the form $aX + b$ (rather than $s(X)$), using again the minimization of the mean squared deviation as a criterion for the "best approximation" of this form? The following result addresses this issue.

Proposition 5.5.3 *Let (X, Y) be a two-dimensional random variable with*

$$E(X) = \mu_X, \quad E(Y) = \mu_Y, \quad V(X) = \sigma_X^2, \quad V(Y) = \sigma_Y^2, \quad \rho_{X,Y} = \rho.$$

Then, the values of a and b that minimize the expected squared deviation

$$E\left[(Y - (aX + b))^2\right]$$

are given by

$$a = \rho \frac{\sigma_Y}{\sigma_X}, \qquad b = \mu_Y - \rho \frac{\sigma_Y}{\sigma_X} \mu_X.$$

The minimum value of the mean squared deviation is then equal to $(1 - \rho^2)\sigma_Y^2$.

Proof: Let us put $Q(a, b)$ for the expected squared deviation in the statement, that is

$$Q(a, b) = E\left[(Y - (aX + b))^2\right].$$

We then have

$$\begin{aligned} Q(a, b) &= E\left[Y^2 + (aX + b)^2 - 2Y(aX + b)\right] \\ &= E\left[Y^2 + a^2X^2 + b^2 + 2abX - 2aXY - 2bY\right] \\ &= E(Y^2) + a^2E(X^2) + 2abE(X) - 2bE(Y) - 2aE(XY) + b^2. \end{aligned}$$

Taking derivatives with respect to a, b we get that

$$\frac{\partial Q(a, b)}{\partial a} = 2aE(X^2) + 2bE(X) - 2E(XY)$$

and

$$\frac{\partial Q(a, b)}{\partial b} = 2aE(X) + 2b - 2E(Y).$$

Setting those two partial derivatives equal to zero, we arrive at the system of two equations

$$E(X^2)a + E(X)b = E(XY),$$

$$E(X)\, a + b = E(Y).$$

The solution of this, with respect to a and b, is easily seen to be the two expressions for a, b in the statement.

Finally, the minimum value of the mean squared deviation is

$$E\left\{\left[Y - \left(\rho \frac{\sigma_Y}{\sigma_X}X + \left(\mu_Y - \rho \frac{\sigma_Y}{\sigma_X}\mu_X\right)\right)\right]^2\right\}$$

$$= E\left[\left((Y - \mu_Y) - \rho \frac{\sigma_Y}{\sigma_X}(X - \mu_X)\right)^2\right]$$

$$= E[(Y - \mu_Y)^2] + \rho^2 \frac{\sigma_Y^2}{\sigma_X^2}E[(X - \mu_X)^2] - 2\rho \frac{\sigma_Y}{\sigma_X}E[(X - \mu_X)(Y - \mu_Y)]$$

$$= \sigma_Y^2 + \rho^2 \frac{\sigma_Y^2}{\sigma_X^2}\sigma_X^2 - 2\rho \frac{\sigma_Y}{\sigma_X}\rho\sigma_X\sigma_Y = \sigma_Y^2(1 - \rho^2). \qquad \square$$

Example 5.5.3 Where does the term *regression* come from? Sir Francis Galton (1822–1911) made an experiment in the late nineteenth century for studying the relationship between the diameter size of sweet pea seeds (mother seeds, whose size is denoted by X) and the diameter size of the seeds produced after self-fertilization (daughter seeds, their size is denoted by Y).

Suppose the means and standard deviations of X and Y are found to be

$$\mu_X = 18, \quad \mu_Y = 17, \quad \sigma_X = 2, \quad \sigma_Y = 2,$$

while the correlation coefficient between X and Y is $\rho = 0.5$ (we expect this to be positive; the larger the seed of the mother, the larger we expect the daughter seed to be). When the value of a mother seed is $X = 23$, the expected size of the daughter seed is

$$\mu_Y + \rho \frac{\sigma_Y}{\sigma_X}(x - \mu_X) = 17 + 0.5(23 - 18) = 19.5,$$

while for a mother seed with $X = 12$, the expected size of the daughter seed is

$$\mu_Y + \rho \frac{\sigma_Y}{\sigma_X}(x - \mu_X) = 17 + 0.5(12 - 18) = 14.$$

We thus see that if a particular seed of a mother is very large, the daughter is also expected to have a seed of large size, **but not as large** as the mother's; similarly, for a mother with very small seed diameter size. Galton, who observed this empirically for sweet peas, called it regression to the mean!

The straight line $y = ax + b$ which, according to Proposition 5.5.3 is the best linear (for X) approximation of the variable Y, called the **regression line** of Y on X. As the coefficients a and b for this straight line are determined by minimizing the mean square of the deviations $Y - (aX + b)$, this method of approximating a random variable by a linear function of another variable is called **least squares method**. The method, which is of paramount importance and everyday use in statistics, is traditionally associated with Carl Friedrich Gauss (1777–1855), one of the greatest mathematicians of all times, who apparently developed the fundamentals of it at the age of 18.

Example 5.5.4 Suppose the pair (X, Y) has joint density function as

$$f(x, y) = \begin{cases} \dfrac{x + 3y}{2}, & \text{if } 0 \leq x \leq 1 \text{ and } 0 \leq y \leq 1, \\ \\ 0, & \text{otherwise.} \end{cases}$$

The marginal densities of X and Y are given by

$$f_X(x) = \int_{-\infty}^{\infty} f(x, y)\, dy = \int_0^1 \frac{x + 3y}{2}\, dy = \frac{2x + 3}{4}, \qquad 0 \leq x \leq 1,$$

and

$$f_Y(y) = \int_{-\infty}^{\infty} f(x, y)\, dx = \int_0^1 \frac{x + 3y}{2}\, dx = \frac{1 + 6y}{4}, \qquad 0 \leq y \leq 1,$$

while the conditional densities are given by

$$f_{X|Y}(x|y) = \frac{f(x,y)}{f_Y(y)} = \frac{2(x+3y)}{1+6y}, \qquad 0 \le x \le 1,$$

and

$$f_{Y|X}(y|x) = \frac{f(x,y)}{f_X(x)} = \frac{2(x+3y)}{2x+3}, \qquad 0 \le y \le 1.$$

Further, the regression curves are given by

$$x = m(y) = E(X|Y = y) = \int_0^1 x\, \frac{2(x+3y)}{1+6y}\, dx = \frac{2+9y}{3(1+6y)}, \qquad 0 \le y \le 1,$$

and

$$y = s(x) = E(Y|X = x) = \int_0^1 y\, \frac{2(x+3y)}{2x+3}\, dy = \frac{x+2}{2x+3}, \qquad 0 \le x \le 1.$$

Clearly, neither of the regression curves is a straight line. Suppose we want to find the regression line of Y on X. This will be a linear function of X which serves as an approximation to Y. For that, we need to calculate the quantities

$$\mu_X = E(X), \quad \mu_Y = E(Y), \quad \sigma_X^2 = Var(X), \quad \sigma_Y^2 = Var(Y), \quad \rho = \rho_{X,Y}.$$

For $r = 1, 2, \ldots$, we readily find

$$E(X^r) = \int_{-\infty}^{\infty} x^r f_X(x)\, dx = \int_0^1 x^r\, \frac{2x+3}{4}\, dx = \frac{1}{4}\left(\frac{2}{r+2} + \frac{3}{r+1}\right),$$

$$E(Y^r) = \int_{-\infty}^{\infty} y^r f_Y(y)\, dy = \int_0^1 y^r\, \frac{6y+1}{4}\, dy = \frac{1}{4}\left(\frac{6}{r+2} + \frac{1}{r+1}\right),$$

and so

$$\mu_X = E(X) = \frac{13}{24}, \quad E(X^2) = \frac{3}{8}, \quad \mu_Y = \frac{5}{8}, \quad E(Y^2) = \frac{11}{24},$$

$$Var(X) = E(X^2) - (E(X))^2 = \frac{47}{576}, \quad Var(Y) = E(Y^2) - (E(Y))^2 = \frac{13}{192}.$$

Moreover,

$$E(XY) = \int_{-\infty}^{\infty} xy f(x,y)\, dx\, dy = \int_0^1 \int_0^1 xy\, \frac{x+3y}{2}\, dx\, dy = \frac{1}{3},$$

and so the covariance between X and Y is

$$Cov(X, Y) = E(XY) - E(X)E(Y) = -\frac{1}{192},$$

and the correlation coefficient is

$$\rho = \rho_{X,Y} = \frac{Cov(X,Y)}{\sqrt{V(X)V(Y)}} = \frac{-\dfrac{1}{192}}{\sqrt{\dfrac{47}{576}\cdot\dfrac{13}{192}}} = -\frac{3}{\sqrt{1833}}.$$

Therefore, the regression line of Y on X is the line $y = ax + b$, where

$$a = \rho\frac{\sigma_Y}{\sigma_X} = -\frac{3}{\sqrt{47\cdot39}}\cdot\frac{\sqrt{\dfrac{13}{192}}}{\sqrt{\dfrac{47}{576}}} = -\frac{3}{47} \cong -0.064$$

and

$$b = \mu_Y - a\mu_X = \frac{5}{8} + \frac{3}{47}\cdot\frac{13}{24} = \frac{31}{47} \cong 0.64.$$

The regression line of X on Y can be found in an analogous manner.

Exercises
Group A

1. Let (X, Y) be a random pair for which we have

$$E(X) = \mu_X, \quad E(Y) = \mu_Y, \quad Var(X) = \sigma_X^2, \quad Var(Y) = \sigma_Y^2.$$

   (i) Write down the equation for the regression line of Y on X.

   (ii) Write down the equation for the regression line of X on Y.

   (iii) Show that the point where these two lines intersect is (μ_X, μ_Y).

   (iv) Assuming $\sigma_X^2 = \sigma_Y^2$, what can we say about the minimum value of the mean squared deviation between the regression curve and the regression line in each of these two cases?

   Application: The regression line of Y on X is

$$y = -\frac{3}{2}x - 2,$$

   while the regression line of X on Y is

$$x = -\frac{3}{5}y - 3.$$

   Calculate the quantities $\mu_X, \mu_Y, \rho, \sigma_X$, and σ_Y.

2. The regression curves for a random pair (X, Y) are given by

$$y = s(x) = E(Y|X = x) = \frac{x+3}{2}$$

and

$$x = m(y) = E(X|Y = y) = \frac{y}{2}.$$

(i) Calculate the expectations $\mu_X = E(X)$, $\mu_Y = E(Y)$.

(ii) What is the value of the correlation coefficient ρ?

(iii) Show that X and Y have equal variances.

3. Let (X, Y) have a joint probability function

$$f(x, y) = \frac{x + y}{27}, \qquad\qquad x = 0, 1, 2 \text{ and } y = 1, 2, 3.$$

(i) Find the regression curve of X on Y, and examine whether it is linear or not.

(ii) Find the regression curve of Y on X.

(iii) Calculate $E(X)$, $E(Y)$, $Var(X)$, $Var(Y)$ and $\rho_{X,Y}$.

(iv) Obtain the regression *line* of Y on X, and that of X on Y.

4. Find the regression curves for the random variables in Exercise 18 of Section 5.2. If the regression curve is not a straight line, obtain the regression line of Y on X, and the corresponding (minimum) mean squared deviation.

5. Suppose the continuous variables X and Y have joint density function as

$$f(x, y) = 24x(1 - y), \qquad\qquad 0 < x < y < 1.$$

(i) Verify that both regression curves (of Y on X and that of X on Y) are straight lines.

(ii) Use Proposition 5.5.2 to find the quantities μ_X, μ_Y and $\rho_{X,Y}$.

6. Let (X, Y) have a joint continuous distribution with density

$$f(x, y) = \begin{cases} 8xy, & 0 \le x \le y \le 1, \\ 0, & \text{otherwise.} \end{cases}$$

(i) Find the regression curve of X on Y and verify that it is linear.

(ii) Find the regression curve of Y on X.

(iii) Obtain the regression *line* of Y on X, and that of X on Y. What do you observe?

7. The joint density function of (X, Y) is

$$f(x, y) = \begin{cases} \dfrac{2}{a^2}, & 0 \le y \le x \le a, \\ \\ 0, & \text{otherwise,} \end{cases}$$

for a suitable constant $a > 0$.

(i) Show that both the regression curves of Y on X and of X on Y are linear.

(ii) Using Proposition 5.5.2, find the correlation coefficient $\rho_{X,Y}$.

(iii) Show that $Var(X) = Var(Y)$.

Group B

8. Let X and Y be two random variables and $s(x) = E(Y|X = x)$. Show that, for any function g, the following inequality holds:

$$E[(Y - g(X))^2] \geq Var(Y) - Var(s(X)).$$

(Hint: In view of Proposition 5.5.1, it suffices to show that

$$E[(Y - s(X))^2] = Var(Y) - Var(s(X)).)$$

9. For a random pair (X, Y), let $y = ax + b$ be the regression line of Y on X. Let $Z = Y - (aX + b)$ be the random variable that expresses the difference between Y and its linear approximation from X.

(i) Obtain the mean and variance of Z.

(ii) Show that the variables X and Z are uncorrelated.

10. Two adjacent stores, A and B, sell television sets in an area. Let X be the number of TV sets of a certain make sold by store A during a day and Y be the number of television sets of the same make sold by store B during the same day. The joint probability function of X and Y has been found to be as follows:

x \ y	1	2	3	4
1	0.10	0.12	0.05	0.03
2	0.16	0.10	0.07	0.04
3	0.04	0.11	0.13	0.05

(i) Find the covariance between X and Y.

(ii) Derive (by the least squares method) the regression line of Y on X, and that of X on Y. Sketch these lines on a graph and find the point where they intersect.

(iii) If it is known that, on a given day, store B has sold two television sets, calculate the mean and variance of the number of sets sold by store A.

5.6   BASIC CONCEPTS AND FORMULAS

Covariance between X and Y	$Cov(X, Y) = \sigma_{XY} = E[(X - \mu_X)(Y - \mu_Y)]$, where $\mu_X = E(X)$, and $\mu_Y = E(Y)$
Properties of the covariance	$Cov(X, Y) = E(XY) - E(X)E(Y)$ $Cov(X, X) = Var(X)$ $Cov(aX + b, cY + d) = ac\, Cov(X, Y)$ $Cov(X + Y, Z) = Cov(X, Z) + Cov(Y, Z)$ $Cov(X, Y + Z) = Cov(X, Y) + Cov(X, Z)$ $Cov\left(\sum_{i=1}^{n} a_i X_i, \sum_{j=1}^{k} b_j Y_j\right) = \sum_{i=1}^{n} \sum_{j=1}^{k} a_i b_j Cov(X_i, Y_j)$
Variance of a sum, difference, and a linear combination	$Var(X \pm Y) = Var(X) + Var(Y) \pm 2Cov(X, Y)$ $Var\left(\sum_{i=1}^{n} a_i X_i\right) = \sum_{i=1}^{n} a_i^2 Var(X_i) + 2 \sum_{j=2}^{n} \sum_{i=1}^{j-1} a_i a_j$ $\qquad\qquad\qquad Cov(X_i, X_j)$ $= \sum_{i=1}^{n} a_i^2 Var(X_i) + \sum_{i=1}^{n} \sum_{j\neq i} a_i a_j$ $\qquad\qquad\qquad Cov(X_i, X_j)$
Correlation coefficient	$\rho = \rho_{X,Y} = \dfrac{Cov(X, Y)}{\sqrt{Var(X)}\sqrt{Var(Y)}} = \dfrac{\sigma_{XY}}{\sigma_X \sigma_Y}$
Properties of the correlation coefficient	$\rho_{aX+b,cY+d} = \rho_{X,Y}$, provided that $ac > 0$ $-1 \le \rho_{X,Y} \le 1$ $\rho_{X,Y} = 1$ if and only if $Y = aX + b$ with $a > 0$ $\rho_{X,Y} = -1$ if and only if $Y = aX + b$ with $a < 0$
Properties of the conditional expectation	$E[E(X|Y)] = E(X), \quad E[E(Y|X)] = E(Y)$ $E[E(g(X)|Y)] = E[g(X)], \quad E[E(h(Y)|X)] = E[h(Y)]$
Calculation of probabilities by conditioning	$P(A) = \displaystyle\sum_{x \in R_X} P(A|X = x) f_X(x)$, if X is discrete $P(A) = \displaystyle\int_{-\infty}^{\infty} P(A|X = x) f_X(x)\, dx$, if X is continuous

Mean and variance for the sum of a random number of random variables (random sum)	$$E\left(\sum_{i=1}^{N} X_i\right) = E(N)\mu,$$ $$Var\left(\sum_{i=1}^{N} X_i\right) = E(N)\sigma^2 + \mu^2 Var(N),$$ where X_i's are i.i.d. with $E(X_i) = \mu$, $Var(X_i) = \sigma^2$ and N is a discrete variable independent of X_i's
Regression curve • of Y on X • of X on Y	$y = s(x) = E(Y|X = x)$ $x = m(y) = E(X|Y = y)$
A property of the regression curve	$E[(Y - g(X))^2] \geq E[(Y - s(X))^2]$ for any function g
Regression curve: linear case	If $E(Y|X = x) = ax + b$, then $$a = \rho\frac{\sigma_Y}{\sigma_X}, \quad b = \mu_Y - a\mu_X = \mu_Y - \rho\frac{\sigma_Y}{\sigma_X}\mu_X.$$ If, in addition, $Var(Y|X = x)$ is a constant, then $Var(Y|X = x) = (1 - \rho^2)\sigma_Y^2$
Regression line	It is the straight line $y = ax + b$ which minimizes the mean squared deviation, $E[(Y - (aX + b))^2]$. The coefficients a and b are given by $$a = \rho\frac{\sigma_Y}{\sigma_X}, \quad b = \mu_Y - a\mu_X = \mu_Y - \rho\frac{\sigma_Y}{\sigma_X}\mu_X$$ and the mean squared deviation is $(1 - \rho^2)\sigma_Y^2$

5.7 COMPUTATIONAL EXERCISES

Consider the discrete pair of variables X and Y from Example 5.2.2, with joint probability function

$$f(x, y) = \begin{cases} \dfrac{x + y}{45}, & \text{for } x = 1, 2, 3 \text{ and } y = 2, 3, 4, \\ 0, & \text{otherwise.} \end{cases}$$

The following command sequence in Mathematica is used to find the expectations $E(X)$, $E(Y)$, and $E(XY)$, the variances of X and Y, the covariance between X and Y and, finally, the correlation coefficient $\rho_{X,Y}$.

```
In[1]:= f[x_, y_] = (x + y)/45
n = 3; m = 3;
xval = {1, 2, 3}; yval = {2, 3, 4};
Print["Joint probability function of two random variables"]
tab = Table[f[xval[[i]], yval[[j]]], {i, 1, n}, {j, 1, m}];
Print[MatrixForm[tab]];
(*Calculation of marginal distribution of X*)
Do[Xmarg[i] = Sum[tab[[i, j]], {j, 1, m}], {i, 1, n}]
(*Calculation of marginal distribution of Y*)
Do[Ymarg[j] = Sum[tab[[i, j]], {i, 1, n}], {j, 1, m}]
(*Calculation of the moments E(X), E(Y), Var(X), Var(Y), E(XY)*)
S = 0;
Do[Do[S = S + xval[[i]] yval[[j]]*tab[[i, j]], {i, 1, n}], {j, 1, m}]
EXY = S;
EX = 0; EY = 0; EX2 = 0 ; EY2 = 0;
Do[ EX = EX + xval[[i]]*Xmarg[i], {i, 1, n}]
Do[ EY = EY + yval[[j]]*Ymarg[j], {j, 1, m}]
Do[ EX2 = EX2 + (xval[[i]]^2)*Xmarg[i], {i, 1, n}]
Do[ EY2 = EY2 + (yval[[j]]^2)*Ymarg[j], {j, 1, m}]
VX = EX2 - EX^2; VY = EY2 - EY^2;
Print[ "E(X) = ", EX, ", E(Y) = ", EY]
Print[ "Var(X)= = ", VX, ", Var(Y) = ", VY]
Print[ "E(XY) = ", EXY]
(*Calculation of the covariance and correlation between X, Y*)
CV = EXY - EX*EY; rho = CV/Sqrt[VX*VY]
Print["Cov(X,Y)= ", CV]
Print["Correlation(X,Y)= ", rho]

Out[1]= Joint probability function of two random variables
```

$$
\begin{pmatrix}
\dfrac{1}{15} & \dfrac{4}{45} & \dfrac{1}{9} \\[2ex]
\dfrac{4}{45} & \dfrac{1}{9} & \dfrac{2}{15} \\[2ex]
\dfrac{1}{9} & \dfrac{2}{15} & \dfrac{7}{45}
\end{pmatrix}
$$

```
E(X) = 32/15, E(Y) = 47/15
Var(X)= = 146/225, Var(Y) = 146/225
E(XY) = 20/3
Cov(X,Y)= -(4/225)
Correlation(X,Y)= -(2/73)
```

For the case when X and Y have a joint continuous distribution, the calculations are similar (obviously sums are replaced by integrals). We give below the commands needed to calculate the same quantities as above for the continuous pair (X, Y) from Example 5.2.3.

```
In[2]:= f[x_, y_] = (1 - x^3*y + x*y^3)/4
(*Marginal density of X*)
fX[X_] = Integrate[f[x, y], {y, -1, 1}]
```

```
Print["Marginal density of X=  " , fX[x]]
(*Marginal density of Y*)
fY[y_] = Integrate[f[x, y], {x, -1, 1}]
Print["Marginal density of Y=  " , fY[y]]
(*Calculation of the moments E(X), E(Y), Var(X), Var(Y), E(XY)*)
EX = Integrate[x*fX[x], {x, -1, 1}];
EY = Integrate[y*fY[y], {y, -1, 1}];
EX2 = Integrate[x^2*fX[x], {x, -1, 1}];
EY2 = Integrate[y^2*fY[y], {y, -1, 1}];
EXY = Integrate[x*y*f[x, y], {x, -1, 1}, {y, -1, 1}];
VX = EX2 - EX^2; VY = EY2 - EY^2;
Print[ "E(X) = ", EX, ", E(Y) = ", EY]
Print[ "Var(X)= = ", VX, ", Var(Y) = ", VY]
Print[ "E(XY) = ", EXY]
(*Calculation of the covariance and correlation between X, Y*)
CV = EXY - EX*EY; rho = CV/Sqrt[VX*VY];
Print["Cov(X,Y)= ", CV]
Print["Correlation(X,Y)= ", rho]

Out[2]= Marginal density of X=  1/2
Marginal density of Y=  1/2
E(X) = 0, E(Y) = 0
Var(X)= = 1/3, Var(Y) = 1/3
E(XY) = 0
Cov(X,Y)= 0
Correlation(X,Y)= 0
```

1. Working in a similar way as above, find $Cov(X, Y)$ and the correlation coefficient $\rho_{X,Y}$ for each pair (X, Y) of discrete variables whose probability functions are as follows:

   (i) $f(x, y) = \dfrac{2^x 3^y}{72\, x!\, y!}$, $x = 1, 2, 3, 4,\ y = 1, 2, 3$;

   (ii) $f(x, y) = \dfrac{2}{81} \cdot \dfrac{2^{x+1} + 3^y}{x!\, y!}$, $x = 1, 2, 3, 4,\ y = 1, 2, 3$;

   (iii) $f(x, y) = \dfrac{7}{22} \cdot \dfrac{|x - y|}{x + y}$, $x = 1, 2, 3,\ y = 1, 2, 3, 4$.

2. For the following probability densities of (X, Y) with joint continuous distributions, first identify the value of c in the density function and then use it to calculate $Cov(X, Y)$ and $\rho_{X,Y}$:

   (i) $f(x, y) = \begin{cases} c\,(x^2 + y), & 0 < x < 1, 0 < y < 1, \\ 0, & \text{otherwise}; \end{cases}$

   (ii) $f(x, y) = \begin{cases} c\,x^2 y, & 0 < x < y < 1, \\ 0, & \text{otherwise}; \end{cases}$

   (iii) $f(x, y) = \begin{cases} c\,(x^2 + y), & 0 < x < y < 1, \\ 0, & \text{otherwise}; \end{cases}$

(iv) $f(x, y, z) = \begin{cases} c\,x^3(1 - y^2), & 0 < x < y < 1, \\ 0, & \text{otherwise.} \end{cases}$

3. The joint probability function of two variables X and Y is given by

$$f(x, y) = \begin{cases} cx^2(y + 1), & x = 1, 2, \ldots, 15, \ y = 1, 2, \ldots, 10, \\ 0, & \text{otherwise,} \end{cases}$$

for some constant c.

   (i) Find the value of c and thence obtain the marginal probability functions of X and Y.

   (ii) Calculate the covariance between X and Y.

   (iii) Find the correlation coefficient $\rho_{X,Y}$.

4. Three variables X, Y, and Z have a joint continuous distribution with joint density function

$$f(x, y, z) = \begin{cases} cx^2 e^{-x(4+y+z)}, & x, y, z > 0, \\ 0, & \text{otherwise,} \end{cases}$$

for some constant c.

   (i) Find the value of c.

   (ii) For each of the three pairs $(X, Y), (X, Z), (Y, Z)$, find the covariance and correlation coefficient. Which pair exhibits the strongest form of (linear) dependence?

5. Let X be a random variable having uniform distribution over the interval $(0, 1)$, and $Y = X^2$.

   (i) Show that
   $$E(X^r) = \frac{1}{r + 1}, \qquad r = 1, 2, \ldots,$$

   and use it to derive the covariance of X and Y.

   (ii) Calculate the variance of Y and thence verify that

   $$\rho_{X,Y} = \frac{\sqrt{15}}{4} \cong 0.968.$$

   Observe that although the relation between X and Y is nonlinear, their correlation coefficient is close to one (compare it with the result of Example 5.3.1).

   (iii) Generalize the result of Part (ii) by calculating the correlation coefficient, say ρ_r, between the variables X and X^r, for $r = 3, 4, \ldots, 20$. For these values of r, plot ρ_r against r to see how the dependence between X and the power X^r decays as r grows. Comment on this plot.

6. Let X and Y be two independent random variables, each having Gamma distribution, but with different parameters. Specifically, the distribution of X is Gamma with

parameters $n_1 = 3$, $\lambda_1 = 5$, while the distribution of Y is Gamma with parameters $n_2 = 5$, $\lambda_2 = 1$.

(i) Calculate

$$E(X^{10}\, Y^5), \quad E(X^3 + X^2 Y + 3), \quad E[(X + Y)^7].$$

(ii) Find the covariance for each of the following pairs of random variables

    (a) X^5 and Y^5,

    (b) X^{10} and Y^8.

7. Find the value of the constant c, so that each of the following functions is a joint density function of a three-dimensional random variable (X, Y, Z). (The density is assumed to take the value zero outside the range specified below.)

(i) $f(x, y, z) = cx^3 e^{-x(4+y+z)}$, $x, y, z > 0$;

(ii) $f(x, y, z) = ce^{-x-y-z}$, $0 < x < y < z$.

Then, in each case, obtain

    (a) the marginal densities of X, Y, and Z,

    (b) the correlation coefficient for each of (X, Y), (X, Z), and (Y, Z).

With the following commands we do, using Mathematica, all calculations which appeared in Example 5.4.1 for the pair (X, Y) having joint density function

$$f(x, y) = \begin{cases} \dfrac{2(x + 2y)}{3}, & 0 < x < 1,\ 0 < y < 1, \\ 0, & \text{otherwise.} \end{cases}$$

```
In[3]:= f[x_, y_] = 2*(x + 2*y)/3;
(*Calculation of the marginal density of X*)
fX[x_] = Integrate[f[x, y], {y, 0, 1}];
Print["MARGINAL DENSITY OF X = " , fX[x]]
(*CalcuLation of the marginal density of Y*)
fY[y_] := Integrate[f[x, y], {x, 0, 1}];
Print["MARGINAL DENSITY OF Y= " , fY[y]]
(*Calculation of conditional moments*)
Print["Conditional expectation of X given that Y=y: m(y)= ",
FullSimplify[Integrate[x*f[x, y], {x, 0, 1}]/fY[y]]]
Print["Conditional expectation of Y given that X=x: s(x)= ",
FullSimplify[Integrate[y*f[x, y], {y, 0, 1}]/fX[x]]]

Out[3]= MARGINAL DENSITY OF X =  2/3+(2 x)/3
MARGINAL DENSITY OF Y=  1/3+(4 y)/3
Conditional expectation of X given that Y=y: m(y)= (2+6y)/(3+12y)
Conditional expectation of Y given that X=x: s(x)= (4+3x)/(6+6x)

In[4]:= m[y_] = (2 + 6*y)/(3 + 12*y);
mY = Integrate[m[y]*fY[y], {y, 0, 1}];
Print[ "E[m(Y)]=", mY]
Plot[m[y], {y, 0, 1}]
```

```
s[x_] = (4 + 3*x)/(6 + 6*x);
sX = Integrate[s[x]*fX[x], {x, 0, 1}];
Print[ "E[s(X)]=", sX]
Plot[s[x], {x, 0, 1}]

Out[3]= E[m(Y)]=5/9
```

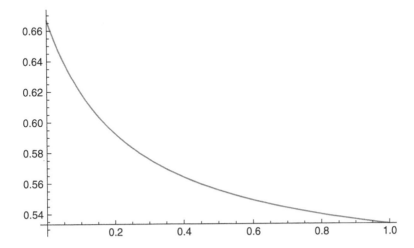

```
E[s(X)]=11/18
```

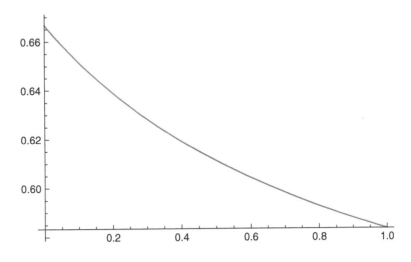

8. In a similar manner, find the conditional expectations

$$m(y) = E(X|Y = y), \qquad s(x) = E(Y|X = x),$$

and then the (unconditional) expectations

$$E[m(Y)], \quad E[s(X)]$$

for each of the pairs (X, Y) with joint density functions $f(x, y)$ as follows:

(i) $f(x, y) = \dfrac{5}{84}(x^4 + 1)$, $0 < x < 2,\ 0 < y < 2$;

(ii) $f(x, y) = \dfrac{15}{62}(x^4 + 1)$, $0 < x < y < 2$;

(iii) $f(x, y) = 24x^5(1 - y)^3$, $0 < x < 1,\ 0 < y < 1$;

(iv) $f(x, y) = 5040\, x^5(1 - y)^3$, $0 < x < y < 1$;

(v) $f(x, y) = \dfrac{60(x^3 + y)}{23}$, $0 < x < y < 1$.

9. Do the same as in Exercise 8 for the pairs (X, Y) of discrete random variables whose joint probability functions are as follows:

(i) $f(x, y) = \dfrac{3x + 7y}{738}$, $x = 1, 3, 5,\ y = 2, 4, 8, 16$;

(ii) $f(x, y) = \dfrac{2x + 4y}{27(x + y)}$, $x = 1, 2, 3,\ y = 1, 2, 3$.

5.8 SELF-ASSESSMENT EXERCISES

We assume below that all the variables considered have a finite nonzero variance.

5.8.1 True–False Questions

1. If for two variables X and Y we know that $Cov(X, Y) < 0$, then their correlation coefficient, $\rho_{X,Y}$, is negative.

2. If $Cov(X, Y) = 0$, then X and Y are *not* independent.

3. Let X and Y be independent standard normal random variables. Then,

$$Cov(2X + 3Y,\ X - 4Y) = 14.$$

4. Let X and Y have a joint continuous distribution with density

$$f(x, y) = \begin{cases} c\, x\, y\, e^{-(x+y)}, & x \geq 0,\ y \geq 0, \\ 0, & \text{otherwise}, \end{cases}$$

for a suitable constant c. Then, $Cov(X, Y) = 0$.

5. Tim, David, and Jonathan, who have met recently, discuss about their birthdays. Let X be the number of persons among them who have their birthday in the winter and Y the number of persons who were born in the summer. Then, $Cov(X, Y) < 0$. (Assume that all four seasons are equally likely for a person to be born.)

6. If $\rho_{X,Y} = 0$, then $Cov(X, Y) = 0$.

7. If $\rho_{X,Y} < 0$, then $Cov(X, Y) < E(X)E(Y)$.

8. If $Var(X + Y) = Var(X) + Var(Y)$, then the random variables X and Y are independent.

9. Steve and Colin, who are archaeologists, are studying the relation between the size of a particular bone of dinosaurs to the size of their foot. Steve has denoted by X the bone size and by Y the foot size, both measured in inches, and has calculated $Cov(X, Y)$ and $\rho_{X,Y}$. Colin, who works independently and has the same information about the joint distribution of the bone and foot sizes, uses centimeters instead. Then, Colin will find the same correlation coefficient, but a different covariance between these two quantities.

10. Let (X, Y) be a two-dimensional random variable. Then, the quantity $E(X|Y)$ is a (one-dimensional) random variable, whose expected value is $E(Y)$.

11. Let X be a random variable and $Y = X^2$. Then, the value of the correlation coefficient between X and Y is either 1 or -1.

12. Let X be a random variable having uniform distribution over $(0, 2)$, and $Y = 2 - X$. Then, $\rho_{X,Y} = 1$.

13. Paula selects two cards at random without replacement from an ordinary deck of cards. Let X be the number of aces in the first card chosen (so that $X = 1$ if the first card is an ace, and $X = 0$ otherwise) and Y be the number of kings in the second card chosen (so that $Y = 1$ if the second card is a king, and $Y = 0$ otherwise). Then, X and Y are positively correlated.

14. If X and Y are two random variables zero mean and unit variance, then $E(XY) \le 1$.

15. Let X and Y be two random variables such that the regression curve of Y on X and that of X on Y are both linear. Then the value of the correlation coefficient $\rho_{X,Y}$ is either 1 or -1.

16. For a two-dimensional random variable (X, Y), it is known that the regression curve of X on Y is given by
$$E(X \mid Y = y) = 4y + 1.$$

It is also known that the correlation coefficient between the two variables is 0.2, while the marginal distribution of Y is exponential with parameter $\lambda = 4$. Then, the variance of X is 25.

17. Let X be a Bernoulli random variable with
$$P(X = 1) = 1 - P(X = 0) = p$$
for $0 < p < 1$, and $Y = X^3$. Then, $Cov(X, Y) = p(1 - p)$.

18. Let X be an exponential random variable with mean 1 and $Y = X^2$. Then, X and Y are uncorrelated.

19. Let $X_1, X_2, \ldots$ be a sequence of independent and identically distributed random variables, each having a standard normal distribution, and

$$S = \sum_{i=1}^{N} X_i,$$

where N is a discrete variable independent of the X_i's. Then, $E(S) = 0$ irrespective of the distribution of N.

20. Suppose X is a random variable having $N(3, 1^2)$ distribution, while Y is independent of X following $N(-4, 2^2)$ distribution. Then, the random variables

$$U = 4X + 2Y \quad \text{and} = -2X + Y$$

are uncorrelated.

21. Suppose X and Y are independent random variables, each having a uniform distribution over the interval $(1, 5)$, and

$$U = X^2 + 2, \qquad\qquad V = 3X - 2Y.$$

Then, $Cov(U, V) = 8$.

5.8.2 Multiple Choice Questions

1. For two variables X and Y, it is known that

$$E(Y) = 1, \quad E(XY) = 10, \quad Cov(X, Y) = -3.$$

Then, the expected value of X is equal to

(a) 3   (b) 10   (c) 13   (d) -7 (e) -13

2. Nick has a green and a red die and he throws them together. Let X be the outcome of the green die and Y be the difference of the two outcomes (the outcome of the green minus the outcome of the red one). Then:

(a) $Cov(X, Y) = Var(X)$

(b) the variables X and Y are independent

(c) the variables X and Y are uncorrelated but not independent

(d) $E(XY) = E(X^2)$

(e) X and Y are negatively correlated

3. Suppose X and Y are independent random variables with $\sigma_X = 1$, and $\sigma_Y = 3$. Then, the standard deviation of the variable $U = X - 3Y$ is equal to

(a) $\sqrt{10}$ (b) 8   (c) -8 (d) 9   (e) $\sqrt{82}$

4. Let X and Y be two discrete random variables with joint probability function

$$f(x, y) = \frac{3x + y}{45}, \qquad\qquad \text{for } x = 1, 2, 3 \text{ and } y = 1, 2.$$

The covariance between X and Y is equal to

(a) $-\dfrac{2}{225}$ (b) $-\dfrac{1}{15}$ (c) $-\dfrac{2}{5}$ (d) $\dfrac{1}{15}$ (e) $\dfrac{2}{5}$

5. For the random variables X and Y, it is given that $\sigma_X = 4$, and $\sigma_Y = 5$, and that their covariance is $Cov(X, Y) = 3$. Let

$$W = 2X + 3(Y - 2).$$

Then, the variance of W is

(a) 79   (b) 325   (c) 41   (d) 73   (e) 307

6. Let X be a discrete random variable with probability function

$$f_X(-1) = f_X(0) = f_X(1) = \frac{1}{3},$$

and define two new variables U and V as $U = 2X - 1$, $V = X^3$. Then, the covariance between U and V is

(a) $4/3$ (b) $2/3$ (c) $1/3$ (d) 0 (e) $-1/3$

7. In a sequence of three independent Bernoulli trials, each with success probability p, we define the variables X and Y as

$$X = \begin{cases} 1, & \text{if the first trial results in a success,} \\ 0, & \text{otherwise,} \end{cases}$$

and

$$Y = \begin{cases} 1, & \text{if there is exactly one success among the three trials,} \\ 0, & \text{otherwise.} \end{cases}$$

Then, the covariance between X and Y is equal to

(a) 0 (b) $2p(1-p)^2$ (c) $-2(1-p)^3$
(d) $p(1-p)^2 - 3p^2(1-p)^2$ (e) $-2p(1-p)^3$

8. Two variables X and Y are independent and have the same distribution with unit variance, while for another variable Z, it is known that $Var(Z) = 9$. If we are given that $Cov(X, Y + Z) = 5/2$, then the correlation coefficient between X and Z equals

(a) $5/6$ (b) $5/18$ (c) $\sqrt{5/6}$ (d) $\sqrt{5/18}$
(e) none of the above

9. For the variables X and Y, it is known that

$$E(X) = 1, \ E(X^2) = 9, \ E(Y) = 2, \ E(XY) = -2, \ \rho_{X,Y} = -1/2.$$

Then, the variance of Y is equal to

(a) $64/9$ (b) 8 (c) $8/3$ (d) $\sqrt{8/3}$ (e) 1

10. For a pair of random variables (X, Y), it is known that X and Y have the same marginal distribution with variance σ^2. If it is known that

$$Var\left(\frac{X}{3} - \frac{Y}{2}\right) = \frac{23\sigma^2}{72},$$

then the correlation coefficient between X and Y is

(a) $\frac{2}{3}$ (b) $\frac{37}{72}$ (c) $\frac{13}{36}$ (d) $\frac{1}{4}$ (e) $\frac{1}{24}$

11. The number of phone calls that Lena makes to her schoolmates during an afternoon is a random variable N, taking values in the set of positive integers according to the probability function

$$P(N = n) = \frac{1}{5} \cdot \left(\frac{4}{5}\right)^{n-1}, \qquad n = 1, 2, \ldots$$

The lengths of these phone calls (in minutes) are assumed to be independent and identically distributed random variables, each having a density

$$f(x) = \frac{1}{4} e^{-x/4}, \qquad x \geq 0.$$

Let S be the *total* time that Lena spends on the phone talking to her schoolmates in an afternoon. Then, the expected value of S, in minutes, is

(a) $\frac{1}{20}$ (b) $\frac{1}{16}$ (c) $\frac{5}{4}$ (d) 16   (e) 20

12. For a random variable X, it is known that $\mu_X = 1$ and $\sigma_X^2 = 9$. Let Y be another variable and assume that the regression line of Y on X is given by

$$y = s(x) = E(Y|X = x) = 3x + 7,$$

while the correlation between the two variables is $\rho_{X,Y} = 1/2$. Then, the values of the mean and variance of Y are given by

(a) $\mu_Y = 4$, $\sigma_Y^2 = 54$ (b) $\mu_Y = 4$, $\sigma_Y^2 = 9/2$

(c) $\mu_Y = 4$, $\sigma_Y^2 = 81/4$ (d) $\mu_Y = 10$, $\sigma_Y^2 = 54$

(e) $\mu_Y = 10$, $\sigma_Y^2 = 324$

13. Let X and Y be two random variables with variances $\sigma_X^2 = 9/4$ and $\sigma_Y^2 = 16$, respectively. If it is known that the regression line of Y on X is

$$y = s(x) = E(Y|X = x) = 2x + 3,$$

then the covariance $Cov(X, Y)$ equals

(a) 16   (b)   8   (c) 4   (d) $\frac{9}{2}$ (e) $\frac{27}{16}$

14. Assume that (X, Y) is a two-dimensional random variable. Then, the conditional expectation

$$E(3Y^2 - 2Y + 1|X)$$

is a random variable whose expected value is

(a) $E(3Y^2 - 2Y + 1)$ (b) $E(3X^2 - 2X + 1)$ (c) $3E(XY^2) - 2E(XY) + 1$

(d) $3E(Y^2) - 2E(Y)$ (e) $3X^2 - 2X + 1$

15. Let $X_1, X_2, \ldots$ be a sequence of independent and identically distributed random variables with mean μ and variance σ^2, and assume that N is a random variable independent of X_i's which follows the Poisson distribution with parameter λ. Then, the variance of the random sum

$$\sum_{i=1}^{N} X_i$$

is equal to

(a) $\lambda^2 \mu^2 + \lambda \sigma^2$ (b) $(\lambda + \mu)\sigma^2$ (c) $\lambda\sqrt{\mu^2 + \sigma^2}$

(d) $\lambda^2(\mu + \sigma^2)$ (e) $\lambda(\mu^2 + \sigma^2)$

16. The number of CDs sold daily by a store is a random variable N with mean $E(N) = 150$ and variance $Var(N) = 25$. The profit (in dollars) that a store makes for each CD sold is a random variable having a normal distribution with $\mu = 3$ and $\sigma = 1$. Then, the variance of the store's daily profit equals

(a) 285   (b) 375   (c) 450   (d) 753   (e) 3759

17. An extract from the joint probability function of two discrete random variables X and Y is given in the following table:

x \ y	-1	0	1
0	0.15	0.30	a
1	0.10	b	0.05
2	0.05	0.10	0.05

The entries a and b in the table are some suitable constants (which should be identified). If it is known that

$$P(Y = 0 \mid X = 1) = \frac{1}{2},$$

then the covariance between X and Y is

(a) 0.055   (b) 0.015   (c) 0   (d) -0.155 (e) -0.055

5.9   REVIEW PROBLEMS

1. Consider a sequence of Bernoulli trials with success probability p. Show that the expected number of trials until k consecutive successes occur equals $(1 - p^k)/(qp^k)$.

   Application: A computer selects digits at random. What is the expected number of digits selected until three successive zeros appear?

   (Hint: Proceed as in Example 5.4.3.)

2. The number of persons arriving at a platform in an underground station in the time interval $(0, t]$ has a Poisson distribution with parameter λt; here, we take as time zero the instant that the previous train departed from the platform. The time until the next train arrives at the platform has a Uniform distribution over the interval (a, b), where a and b are given positive constants with $a < b$.

   (i) Show that the expected number of persons embarking on the next arriving train is $\lambda(a + b)/2$.

   (ii) Prove that the variance of the number of persons embarking on the arriving train is given by
   $$\frac{\lambda[6(a + b) + \lambda(b - a)^2]}{12}.$$

3. A random pair (X, Y) has range $R_{X,Y}$, defined by the triangle on the plane with vertices at the points $(0, 0), (0, 1)$, and $(1, 0)$.

   (i) Find the joint density function, $f(x, y)$, of (X, Y) and the marginal densities $f_X(x)$ and $f_Y(y)$.

   (ii) Calculate $E(X), E(Y), Var(X), Var(Y), Cov(X, Y)$, and $\rho_{X,Y}$.

   (iii) Obtain the conditional expectations $E(X|Y = y)$ and $E(Y|X = x)$, and draw a graph of the functions $m(y) = E(X|Y = y)$ and $s(x) = E(Y|X = x)$. What do you observe?

   (iv) Find the variances of the random variables $E(X|Y)$ and $E(Y|X)$, and verify that these variables are homoscedastic (that is, they have the same variance).

4. Two variables X and Y have variances σ_X^2 and σ_Y^2, respectively, and their correlation coefficient is ρ. Show that the value of a that minimizes the variance of the sum $aX + Y$ is
   $$a = -\rho \frac{\sigma_Y}{\sigma_X}.$$

5. The variables X and Y have joint density function as
   $$f(x, y) = \begin{cases} 6, & \text{if } 0 \le x^2 \le y \le x \le 1, \\ 0, & \text{otherwise} \end{cases}$$

   (that is, the pair (X, Y) is uniformly distributed in the region A shown below).

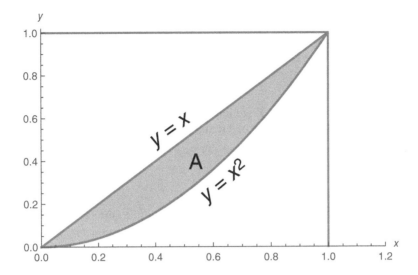

(i) Find the marginal density function of X.

(ii) Show that the conditional distribution of Y, given $X = x$, is uniform distribution over the interval (x^2, x).

(iii) Calculate the regression curve

$$s(x) = E(Y|X = x), \qquad 0 \leq x \leq 1,$$

and show that it is not linear.

(iv) Find the mean and variance of $s(X) = E(Y|X)$.

6. A stick of unit length is broken at an arbitrary point. Let X and Y be the lengths of the two segments formed. Show that

$$Cov(X, Y) = 12\rho_{X,Y}.$$

7. Let X and Y be two random variables and $a, b, c, d \in \mathbb{R}$. Prove that

$$\rho_{aX+b,cY+d} = sgn(ac)\rho_{X,Y},$$

where the function $sgn(z)$ is defined by

$$sgn(z) = \begin{cases} -1, & \text{if } z < 0, \\ 0, & \text{if } z = 0, \\ 1, & \text{if } z > 0. \end{cases}$$

8. For two random variables X and Y and any real numbers $a, b, c,$ and d, let us define

$$U = aX + bY, \quad V = cX + dY.$$

(i) Show that

$$Cov(U, V) = ac\ Var(X) + bd\ Var(Y) + (ad + bc)\ Cov(X, Y).$$

(ii) If X and Y have the same variance, verify that the variables $X + Y$ and $X - Y$ are uncorrelated.

(iii) Assuming that X and Y are uncorrelated and have the same variance, show that U and V are uncorrelated if and only if $ac + bd = 0$.

(iv) In the special case where $a = b = c = 1, d = -1$, demonstrate that it holds

$$Cov(X + Y, X - Y) = Var(X) - Var(Y)$$

and deduce that, if X, Y have the same variance, than the variables $X + Y, X - Y$ are uncorrelated (even if the original variables X, Y are correlated).

9. Let X, Y, and Z be three independent random variables with $Var(X) = Var(Y) = Var(Z) = \sigma^2$. Define

$$U = X, \quad V = aX + Y, \quad W = bY + Z.$$

Find the values of the constants $a, b \in \mathbb{R}$, such that

$$\rho_{U,V} = \rho_{V,W} = \frac{1}{2}.$$

10. In a class of n students wearing uniforms, each student leaves their uniform jacket at the entrance when entering the class. At the end of the class, each student selects a jacket to wear at random (assuming all jackets are identical). Let X be the number of persons who picked their own jacket. Show that

$$E(X) = Var(X).$$

(Hint: Express X as $X_1 + \ldots + X_n$ for suitable variables X_i, and then find $E(X_i)$, $Var(X_i)$ and $Cov(X_i, X_j)$ for $i \neq j$.)

11. Suppose X and Y are independent Bernoulli variables with the same mean $1/2$. Show that the variables $U = X + Y$ and $V = |X - Y|$ are uncorrelated, but not independent.

12. Let X be a standard normal random variable and assume that Y is independent of X and follows t distribution with 5 degrees of freedom. Calculate the covariance between the variables U and V, where

$$U = X - Y, \qquad V = 4X + 3Y.$$

13. Assume that X and Y are independent random variables with uniform distribution over $[0, 1]$. Let

$$U = min\{X, Y\}, \quad V = max\{X, Y\}.$$

    (i) Obtain the distribution of the product UV and its expected value.

    (ii) Calculate the correlation coefficient between U and V.

(Hint: Observe that $UV = XY$.)

14. We ask a computer to select a (decimal) number Y from the interval $(0, \theta)$. If the number selected is y (that is, $Y = y$), we ask the computer to select a second number, X, from the interval $(0, \sqrt{y})$. Calculate the mean and variance of X

    (i) by finding first the distribution of X;

    (ii) using Propositions 5.4.1 and 5.4.3.

15. Let X be a random variable having the uniform distribution over $(0, a)$, while Y is another variable, independent of Y, which has the uniform distribution over $(0, b)$, for some $a > 0, b > 0$. It is given that

$$Var(X + Y) = \frac{5}{6}$$

and

$$Cov(3X + Y, \ 3X - Y) = 0.$$

    (i) Find the values of a and b.

    (ii) Calculate

$$E[(X - Y)^2], \qquad E(X^4 Y^3).$$

16. In statistics, when a random sample is drawn from a population, it is often convenient to assume that this population has infinitely many elements. On the other hand, suppose we have a population consisting of N units (e.g. representing the heights of N students present in a class) and we select (without replacement) n of those units. Let $X_1, \ldots, X_n$ be the random variables representing the values of the chosen units (e.g. the heights of the students selected). In this case, define $S_n = X_1 + \cdots + X_n$, and let

$$\overline{X}_n = \frac{S_n}{n}$$

denote the sample mean. Assume that the population variance is σ^2, so that $Var(X_i) = \sigma^2$ for all i.

    (i) Show that

$$Var(S_n) = n\sigma^2 + n(n - 1) \ Cov(X_1, X_2).$$

    (ii) Prove that

$$Cov(X_1, X_2) = -\frac{\sigma^2}{N - 1},$$

    and hence show that

$$Var\left(\overline{X}_n\right) = \frac{\sigma^2}{n} \cdot \frac{N - n}{N - 1}.$$

    Compare your answer to the result of Example 5.2.4. The term $(N - n)/(N - 1)$ is known as the *finite population correction*.

17. (Variance of the product between two variables) Let X_1 and X_2 be two random variables with

$$E(X) = \mu_1, \quad E(Y) = \mu_2, \quad Var(X) = \sigma_1^2, \quad Var(Y) = \sigma_2^2$$

and

$$Cov(X_1, X_2) = \sigma_{12}, \quad Cov(X_1^2, X_2^2) = s_{12}.$$

Show that

$$Var(X_1 X_2) = s_{12} - \sigma_{12}^2 - 2\sigma_{12}^2 \mu_1 \mu_2 + \sigma_1^2 \sigma_2^2 + \mu_1^2 \sigma_2^2 + \mu_2^2 \sigma_1^2.$$

How does the above formula simplify when X and Y are independent?

18. Let X and Y be two random variables, and Z be another variable defined as

$$Z = \begin{cases} X, & \text{with probability } p, \\ Y, & \text{with probability } 1 - p. \end{cases}$$

Express the expected value of Z in terms of $E(X)$ and $E(Y)$.

(Hint: Consider the variable

$$W = \begin{cases} 1, & \text{if } Z = X, \\ 0, & \text{if } Z = Y, \end{cases}$$

and use the fact that $E(Z) = E[E(Z|W)]$.)

19. Let X_1, X_2, and X_3 be a random sample from uniform distribution over the interval $(0, 1)$, and $X_{(i)}$ be the ith order statistics from this sample, for $i = 1, 2, 3$. Calculate the covariance between

(a) $X_{(1)}$ and $X_{(2)}$,

(b) $X_{(1)}$ and $X_{(3)}$.

20. A discrete variable X has Poisson distribution with parameter Λ, where Λ is a continuous random variable having exponential distribution with parameter $\theta > 0$. Show that the probability function of X is given by

$$P(X = x) = \frac{\theta}{(1 + \theta)^{x+1}}, \qquad x = 0, 1, 2, \ldots$$

Identify the distribution of X, and thence find its mean and variance.

(Hint: Use the formula [Part (ii) of Proposition 5.4.2]

$$P(X = x) = \int_0^\infty P(X = x | \Lambda = \lambda) f_\Lambda(\lambda) d\lambda.$$

21. At the inspection control in a production line of a factory, a random sample of n items is selected and the number, X, of defective items is recorded. The probability

that a randomly selected item is defective is unknown, and so we regard it as a random variable having uniform distribution over an interval $(0, \theta)$, where $\theta \in (0, 1)$ is a known number. Calculate the mean and variance of X.

(Hint: Let P_0 be the unknown probability that an item is defective. Then, use the formula [see Part (ii) of Proposition 5.4.2]

$$P(X = x) = \int_{-\infty}^{\infty} P(X = x | P_0 = p) f_{P_0}(p) dp.)$$

22. Let Y be a continuous random variable having Gamma distribution with parameters a, λ, where a is a positive integer and λ is a positive real number. For another, discrete random variable X it is known that, given $Y = y$, the distribution of X is Poisson with parameter y.

(i) Show that the probability function of X is given by

$$P(X = x) = \binom{x + a - 1}{a - 1} \left(\frac{1}{\lambda + 1}\right)^x \left(\frac{\lambda}{\lambda + 1}\right)^a, \qquad x = 0, 1, 2, \ldots$$

(ii) Find the mean and variance of X in two different ways.

Application: The number of traffic violations by a driver during a year is a discrete random variable X. The distribution of X depends on a parameter y whose value differs from one driver to another, as it is affected by a number of factors (e.g. the number of miles that the driver covers during a year, the driver's age, driving skills, etc.). We assume that the parameter y for each driver is the observed value of a random variable Y, following a Gamma distribution with parameters $a = 3$ and $\lambda = 5/3$. Assuming further that, given $Y = y$, the (conditional) distribution of X is Poisson with parameter y, derive the mean and variance of the number of traffic violations that a driver makes during a year.

(Hint: Use the result of Part (i) in Proposition 5.4.2.)

23. Nick rolls a die and let X be the random variable that represents the outcome of the throw. If this outcome takes the value $X = x$, for $x \in \{1, 2, 3, 4, 5, 6\}$, Nick asks a computer to generate x numbers at random from the interval $(0, 1)$. Let Y be the largest number that the computer has generated.

(i) Write down the range of the two-dimensional random variable (X, Y).

(ii) Obtain the conditional distribution function of Y, given $X = x$ and thence find the conditional density function $f_{Y|X}(y|x)$.

(iii) Derive the marginal density function of Y.

24. From the records of a company that issues a credit card, it has been estimated that, if the primary cardholder is a man, the number of weekly transactions is a random variable having Poisson distribution with mean 5, while if the primary cardholder is a woman, the number of weekly transactions has Poisson distribution with a mean of eight transactions. If 55% of the primary cardholders for this particular credit card are women, calculate the mean and variance of the number of transactions for a randomly selected cardholder.

25. A roulette wheel has a proportion r of its numbers red while a proportion b are black; assume that $r + b < 1$ (in a standard wheel, $r = b = 18/37$, as there are 18 numbers of each color and also a zero number on the wheel which is neither red or black). Suppose we spin the wheel m times, and let N_R be the number of reds and N_B be the number of blacks that appear. Set $N = N_R + N_B$.

   (i) Explain why N has a binomial distribution and identify the parameters of that distribution. Thence, find the mean and variance of N.

   (ii) Show that
   $$Cov(N_R, N_B) = -mrb.$$

   (iii) Calculate the correlation coefficient between N_R and N_B. Give a numerical answer for this coefficient in the case when $r = b = 18/37$.

   (Hint: For Part (ii), use the formula
   $$Var(N) = Var(N_R + V_B) = Var(N_R) + Var(N_B) + 2Cov(N_R, N_B)$$

   and the result from Part (i).)

5.10 APPLICATIONS

5.10.1 Portfolio Optimization Theory

Modern Portfolio theory is a mathematical theory which provides a framework for investors to maximize their expected return for a given level of risk (Markowitz 1952). The theory, also known as mean-variance analysis, was introduced by the economist Harry Markowitz in an essay he wrote in 1952, and for which he later received the Nobel Prize in Economics.

Suppose we have two assets (securities) A_1, A_2 and let R_1, R_2 be the random variables representing the returns from these two assets, in some monetary unit. The distribution of these two variables is not known, however we know their means and variances, which are as follows:

$$E(R_1) = \mu_1, \quad E(R_2) = \mu_2, \quad Var(R_1) = \sigma_1^2, \quad Var(R_2) = \sigma_2^2.$$

Moreover, we write ρ for the correlation coefficient between R_1 and R_2. A person wants to invest in these two assets and seeks the "optimal proportion" he should invest in each of the two. We write a for the proportion he will invest in A_1, so that $1 - a$ is the proportion he is going to invest in A_2. His return will then be

$$R_P = aR_1 + (1 - a)R_2$$

and his expected return is

$$\mu = E(R_P) = a\mu_1 + (1 - a)\mu_2.$$

We denote by σ_P^2 the variance of the variable R_p. In finance, this is known as the **volatility** of the risk associated with this portfolio. A key question in this context is: find the value of a such that, for given $\mu_1, \mu_2, \sigma_1^2, \sigma_2^2$, the volatility of the portfolio risk is minimized.

To begin with, we express the volatility σ_P^2 in terms of the variances σ_1^2, σ_2^2 and the correlation coefficient ρ between the individual investments. More explicitly, we have

$$
\begin{aligned}
\sigma_P^2 &= Var(aR_1 + (1-a)R_2) \\
&= a^2 Var(R_1) + (1-a)^2 Var(R_2) + 2a(1-a)Cov(R_1, R_2) \\
&= a^2 \sigma_1^2 + (1-a)^2 \sigma_2^2 + 2a(1-a)\rho \sigma_1 \sigma_2.
\end{aligned}
\tag{5.12}
$$

We treat the quantity above as a function of a, and we denote it by $g(a)$. In order to obtain the minimum of this function with respect to a, we find its first and second derivatives, which are given by

$$
g'(a) = 2a\sigma_1^2 - 2(1-a)\sigma_2^2 + (2-4a)\rho \sigma_1 \sigma_2
\tag{5.13}
$$

and

$$
g''(a) = 2\sigma_1^2 + 2\sigma_2^2 - 4\rho \sigma_1 \sigma_2.
$$

We note in particular that the second derivative does not depend on a and that, from the above expression,

$$
g''(a) = 2Var(R_1 - R_2) > 0.
$$

We now look for the value of a such that $g'(a) = 0$ (the above result guarantees that this will automatically be a minimum). In view of (5.13), we see that a must satisfy

$$
a\left(\sigma_1^2 + \sigma_2^2 - 2\rho \sigma_1 \sigma_2\right) = \sigma_2^2 - \rho \sigma_1 \sigma_2,
$$

which gives the value of a as

$$
a = \frac{\sigma_2^2 - \rho \sigma_1 \sigma_2}{\sigma_1^2 + \sigma_2^2 - 2\rho \sigma_1 \sigma_2} = \frac{Var(R_2) - Cov(R_1, R_2)}{Var(R_1 - R_2)}.
\tag{5.14}
$$

This is the proportion that has to be invested in asset A_1 so that the volatility is minimized. The proportion that is then invested to asset A_2 is

$$
1 - a = \frac{\sigma_1^2 - \rho \sigma_1 \sigma_2}{\sigma_1^2 + \sigma_2^2 - 2\rho \sigma_1 \sigma_2} = \frac{Var(R_1) - Cov(R_1, R_2)}{Var(R_1 - R_2)}.
$$

For this value of a, and $1 - a$, the investor's expected return from the portfolio is

$$
\mu = E(R_p) = a\mu_1 + (1-a)\mu_2 = \frac{\sigma_2(\sigma_2 - \rho \sigma_1)\mu_1 + \sigma_1(\sigma_1 - \rho \sigma_2)\mu_2}{\sigma_1^2 + \sigma_2^2 - 2\rho \sigma_1 \sigma_2}.
$$

Moreover, substituting the value of a in (5.12), we see that the minimum volatility of this portfolio is given by

$$\sigma_P^2 = \frac{\sigma_1^2 \sigma_2^2 (1 - \rho^2)}{\sigma_1^2 + \sigma_2^2 - 2\rho\sigma_1\sigma_2}.$$

Note, however, that a denotes the *proportion* that should be invested to asset A_1, so that its value must be in the interval $[0, 1]$. Therefore, we see from (5.14) that the values of $\sigma_1, \sigma_2,$ and ρ must satisfy

$$0 \leq \frac{\sigma_2^2 - \rho\sigma_1\sigma_2}{\sigma_1^2 + \sigma_2^2 - 2\rho\sigma_1\sigma_2} \leq 1.$$

The left-hand inequality gives

$$\sigma_2^2 - \rho\sigma_1\sigma_2 \geq 0$$

while from the right-hand inequality we see that we must have

$$\sigma_2^2 - \rho\sigma_1\sigma_2 \leq \sigma_1^2 + \sigma_2^2 - 2\rho\sigma_1\sigma_2.$$

The last two conditions yield

$$\rho\sigma_1 \leq \sigma_2 \quad \text{and} \quad \rho\sigma_2 \leq \sigma_1,$$

which indicates that the following inequality

$$\rho \leq \min\left\{\frac{\sigma_1}{\sigma_2}, \frac{\sigma_2}{\sigma_1}\right\}$$

must hold in order to have a feasible (optimal) allocation for the two investments.

As a final consideration, we may ask ourselves: is there a general rule which decides which of the two assets should be given a greater weight than the other? To answer this, we simply compare the values of a and $1 - a$ found above. By doing so, we see that $a > 1 - a$, so that $a > 1/2$ if and only if

$$\sigma_2^2 - \rho\sigma_1\sigma_2 > \sigma_1^2 - \rho\sigma_1\sigma_2,$$

which is the same as

$$\sigma_2^2 > \sigma_1^2.$$

This shows that, to achieve minimum volatility, we should invest more on the asset with the smaller variance, i.e. to the less risky between the two assets. This is further illustrated by our empirical investigation in the following table. The table presents the values of $E(R_P)$ and σ_P^2 when μ_1, μ_2 are fixed, but we allow the correlation coefficient ρ and the variances σ_1^2, σ_2^2 (or, equivalently the standard deviations σ_1, σ_2) to vary. More precisely, we have used $\mu_1 = 50, \mu_2 = 100$ and various choices for ρ, σ_1, σ_2.

σ_1	σ_2	ρ	$E(R_P)$	σ_P
5	10	−0.8	65.85	4.39
		−0.4	63.64	12.73
		0	60	20
		0.4	52.94	24.71
20	10	−0.8	84.14	17.57
		−0.4	86.36	50.91
		0	90	80
		0.4	97.05	98.82
20	30	−0.8	69.47	57.35
		−0.4	67.98	169.89
		0	65.38	276.92
		0.4	59.76	368.78

The values on the table demonstrate that the value of $E(R_P)$ is mostly affected by the relative magnitude of the standard deviations, while the value of ρ seems to have a less drastic effect. Further, for $\sigma_1 = \sigma_2$, it is easily checked that the value of a is $1/2$, so that $E(R_P)$ is the arithmetic average of μ_1, μ_2 and this is irrespective of the value of ρ. On the other hand, the volatility σ_P^2 increases rapidly both with the values of ρ and those of the standard deviations σ_1, σ_2.

KEY TERMS

correlation coefficient

covariance

linear regression

least squares method

mean squared deviation

negatively correlated random variables

positively correlated random variables

regression curve

regression line

uncorrelated random variables

variance–covariance inequality

CHAPTER 6

IMPORTANT MULTIVARIATE DISTRIBUTIONS

Harold Hotelling (Fulda, Minnesota 1895 – Chapel Hill, North Carolina 1973)

Harold Hotelling was born on 29 September 1895, in Fulda, Minnesota, and died on 26 December 1973, at the age of 78, in Chapel Hill, North Carolina. After completing his BA and MA in 1919 and 1921, respectively, at the University of Washington, he went to Princeton University and finished his PhD in 1924 under the supervision of Professor Oswald Veblen. He was much influenced by the book *Statistical Methods for Research Workers* by Ronald Fisher with whom he maintained a good professional relationship. He developed Hotelling's T-squared distribution, a generalization of Student's t-distribution to the multivariate case, and demonstrated its applications in tests of hypotheses and confidence regions. He also developed canonical correlation analysis and principal component analysis which are some of the most commonly used statistical techniques. He was Associate Professor at Stanford University during 1927–1931, then at Columbia University from 1931 to 1946, and finally served as Professor at the University of North Carolina at Chapel Hill from 1946 until his death in 1973.

6.1   INTRODUCTION

Thus far, we have seen various concepts and results relating to the probabilistic behavior of two, or more, random variables, as they vary simultaneously. In this chapter, we describe some of the most important families of multivariate probability distributions. For each family, we discuss its main characteristics (mean, variance, covariance, and correlation) and also the marginal and conditional distributions associated with it. These would facilitate the analysis in problems where these distributions are used in modeling random quantities in real-life situations.

6.2   MULTINOMIAL DISTRIBUTION

The multinomial distribution is a natural generalization of the binomial distribution. Recall that a binomial distribution arises when n independent trials are performed, each with two possible outcomes (Bernoulli trials). We typically identify one of these outcomes as "success", and we further assume that the success probability is the same for all trials. Although it is conceptually possible to think of this as a bivariate problem (we can define two variables, X and Y, representing the numbers of successes and "failures", respectively), there is no need to do so, as the distribution of Y is simply that of $Y = n - X$.

Now, let us suppose we have n trials that are independent and each has $k + 1$ possible outcomes ($k \geq 2$). In analogy with the experiment associated with the binomial, we may term k among these outcomes as "successes". Specifically, we label the $k + 1$ outcomes by $S_1, S_2, \ldots, S_k$ and F (as with the binomial distribution, the correspondence between the outcomes and these labels is arbitrary, but it should be kept in mind that the calculations and interpretation of results need to be carried out in a consistent manner).

As we now have more than one type of "success", we refer to the outcome denoted by S_r as "success of type r", for $r = 1, 2, \ldots, k$, while F stands for a failure outcome. We assume further that the probabilities of these events remain constant throughout the experiment (i.e. the n trials may simply be considered as n independent repetitions of the same experiment). For $r = 1, 2, \ldots, k$, let p_r be the probability that the event S_r occurs in a trial and q be the probability that a trial results in a failure. Thus, for each trial, we have $P(S_r) = p_r$ and $P(F) = q$, and since exactly only one of the events $S_1, S_2, \ldots, S_k, F$ could occur, we must have

$$q + p_1 + p_2 + \cdots + p_k = 1. \tag{6.1}$$

We then have the following definition.

Definition 6.2.1 Let $X_1, \ldots, X_k$ be the numbers of appearances of events $S_1, S_2, \ldots, S_k$ in n independent trials, each with $k + 1$ possible outcomes, viz. $S_1, S_2, \ldots, S_k$ and F. We assume that, in each trial, the event S_r has probability p_r, while the event F has probability q (so that (6.1) holds) and that these probabilities are the same for all trials. Then, the k-dimensional random variable $(X_1, \ldots, X_k)$ is said to follow the **multinomial distribution** with parameters n and $p_1, p_2, \ldots, p_k$.

Observe that the parameter q is not explicitly listed as one of the parameters of the multinomial distribution, since its value is determined by $p_1, p_2, \ldots, p_k$ (in the same way that the probability of failure, $q = 1 - p$, is not one of the parameters for the binomial distribution). Clearly, $k = 1$ in the above definition corresponds to the binomial distribution, but this case will not be treated here, as it has been dealt with in Volume I.

When we have $k + 1$ possible outcomes, we consider k rather than $k + 1$ variables, just as we do in the case of binomial with one variable when there are two possible outcomes. Also, observe that if X_{k+1} denotes the number of appearances of event F in the n trials, then

$$X_{k+1} = n - \left(X_1 + X_2 + \cdots + X_k \right).$$

As a first example, assume that n patients visit a surgery clinic on a specific day. Let X_1 be the number of patients who leave the clinic, after seeing a doctor, within three hours, and X_2 be the number of patients who are released after three hours but by the end of the day (while the remaining $n - (X_1 + X_2)$ have to stay overnight). Then, (X_1, X_2) have a multinomial distribution; in this case, i.e. the case when $k = 2$, the resulting distribution is often called the **trinomial distribution**.

In the following example, we consider n rolls of a die and introduce the multinomial distribution in such an experiment.

Example 6.2.1 Suppose we roll a die n times, and let X_i denote the number of times i appears as outcome of a throw, for $i = 1, 2, 3, 4, 5, 6$. Then, the random variable $(X_1, X_2, \ldots, X_5)$ follows the multinomial distribution with parameters n and $p_1 = p_2 = \cdots = p_5 = 1/6$. Evidently, the number of times six appears will be $X_6 = n - (X_1 + X_2 + \cdots + X_5)$.

In the same experiment, let us define the following variables:

Y_1:   the number of times the outcome of a throw is either 1 or 2,
Y_2:   the number of times the outcome of a throw is 3, 4, or 5,
Y_3:   the number of times the outcome of a throw is 6.

Then:

- the bivariate random variable (Y_1, Y_2) follows the trinomial distribution with parameters n and $(2/6, 3/6)$;

- the bivariate random variable (Y_3, Y_1) follows the trinomial distribution with parameters n and $(1/6, 2/6)$.

Example 6.2.2 Suppose we know (from data of previous years) that among the students who register at a University Sports Club, a proportion p_1 choose to play baseball, a proportion p_2 select basketball, a proportion p_3 prefer hockey, while the proportions of students who choose rugby and volleyball are p_4 and p_5, respectively. Finally, the remaining proportion $q = 1 - (p_1 + p_2 + p_3 + p_4 + p_5)$ choose other sports. We assume that no student can opt for more than one sport.

We select a sample of n students registered with the Sports Club, and let X_i (for $i = 1, 2, 3, 4, 5$) be the number of students in this sample who have chosen baseball, basketball, hockey, rugby, and volleyball, respectively. Finally, let the variable X_6 represent the number of students who do other sports. Then clearly $X_6 = n - (X_1 + X_2 + X_3 + X_4 + X_5)$ and the 5-dimensional random variable $(X_1, X_2, X_3, X_4, X_5)$ follows the multinomial distribution with parameters n and p_i (for $1 \leq i \leq 5$).

We first now focus on some results for the special case of a trinomial distribution, corresponding to $k = 2$ in Definition 6.2.1.

Proposition 6.2.1 *Let (X_1, X_2) follow a trinomial distribution with parameters n and p_1, p_2. Then, the joint probability function of (X_1, X_2) is given by*

$$f(x_1, x_2) = P(X_1 = x_1, X_2 = x_2) = \frac{n!}{x_1! x_2! x_3!} p_1^{x_1} p_2^{x_2} q^{x_3},$$

for any $0 \leq x_1, x_2 \leq n$ such that $x_1 + x_2 \leq n$, where $x_3 = n - x_1 - x_2$ and $q = 1 - p_1 - p_2$.

Proof: Any possible outcome of the experiment under consideration here, i.e. performing a series of n trials, is an n-tuple consisting of successes (S_1, S_2) and failures (F), having the form

$$\underbrace{S_1 \ S_2 \ S_2 \ F \ S_2 \ F \ F \ S_1 \ F \ S_2 \ S_1 \dots F \ F \ S_2}_{n \text{ successes and failures in total}}.$$

Among these outcomes, the ones which are favorable to the event $\{X_1 = x_1, \ X_2 = x_2\}$ are those consisting of exactly x_1 individual outcomes that are of type S_1 and x_2 individual outcomes which are of type S_2. Because the trials are assumed independent, any such n-tuple that is favorable for the event under consideration has probability

$$p_1^{x_1} p_2^{x_2} q^{n - x_1 - x_2}.$$

All we need now is to calculate the number of distinct ways (trial outcomes) in which the event $\{X_1 = x_1, \ X_2 = x_2\}$ occurs. To find this, consider the following two-step procedure:

Step 1: We select the x_1 places at which the successes of first type (S_1) will occur.

Step 2: From the remaining $n - x_1$ places, we select x_2 places at which the successes of second type (S_2) will occur.

It is evident that in Step 1 there are $\binom{n}{x_1}$ distinct choices, while in Step 2 there are $\binom{n - x_1}{x_2}$ choices. Consequently, the total number of favorable ways in which the event of interest $\{X_1 = x_1, \ X_2 = x_2\}$ occurs is given by

$$\binom{n}{x_1} \binom{n - x_1}{x_2} = \frac{n!}{x_1!(n - x_1)!} \cdot \frac{(n - x_1)!}{x_2!((n - x_1) - x_2)!} = \frac{n!}{x_1! x_2! x_3!},$$

where $x_3 = n - x_1 - x_2$, as required. $\square$

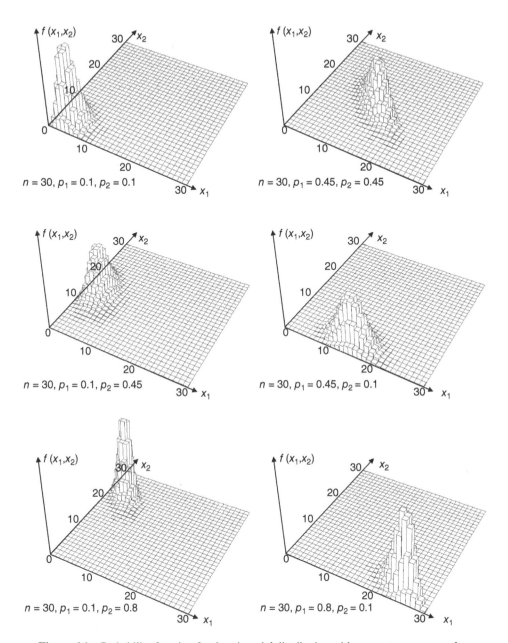

Figure 6.1 Probability function for the trinomial distribution with parameters n, p_1, and p_2.

Figure 6.1 shows the probability function of the trinomial distribution for different values of parameters n, p_1, and p_2.

As the trinomial distribution is a bivariate distribution, it is reasonable to study its marginal distributions and also conditional distributions.

Proposition 6.2.2 *Suppose the pair* (X_1, X_2) *has the trinomial distribution with parameters* n *and* p_1, p_2. *Then, the following hold:*

(i) *The marginal distribution of* X_1 *is the binomial distribution with parameters* n *and* p_1, *while the marginal distribution of* X_2 *is binomial distribution with parameters* n *and* p_2.

(ii) *The conditional distribution of* X_1, *given* $X_2 = x_2$, *is the binomial distribution with parameters* $n - x_2$ *and* $p_1/(p_1 + q)$, *while the conditional distribution of* X_2, *given* $X_1 = x_1$, *is the binomial distribution with parameters* $n - x_1$ *and* $p_2/(p_2 + q)$.

Proof: Let $f(x_1, x_2)$ denote the joint probability function of (X_1, X_2).

(i) For the marginal probability function of X_1, we have

$$f_{X_1}(x_1) = \sum_{x_2} f(x_1, x_2) = \sum_{x_2} \frac{n!}{x_1! x_2! x_3!} p_1^{x_1} p_2^{x_2} q^{x_3}, \qquad 0 \leq x_1 \leq n,$$

where $x_3 = n - x_1 - x_2$ and the summation extends over all x_2 such that

$$0 \leq x_2 \leq n \quad \text{and} \quad x_1 + x_2 \leq n.$$

These two conditions can be written in a concise from as $0 \leq x_2 \leq n - x_1$, so that the marginal probability function can be written as

$$f_{X_1}(x_1) = \frac{n!}{x_1!(n - x_1)!} p_1^{x_1} \sum_{x_2=0}^{n-x_1} \frac{(n - x_1)!}{x_2!((n - x_1) - x_2)!} p_2^{x_2} q^{(n-x_1)-x_2}.$$

Now, using the binomial formula, we readily see that the above sum equals $(p_2 + q)^{n-x_1}$, which then yields

$$f_{X_1}(x_1) = \frac{n!}{x_1!(n - x_1)!} p_1^{x_1} (p_2 + q)^{n-x_1} = \binom{n}{x_1} p_1^{x_1}(1 - p_1)^{n-x_1}.$$

This shows that the distribution of X_1 is indeed $b(n, p_1)$, as asserted.

It should be mentioned that a much easier way to obtain the same result is by observing that X_1 is the number of successes in a sequence of n Bernoulli trials, each with two possible outcomes, namely, "S_1" and "not S_1".

The result for the distribution of X_2 follows in an analogous manner.

(ii) From the definition of a conditional probability function, we find

$$f_{X_1 | X_2}(x_1 | x_2) = \frac{f(x_1, x_2)}{f_{X_2}(x_2)} = \frac{\dfrac{n!}{x_1! x_2! x_3!} p_1^{x_1} p_2^{x_2} q^{x_3}}{\dfrac{n!}{x_2!(n - x_2)!} p_2^{x_2}(1 - p_2)^{n-x_2}}$$

$$= \frac{(n - x_2)!}{x_1! x_3!} \frac{p_1^{x_1} q^{x_3}}{(p_1 + q)^{n-x_2}},$$

which can be rewritten as

$$f_{X_1|X_2}(x_1|x_2) = \binom{n-x_2}{x_1} \left(\frac{p_1}{p_1+q}\right)^{x_1} \left(1 - \frac{p_1}{p_1+q}\right)^{(n-x_2)-x_1}.$$

The right-hand side term is precisely the probability function of the binomial distribution with parameters $n - x_2$ and $p_1/(p_1 + q)$. The result for the conditional distribution of X_2, given $X_1 = x_1$, follows similarly.   □

From the result in Part (ii) of Proposition 6.2.2, the conditional expectation of X_2, given $X_1 = x_1$, is

$$E(X_2 \mid X_1 = x_1) = (n - x_1)\frac{p_2}{p_2+q} = \left(-\frac{p_2}{p_2+q}\right)x_1 + \left(\frac{np_2}{p_2+q}\right) = ax_1 + b,$$

with

$$a = -\frac{p_2}{p_2+q}, \quad b = \frac{np_2}{p_2+q}.$$

Thus, the regression curve of X_2 on X_1 is linear which, in view of Proposition 5.5.2, implies that

$$a = \rho_{X_1,X_2}\frac{\sigma_{X_2}}{\sigma_{X_1}} = \frac{Cov(X_1,X_2)}{\sigma_{X_1}^2}.$$

But, from Part (i) of Proposition 6.2.2, we have

$$\sigma_{X_1}^2 = Var(X_1) = np_1(1 - p_1) = np_1(p_2 + q),$$

and thence

$$Cov(X_1,X_2) = a\sigma_{X_1}^2 = \left(-\frac{p_2}{p_2+q}\right)np_1(p_2 + q) = -np_1p_2.$$

We have therefore shown the following proposition.

Proposition 6.2.3 *Let* (X_1,X_2) *follow trinomial distribution with parameters n and* p_1,p_2. *Then, it holds that*
$$Cov(X_1,X_2) = -np_1p_2.$$

Example 6.2.3 The owner of an on-line bookstore estimates that 50% of her customers prefer books, 30% prefer e-books, while the remaining 20% prefer audio books. What is the probability that among the next five customers,

(i) two of them will buy books and two will buy e-books?

(ii) exactly one customer will buy e-books and at most one customer will buy audio books?

(iii) the same number of customers will buy books and e-books?

Assume that each customer buys one type of product only. Moreover, calculate the correlation coefficient between the number of customers buying books and the number of those who will buy e-books.

SOLUTION   Let X_1 be the number of customers who will buy books and X_2 be the number of customers, among the next *five* customers of the bookstore, who will purchase e-books. Let also X_3 be the number of customers who will purchase audio books. Then, the joint distribution of (X_1, X_2) is trinomial with parameters $n = 5$ and $p_1 = 0.5$, $p_2 = 0.3$. Their joint probability function is

$$f(x_1, x_2) = \frac{5!}{x_1! x_2! (n - x_1 - x_2)!} (0.5)^{x_1} (0.3)^{x_2} (0.2)^{n - x_1 - x_2},$$

for $0 \leq x_1, x_2 \leq 5$ and $x_1 + x_2 \leq 5$. Moreover, we have $X_3 = 5 - X_1 - X_2$.

(i) We require the probability $P(X_1 = 2, X_2 = 2)$, which is given by

$$P(X_1 = 2, X_2 = 2) = \frac{5!}{2! 2! 1!} (0.5)^2 (0.3)^2 (0.2)^1 = 0.135 = 13.5\%.$$

(ii) The required probability $P(X_2 = 1, X_3 \leq 1)$ is evidently $P(X_2 = 1, X_3 = 0) + P(X_2 = 1, X_3 = 1)$. But, the first of these probabilities is the same as $P(X_1 = 4, X_2 = 1)$, while the second one is $P(X_1 = 3, X_2 = 1)$. Thus, the required probability is

$$P(X_2 = 1, X_3 \leq 1) = f(4, 1) + f(3, 1)$$

$$= \frac{5!}{4! 1! 0!} (0.5)^4 (0.3)^1 (0.2)^0 + \frac{5!}{3! 1! 1!} (0.5)^3 (0.3)^1 (0.2)^1$$

$$= 0.2438 \cong 24.4\%.$$

(iii) For this case, we have

$$P(X_1 = X_2) = f(0, 0) + f(1, 1) + f(2, 2)$$

$$= (0.2)^5 + \frac{5!}{1! 1! 3!} (0.5)^1 (0.3)^1 (0.2)^3 + \frac{5!}{2! 2! 1!} (0.5)^2 (0.3)^2 (0.2)^1$$

$$= 0.159\ 32 \cong 15.9\%.$$

Moreover, the correlation coefficient between X_1 and X_2 is

$$\rho_{X_1, X_2} = \frac{Cov(X_1, X_2)}{\sqrt{Var(X_1) Var(X_2)}}.$$

As

$$Cov(X_1, X_2) = -np_1 p_2 = -5(0.5)(0.3) = -0.75$$

and the variances of X_1 and X_2 are

$$Var(X_1) = np_1(1 - p_1) = 5(0.5)(0.5) = 1.25,$$

$$Var(X_2) = np_2(1 - p_2) = 5(0.3)(0.7) = 1.05,$$

we find

$$\rho_{X_1,X_2} = \frac{-0.75}{\sqrt{(1.25)(1.05)}} = -0.6547 \cong -65.5\%.$$

Results similar to those for the trinomial distribution hold in the general case of the multinomial distribution. The key properties of the multinomial distribution are stated in the following proposition and their proofs are left as an exercise for the reader.

Proposition 6.2.4 *Let* $(X_1, \ldots, X_k)$ *be a k-dimensional random variable having multinomial distribution with parameters n and* $p_1, p_2, \ldots, p_k$. *Then, the following hold:*

(i) *The joint probability function of* $X_1, \ldots, X_k$ *is given by*

$$f(x_1, \ldots, x_k) = \frac{n!}{x_1! \ldots x_k! x_{k+1}!} p_1^{x_1} \ldots p_k^{x_k} q^{x_{k+1}}$$

for $x_1, \ldots, x_k$ *in the set* $\{0, 1, 2, \ldots, n\}$ *such that* $x_1 + \cdots + x_k \leq n$ *(for all other values, the probability function vanishes), where*

$$x_{k+1} = n - x_1 - \cdots - x_k, \qquad q = 1 - p_1 - \cdots - p_k.$$

(ii) *The marginal distribution of* X_i, *for* $i = 1, \ldots, k$, *is binomial distribution with parameters n and* p_i.

(iii) *The marginal distribution of any collection of* r $(2 \leq r \leq k)$ *among* $X_1, \ldots, X_k$ *is multinomial with parameters n and success probabilities* p_i *that are precisely the probabilities corresponding to the chosen variables.*

(iv) *For any* $i = 1, \ldots, k$, *the conditional joint distribution of all* X_j'*s, for* $j \neq i$, *given* $X_i = x_i$, *is multinomial distribution with parameters* $n - x_i$ *and* $p_j/(1 - p_i)$, *for* $j \neq i$.

(v) *For any* $i \neq j$, *we have*

$$Cov(X_i, X_j) = -np_i p_j.$$

Example 6.2.4 Nick rolls a die 12 times. Calculate the probability that

(i) each of the six faces appears twice;

(ii) three 1's and four 2's appear;

(iii) three 1's and four 2's appear, given that there is exactly one outcome as 4.

SOLUTION For $i = 1, 2, 3, 4, 5$, let us define the variables

X_i: the number of times the outcome i appears in the 12 rolls.

Then, the *five*-dimensional random variable $(X_1, X_2, X_3, X_4, X_5)$ has multinomial distribution with parameters 12 and $p_i = 1/6$, for $i = 1, 2, 3, 4, 5$. The associated joint probability function is

$$f(x_1, x_2, x_3, x_4, x_5) = \frac{12!}{x_1! \dots x_5! x_6!} \left(\frac{1}{6}\right)^{x_1} \dots \left(\frac{1}{6}\right)^{x_5} \left(\frac{1}{6}\right)^{x_6}$$

$$\text{for } 0 \le x_1, \dots, x_5 \le 12, \; x_1 + \dots + x_5 \le 12,$$

where $x_6 = 12 - (x_1 + x_2 + x_3 + x_4 + x_5)$ or, equivalently,

$$f(x_1, x_2, x_3, x_4, x_5) = \frac{12!}{x_1! \cdots x_5! (12 - x_1 - \cdots - x_5)!} \cdot \left(\frac{1}{6}\right)^{12}.$$

(i) The probability that each face appears twice is

$$P(X_1 = 2, X_2 = 2, X_3 = 2, X_4 = 2, X_5 = 2) = \frac{12!}{(2!)^6} \cdot \left(\frac{1}{6}\right)^{12} = 0.0034.$$

(ii) By Part (iii) of Proposition 6.2.4, we have the joint distribution of (X_1, X_2) to be multinomial (in fact, trinomial) with parameters $n = 12$ and $p_1 = p_2 = 1/6$. So, we find

$$P(X_1 = 3, X_2 = 4) = \frac{12!}{3!4!5!} \cdot \left(\frac{1}{6}\right)^3 \cdot \left(\frac{1}{6}\right)^4 \cdot \left(\frac{4}{6}\right)^{12-3-4}$$

$$= 277\,20 \cdot \left(\frac{1}{6}\right)^7 \cdot \left(\frac{4}{6}\right)^5$$

$$= 0.013.$$

(iii) In this case, we need to find the conditional distribution of (X_1, X_2), given $X_4 = 1$. By Part (iv) of Proposition 6.2.4, this is multinomial with parameters $n - 1 = 11$ and success probabilities

$$\frac{p_1}{1 - p_4} = \frac{1/6}{1 - 1/6} = \frac{1}{5}, \quad \frac{p_2}{1 - p_4} = \frac{1/6}{1 - 1/6} = \frac{1}{5}.$$

Hence,

$$P(X_1 = 3, X_2 = 4 | X_4 = 1) = \frac{11!}{3!4!4!} \cdot \left(\frac{1}{5}\right)^3 \cdot \left(\frac{1}{5}\right)^4 \cdot \left(\frac{3}{5}\right)^{11-3-4}$$

$$= 11550 \cdot \left(\frac{1}{5}\right)^7 \cdot \left(\frac{3}{5}\right)^4 = 0.019.$$

EXERCISES

Group A

1. In a recent survey among teenagers about their movie preferences, it was estimated that 35% among them prefer to watch adventure movies, 25% prefer comedies, 20% like thriller movies, and the remaining 20% prefer other types of movies. If we ask 20 students in a class about their movie preferences, what is the probability that there will be

   (i) more than two, but less than five, students preferring comedies?

   (ii) no one who likes adventure movies and at least three who like thriller movies?

2. In Example 6.2.4 involving 12 rolls of a die, calculate the probability that

   (i) faces $1, 2, 3, 4$ appear three times each;

   (ii) faces $1, 2, 3, 4$ appear three times each, given that no 6 appears.

3. Suppose Nick rolls a die $6r$ times, for some positive integer r. Let X_1 be the number of times that outcome 1 appears and X_5 be the number of times outcome 5 appears.

   (i) Find the probability of the event

   $$P(X_1 = r, \ X_5 = r).$$

   (ii) Determine the variances of X_1 and X_5, and the covariance between them.

   (iii) Calculate the correlation coefficient between X_1 and X_5, and thence show that its value does not depend on r.

4. In Example 6.2.2, suppose the proportions p_i for $1 \leq i \leq 5$, of students choosing to do different sports are as follows:

   $$p_1 = 0.15, \quad p_2 = 0.12, \quad p_3 = 0.10, \quad p_4 = 0.08, \quad p_5 = 0.06.$$

   Find the probability that among the next 10 students who will register at the Sports Club,

   (i) exactly five will select basketball and no more than two will choose volleyball;

   (ii) no one will select rugby and at least three will choose baseball;

   (iii) the total number of students who will select either hockey or rugby is 4, if we know that three students have selected other sports.

5. A shooter aims at a target and with each shot, he earns some points, provided his shot finds the target. The chances that in any given shot, he earns $100, 50$, and 30 points are, respectively, 15%, 30%, and 45%.

   (i) If he shoots at the target 10 times, what is the probability that he scores 100 points 3 times, 50 points twice, and he does not find the target once?

(ii) What is the expected number of points he accumulates in the 10 shots?

(iii) In a series of 10 shots, let X be the number of times he scores 100 points and Y be the number of times he does not hit the target at all. Write down the joint probability function of X and Y. Use it then to calculate the correlation coefficient between X and Y.

6. A company dinner will be attended by 20 persons. Let X_1 and X_2 be the number of persons who will select white and red wine with their meal, respectively. Further, let $X_3 = n - X_1 - X_2$ be the number of persons who will not drink wine with their meal. Assume that the distribution of (X_1, X_2) is trinomial with parameters 20 and p_1, p_2.

(i) After identifying the distribution of X_3, find the mean and variance of the number of persons who will *not* drink wine with their meal.

(ii) Explain what is wrong in the following line of reasoning:

$$Var(X_3) = Var(n - X_1 - X_2) = Var(X_1 + X_2) = Var(X_1) + Var(X_2).$$

(iii) Suppose we know that eight persons will select red wine with their meals. Then, find the conditional distribution of X_3, the number of persons who will not drink wine, and thence obtain the associated conditional expectation and variance.

7. Let (X_1, X_2) follow trinomial distribution with parameters n and p_1, p_2.

(i) Show that $E(X_1 X_2) = n(n-1)p_1 p_2$, and thence explain whether X_1 and X_2 are independent.

(ii) Let $X_3 = n - X_1 - X_2$. Then, calculate the covariance between the random variables

(a) X_1 and X_3;

(b) X_2 and X_3.

[Hint: Use the results in Part (i) of Propositions 6.2.2 and 6.2.3.]

Group B

8. A gambler participates in the following game: she rolls a die and if the outcome is at most 3, she wins US$30; if it is 4 or 5, she loses US$20, while if the outcome is 6, she neither wins nor loses.

Suppose she rolls the die 6 times, and X_1 denotes the number of times she wins and X_2 denotes the number of times she loses.

(i) Calculate the covariance and correlation coefficient between X_1 and X_2.

(ii) Find the expectation and variance of her profit in the 6 rolls.

(iii) Calculate the probability that, after 6 rolls, she

(a) has the same amount of money that she entered the game with;

(b) has won a total of US$50 from this game.

9. For the shooter in Exercise 5, calculate

   (i) the variance of the number of points he accumulates when he shoots 10 times at the target;

   (ii) the correlation coefficient between his total score in the 10 shots and the *number of times* that

      (a) he finds the center of the target (i.e. he scores 100 points);

      (b) he does not find the target at all.

10. Among the customers entering a large store, 20% do not make a purchase, 40% pay with cash, 30% use a credit card for payment, while the remaining 10% of the customers pay using the reward card offered by the store to its customers. The total number of customers entering the store within a minute follows Poisson distribution with parameter $\lambda = 2$. Find the probability that, among the customers entering the store within the next five minutes, there will be 3 who will pay cash, 4 who will use their credit card, and 1 who will use the reward card.

11. A person's blood type can be one of four types: O, A, B, or AB. In a certain country, the percentages of people with each of these four groups in the population are, respectively, 47%, 38%, 11%, and 4%. Suppose a sample of 20 persons is chosen from that population and tested for their blood type.

   (i) What is the probability that none of them has blood type B or AB?

   (ii) What is the probability that exactly three persons have blood type A and two persons have blood type B?

   (iii) Let $X_1, X_2,$ and X_3 be the number of persons in the sample whose blood type is found to be O, A, and B, respectively. Then, calculate the probabilities

      (a) $P(X_1 + X_2 = 9)$,

      (b) $P(X_1 + X_2 + X_3 \leq 18)$,

      (c) $P(4X_2 + 3X_3 = 16)$

      (d) $P(X_2 + X_3 = 10|X_1 = 9)$.

12. The lifetimes for a certain type of batteries (in thousands of hours) follow exponential distribution with mean 5. If we buy 12 such batteries, find the probability that 2 of them will fail before 3000 hours, 5 batteries will have a lifetime between 3000 and 5000 hours, while the remaining 5 will last for more than 5000 hours.

13. Consider a sequence of $n = 3r$ trials (r is a positive integer) with three possible outcomes in each trial. Denote these outcomes by $A, B,$ and C, respectively. Assume that in each trial, A occurs with probability p^2 and B occurs with probability $2p - 2p^2$, for some $0 < p < 1$.

   (i) Find the value of p that maximizes the probability that in the n trials, all three outcomes appear the same number of times.

   (ii) For the value of p found in Part (i), verify that the probability that B occurs in each trial is twice the probability that A occurs, while the probabilities of the outcomes A and C are equal.

14. Let $X_{(1)} \leq \cdots \leq X_{(n)}$ be the ordered sample obtained from a random sample $X_1, \ldots, X_n$ from a continuous distribution with density function $f(x)$ and distribution function $F(x)$. Consider the ordered variables $X_{(i)}$ and $X_{(j)}$, for $1 \leq i < j \leq n$, and let x and y be two real numbers, with $x < y$.

(i) Let Z denote the number of X_r, for $r = 1, 2, \ldots, n$, that are less than or equal to x, and W denote the number of X_r, for $r = 1, 2, \ldots, n$, for which $x < X_r \leq y$. Then, show that the joint probability function of Z and W is given by

$$P(Z = z,\ W = w) = \frac{n!}{z!\,w!\,(n - z - w)!}(F(x))^z (F(y) - F(x))^w (1 - F(y))^{n-z-w},$$

$$\text{for } 0 \leq z, w \leq n \quad \text{and } z + w \leq n.$$

(ii) Give an expression for the joint distribution function of $X_{(i)}$ and $X_{(j)}$ in terms of the joint probability function of Z and W.

(iii) Thence, deduce that

$$F_{X_{(i)}, X_{(j)}}(x, y) = \sum_{s=j}^{n} \sum_{r=i}^{s} \frac{n!}{r!\,(s-r)!\,(n-s)!}$$
$$\times (F(x))^r (F(y) - F(x))^{s-r} (1 - F(y))^{n-s}.$$

(iv) Differentiating the above expression with respect to x and y, verify that the joint density function of $X_{(i)}$ and $X_{(j)}$ is the same as the one presented in Proposition 3.6.2.

6.3 MULTIVARIATE HYPERGEOMETRIC DISTRIBUTION

The multinomial distribution, discussed in Section 6.2, is a generalization of the binomial distribution. An analogous generalization for the hypergeometric distribution to the multivariate case is presented next.

Definition 6.3.1 Suppose an urn contains balls of $k + 1$ different colors, for $k \geq 1$. Let the number of balls with color of type r, for $r = 1, \ldots, k$, be a_r, and that b balls have color of the last type, $k + 1$. Suppose we select from the urn at random and without replacement n balls, where $n \leq a_1 + \cdots + a_k + b$, and $X_1, \ldots, X_k$ denote the number of balls of colors $1, \ldots, k$, respectively, selected. Then, the k-dimensional random variable $(X_1, \ldots, X_k)$ is said to have the multivariate hypergeometric distribution with parameters $a_1, a_2, \ldots, a_k, b$, and n.

It is evident that for $k = 1$, we get the hypergeometric distribution seen in the first volume. Note that another random variable which appears implicitly in the experiment above is X_{k+1} denoting the number of balls selected with color of type $k + 1$. But this is determined by the values of the variables $X_1, \ldots, X_k$ since

$$X_{k+1} = n - (X_1 + \cdots + X_k),$$

and for this reason it does not get included in the set of variables describing the distribution.

In the special case of $k = 2$, the distribution is called the **double hypergeometric distribution**, which is discussed next.

Proposition 6.3.1 *Let (X_1, X_2) follow a double hypergeometric distribution. Then, with the notation of Definition 6.3.1, the joint probability function of (X_1, X_2) is given by*

$$f(x_1, x_2) = P(X_1 = x_1, \ X_2 = x_2) = \frac{\binom{a_1}{x_1}\binom{a_2}{x_2}\binom{b}{x_3}}{\binom{a_1 + a_2 + b}{n}}, \qquad (6.2)$$

for $0 \leq x_1 \leq a_1$, $0 \leq x_2 \leq a_2$, and $0 \leq x_3 \leq b$, where $x_3 = n - x_1 - x_2$.

Proof: The number of possible outcomes in the random experiment of selecting n balls out of a total of $a_1 + a_2 + b$ balls from the urn is simply $\binom{a_1 + a_2 + b}{n}$.

Considering now the favorable outcomes for the event $\{X_1 = x_1, \ X_2 = x_2\}$, we see that these are exactly the selections that contain x_1 balls of the first color (out of a total of a_1 balls available in the urn), x_2 balls of the second color (from a total of a_2 balls available), and finally, $x_3 = n - x_1 - x_2$ balls of the third color (out of b such balls in the urn). An application of the multiplicative rule yields the number of different ways in which the three distinct selections above can be made to be

$$\binom{a_1}{x_1}\binom{a_2}{x_2}\binom{b}{x_3},$$

and the result in (6.2) then follows. □

Figure 6.2 shows the probability function of the double hypergeometric distribution for different choices of its parameters.

Example 6.3.1 In a class of 20 students, 4 come to the University every morning by car, 6 arrive with their bicycle, while 10 use public transport. Suppose we select two students at random, and let X_1 and X_2 be the number of students arriving by car and bicycle, respectively. The distribution of the pair (X_1, X_2) is double hypergeometric with parameters $a_1 = 4, a_2 = 6, b = 10$, and $n = 2$, so that the joint probability function of (X_1, X_2) is given by

$$f(x_1, x_2) = P(X_1 = x_1, \ X_2 = x_2) = = \frac{\binom{4}{x_1}\binom{6}{x_2}\binom{10}{2 - x_1 - x_2}}{\binom{20}{2}},$$

for $0 \leq x_1 \leq 4$, $0 \leq x_2 \leq 6$, and $0 \leq 2 - x_1 - x_2 \leq 10$. These conditions are in fact equivalent to the obvious requirement that both x_1 and x_2, but also their sum, is nonnegative and less than or equal to two. The values of $f(x_1, x_2)$ are readily seen to be as follows:

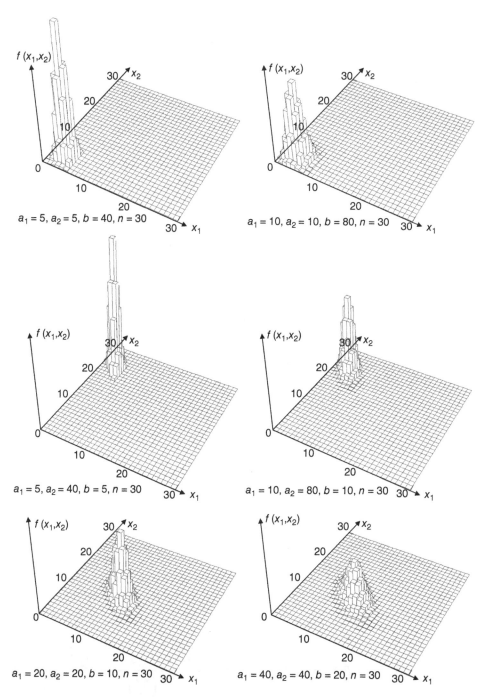

Figure 6.2 Probability function of the double hypergeometric distribution for different choices of parameters a_1, a_2, b, and n.

x_1 \ x_2	0	1	2
0	0.2368	0.3158	0.0790
1	0.2105	0.1263	0
2	0.0316	0	0

The values in the table enable us to calculate any probability associated with X_1, X_2 (and X_3, the number of students who use public transport). For instance, the probability that, from the two students selected, one uses bicycle and the other uses public transport to arrive at the University is simply

$$f(0, 1) = P(X_1 = 0, \ X_2 = 1) = 0.3158.$$

Similarly, we find,

$$P(X_1 \leq 1, \ X_2 \geq 1) = f(0, 1) + f(0, 2) + f(1, 1) = 0.3158 + 0.0790 + 0.1263$$

$$= 0.5211,$$

$$P(X_1 + X_2 = 1) = f(0, 1) + f(1, 0) = 0.3158 + 0.2105 = 0.5263,$$

and so on.

The following proposition gives the marginal probability functions and the corresponding conditional distributions for the double hypergeometric distribution.

Proposition 6.3.2 *Suppose* $(X_1, \ X_2)$ *follows double hypergeometric distribution with parameters* $a_1, a_2, b,$ *and* n. *Then, the following hold:*

(i) *The marginal distribution of* X_1 *is hypergeometric with parameters* $a_1, a_2 + b,$ *and* n, *while the marginal distribution of* X_2 *is hypergeometric with parameters* $a_2, a_1 + b,$ *and* n.

(ii) *The conditional distribution of* X_1, *given* $X_2 = x_2$, *is hypergeometric with parameters* $a_1, b,$ *and* $n - x_2$, *while the conditional distribution of* X_2, *given* $X_1 = x_1$, *is hypergeometric with parameters* $a_2, b,$ *and* $n - x_1$.

Proof:

(i) The marginal probability function of X_1 is given by

$$f_{X_1}(x_1) = P(X_1 = x_1) = \sum_{x_2} P(X_1 = x_1, X_2 = x_2)$$

$$= \sum_{x_2} \frac{\binom{a_1}{x_1}\binom{a_2}{x_2}\binom{b}{n-x_1-x_2}}{\binom{a_1+a_2+b}{n}}$$

$$= \frac{\binom{a_1}{x_1}}{\binom{a_1+a_2+b}{n}} \sum_{x_2} \binom{a_2}{x_2}\binom{b}{n-x_1-x_2},$$

where the summation extends over all x_2 for which

$$0 \le x_2 \le a_2, \quad 0 \le n - x_1 - x_2 \le b.$$

But, in view of the Cauchy combinatorial formula (see Volume I), it holds that

$$\sum_{x_2} \binom{a_2}{x_2}\binom{b}{(n-x_1)-x_2} = \binom{a_2+b}{n-x_1},$$

and so we obtain

$$f_{X_1}(x_1) = \frac{\binom{a_1}{x_1}\binom{a_2+b}{n-x_1}}{\binom{a_1+(a_2+b)}{n}}, \qquad 0 \le x_1 \le a_1 \text{ and } 0 \le n - x_1 \le a_2 + b.$$

This shows that X_1 follows hypergeometric distribution with parameters $a_1, a_2 + b$, and n.

A much simpler proof of the result for this part could be obtained by viewing that the urn initially contains a_1 white balls (those with color type 1) and $a_2 + b$ black balls (the ones which have in fact color types 2 or 3), and then noting that X_1 represents the number of white balls selected in a sample of size n (without replacement).

The marginal distribution of X_2 can be obtained in an analogous manner.

(ii) From the definition of conditional probability, we readily find

$$f_{X_1|X_2}(x_1|x_2) = \frac{f(x_1,x_2)}{f_{X_2}(x_2)} = \frac{\binom{a_1}{x_1}\binom{a_2}{x_2}\binom{b}{n-x_1-x_2}\binom{a_1+a_2+b}{n}^{-1}}{\binom{a_2}{x_2}\binom{a_1+b}{n-x_2}\binom{a_1+a_2+b}{n}^{-1}}$$

$$= \frac{\binom{a_1}{x_1}\binom{b}{(n-x_2)-x_1}}{\binom{a_1+b}{n-x_2}},$$

which is the hypergeometric distribution with parameters a_1, b, and $n - x_2$. Again, an alternative simple argument which requires no calculations is as follows: under the condition $X_2 = x_2$, we end up with an urn containing a_1 balls of color 1 and b balls of color 3. So the number X_1 of color 1 balls chosen is indeed hypergeometric with the given parameters.

The conditional distribution of X_2, given $X_1 = x_1$, can be obtained in an analogous manner.   □

From Proposition 6.3.2 and the formula for the mean of hypergeometric distribution, we immediately have

$$E(X_2 | X_1 = x_1) = (n - x_1)\frac{a_2}{a_2 + b} = \left(-\frac{a_2}{a_2 + b}\right) x_1 + \frac{na_2}{a_2 + b}.$$

As the regression curve of X_2 on X_1 is linear, we can apply Proposition 5.5.2 to obtain

$$\rho_{X_1, X_2}\frac{\sigma_{X_2}}{\sigma_{X_1}} = \frac{Cov(X_1, X_2)}{\sigma_{X_1}^2} = -\frac{a_2}{a_2 + b}.$$

In addition, from Proposition 6.3.2 we have the distribution of X_1 to be hypergeometric with parameters $a_1, a_2 + b$, and n, so that the variance of X_1 is (as seen in Volume I)

$$\sigma_{X_1}^2 = n\frac{a_1(a_2 + b)(a_1 + a_2 + b - n)}{(a_1 + a_2 + b)^2(a_1 + a_2 + b - 1)}.$$

Using the above two equations, we obtain the following proposition for the covariance between X_1 and X_2.

Proposition 6.3.3 *For a random pair (X_1, X_2) following double hypergeometric distribution with parameters a_1, a_2, b and n, the covariance between X_1 and X_2 is*

$$Cov(X_1, X_2) = -n\frac{a_1 a_2(a_1 + a_2 + b - n)}{(a_1 + a_2 + b)^2(a_1 + a_2 + b - 1)}.$$

Example 6.3.2 Suppose items manufactured in a production line are packaged in batches of 30 units. When an item is produced, it is classified (according to certain measurements of characteristics) into one of the three following categories: Good (G), Satisfactory (S), and Unacceptable (U). The production line manager takes samples of size $n = 6$ from each batch of items and records the number of items that fall into each of these three categories. Let X_1 be the number of items that are good and X_2 the number of items classified as satisfactory in such a random sample of size 6. The number of unacceptable items will then be $X_3 = 6 - X_1 - X_2$.

Suppose we know in a batch of 30 items produced, there are $a_1 = 16$ items of class $G, a_2 = 11$ items of class S, and $b = 3$ items of class U. Then, the distribution of (X_1, X_2)

is double hypergeometric with joint probability function

$$f(x_1, x_2) = \frac{\binom{16}{x_1}\binom{11}{x_2}\binom{3}{6 - x_1 - x_2}}{\binom{30}{6}},$$

$$0 \le x_1 \le 16, \quad 0 \le x_2 \le 11, \quad 3 \le x_1 + x_2 \le 6.$$

Moreover, from Proposition 6.3.2, we have the following:

(i) The number of items classified as good in the sample follows hypergeometric distribution with probability function

$$f_{X_1}(x_1) = \frac{\binom{16}{x_1}\binom{14}{6 - x_1}}{\binom{30}{6}}, \qquad 0 \le x_1 \le 6,$$

while the mean and variance of the number of items classified as good in the sample are

$$E(X_1) = 6 \cdot \frac{16}{16 + 14} = \frac{16}{5}, \quad Var(X_1) = 6 \cdot \frac{16 \cdot 14 \cdot (30 - 6)}{30^2 \cdot (30 - 1)} = \frac{896}{725}.$$

(ii) The number of items classified as satisfactory in the sample of 6 items follows the hypergeometric distribution with parameters $11, 19$, and 6, and so its mean and variance are

$$E(X_2) = 6 \cdot \frac{11}{11 + 19} = \frac{11}{5}, \quad Var(X_2) = 6 \cdot \frac{11 \cdot 19 \cdot (30 - 6)}{30^2 \cdot (30 - 1)} = \frac{836}{725}.$$

(iii) Suppose it is known that a sample contains 4 items classified as S. In this case, the (conditional) distribution of the number of items classified as G is hypergeometric with parameters $16, 3$ and $n - 4 = 2$, and so

$$f_{X_1 | X_2}(x_1 | 4) = \frac{\binom{16}{x_1}\binom{3}{2 - x_1}}{\binom{19}{2}}, \qquad 0 \le x_1 \le 2.$$

Thus, for example, if 4 items are classified as S, the probability that both remaining items are G is

$$f_{X_1 | X_2}(2 | 4) = \frac{\binom{16}{2}\binom{3}{0}}{\binom{19}{2}} = \frac{40}{57} \cong 70\%.$$

(iv) From Proposition 6.3.3, we see that the covariance between X_1 and X_2 is

$$Cov(X_1, X_2) = -6 \cdot \frac{16 \cdot 11 \cdot (30 - 6)}{30^2 \cdot (30 - 1)} = -\frac{704}{725}.$$

The expected number of items in the sample that are acceptable (i.e. they are either good or satisfactory) is

$$E(X_1 + X_2) = E(X_1) + E(X_2) = \frac{16}{5} + \frac{11}{5} = \frac{27}{5},$$

while the corresponding variance can be found as

$$Var(X_1 + X_2) = Var(X_1) + Var(X_2) + 2Cov(X_1, X_2)$$

$$= \frac{896}{725} + \frac{836}{725} - 2 \cdot \frac{704}{725} = \frac{324}{725}.$$

It may seem somewhat strange that $X_1 + X_2$ has a smaller variance than either X_1 or X_2. Mathematically, this is due to the fact that X_1 and X_2 are negatively correlated; a simpler explanation can be provided from the observation that the variance of acceptable items in the sample is equal to the variance of *unacceptable items* in the sample. These two variables are in fact perfectly (negatively) correlated, since their sum is fixed and equal to 6 in this case. Note also that the distribution of X_3 is hypergeometric with parameters $3, 27$, and 6, and so an alternative way to get the above obtained results is as follows:

$$E(X_1 + X_2) = 6 - E(X_3), \qquad Var(X_1 + X_2) = Var(X_3).$$

Let us now assume that in the urn model considered in Definition 6.3.1, the selection of n balls is made *with replacement*; that is, when a ball is selected, we record its color and put it back into the urn so that it is available for the next selection as well. For $r = 1, 2, \ldots, k$, we call a "success of type r" the selection of a ball whose color is of type r, while the event that a ball of type $k + 1$ is chosen is called a "failure". In such a sampling scheme, the successive selection of balls from the urn represents a sequence of independent trials, each with $k + 1$ possible outcomes and constant probabilities of appearance for each of these outcomes throughout the experiment; this is so because, once a selected ball is put back into the urn, the composition of the urn remains the same for each trial. Specifically, the probabilities for the $k + 1$ different outcomes are given by

$$p_r = \frac{a_r}{a_1 + + \cdots + a_k + b}, \qquad r = 1, \ldots, k,$$

and

$$q = \frac{b}{a_1 + \cdots + a_k + b}.$$

Therefore, when sampling from the urn takes place with replacement, the k- dimensional random variable $(X_1, \ldots, X_k)$ follows the multinomial distribution (as opposed to the hypergeometric when there is no replacement).

For the special case when $k = 2$, the distribution of (X_1, X_2) will be the trinomial distribution with probability function

$$P(X_1 = x_1, \ X_2 = x_2) = \frac{n!}{x_1! x_2! (n - x_1 - x_2)!} \left(\frac{a_1}{a_1 + a_2 + b} \right)^{x_1} \left(\frac{a_2}{a_1 + a_2 + b} \right)^{x_2}$$
$$\cdot \left(\frac{b}{a_1 + a_2 + b} \right)^{x_3}$$

for $0 \leq x_1, x_2 \leq n$, and $x_1 + x_2 \leq n$, with $x_3 = n - x_1 - x_2$.

Intuitively, when the urn contains a large number of balls of any type, putting or not a selected ball back into the urn would not make much difference. The precise mathematical formulation of this fact is stated in the following proposition, which generalizes an analogous result for the univariate case providing an approximation to the hypergeometric by the binomial distribution.

Proposition 6.3.4 (Approximation of double hypergeometric by trinomial distribution) *Let (X_1, X_2) follow double hypergeometric distribution with parameters a_1, a_2, b, and n. Suppose $a_1 \to \infty$, $a_2 \to \infty$, $b \to \infty$ in such a way that the ratios of a_1 and a_2 over the sum $a_1 + a_2 + b$ tend to finite constants $p_1, p_2 \in (0, 1)$, respectively; that is,*

$$\lim \frac{a_1}{a_1 + a_2 + b} = p_1, \qquad \lim \frac{a_2}{a_1 + a_2 + b} = p_2.$$

Then, we have

$$\lim f(x_1, x_2) = \lim P(X_1 = x_1, \ X_2 = x_2) = \frac{n!}{x_1! x_2! x_3!} p_1^{x_1} p_2^{x_2} q^{x_3}$$

for $0 \leq x_1, x_2 \leq n$ and $x_1 + x_2 \leq n$, where $x_3 = n - x_1 - x_2$ and $q = 1 - p_1 - p_2$.

Proof: We first recall that, for a positive integer n and another integer k with $1 \leq k \leq n$, the symbol $(n)_k$ is defined as the product[1]

$$(n)_k = n(n - 1)(n - 2) \cdots (n - k + 1).$$

Using this notation, we can express the probability function of the double hypergeometric distribution in (6.2) as

$$f(x_1, x_2) = \frac{\dfrac{(a_1)_{x_1}}{x_1!} \cdot \dfrac{(a_2)_{x_2}}{x_2!} \cdot \dfrac{(b)_{x_3}}{x_3!}}{\dfrac{(a_1 + a_2 + b)_n}{n!}} = \frac{n!}{x_1! x_2! x_3!} \cdot \frac{(a_1)_{x_1} (a_2)_{x_2} (b)_{x_3}}{(a_1 + a_2 + b)_n}$$

$$= \frac{n!}{x_1! x_2! x_3!} \cdot \frac{(a_1)_{x_1}}{(a_1 + a_2 + b)^{x_1}} \cdot \frac{(a_2)_{x_2}}{(a_1 + a_2 + b)^{x_2}} \cdot \frac{(b)_{x_3}}{(a_1 + a_2 + b)^{x_3}}$$
$$\cdot \frac{(a_1 + a_2 + b)^n}{(a_1 + a_2 + b)_n}.$$

---

[1] In Combinatorics, $(n)_k$ is usually defined as the number of k-element permutations of a given set of n elements; see Chapter 2 of Volume I.

But,

$$\frac{(a_1)_{x_1}}{(a_1 + a_2 + b)^{x_1}} = \frac{a_1(a_1 - 1) \cdots (a_1 - x_1 + 1)}{(a_1 + a_2 + b)^{x_1}}$$

$$= \frac{a_1}{a_1 + a_2 + b} \cdot \frac{a_1 - 1}{a_1 + a_2 + b} \cdots \frac{a_1 - x_1 + 1}{a_1 + a_2 + b}$$

$$= \frac{a_1}{a_1 + a_2 + b} \left(\frac{a_1}{a_1 + a_2 + b} - \frac{1}{a_1 + a_2 + b} \right) \cdots$$

$$\left(\frac{a_1}{a_1 + a_2 + b} - \frac{x_1 - 1}{a_1 + a_2 + b} \right).$$

Taking now the limits as $a_1 \to \infty, a_2 \to \infty, b \to \infty$, we obtain

$$\lim \frac{(a_1)_{x_1}}{(a_1 + a_2 + b)^{x_1}} = p_1(p_1 - 0) \cdots (p_1 - 0) = p_1^{x_1}.$$

Similarly, we find

$$\lim \frac{(a_2)_{x_2}}{(a_1 + a_2 + b)^{x_2}} = p_2^{x_2}, \qquad \lim \frac{(b)_{x_3}}{(a_1 + a_2 + b)^{x_3}} = q^{x_3},$$

and that

$$\lim \frac{(a_1 + a_2 + b)^n}{(a_1 + a_2 + b)_n} = 1.$$

So, upon taking the limit in the expression for $f(x_1, x_2)$, above, we get

$$\lim f(x_1, x_2) = \frac{n!}{x_1! x_2! x_3!} p_1^{x_1} p_2^{x_2} q^{x_3} \cdot 1 = \frac{n!}{x_1! x_2! x_3!} p_1^{x_1} p_2^{x_2} q^{x_3},$$

as required.   □

Example 6.3.3 In a student election at an University, there are three candidates, Bob, Nicky, and Paul, for the post of the President. After the voting got completed, it is found that 1000 valid votes were cast. Bob received 25% of them, Nicky received 45%, and Paul the remaining 30%. If we select 15 students at random, among those who voted, what is the probability that 5 of them voted for each of the three candidates?

SOLUTION Let $X_1, X_2,$ and X_3 be the numbers of students, among the 15 students chosen, who voted for Bob, Nicky, and Paul, respectively. The distribution of the pair (X_1, X_2) is double hypergeometric with parameters $a_1 = 250, a_2 = 450, b = 300$, and $n = 15$. Thus, the joint probability function of X_1 and X_2 is

$$f(x_1, x_2) = P(X_1 = x_1, \ X_2 = x_2) = \frac{\binom{250}{x_1} \binom{450}{x_2} \binom{300}{15 - x_1 - x_2}}{\binom{1000}{15}},$$

for $0 \leq x_1 \leq 250, 0 \leq x_2 \leq 450$, and $x_1 + x_2 \leq 15$. So, the required probability is simply

$$f(5,5) = P(X_1 = 5, \ X_2 = 5) = \frac{\binom{250}{5}\binom{450}{5}\binom{300}{5}}{\binom{1000}{15}}.$$

This is an exact expression and, although it is possible to calculate the desired probability explicitly, the calculation is rather cumbersome, as it involves large numbers. Instead, we may take advantage of the fact that the parameters $a_1, a_2,$ and b of the double hypergeometric distribution are large, and so make use of the trinomial approximation presented in Proposition 6.3.4. The parameters of the trinomial in this approximation are $n = 15$,

$$p_1 = \frac{a_1}{a_1 + a_2 + b} = \frac{250}{1000} = \frac{1}{4}, \qquad p_2 = \frac{a_1}{a_1 + a_2 + b} = \frac{450}{1000} = \frac{9}{20},$$

and the associated probability function is

$$f(x_1, x_2) = \frac{15!}{x_1! x_2! (15 - x_1 - x_2)!} \cdot \left(\frac{1}{4}\right)^{x_1} \cdot \left(\frac{9}{20}\right)^{x_2} \cdot \left(\frac{3}{10}\right)^{15 - x_1 - x_2}$$

for $0 \leq x_1, x_2 \leq 15$ and $x_1 + x_2 \leq 15$.

As a result, we get an approximation for the required probability to be

$$f(5,5) \cong \frac{15!}{5! \, 5! \, 5!} \cdot \left(\frac{1}{4}\right)^{5} \cdot \left(\frac{9}{20}\right)^{5} \cdot \left(\frac{3}{10}\right)^{5} = 0.0331 \cong 3.3\%.$$

On the other hand, the exact value can be computed from the above double hypergeometric probability expression to be 0.0334 (see also Exercise 6 in Section 6.6).

The results for double hypergeometric distribution presented thus far can be generalized for the general case of multivariate hypergeometric distribution, as stated in the following proposition.

Proposition 6.3.5 *Let* $(X_1, \ldots, X_k)$ *be a k-dimensional random variable having multivariate hypergeometric distribution with parameters* $a_1, a_2, \ldots, a_k, b$ *and n. Then, the following hold:*

(i) *The joint probability function of* $X_1, \ldots, X_k$ *is given by*

$$f(x_1, \ldots, x_k) = \frac{\binom{a_1}{x_1} \cdots \binom{a_k}{x_k}\binom{b}{x_{k+1}}}{\binom{a_1 + \cdots + a_k + b}{n}}$$

for $0 \leq x_i \leq a_i, \ i = 1, \ldots, k,$ *and* $0 \leq x_{k+1} \leq b,$ *where* $x_{k+1} = n - x_1 - x_2 - \cdots x_k.$

(ii) *The marginal distribution of X_i, for $i = 1, \ldots, k$, is hypergeometric with parameters $a_i, b + \sum_{j \neq i} a_j$ and n.*

(iii) *The marginal distribution of any collection of r $(2 \leq r \leq k)$ among $X_1, \ldots, X_k$ is an r-dimensional hypergeometric with parameters corresponding to those of the chosen variables (in place of b, we put the sum of all parameters associated with the variables not chosen).*

(iv) *For any $i = 1, \ldots, k$, the conditional joint distribution of all X_j's, for $j \neq i$, given $X_i = x_i$, is a $(k - 1)$-dimensional hypergeometric with parameters a_j for $j \neq i$, b and $n - x_i$.*

(v) *For any $i \neq j$, we have*

$$Cov(X_i, X_j) = -n \frac{a_i a_j \left(\sum_{j=1}^{k} a_j + b - n \right)}{\left(\sum_{j=1}^{k} a_j + b \right)^2 \left(\sum_{j=1}^{k} a_j + b - 1 \right)}.$$

(vi) *Suppose $a_i \rightarrow \infty$, for $i = 1, \ldots, k$, and $b \rightarrow \infty$ in such a way that*

$$\lim \frac{a_i}{\sum_{j=1}^{k} a_j + b} = p_i, \qquad i = 1, \ldots, k.$$

Then, the probability function of multivariate hypergeometric distribution converges to the probability function of multinomial distribution with parameters n and $p_1, p_2, \ldots, p_k$.

Example 6.3.4 A tea chest box contains 40 tea bags of four different flavors: raspberry, lemon, peppermint, and forest fruits, with 8 bags having raspberry flavor, 10 lemon flavored, 7 peppermint flavored, and the rest have forest fruits flavor. If we select at random 8 bags from the chest box, find the probability that

(i) there are two bags from each flavor;

(ii) all selected bags have forest fruits flavor.

SOLUTION Let X_1, X_2, and X_3 be the numbers of bags selected having a raspberry, lemon, and peppermint flavors, respectively. The distribution of the random vector (X_1, X_2, X_3) is 3-dimensional hypergeometric with parameters $a_1 = 8$, $a_2 = 10$, $a_3 = 7$, $b = 15$, and $n = 8$. So, the joint probability function of (X_1, X_2, X_3) is

$$f(x_1, x_2, x_3) = P(X_1 = x_1, X_2 = x_2, X_3 = x_3)$$

$$= \frac{\binom{8}{x_1}\binom{10}{x_2}\binom{7}{x_3}\binom{15}{8-x_1-x_2-x_3}}{\binom{40}{8}},$$

where $0 \le x_1 \le 8$, $0 \le x_2 \le 10$, $0 \le x_3 \le 7$, and $x_1 + x_2 + x_3 \le 8$. We then obtain the required probabilities as follows:

(i) $P(X_1 = 2, X_2 = 2, X_3 = 2) = \dfrac{\binom{8}{2}\binom{10}{2}\binom{7}{2}\binom{15}{2}}{\binom{40}{8}} = 0.0361;$

(ii) $P(X_1 = 0, X_2 = 0, X_3 = 0) = \dfrac{\binom{15}{8}}{\binom{40}{8}} = 0.000084.$

EXERCISES

Group A

1. From an ordinary pack of cards, we select 6 cards at random. What is the probability that among these cards

    (i) there are 4 aces and 2 kings?

    (ii) there are 3 aces and 2 kings?

    (iii) there are two diamonds and one spade?

2. In a company with 100 employees, 20 are over 50 years of age, 50 are between 35 and 50 years of age, while the rest are less than 35 years old. Six persons are chosen at random to form a committee.

    (i) What is the probability that the number of persons from each age group in the committee is the same?

    (ii) Find the probability that the committee contains at least five persons who are over 50 years old.

    (iii) Calculate the probability that the committee contains at least three persons who are over 50 years old and at most two persons between 35 and 50 years old.

    (iv) If it is known that the committee has exactly three members whose age is less than 35 years, find the distribution and the mean and variance of the number of committee members whose age is over 50 years.

3. In the Olympic games final of a track and field event, among the 8 athletes who have qualified, there are 3 from the US, 2 from Russia, 2 from France, and 1 from Australia. Assuming that all athletes are equally likely to win a medal, what is the probability that

(i) at least two US athletes get in the podium?

(ii) no US athlete wins a medal, if we know that the winner of the event is French?

4. A farmer has 4 cows, 15 sheep, and 21 chickens. One morning he counts his animals and finds out that 5 are missing. Let $X_1, X_2,$ and X_3 be the number of cows, sheep, and chickens, respectively, who are missing.

   (i) Write down the joint probability function of (X_1, X_2) and of (X_2, X_3).

   (ii) Calculate the covariance and correlation coefficient for the two pairs in Part (i).

   (iii) Find the mean and variance of

$$Y_1 = X_1 + X_2 + X_3 \quad \text{and} \quad Y_2 = X_1 + 2X_2 + 3X_3.$$

5. A multiple choice test contains 20 questions: 12 of them have four possible answers each, 3 of them have three possible answers each, while the remaining 5 questions are "True/False" questions. A student who takes this test selects 9 questions at random to answer.

   (i) What is the probability that she has chosen the same number of questions for each of the three types?

   (ii) Let X be the number of all possible answers, in total, for the 9 questions chosen. Calculate the probability that $X = 36$.

   (iii) Find the mean and variance of X.

6. An ordinary deck of cards contains 13 cards of each suit (clubs, diamonds, spades, and hearts). Linda selects 12 cards at random. What is the probability that there are 3 cards from each suit, if the selection of cards is made

   (i) with replacement?

   (ii) without replacement?

7. Let (X_1, X_2) follow double hypergeometric distribution with parameters $a_1, a_2, b,$ and n.

   (i) Write down the distribution of the random variable $X_3 = n - X_1 - X_2$. Find the mean and variance of X_3, and explain what is wrong in the following line of reasoning:

$$Var(X_3) = Var(n - X_1 - X_2) = Var(X_1 + X_2) = Var(X_1) + Var(X_2).$$

   (ii) Write down the conditional probability function of X_3, given $X_1 = x_1$, and thence find the associated conditional mean and variance of X_3.

Group B

9. In each draw of a lottery, 6 balls are selected at random from a set of 49 balls numbered $1, \ldots, 49$. Let X_1 be the number of one-digit numbers selected, and

X_2, X_3, X_4 denote the amount of numbers in the ranges $10 - 19, 20 - 29$, and $30 - 39$, respectively, selected in this week's draw.

(i) Find the joint distribution of (X_1, X_2, X_3, X_4).

(ii) Find the marginal distribution of X_1, and derive its mean and variance. Do the same for variable X_2.

(iii) Write down the marginal probability function of the pair (X_1, X_2) and determine the covariance between X_1 and X_2.

(iv) Calculate the expectation of $aX_1 + bX_2$, where a and b are two given positive constants.

10. Among the students who train regularly with the basketball team of a College, there are 6 freshmen, 4 sophomores, 3 juniors, and 7 seniors. The team coach wants to select 12 students for the next basketball game. Assuming that all students are equally likely to be selected, find the probability that

(i) exactly 2 juniors and one sophomore student are selected;

(ii) more juniors than sophomores are selected;

(iii) all four years are represented in the selection.

11. With the assumptions of Example 6.3.4 wherein 8 bags are selected from the chest box, what is the probability that

(i) at least one peppermint and one forest fruits tea bag is selected?

(ii) at least one peppermint and at least two lemon tea bags are selected?

12. A store has a stock of 1000 light bulbs in total, which are of four types: Type I bulbs have an expected lifetime of 1, Type II bulbs have an expected lifetime of 2, Type III bulbs have an expected lifetime of 4 and, finally, Type IV bulbs have an expected lifetime of 8, where all lifetimes are in thousands of hours. A person selects 10 bulbs from this stock at random for lighting the reception of a building. Calculate the mean and variance of the *total lifetime* of these 10 bulbs.

6.4   BIVARIATE NORMAL DISTRIBUTION

In the last two sections, we considered multivariate extensions of the binomial and hypergeometric distributions. The current section deals with bivariate generalization of the normal distribution which, along with its univariate counterpart, has found numerous applications.

The bivariate normal distribution may be introduced in a number of ways. Perhaps the most common one is to give its density function directly. This has the disadvantage of being rather technical and does not really reveal how the distribution arises naturally. For this reason, here we take a different route by introducing the bivariate normal model in terms of the associated marginal and conditional distributions, which we stipulate to be (univariate) normal. Upon using them, we obtain the density function of the bivariate normal.

Definition 6.4.1 Suppose for a two-dimensional random variable (X, Y) having a joint continuous distribution, the following hold:

(i) X has a normal distribution.

(ii) For any $x \in \mathbb{R}$, the conditional distribution of Y given $X = x$ is normal with mean $E(Y|X = x) = ax + b$ and variance $Var(Y|X = x) = c^2$, for some constants a, b, c.

Then, we say that (X, Y) follows the **bivariate normal distribution**.

Condition (ii) in Definition 6.4.1 suggests that the conditional distribution of Y, given $X = x$, is normal with parameters $\mu = ax + b$ and $\sigma^2 = c^2$. This is shown graphically in Figure 6.3 wherein the normal curve appears to follow a parallel movement, keeping however its peak (the point with the highest value for the density) on the line with equation $y = ax + b$.

We now proceed to find the density function $f(x, y)$ of the bivariate normal distribution. For convenience, we set

$$\mu_X = E(X), \quad \sigma_X^2 = Var(X),$$

and we then see from condition (i) in Definition 6.4.1 that the (marginal) density of X is given by

$$f_X(x) = \frac{1}{\sigma_X \sqrt{2\pi}} \exp\left\{ -\frac{(x - \mu_X)^2}{2\sigma_X^2} \right\}, \qquad x \in \mathbb{R}.$$

Further, the conditional density of Y, given $X = x$, is given by

$$f_{Y|X}(y|x) = \frac{1}{c\sqrt{2\pi}} \exp\left\{ -\frac{[y - (ax + b)]^2}{2c^2} \right\}, \qquad y \in \mathbb{R}.$$

Combining the above two expressions, we get, for any $x, y \in \mathbb{R}$,

$$f(x, y) = f_X(x) f_{Y|X}(y|x) = \frac{1}{2\pi c \sigma_X} \exp\left\{ -\frac{1}{2} \left[\frac{(x - \mu_X)^2}{\sigma_X^2} + \frac{(y - ax - b)^2}{c^2} \right] \right\}. \qquad (6.3)$$

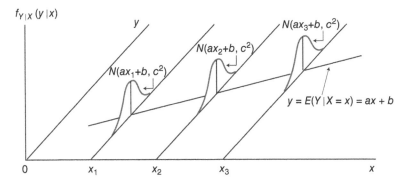

Figure 6.3 The conditional density function $f_{Y|X}(y|x)$, as a function of x, y.

Let us now introduce the notation

$$\mu_Y = E(Y), \quad \sigma_Y^2 = Var(Y), \quad \rho = \rho_{X,Y} = \frac{Cov(X, Y)}{\sigma_X \sigma_Y}.$$

Since the equality $E(Y|X = x) = ax + b$ holds true for all $x \in \mathbb{R}$, we may write $E(Y|X) = aX + b$ and, consequently, $XE(Y|X) = aX^2 + bX$. Next, applying the mean and variance operator in the first equality and the mean operator in the second one we get

$$E(E(Y|X)) = aE(X) + b, \quad Var(E(Y|X)) = a^2 Var(X), \quad E(XE(Y|X)) = aE(X^2) + bE(X)$$

and, in view of Propositions 5.4.1 and 5.4.3, we derive the following formulas

$$\mu_Y = a\mu_X + b, \quad \sigma_X^2 - c^2 = a^2 \sigma_Y^2, \quad E(XY) = aE(X^2) + b\mu_X.$$

Combining all these relations, we obtain

$$a = \rho \frac{\sigma_Y}{\sigma_X}, \quad b = \mu_Y - a\mu_X,$$

while from condition (iv) in Definition 6.4.1, we see that

$$c^2 = (1 - \rho^2)\sigma_Y^2.$$

Substituting these quantities in the expression for the exponent of the function $f(x, y)$ in (6.3), we obtain after a little algebra that

$$\frac{(x - \mu_X)^2}{\sigma_X^2} + \frac{(y - ax - b)^2}{c^2} = \frac{(x - \mu_X)^2}{\sigma_X^2} + \frac{\left[(y - \mu_Y) - \rho \frac{\sigma_Y}{\sigma_X}(x - \mu_X)\right]^2}{(1 - \rho^2)\sigma_Y^2}$$

$$= \frac{1}{1 - \rho^2}\left\{\left(\frac{x - \mu_X}{\sigma_X}\right)^2 - 2\rho \cdot \frac{x - \mu_X}{\sigma_X} \cdot \frac{y - \mu_Y}{\sigma_Y} + \left(\frac{y - \mu_Y}{\sigma_Y}\right)^2\right\}.$$

Thus, we finally obtain the joint density function of the bivariate normal distribution as

$$f(x, y) = \frac{1}{2\pi\sigma_X\sigma_Y\sqrt{1 - \rho^2}} \exp\left[-\frac{1}{2(1 - \rho^2)}Q(x, y)\right], \qquad x, y \in \mathbb{R}, \qquad (6.4)$$

where the function $Q(x, y)$ is given by

$$Q(x, y) = \left(\frac{x - \mu_X}{\sigma_X}\right)^2 - 2\rho \cdot \frac{x - \mu_X}{\sigma_X} \cdot \frac{y - \mu_Y}{\sigma_Y} + \left(\frac{y - \mu_Y}{\sigma_Y}\right)^2. \qquad (6.5)$$

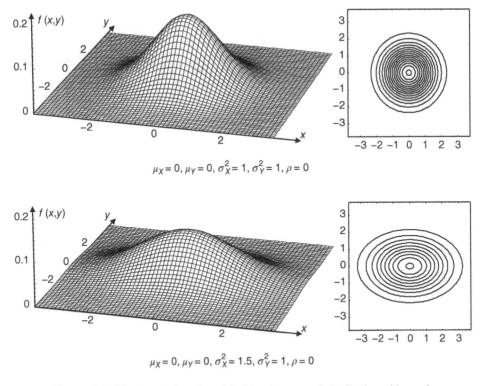

$\mu_X = 0$, $\mu_Y = 0$, $\sigma_X^2 = 1$, $\sigma_Y^2 = 1$, $\rho = 0$

$\mu_X = 0$, $\mu_Y = 0$, $\sigma_X^2 = 1.5$, $\sigma_Y^2 = 1$, $\rho = 0$

Figure 6.4   The density function of the bivariate normal distribution with $\rho = 0$.

The distribution with this density is referred to as the "normal distribution with parameters $\mu_X, \mu_Y, \sigma_X^2, \sigma_Y^2$ and ρ". Figures 6.4 and 6.5 show the shape of the density function $f(x, y)$ for various values of these parameters. The right part of each graph shows the form of the horizontal intersections of the *three*-dimensional graph with planes parallel to the Oxy plane.

It has to be noted that the process we have followed above can be reversed. To be specific, suppose the joint density function $f(x, y)$ of (X, Y) is given by (6.4), where the function $Q(x, y)$ is as in (6.5). Then, we can express the joint density as $f(x, y) = f_1(x)f_2(x, y)$, where

$$f_1(x) = \frac{1}{\sigma_X \sqrt{2\pi}} \exp\left\{ -\frac{(x - \mu_X)^2}{2\sigma_X^2} \right\},$$

and

$$f_2(x, y) = \frac{1}{\sigma_Y \sqrt{2\pi}\sqrt{1 - \rho^2}} \exp\left\{ \frac{\left[y - \left(\rho \dfrac{\sigma_Y}{\sigma_X}(x - \mu_X) + \mu_Y \right) \right]^2}{2(1 - \rho^2)\sigma_Y^2} \right\}.$$

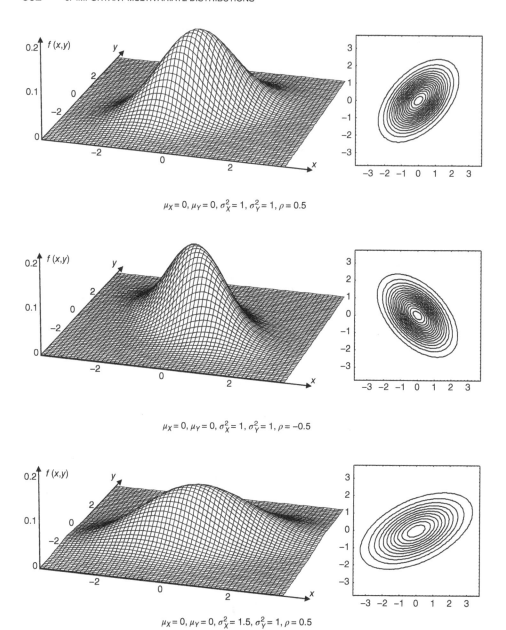

$\mu_X = 0,\ \mu_Y = 0,\ \sigma_X^2 = 1,\ \sigma_Y^2 = 1,\ \rho = 0.5$

$\mu_X = 0,\ \mu_Y = 0,\ \sigma_X^2 = 1,\ \sigma_Y^2 = 1,\ \rho = -0.5$

$\mu_X = 0,\ \mu_Y = 0,\ \sigma_X^2 = 1.5,\ \sigma_Y^2 = 1,\ \rho = 0.5$

Figure 6.5   The density function of the bivariate normal distribution with $\rho \neq 0$.

We, therefore, obtain the marginal density of X as

$$f_X(x) = \int_{-\infty}^{\infty} f(x, y)\, dy = \int_{-\infty}^{\infty} f_1(x) f_2(x, y)\, dy = f_1(x) \int_{-\infty}^{\infty} f_2(x, y)\, dy = f_1(x),$$

(explain why!), while the conditional density function of Y, given $X = x$, is simply

$$f_{Y|X}(y|x) = \frac{f(x, y)}{f_X(x)} = f_2(x, y)$$

$$= \frac{1}{\sqrt{(1 - \rho^2)\sigma_Y^2}\sqrt{2\pi}} \exp\left\{\frac{\left[y - \left(\rho\frac{\sigma_Y}{\sigma_X}(x - \mu_X) + \mu_Y\right)\right]^2}{2(1 - \rho^2)\sigma_Y^2}\right\}.$$

We have thus proved the following result.

Proposition 6.4.1 *Assume that the joint density function of (X, Y) is given by (6.4). Then, we have the following:*

(i) *The marginal distribution of X is normal with parameters μ_X and σ_X^2.*

(ii) *The conditional distribution of Y, given $X = x$, is normal. The associated (conditional) expectation is*

$$E(Y|X = x) = \rho\frac{\sigma_Y}{\sigma_X}(x - \mu_X) + \mu_Y,$$

while the (conditional) variance is

$$Var(Y|X = x) = (1 - \rho^2)\sigma_Y^2.$$

It should be mentioned that some textbooks use the form of the density function in (6.4) to define the bivariate normal distribution; this, as we have seen above, is equivalent to Definition 6.4.1.

Further, it should be clear by symmetry considerations that if the pair (X, Y) follows bivariate normal distribution with parameters μ_X, μ_Y, σ_X^2, σ_Y^2, and ρ, then the following hold:

(a) the marginal distribution of Y is normal with parameters μ_Y and σ_Y^2, and

(b) the conditional distribution of X given that $Y = y$ is normal. The associated conditional expectation is

$$E(X|Y = y) = \rho\frac{\sigma_X}{\sigma_Y}(y - \mu_Y) + \mu_X,$$

and the conditional variance of X, given $Y = y$, is

$$Var(X|Y = y) = (1 - \rho^2)\sigma_X^2.$$

Example 6.4.1 For a certain population of families, it is assumed that (X, Y), where X denotes the height of the father in the family and Y denotes the height of the eldest son, follows the bivariate normal distribution with parameters $\mu_X = 1.70$ m, $\mu_Y = 1.75$ m, $\sigma_X = 0.05$ m, $\sigma_Y = 0.03$ m, and $\rho = 0.5$. Suppose in a specific family, the father has a height of 1.75 m. What is the probability that his eldest son will be taller than his father?

SOLUTION   Proposition 6.4.1 gives the conditional distribution of Y, given that $X = 1.75$, to be normal with

$$E(Y|X = 1.75) = \rho_{X,Y} \frac{\sigma_Y}{\sigma_X}(1.75 - \mu_X) + \mu_Y$$

$$= \frac{(0.5)(0.03)}{0.05}(1.75 - 1.70) + 1.75$$

$$= 1.765,$$

and

$$Var(Y|X = 1.75) = (1 - \rho^2)\sigma_Y^2 = (0.75)(0.0009) = 0.000675.$$

As a result, the required probability is

$$P(Y > 1.75 \mid X = 1.75) = P\left(\frac{Y - 1.765}{\sqrt{0.000\,675}} > \frac{1.75 - 1.765}{\sqrt{0.000\,675}} \mid X = 1.75\right)$$

$$= P(Z > -0.577) = 1 - \Phi(-0.577) = \Phi(0.577)$$

$$= 0.719 \cong 72\%.$$

Here, Z denotes a standard normal random variable, with distribution function Φ.

For comparison, we observe that, since $Y \sim N(1.75, (0.03)^2)$, the (unconditional) probability that the son's height exceeds 1.75m is

$$P(Y > 1.75) = P\left(\frac{Y - 1.75}{0.03} > \frac{1.75 - 1.75}{0.03}\right) = P(Z > 0) = 1 - \Phi(0) = 0.5.$$

How do you explain the difference between this and the previous value?

As we have explained in the last chapter, if two variables are independent they are also uncorrelated, but the converse is not true. The following result shows that for normal random variables, the concepts of independent and uncorrelatedness are equivalent.

Proposition 6.4.2 *Let (X, Y) follow bivariate normal distribution. Then, if the variables X and Y are uncorrelated, they are also independent.*

Proof: As X and Y are uncorrelated, we have their correlation coefficient, $\rho_{X,Y}$, to be zero. Then the density function in (6.4) becomes

$$f(x,y) = \frac{1}{2\pi\sigma_X\sigma_Y} \exp\left\{ -\frac{1}{2}\left[\left(\frac{x-\mu_X}{\sigma_X}\right)^2 + \left(\frac{y-\mu_Y}{\sigma_Y}\right)^2 \right] \right\}, \qquad x,y \in \mathbb{R},$$

or, equivalently,

$$f(x,y) = \frac{1}{\sigma_X\sqrt{2\pi}} \exp\left[-\frac{(x-\mu_X)^2}{2\sigma_X^2} \right] \cdot \frac{1}{\sigma_Y\sqrt{2\pi}} \exp\left[-\frac{(y-\mu_Y)^2}{2\sigma_Y^2} \right], \ x,y \in \mathbb{R}.$$

The independence between X and Y now follows immediately from the above factorization of the joint density, by an appeal to Proposition 3.2.4.   □

For a one-dimensional random variable X following normal distribution, it is known that any linear transformation of it, $aX + b$ (with $a \neq 0$), is also normal. The following result, which is quite useful in many applications, presents a similar result for the bivariate normal distribution.

Proposition 6.4.3 *Let* (X,Y) *follow bivariate normal distribution, and assume that* $a_1, a_2, b_1,$ *and* b_2 *are four real numbers for which*

$$\begin{vmatrix} a_1 & b_1 \\ a_2 & b_2 \end{vmatrix} = a_1 b_2 - a_2 b_1 \neq 0.$$

Then, the distribution of (U, W), *where*

$$U = a_1 X + b_1 Y,$$

$$W = a_2 X + b_2 Y,$$

is also bivariate normal.

Proof: A standard way of proving this result is by finding the joint density function of (U, W) and then verifying that it can be written in the form (6.4). Here we present a different proof, based on Definition 6.4.1.

To start with, we make the following observations:

(a) For the random variable X, we have $X \sim N(\mu_X, \sigma_X^2)$, since the distribution of the pair (X, Y) is bivariate normal.

(b) The conditional distribution of $U = a_1 X + b_1 Y$, given $X = x$, is exactly the distribution of the random variable $a_1 x + b_1 Y = cY + d$ (where we have set $c = b_1$, $d = a_1 x \in \mathbb{R}$). But, since $Y \sim N(\mu_Y, \sigma_Y^2)$, the distribution of the variable $cY + d$ is

$N(c\mu_Y + d, c^2\sigma_Y^2)$. Consequently,

$$E(U \mid X = x) = c\mu_Y + d = b_1\mu_Y + a_1 x$$

(a linear function of x), while the conditional variance of U is

$$Var(U \mid X = x) = c^2\sigma_Y^2 = b_1^2\sigma_Y^2$$

(a constant, that is a quantity not depending on x).

In view of (a), and (b), the distribution of (X, U) is bivariate normal. For reasons of symmetry, this is also true for the pair (U, Y). Thus, it has been established that, if we replace one of the variables in the pair (X, Y) by a linear combination of X and Y, the resulting pair has a bivariate normal distribution.

By the same reasoning, starting from (U, X), we conclude that the pair

$$\left(U, \left(\frac{b_2}{b_1}\right) U + \left(\frac{a_2 b_1 - a_1 b_2}{b_1}\right) X \right),$$

also has a bivariate normal distribution.

However,

$$\frac{b_2}{b_1} U + \frac{a_2 b_1 - a_1 b_2}{b_1} X = \frac{b_2}{b_1}(a_1 X + b_1 Y) + \frac{a_2 b_1 - a_1 b_2}{b_1} X = a_2 X + b_2 Y = W,$$

and this completes the proof of the proposition. Note that we have implicitly assumed above that $b_1 \neq 0$. If $b_1 = 0$, the proof is simpler and it is established using an argument as in (a) and (b) above.   □

We mention that the parameters of the normal distribution for the pair (U, W) in Proposition 6.4.3 can be found easily using standard properties for the expectation and (co)variance of random variables. For example, we have that

$$E(U) = E(a_1 X + b_1 Y) = a_1\mu_X + b_1\mu_Y,$$

$$Var(U) = Var(a_1 X + b_1 Y) = a_1^2\sigma_X^2 + b_1^2\sigma_Y^2 + 2a_1 b_1\rho_{X,Y}\sigma_X\sigma_Y,$$

$$Cov(U, W) = a_1 a_2\sigma_X^2 + b_1 b_2\sigma_Y^2 + (a_1 b_2 + a_2 b_1)\rho_{X,Y}\sigma_X\sigma_Y,$$

and so on.

From Proposition 6.4.3 and the fact that the marginal distributions of bivariate normal distribution are (univariate) normal, we readily obtain the following corollary.

Corollary 6.4.1

(i) *Assume that the pair (X, Y) follows bivariate normal distribution. Then, the distribution of the linear combination $U = a_1 X + b_1 Y$, with $|a_1| + |b_1| \neq 0$, is normal.*

(ii) *(ii) Let $X_1, \ldots, X_n$ be a set of independent and normally distributed random variables, and $a_1, \ldots, a_n$ be real numbers such that not all a_i's are zero. Then, the random variable*

$$S = \sum_{i=1}^{n} a_i X_i,$$

follows a normal distribution.

It is of interest to note that the result in Part (ii) of Corollary 6.4.1 also holds in the case when X_i's are *not independent*. To see this, however, one needs the notion and general properties of a multivariate normal distribution, which are not discussed in this book.

Example 6.4.2 Let (X, Y) follow bivariate normal distribution, and assume that the variables X, Y are homoscedastic, that is, they have the same variance, $Var(X) = Var(Y)$. Then, the random variables $X - Y$ and $X + Y$ are independent.

SOLUTION From Proposition 6.4.3, we see that the pair (U, V), where

$$U = X + Y = 1 \cdot X + 1 \cdot Y,$$
$$V = X - Y = 1 \cdot X + (-1) \cdot Y,$$

follows a bivariate normal distribution. Next, note that

$$Cov(U, V) = Cov(X + Y, X - Y) = Var(X) - Cov(X, Y) + Cov(X, Y) - Var(Y)$$
$$= Var(X) - Var(Y) = 0,$$

and since U and V are uncorrelated and normally distributed, they are also independent by Proposition 6.4.2.

Example 6.4.3 Anne, who is a student at University, wants to attend a seminar next week. The seminar is scheduled to finish at 11 am. However, from past experience it is estimated that the actual time the seminar finishes is a random variable having normal distribution with mean time of 11:06 am and a standard deviation of 1.5 minutes. Anne wants also to attend a lecture that starts at 11:15 am; in fact, it is assumed that the start time of the lecture is again a normal random variable with mean time of 11:15 am and a standard deviation of one minute. Finally, the time that Anne needs to go from the seminar venue to the room where the lecture takes place is a normal random variable with mean of six minutes and a standard deviation of one minute. Assuming that these three variables are independent, what is the probability that she manages to arrive at the lecture room in time for the lecture?

SOLUTION Let us define the variables:

X_1: the time, in minutes after 11 am, at which the seminar ends;

X_2: the time that Anne will need to walk from the seminar room to the room where the lecture takes place;

X_3: the time, in minutes after 11 am, at which the lecture starts.

Then, the required probability is given by

$$P(X_1 + X_2 \leq X_3) = P(X_1 + X_2 - X_3 \leq 0) = P(S \leq 0),$$

where we have set $S = X_1 + X_2 - X_3$. But, since $X_1, X_2,$ and X_3 are assumed independent, each having a normal distribution, the distribution of S will also be normal with mean and variance, respectively, given by

$$E(S) = E(X_1) + E(X_2) - E(X_3) = 6 + 6 - 15 = -3,$$

and

$$Var(S) = Var(X_1) + Var(X_2) + Var(X_3) = (1.5)^2 + 1^2 + 1^2 = 4.25.$$

Thence,

$$P(S \leq 0) = P\left(\frac{S - (-3)}{\sqrt{4.25}} \leq \frac{0 - (-3)}{\sqrt{4.25}} \right) = P(Z \leq 1.455) \cong 0.927 = 92.7\%,$$

and so Anne has a pretty good chance of being in time for the lecture!

EXERCISES

Group A

1. Let (X, Y) follow bivariate normal distribution with $\mu_X = -2$, $\mu_Y = 5$, $\sigma_X^2 = 16$, $\sigma_Y^2 = 9$, and $\rho_{X,Y} = 2/5$. Then, compute the probabilities

   (i) $P(-5 < X \leq 1)$, $P(-5 < X \leq 1 \mid Y = 4)$;

   (ii) $P(-1 < Y \leq 5)$, $P(-1 < Y \leq 5 \mid X = -1)$.

2. Assume that the distribution of (X, Y) is bivariate normal with parameters

$$\mu_X = \mu_Y = 10, \quad \sigma_X^2 = 4, \quad \sigma_Y^2 = 16, \quad \rho_{X,Y} = 1/5.$$

   Calculate each of the quantities below:

   (i) $E(X|Y = 11)$, $Var(X|Y = 11)$;

   (ii) $E(Y|X = 12)$, $Var(Y|X = 12)$;

   (iii) $E(X^2|Y = 11)$, $E(Y^2|X = 12)$;

   (iv) $Cov(2X + 3Y, 3X - 4Y)$.

3. Let X represent a person's height, measured in cm, and Y be his/her weight, measured in kilograms. We assume that the joint distribution of (X, Y) is normal with parameters

$$\mu_X = 175, \quad \mu_Y = 73, \quad \sigma_X^2 = 81, \quad \sigma_Y^2 = 49, \quad \rho_{X,Y} = 3/7.$$

(i) What is the weight distribution for persons whose height is 179 cm?

(ii) For a person whose weight is 67 kg, find the probability that his/her height is between 170 and 176 cm.

4. A psychology test consists of two parts: verbal and quantitative. Each part is assessed on a scale $0 - 100$, and let X and Y denote, respectively, a person's scores in the verbal and quantitative parts. It is assumed that the distribution of (X, Y) is bivariate normal with $\mu_X = 65$, $\mu_Y = 57$, $\sigma_X^2 = 49$, $\sigma_Y^2 = 64$, and $\rho_{X,Y} = 3/5$.

(i) What is the probability that a person who takes this test achieves a score higher than 57 in the quantitative part?

(ii) What is the probability that a person who takes this test achieves a score higher than 57 in the quantitative part, if the person scored 55 in the verbal part?

(iii) What is the probability that a person's scores in the two parts differ, in absolute value, at most by 4?

5. Let X and Y be random variables with $Var(X) = \sigma_X^2$ and $Var(Y) = \sigma_Y^2$. Assuming that the distribution of (X, Y) is bivariate normal and that $a\sigma_X^2 = b\sigma_X^2$ for two constants $a, b > 0$, show that the random variables $aX + bY$ and $aX - bY$ are independent.

[Hint: It suffices to verify that these two variables are uncorrelated and their joint distribution is bivariate normal.]

6. The length of screws produced by a machine follows normal distribution with a mean value of 2.5 in. and a standard deviation of 0.2 in. Find the probability that for a sample of 50 screws examined, their average is between 2 and 2.4 in.

7. The amount of nicotine per cigarette, for a certain cigarette brand, can be described by a normal distribution with mean 0.4 mg and a standard deviation 0.08 mg. If a person smokes 10 cigarettes per day, what is the probability that, during a week, he will be exposed to more than 32 mg of nicotine?

8. Using data from a large population, it is known that the heights of fathers and sons in this population follow a bivariate normal distribution with correlation coefficient $\rho = 0.5$. Both the fathers' and sons' heights have mean 5 ft 8 in. and standard deviation 2 in.

Among the sons who are above average in height, what percentage are shorter than their fathers?

9. The numbers of calories that a person gets during the three main meals of the day are random variables with a mean of 500 cal for breakfast, 950 cal for lunch, and 1100 cal during dinner. The standard deviations of these variables are, respectively, 180, 280, and 310 cal. Assuming that the three amounts of calories are independent

random variables, find the probability that, in a given month (30 days), the average daily income of calories for a person will be between 2520 and 2620.

Group B

10. A darts player aims at a target. The coordinates, X and Y, at a coordinate system whose origin is at the center of the darts board, of the point that the darts hit the board have a bivariate normal distribution with $E(X) = 0, E(Y) = 0, Var(X) = Var(Y) = 9$ and the correlation coefficient between X and Y is zero.

    (i) Write down the distributions of X and Y.

    (ii) Calculate $P(X \le 3, Y \le 3)$ and $P(X > 3 \text{ or } Y > 3)$.

    (iii) Find the distributions of the variables $T_1 = X^2/9$ and $T_2 = Y^2/9$.

    (iv) What is the distribution of $T_1 + T_2$?

    (v) Calculate the probability that a dart lands at the interior circle of the board having radius 3.

11. Let X and Y be two uncorrelated random variables and define two new variables U and V as $U = X - Y$ and $V = X + Y$.

    (i) Prove that the correlation coefficient between U and V is

    $$\rho_{U,V} = \frac{Var(X) - Var(Y)}{Var(X) + Var(Y)}.$$

    (ii) Assume now that X and Y have the same variance and that their joint distribution is bivariate normal. If it is known in addition that

    $$P(X - Y \le a) = p, \quad P(X + Y \le b) = q,$$

    express, in terms of p and q, the probabilities

    $$P(U \le a, W \le b) \quad \text{and} \quad P(U \le a \text{ or } W \le b).$$

12. Let (X, Y) have a joint continuous distribution with joint density function

    $$f(x, y) = \phi(x, y) \left[1 + 2\pi \, e \, xy \, \phi(x, y) \right], \qquad x, y \in \mathbb{R},$$

    where the function $\phi(x, y)$ is defined, for any real x, y, as

    $$\phi(x, y) = (2\pi)^{-1} \exp \left[-\frac{1}{2} \left(x^2 + y^2 \right) \right].$$

    Verify that each of the variables X, Y follows standard normal distribution. (This shows that if the marginal distributions of two random variables are normal, it does not follow that their joint distribution is bivariate normal.)

13. We want to measure the pH of a chemical substance. Let μ be the true value of pH, which is assumed unknown. Using an instrument, we make n independent

measurements. Suppose each measurement is a normal random variable with mean μ and variance 0.01. Find how many measurements are needed so that, with probability 0.95, the average of the measurements differs from μ at most by 0.02 in absolute value.

14. Let X and Y be two independent standard normal random variables, while U and W are related to X and Y as

$$U = \sigma_1 X + \mu_1, \qquad W = \sigma_2 \left[\rho X + \sqrt{1 - \rho^2} Y \right] + \mu_2,$$

where $\mu_1, \mu_2 \in \mathbb{R}$, $\sigma_1 > 0$, $\sigma_2 > 0$, $-1 < \rho < 1$ are given real numbers. Show that (U, W) has bivariate normal distribution with parameters

$$E(U) = \mu_1, \quad E(W) = \mu_2, \quad Var(U) = \sigma_1^2, \quad Var(W) = \sigma_2^2, \quad \rho_{U,W} = \rho.$$

15. Prove that if (U, W) follows a bivariate normal distribution, then U and W can be expressed in the form

$$U = aX + b, \quad W = cX + dY + h,$$

where X and Y are independent variables with standard normal distribution, for some suitable real constants $a, b, c, d,$ and h.

[Hint: Set $E(U) = \mu_1, E(W) = \mu_2, Var(U) = \sigma_1^2, Var(W) = \sigma_2^2$ and $\rho_{U,W} = \rho$. Then, use the result of Exercise 14 above to show that X and Y are given by

$$X = \frac{U - \mu_1}{\sigma_1}, \quad Y = \frac{1}{\sqrt{1 - \rho^2}} \left(-\frac{\rho(U - \mu_1)}{\sigma_1} + \frac{W - \mu_2}{\sigma_2} \right).]$$

6.5   BASIC CONCEPTS AND FORMULAS

Multinomial distribution	The joint distribution of the numbers that the outcomes $S_1, \ldots, S_k$ occur in n independent repetitions of an experiment with $k + 1$ possible outcomes, $S_1, \ldots, S_k, F$, with respective probabilities $p_1, \ldots, p_k, q$
Trinomial distribution with parameters n and p_1, p_2	The special case of $k = 2$ in the multinomial distribution • $f(x_1, x_2) = P(X_1 = x_1, \ X_2 = x_2) = \dfrac{n!}{x_1! x_2! x_3!} p_1^{x_1} p_2^{x_2} q^{x_3}$ for any $0 \leq x_1, x_2 \leq n$ such that $x_1 + x_2 \leq n$, where $x_3 = n - x_1 - x_2$ and $q = 1 - p_1 - p_2$ • $Cov(X_1, X_2) = -np_1 p_2$ • The marginal distribution of X_1 is binomial with parameters n and p_1, while that of X_2 is $b(n, p_2)$

	• The conditional distribution of X_1, given $X_2 = x_2$, is $b(n - x_2, p_1/(p_1 + q))$ $$E(X_1 \mid X_2 = x_2) = (n - x_2)\frac{p_1}{p_1 + q}$$ Similarly for the conditional distribution of X_2, given $X_1 = x_1$
Multivariate hypergeometric distribution	The joint distribution for the numbers of balls with colors $1, \ldots, k$ that are included in a sample of n balls, taken without replacement from an urn containing a_r balls of color r (for $r = 1, \ldots, k$) and b balls of color $k + 1$
Double hypergeometric distribution with parameters a_1, a_2, b and n	The special case of $k = 2$ in the multivariate hypergeometric distribution • $f(x_1, x_2) = P(X_1 = x_1, \ X_2 = x_2) =$ $$\frac{\binom{a_1}{x_1}\binom{a_2}{x_2}\binom{b}{x_3}}{\binom{a_1 + a_2 + b}{n}}$$ for $0 \le x_1 \le a_1$, $0 \le x_2 \le a_2$ and $0 \le x_3 \le b$, where $x_3 = n - x_1 - x_2$ • $Cov(X_1, X_2) = -n\dfrac{a_1 a_2 (a_1 + a_2 + b - n)}{(a_1 + a_2 + b)^2 (a_1 + a_2 + b - 1)}$ • The marginal distribution of X_1 is hypergeometric with parameters a_1, $a_2 + b$, and n, while that of X_2 is hypergeometric with parameters a_2, $a_1 + b$, and n • The conditional distribution of X_1, given $X_2 = x_2$, is hypergeometric with parameters a_1, b, and $n - x_2$, while the conditional distribution of X_2, given $X_1 = x_1$, is hypergeometric with parameters a_2, b, and $n - x_1$
Approximation of the double hypergeometric by the trinomial distribution	Suppose $a_1 \to \infty$, $a_2 \to \infty$, $b \to \infty$ in such a way that $$\lim \frac{a_1}{a_1 + a_2 + b} = p_1, \quad \lim \frac{a_2}{a_1 + a_2 + b} = p_2$$ Then, the joint probability function of double hypergeometric with parameters a_1, a_2, b, and n converges to that of trinomial distribution with parameters n and p_1, p_2

Bivariate normal distribution with parameters μ_X, μ_Y, σ_X^2, σ_Y^2, and $\rho_{X,Y}$	$$f(x,y) = \frac{1}{2\pi\sigma_X\sigma_Y\sqrt{1-\rho^2}} \exp\left[-\frac{1}{2(1-\rho^2)}Q(x,y)\right]$$ for $x,y \in \mathbb{R}$, where $$Q(x,y) = \left(\frac{x-\mu_X}{\sigma_X}\right)^2 - 2\rho \cdot \frac{x-\mu_X}{\sigma_X} \cdot \frac{y-\mu_Y}{\sigma_Y} + \left(\frac{y-\mu_Y}{\sigma_Y}\right)^2$$
Properties of a random pair (X, Y) following the bivariate normal distribution	• The marginal distribution of X is normal with parameters μ_X and σ_X^2, while the marginal distribution of Y is normal with parameters μ_Y and σ_Y^2 • The conditional distribution of Y, given $X = x$, is normal with (conditional) expectation $$E(Y \mid X = x) = \rho\,\frac{\sigma_Y}{\sigma_X}\,(x-\mu_X) + \mu_Y$$ and (conditional) variance $$Var(Y \mid X = x) = (1-\rho^2)\sigma_Y^2$$ Similar results hold for the conditional distribution, mean and variance of X, given $Y = y$ • If the variables X and Y are uncorrelated, then they are independent • If $U = a_1X + b_1Y$, and $V = a_2X + b_2Y$ with $a_1b_2 - a_2b_1 \neq 0$, then (U, V) follows bivariate normal distribution • If $U = a_1X + b_1Y$ with $|a_1| + |b_1| \neq 0$, the distribution of U is (univariate) normal • Every linear combination of independent normal random variables has a normal distribution

6.6 COMPUTATIONAL EXERCISES

As with univariate distributions, Mathematica has built-in functions for the most important multivariate distributions, and in particular the distributions discussed in this chapter. The syntax for calling these functions, for the three distributions discussed so far in this chapter, is given in Table 6.1.

Mathematica function	Distribution
`MultinormalDistribution` $[\{\mu_X, \mu_Y\}, r]$	Bivariate normal distribution of (X, Y) with means μ_X, μ_Y. The list r has the remaining characteristics of the distribution, namely, $$r = \{\{\sigma_X^2, \rho_{X,Y}\sigma_X\sigma_Y\}, \{\rho_{X,Y}\sigma_X\sigma_Y, \sigma_Y^2\}\}$$
`MultinomialDistribution` $[n, \{p_1, p_2, p_3\}]$	Trinomial distribution with parameters n and p_1, p_2 ($p_3 = 1 - p_1 - p_2$)

`MultivariateHypergeometric`
`Distribution`
$[n, \{m_1, m_2, m_3\}]$
	Double hypergeometric distribution with parameters m_1, m_2, m_3 and n (sampling of n units without replacement from an urn containing m_i items of color i, $i = 1, 2, 3$)

Table 6.1   Mathematica functions for multivariate distributions.

The following program calculates various probabilities associated with the bivariate normal distribution of (X, Y) with respective means $\mu_X = 1, \mu_Y = 2$, variances $\sigma_X^2 = 4$, $\sigma_Y^2 = 9$, and correlation coefficient $\rho_{X,Y} = 5/6$.

```
In[1]:= r:= {{4, 5}, {5, 9}};
ndist:= MultinormalDistribution[{1, 2}, r];
Print["Joint Density Function of a Bivariate Normal Distribution"]
Plot3D[PDF[ndist, {x, y}], {x, -5, 7}, {y, -4, 8}]
Print["Contour Plot for the Joint Density of a Bivariate Normal Distribu-
tion"]
ContourPlot[PDF[ndist, {x, y}], {x, -5, 7}, {y, -4, 8}]
Print["Joint Distribution Function of a Bivariate Normal Distribution"]
Plot3D[CDF[ndist, {x, y}], {x, -5, 7}, {y, -4, 8}]
Print["Calculation of Probabilities for a Bivariate Normal Distribution"]
Print["P(X<3,Y<5)= ", N[CDF[ndist, {3, 5}]]]
Print["P(X<3)= ", N[CDF[ndist, {3, Infinity}]]]
Print["P(Y<4)= ", N[CDF[ndist, {Infinity, 4}]]]
Print["P(X<3|Y<4)= ",
N[CDF[ndist, {3, 4}]]/N[CDF[ndist, {Infinity, 4}]]]
Print["P(2<X<3, 4<Y<5)= ",
N[CDF[ndist, {3, 5}]] - N[CDF[ndist, {2, 5}]] -
N[CDF[ndist, {3, 4}]] + N[CDF[ndist, {2, 4}]]]
```

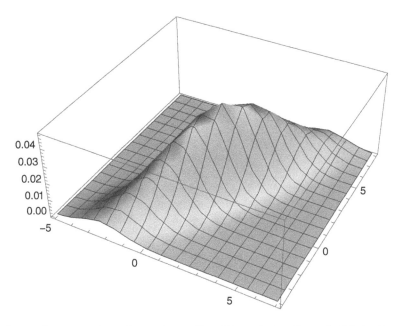

Out[4]= Joint Density Function of a Bivariate Normal Distribution
Out[6]= Contour Plot for the Joint Density of a Bivariate Normal Distribution

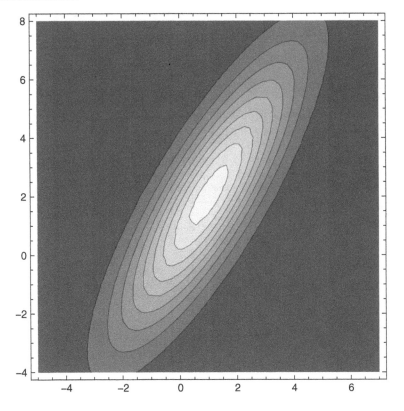

Out[8]= Joint Distribution Function of a Bivariate Normal Distribution

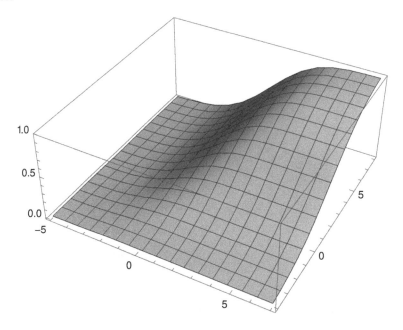

Calculation of Probabilities for a Bivariate Normal Distribution
P(X<3,Y<5)= 0.785633
P(X<3)= 0.841345
P(Y<4)= 0.747507
P(X<3|Y<4)= 0.961653
P(2<X<3, 4<Y<5)= 0.0322392

Similarly, we now carry out the calculations for Example 6.2.3, associated with trinomial distribution with parameters $n = 5$ and $p_1 = 0.5, p_2 = 0.3$ (so that $p_3 = 1 - p_1 - p_2 = 0.2$).

```
In[2]:= n:= 5; p1:= 0.5; p2:= 0.3; p3:= 1 - p1 - p2;
ndist:= MultinomialDistribution[n, {p1, p2, p3}];
Print["Calculation of probabilities for the trinomial distribution"]
f[x_, y_]:= PDF[ndist, {x, y, n - x - y}]
Print["P(X=2, Y=2) = ", f[2, 2]];
Print["P(Y=1, Z<=1) = ", f[3, 1] + f[4, 1]];
prob:= Sum[f[i, i], {i, 0, 2}];
Print["P(X=Y) = ", prob]
```

Out[2]= Calculation of probabilities for the trinomial distribution
P(X=2, Y=2) = 0.135
P(Y=1, Z<=1) = 0.24375
P(X=Y) = 0.15932

Use similar ideas to solve the following exercises.

1. 12 couples attend a company dinner. For the main course of the meal, there are four options: chicken, lamb, fish, or vegetarian. Assuming all options are equally likely, use Mathematica to calculate the probabilities that

    (i) more persons choose lamb than fish;

    (ii) the same number of persons choose chicken and vegetarian dish;

    (iii) at least four persons choose lamb and at least seven choose fish, given that exactly three persons choose the vegetarian dish.

2. For the position of the President in a Student election, there are three candidates: A, B, and C. Suppose after the voting is complete, 3000 valid votes have been cast with 900 of them for candidate A, 1400 for B, and the rest for the third candidate.

    (i) If we pick 30 persons and ask them who they voted for, what is the probability that 10 persons among them voted for each candidate?

    (ii) How does the result in Part (i) change if the 30 persons are chosen with replacement (so that we may ask the same person more than once)? What conclusion can we draw from this?

3. For a certain football team, we assume that the team wins a game with probability 60%, loses a game with a probability 10%, while the chance of drawing a game is 30%.

    (i) In the first 30 games in a season, what is the probability that the team has the same number of wins, draws, and losses?

    (ii) In these 30 games, let X be the number of wins and Y be the number of draws. Write down the joint probability function of X and Y. Use this to calculate the correlation coefficient between X and Y.

4. Helen plays the following game with dice: she rolls a die and, on each roll, she receives US\$5 if the outcome is 1, 3, or 5; if the outcome is 2 or 6, she loses US\$4, while if the outcome is 4 she neither wins nor loses. Calculate the probability that after 100 rolls,

    (i) she has won a total of US\$10;

    (ii) she has won at least US\$10;

    (iii) she has the same amount of money as what she entered the game with.

5. The lifetimes X and Y, in thousands of hours, of two machines used as parts of the same electric system follow bivariate normal distribution with $\mu_X = 100$, $\mu_Y = 200$, $\sigma_X^2 = 900$, $\sigma_Y^2 = 1600$, and $\rho_{X,Y} = 3/8$. Calculate the probability that

    (i) both machines operate after 150 000 hours;

    (ii) neither machine operates after 170 000 hours;

    (iii) the life duration of the first machine is between 70 000 and 90 000 hours while that of the second one is between 170 000 and 190 000 hours;

    (iv) the first machine is still working after 150 000 hours given that the other machine also works at that time.

6. Use Mathematica to obtain the exact probability $f(5, 5)$ in Example 6.3.3, that is, all three candidates received five votes in the sample of 15 students. Thence, assess the accuracy of the trinomial approximation used in the solution of that example.

7. A person who flies with a parachute wants to land at a specific point, which can be considered as the origin of a Cartesian plane. The coordinates (X, Y) of the landing point on that plane are assumed to follow a bivariate normal distribution with parameters

$$\mu_X = 10, \quad \mu_Y = 30, \quad \sigma_X^2 = 9, \quad \sigma_Y^2 = 25, \quad \rho_{X,Y} = 2/5.$$

Calculate the following probabilities related to the coordinates of the point that he lands at:

(i) $P(X < 10, Y < 30)$;

(ii) $P(5 < X < 15, Y < 30)$;

(iii) $P(5 < X < 15, 20 < Y < 35)$;

(iv) $P(5 < X < 15, Y < 30 \mid X < 18, Y < 38)$.

8. A psychology test consists of two parts: verbal and written. Every person who takes the test receives a score, on a scale $0 - 100$, for each of the two parts. Let X and Y denote a person's score in the verbal and written parts, respectively. Assuming that the pair (X, Y) has a bivariate normal distribution with

$$\mu_X = 60, \quad \mu_Y = 70, \quad \sigma_X^2 = 81, \quad \sigma_Y^2 = 100, \quad \rho_{X,Y} = 1/2,$$

answer the following:

(i) Find the percentage of persons taking on this test who will

- have a score less than 50 in the verbal part and at least 75 in the written part,
- have a score higher than 70 in the verbal part and at least 75 in the written part,
- have a score higher than 70 in the verbal part, given that the score in the written part is at least 75.
- have a score higher than 75 in the written part, given that the score in the verbal part is at least 70.

(ii) If 60 persons take the test on a given day, calculate the probability that their average score is at least 64 in the verbal part and at least 68 in the quantitative part.

6.7 SELF-ASSESSMENT EXERCISES

6.7.1 True–False Questions

1. Ten persons are sitting in a room. Let $X_1, X_2,$ *and* X_3 be the number of persons, among them, who were born in January, February, and March, respectively (assume that all months are equally likely). Then, the distribution of (X_1, X_2, X_3) is trinomial.

2. Nick rolls a die *four* times. The probability that exactly two sixes and one ace will appear is $6!(1/6)^4$.

3. In a poker hand, a player receives *five* cards. Let X_1 be the number of spades and X_2 be the number of hearts he receives in a given hand. Then, the joint distribution of (X_1, X_2) is trinomial.

4. Steve, Tom, and Laura receive five cards each in a poker hand. Let $X, Y,$ *and* Z be the number of aces that Steve, Tom, and Laura receive, respectively, in a given hand. Then, the joint distribution of (X, Y) is double hypergeometric.

5. Assume that (X, Y) follows a trinomial distribution with parameters n and p_1, p_2. Then,
$$P(Y = k) = \binom{n}{k} p_2^k p_1^{n-k}, \qquad k = 0, 1, \ldots, n.$$

6. We roll a die 20 times. Let X be the number of ones and Y be the number of sixes that appear in these 20 rolls. Then,
$$Cov(X, Y) = -\frac{5}{9}.$$

Questions 7–11 below refer to the following scenario:

In a group of 50 persons, there are 21 persons with blood type O, 17 persons with blood type A, 8 persons with blood type B, and 4 persons with blood type AB. We select five persons at random and let $X_1, X_2, X_3,$ *and* X_4 be the number of persons with blood types $O, A, B,$ and AB, respectively, in that sample.

Consider now the following statements and answer whether they are true or not.

7. The distribution of $X_1 + X_2$ is hypergeometric.

8. The conditional distribution of (X_2, X_3), given $X_1 = 3$, is double hypergeometric.

9. The conditional probability $P(X_1 = 2, X_2 = 1 | X_3 = 0)$ is equal to
$$\frac{\binom{21}{2}\binom{17}{1}}{\binom{42}{5}}.$$

10. The probability that *two* persons in the sample have a blood type A, *one* person has blood type B and none has blood type AB is equal to
$$\frac{\binom{21}{2}\binom{17}{2}\binom{8}{1}}{\binom{50}{5}}.$$

11. The probability that all five persons selected have a blood type that is either O or AB is equal to
$$\frac{\binom{25}{5}}{\binom{50}{5}}.$$

12. Suppose (X, Y) follows a bivariate normal distribution with parameters

$$\mu_X = 1, \quad \mu_Y = 10, \quad \sigma_X^2 = 9, \quad \sigma_Y^2 = 81, \quad E(XY) = 10.$$

Then, X and Y are independent.

13. If the distribution of (X, Y) is bivariate normal, then the marginal distributions of X and Y are normal.

14. Suppose the pair (X, Y) follows a bivariate normal distribution. Then, the variable $5X - 3Y^2$ also follows normal distribution.

15. Suppose the variables X and Y have a joint continuous distribution such that the marginal distributions of X and Y are normal. Then, the distribution of (X, Y) is bivariate normal.

16. Assume that (X, Y) have a bivariate normal distribution with parameters

$$\mu_X = 1, \quad \mu_Y = 2, \quad \sigma_X^2 = 4, \quad \sigma_Y^2 = 16, \quad \rho_{X,Y} = 1/3.$$

Then,

$$E(Y \mid X = 2) = \frac{8}{3}.$$

17. Let (X, Y) have a bivariate normal distribution with parameters

$$\mu_X = 4, \quad \mu_Y = -3, \quad \sigma_X^2 = 25, \quad \sigma_Y^2 = 25, \quad \rho_{X,Y} = 1/4.$$

Then

$$P(|Y| \leq 1) = \Phi\left(\frac{4}{5}\right) - \Phi\left(\frac{2}{5}\right),$$

where Φ denotes the (cumulative) distribution function of the standard normal distribution.

18. We select three digits from 1 to 9 with replacement. The probability that there is exactly one 4 and one 5 among the three selected digits is

$$3\left(\frac{1}{9}\right)^2 \left(\frac{7}{9}\right).$$

19. We select three digits from 1 to 9 without replacement. The probability that both 4 and 5 are among the selected digits is $1/12$.

6.7.2 Multiple Choice Questions

1. Nick rolls a die 12 times. The probability that each of the outcomes 3 and 5 appear exactly twice is

(a) $\dfrac{12!}{8!} \cdot \left(\dfrac{2}{3}\right)^{12}$

(b) $\dfrac{12!}{4! \, 8!} \cdot \dfrac{4^8}{6^{12}}$

(c) $\dfrac{12!}{8!} \cdot \dfrac{4^7}{6^{12}}$

(d) $\dfrac{12!}{4! \cdot 8!} \cdot \left(\dfrac{1}{3}\right)^2 \left(\dfrac{2}{3}\right)^8$

(e) $\dfrac{12!}{8!} \cdot \dfrac{4^{10}}{6^{12}}$

2. When rolling a die five times, the probability that exactly one six appears given that two aces appear, is

(a) $\dfrac{16}{25}$ (b) $\dfrac{25}{72}$ (c) $\dfrac{48}{125}$ (d) $\dfrac{480}{6^5}$ (e) $\dfrac{125}{216}$

3. Let (X, Y) follow trinomial distribution with $n = 30$ and $p_1 = 1/3, p_2 = 1/2$. Then, the correlation coefficient between X and Y is equal to

(a) $\dfrac{1}{5}$ (b) $-\dfrac{1}{\sqrt{18}}$ (c) $-\dfrac{1}{5}$ (d) $-\dfrac{\sqrt{2}}{2}$ (e) $-\dfrac{\sqrt{18}}{3}$

4. A box contains 5 red, 7 blue, and 4 yellow balls. A friend of Lucy selects four balls from the box at random, without replacement, and she tells Lucy that exactly one of them is red. Then, the distribution of yellow balls selected is hypergeometric with parameters $a, b,$ and n, where

(a) $a = 4, b = 7, n = 3$ (b) $a = 4, b = 5, n = 7$

(c) $a = 7, b = 5, n = 4$ (d) $a = 3, b = 7, n = 4$

(e) $a = 4, b = 5, n = 3$

5. In each hand of bridge, a player receives 13 cards. The probability that, in a given hand, a player receives 4 clubs and 4 hearts is

(a) $\dfrac{\binom{13}{4}\binom{13}{4}}{\binom{52}{13}}$ (b) $\dfrac{\binom{13}{4}\binom{13}{4}\binom{26}{5}}{\binom{52}{13}}$ (c) $\dfrac{\binom{26}{8}\binom{26}{5}}{\binom{52}{13}}$

(d) $\dfrac{52!}{26!26!}\left(\dfrac{1}{4}\right)^8\left(\dfrac{1}{2}\right)^5$ (e) $\dfrac{52!}{13!13!26!}\left(\dfrac{1}{4}\right)^8\left(\dfrac{1}{2}\right)^5$

6. In a given hand of bridge, the probability that a player receives 5 clubs and 3 hearts, given that no spades are received, is

(a) $\dfrac{\binom{13}{5}\binom{13}{3}\binom{13}{5}}{\binom{52}{13}}$ (b) $\dfrac{\binom{13}{5}\binom{13}{3}}{\binom{52}{13}}$ (c) $\dfrac{\binom{13}{5}\binom{13}{3}}{\binom{26}{5}}$

(d) $\dfrac{\binom{13}{5}\binom{13}{3}\binom{26}{5}}{\binom{39}{13}\binom{52}{13}}$ (e) $\dfrac{\binom{13}{5}\binom{13}{3}\binom{13}{5}}{\binom{39}{13}}$

7. A box contains 10 red, 10 blue, and 10 black chips. We select five chips without replacement from the box, and let X_1 be the number of red chips chosen and X_2 be the number of chips which are either blue or black. Then, the *correlation coefficient* between X_1 and X_2 is equal to

(a) 0 (b) 1 (c) $-\dfrac{125}{261}$ (d) $-\dfrac{1}{2}$ (e) -1

8. Suppose (X, Y) have a bivariate normal distribution with parameters

$$\mu_X = 1, \quad \mu_Y = 3, \quad \sigma_X^2 = 4, \quad \sigma_Y^2 = 9, \quad \rho_{X,Y} = \dfrac{1}{3}.$$

Then, the value of $Cov(X, Y)$ is

(a) 3   (b) 2   (c) $\dfrac{1}{3}$ (d) $\dfrac{1}{9}$ (e) $\dfrac{1}{18}$

9. Let (X, Y) have bivariate normal distribution with parameters

$$\mu_X = 1, \quad \mu_Y = 3, \quad \sigma_X^2 = 4, \quad \sigma_Y^2 = 9, \quad \rho_{X,Y} = \frac{1}{3}.$$

Then, the conditional expectation $E(Y|X = 5)$ equals

(a) 1   (b) $\dfrac{13}{9}$ (c) $\dfrac{3}{2}$ (d) $\dfrac{85}{27}$ (e) 5

10. Assume that the distribution of (X, Y) is bivariate normal with parameters

$$\mu_X = 1, \quad \mu_Y = -3, \quad \sigma_X^2 = 4, \quad \sigma_Y^2 = 9, \quad \rho_{X,Y} = -\frac{1}{2},$$

and let $W = 2X - Y$. Then, the distribution of W is

(a) $N(-1, 1)$ (b) $N(-1, 37)$ (c) $N(5, 1)$

(d) $N(5, 13)$ (e) $N(5, 37)$

11. Let (X, Y) be as Exercise 10 above, and define a new variable U as $U = 5X + 2Y$. Then, the conditional distribution of U, given $X = 3$, is normal with

(a) $\mu_U = 9$, $\sigma_U^2 = 36$ (b) $\mu_U = 21$, $\sigma_U^2 = 18$ (c) $\mu_U = 9$, $\sigma_U^2 = 18$

(d) $\mu_U = 6$, $\sigma_U^2 = 27$ (e) $\mu_U = 15$, $\sigma_U^2 = 36$

12. Assume that the pair (X, Y) follows a bivariate normal distribution. Both X and Y have zero means, while $\sigma_X^2 = 2, \sigma_Y^2 = 8$. If it is known further that $Var(Y|X = 4) = 6$, then the value of $E(XY)$ is equal to

(a) 0   (b) 2   (c) 4   (d) $4 - 2\sqrt{3}$ (e) 8

13. A high school offers four language courses: French, Spanish, German, and Italian. Each student has to select exactly two of these courses. Lena, Mary, and Linda have just made their choices. Assuming all choices are equally likely, the probability that all three are going to attend the French course and *at least one* of them is going to take an Italian course equals

(a) $\dfrac{7}{216}$ (b) $\dfrac{19}{216}$ (c) $\dfrac{1}{6}$ (d) $\dfrac{37}{216}$ (e) $\dfrac{91}{216}$

14. A canteen offers three choices for ice cream: strawberry, vanilla, and chocolate. Four friends have just arrived at the canteen and they all want to order ice cream. Assuming the three flavors to be equally likely, the probability that at least two friends select strawberry, and no one selects vanilla flavor is

(a) $\dfrac{1}{9}$ (b) $\dfrac{2}{27}$ (c) $\dfrac{7}{81}$ (d) $\dfrac{11}{81}$ (e) $\dfrac{17}{81}$

15. With the assumptions of the above exercise, the probability that no more than two friends order chocolate ice cream, given that exactly one has chosen vanilla ice cream, equals

(a) $\dfrac{23}{27}$ (b) $\dfrac{26}{27}$ (c) $\dfrac{7}{8}$ (d) $\dfrac{15}{16}$ (e) $\dfrac{33}{81}$

6.8 REVIEW PROBLEMS

1. In each hand in the game of poker, a player receives 5 cards. Find the probability that in a given hand, a player receives

   (i) three aces;

   (ii) three aces, but no kings;

   (iii) three aces and one face card;

   (iv) three hearts and one spade;

   (v) three hearts, given that she has exactly one spade;

   (vi) three hearts, given that the first card she receives is a spade.

2. In a US state, a committee has been formed in order to assess the academic standards of high schools in that state. The standards are classified as Excellent (E), Very Good (V), Good (G), and Poor (P). From previous experience, it is known that 40% of the schools are in class E, 30% in class V, 20% in class G, and 10% in class P, and it is assumed that these percentages still represent the classification of schools in this state. If the committee visits 12 schools at random, find the probability that

   (i) 5 schools will be classified E and 3 will be classified V;

   (ii) the number of schools that will be classified E or V will be 8 and 2 schools will be classified P;

   (iii) the number of schools that will be classified E or V will be 8, given that 2 schools will receive P;

   (iv) 5 schools will be classified E and 3 will be classified V, given that 2 schools will receive G.

3. At a certain University, based on data from previous years, it is known that the marks of students in the Mathematics and Statistics courses, respectively, follow a bivariate normal distribution. Let X denote the mark of a student in Mathematics and Y the mark of the same student in Statistics. The parameters of the bivariate normal distribution are

$$\mu_X = 76, \quad \sigma_X^2 = 16, \quad \mu_Y = 81, \quad \sigma_Y^2 = 25, \quad \rho_{X,Y} = \frac{1}{4}.$$

   Sophie is taking both exams this semester.

   (i) What is the probability that she performs better in Mathematics than in Statistics?

   (ii) Find the probability that her mark in Statistics is at least 10 points higher than that in Mathematics.

   (iii) Sophie is confident that her average mark in the two courses will be higher than 80. How likely is this to happen?

4. In each hand of bridge, a player receives 13 cards. What is the probability that, in a given hand, a player receives

   (i) 7 cards of the same suit (clubs, diamonds, hearts, or spades)?

   (ii) 5 cards of one suit and 5 cards of another suit?

   (iii)   5 cards of the same suit, given exactly two diamonds are received?

   (iv) 3 aces and 4 spades, but not the ace of spades?

5. A roulette wheel has 37 pockets, numbered from 0 to 36. In the number range 1–36, 18 pockets are red and 18 are black. The pocket numbered 0 is green. In each spin of the wheel, the ball lands on any of the 37 pockets with the same probability, namely, $1/37$.

Suppose we spin the wheel n times, and let X_1, X_2 denote, respectively, the number of times the ball lands on red and black pockets, respectively.

   (i) Write down the probability function of (X_1, X_2).

   (ii) For each of X_1 and X_2, find its mean and variance. Calculate also the covariance between X_1 and X_2.

   (iii) Suppose that before each spin, Jimmy selects a color (red or black) at random and bets on it. What is the probability that, in a particular spin, the winning color is red and Jimmy wins?

   (iv) Suppose now that Jimmy, when he decides to bet on red, he bets US\$5 and each time he chooses black, he bets US\$10. If he wins, he earns an amount equal to his bet, while if the winning number is zero, he always loses his bet. Calculate the expected value and variance of Jimmy's total earnings after 10 bets.

6. (**Negative trinomial distribution**) Consider a sequence of independent trials, each with three possible outcomes S_1, S_2, and F with respective probabilities p_1, p_2, and q, where $q = 1 - p_1 - p_2$. Let X_1, X_2 be the number of times the outcomes S_1, S_2 have appeared until F occurs for the r-th time; here r is a given positive integer.

   (i) Verify that the joint probability function of (X_1, X_2) is given by

$$f(x_1, x_2) = \frac{(r + x_1 + x_2 - 1)!}{x_1! x_2! (r-1)!} p_1^{x_1} p_2^{x_2} q^r, \qquad x_1 = 0, 1, \dots, \ x_2 = 0, 1, \dots.$$

   (ii) Show that the marginal probability function of X_1 is given by

$$f_{X_1}(x_1) = \binom{r + x_1 - 1}{r - 1} \left(\frac{p_1}{p_1 + q}\right)^{x_1} \left(\frac{q}{p_1 + q}\right)^{r}, \quad x_1 = 0, 1, \dots.$$

Use this result to find the mean and variance of X_1.

   (iii) Prove that the conditional probability function of X_1, given $X_2 = x_2$, is given by

$$f_{X_1 | X_2}(x_1 | x_2) = \binom{r + x_1 + x_2 - 1}{r + x_2 - 1} p_1^{x_1}(1 - p_1)^{r + x_2}, \quad x_1 = 0, 1, \dots.$$

   (iv) Conclude that the conditional distribution of $X_1 + X_2 + r$, given $X_2 = x_2$, is negative binomial with parameters $x_2 + r$ and $1 - p_1$. Using this, prove that

$$E(X_1 \mid X_2 = x_2) = \frac{(r + x_2)p_1}{1 - p_1} = \frac{p_1}{1 - p_1} x_2 + \frac{r p_1}{1 - p_1}.$$

(v) Making use of the fact that the regression curve of X_1 on X_2 is linear, show that

$$Cov(X_1, X_2) = \frac{rp_1p_2}{q^2}.$$

[Hint: The event $\{X_1 = x_1, X_2 = x_2\}$ occurs when a sequence of outcomes

$$\underbrace{S_1 \; F \; F \; S_2 \; F \; S_2 \; S_1 \; F \; F \; S_1 \; F \; S_2 \; \dots \; F \; S_2 \; S_1 \; F}_{x_1 + x_2 + r \text{ trials}}$$

occurs, where the last outcome is F, while in the previous $x_1 + x_2 + r - 1$ trials, there are exactly x_1 outcomes of type S_1 and x_2 outcomes of type S_2.]

7. In the game of roulette (see Exercise 5 above), suppose a player stops his game when a zero winning number has appeared exactly r times, where r is a given positive integer. Let X_1 be the number of times the winning number is red and X_2 be the number of times the winning number is black during the game.

   (i) Write down the joint probability function of (X_1, X_2).

   (ii) For each of X_1 and X_2, find the mean and variance.

   (iii) Calculate the covariance and correlation coefficient between X_1 and X_2.

   (iv) Given two real numbers a_1 and a_2, calculate the expected value and variance of the random variable $Y = a_1X_1 + a_2X_2$.

   (v) Find the covariance between X_1, Y and between X_2, Y.

   Application: Suppose the game stops when a zero winning number occurs for the fourth time. Give numerical answers to the results above, assuming that the player of the game wins US\$4 each time the winning number is red, loses US\$4 when the winning number is black, and loses US\$12 when a zero outcome occurs.

   [Hint: You may use the results of Exercise 6 above.]

8. Let (X, Y) have bivariate normal distribution with parameters μ_X, μ_Y, σ_X^2, σ_Y^2, and $\rho = \rho_{X,Y}$. We consider the random variable

$$W = Y - aX,$$

where $a \in \mathbb{R}$ is a constant that minimizes the variance of W. Show that the distribution of W is the same as the conditional distribution of Y, given $X = 0$.

9. The magnitude Y of a signal received from a satellite can be expressed as

$$Y = a + bV + W,$$

where V is the voltage which the satellite is measuring, a and b are constants and W is a "noise term", independent of V. We assume that

$$V \sim N(0, \sigma_V^2), \quad W \sim N(0, \sigma_W^2).$$

(i) Calculate the correlation coefficient between Y and V.

(ii) Given that $Y = y$, what is the distribution of V?

10. Let Z be a standard normal random variable and W be another variable such that $P(W = 1) = P(W = -1) = 1/2$. Define a new variable X as $X = ZW$.

(i) Establish that X follows a normal distribution with zero mean.

(ii) Show that X and Z are
   • uncorrelated

   • *not* independent
   random variables.

(iii) Explain whether or not the joint distribution of X and Z is bivariate normal.

11. Let $g_1(x, y)$ and $g_2(x, y)$ be two joint density functions of two pairs of random variables, each of which has a bivariate normal distribution with zero mean and unit variance. However, the two pairs have different correlation coefficients. Show then that

(i) the function defined by

$$f(x, y) = \frac{1}{2} g_1(x, y) + \frac{1}{2} g_2(x, y)$$

is the joint density function of some two-dimensional random variable (X, Y);

(ii) although the pair (X, Y) in Part (i) does *not* follow a bivariate normal distribution, the marginal distributions of X and Y are standard normal.

12. Let $g : \mathbb{R} \mapsto \mathbb{R}$ be an odd function (i.e. $(g(-x) = -g(x)$ for all x) such that $g(x) = 0$ for any $x \notin (-1, 1)$ and for some $c > 0$,

$$|g(x)| \le c, \qquad \text{for } x \in (-1, 1).$$

Now, let
$$\phi(x) = \frac{1}{\sqrt{2\pi}} \exp\left(-\frac{1}{2} x^2\right), \qquad x \in \mathbb{R}$$

be the density function of the standard normal distribution. Prove that

(i) the function, defined for any real x, y, as

$$f(x, y) = \phi(x)\phi(y) + g(x)g(y)$$

is the joint density function of a random pair (X, Y);

(ii) the pair (X, Y) in Part (i) does *not* follow a bivariate normal distribution, but the marginal distributions of X and Y are standard normal.

13. Let $X_1, \ldots, X_n$ be independent random variables, each having standard normal distribution, and define

$$Y = \sum_{i=1}^{n} a_i X_i, \qquad Z = \sum_{i=1}^{n} b_i X_i,$$

where a_i, b_i are given real constants, for $i = 1, \ldots, n$. Prove that the variables Y and Z are independent if and only if

$$\sum_{i=1}^{n} a_i b_i = 0.$$

6.9 APPLICATIONS

6.9.1 The Effect of Dependence on the Distribution of the Sum

In financial modeling, a common assumption is that investment returns follow the normal distribution. Moreover, it is often assumed that returns from different investments (or, from different time periods of the *same investment*) are independent. Although these assumptions facilitate the analysis to a large extent, their validity in practice is often questionable. In this section, we investigate the effect of dependence (more precisely, the correlation) between two normally distributed random variables on the distribution of the sum between these variables.

Suppose someone has made two investments, say A and B, and let X and Y denote the returns from these investments. We assume that the joint distribution of X, Y is bivariate normal with parameters

$$\mu_X = \mu_Y = 400, \quad \sigma_X = \sigma_Y = 50, \quad \rho_{X,Y} = 0.5.$$

Here the means and standard deviations are given in some monetary unit, e.g. US dollar. The investor is interested in the total return from this investment. More precisely, she wants to calculate the probability that the total return is at least US\$900. We therefore need to find

$$P(X + Y \geq 900). \tag{6.6}$$

A first key observation in tackling the problem above is that, once X, Y are individually normally distributed, their sum also follows the normal distribution. The mean of $X + Y$, denoted by μ, is simply

$$\mu = \mu_X + \mu_Y = 800,$$

while the variance of $X + Y$, denoted by σ^2, can be found using the formula

$$\sigma^2 = Var(X + Y) = Var(X) + Var(Y) + 2Cov(X, Y)$$

$$= \sigma_X^2 + \sigma_Y^2 + 2\rho_{X,Y} \cdot \sigma_X \cdot \sigma_Y$$

$$= 50^2 + 50^2 + 2 \cdot \frac{1}{2} \cdot 50 \cdot 50 = 7500.$$

Therefore, we have shown that $X + Y \sim N(800, 7500)$ and the required probability is found to be

$$P(X + Y \geq 900) = P\left(\frac{(X + Y) - \mu}{\sigma} \geq \frac{900 - \mu}{\sigma}\right)$$

$$= P\left(Z \geq \frac{900 - 800}{\sqrt{7500}}\right)$$

$$= 1 - \Phi(1.155) \cong 0.1241.$$

How does that compare with the probability in (6.6) if one assumes instead that the variables X, Y are independent? (Perhaps you should try a guess before reading the answer below!)

In the case of independence, the mean of $X + Y$ of course remains unaltered while the variance is now

$$Var(X + Y) = Var(X) + Var(Y) + 2Cov(X, Y)$$

$$= 50^2 + 50^2 + 0 = 5000.$$

The distribution of $X + Y$ is then $N(800, 5000)$ and the probability that the total return is at least 900 is equal to

$$P(X + Y \geq 900) = P\left(Z \geq \frac{900 - 800}{\sqrt{5000}}\right)$$

$$= 1 - \Phi(1.414) \cong 0.079.$$

We thus see that the probability has been reduced by more than a third, compared to the previous answer. Trying to understand why this happens, we note that the distribution of $X + Y$ is in the latter case more concentrated around the mean, so large values are less likely.

This observation becomes more pronounced if one assumes that X, Y are negatively correlated. Assume, for instance that $\rho_{X,Y} = -0.5$; then the variance of the sum becomes

$$Var(X + Y) = 50^2 + 50^2 - 2 \cdot \frac{1}{2} \cdot 50 \cdot 50 = 2500,$$

and we obtain that, in this case,

$$P(X + Y \geq 900) = P\left(Z \geq \frac{900 - 800}{\sqrt{2500}}\right) = 1 - \Phi(2) \cong 0.023.$$

Roughly speaking, negative correlation means that when X is large then Y is small and vice versa, so that large values of $X + Y$ are even less likely.

Following the above, one may wonder: what is the value of $\rho_{X,Y}$ that maximizes the probability in (6.6)? In view of the preceding discussion, it is not difficult to see that the probability can be expressed as

$$P(X + Y \geq 900) = P\left(Z \geq \frac{900 - 800}{\sqrt{50^2 \cdot 2 \cdot (1 + \rho_{X,Y})}}\right)$$

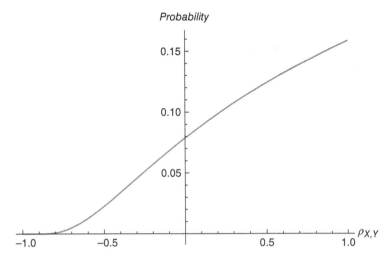

Figure 6.6 The probability $P(X + Y \geq 900)$, as the correlation coefficient $\rho_{X,Y}$ varies from -1 to $+1$.

$$= 1 - \Phi\left(\sqrt{\frac{2}{1 + \rho_{X,Y}}}\right) = g(\rho_{X,Y}).$$

It is clear that g is an increasing function of $\rho_{X,Y}$ attaining its maximum value at $\rho_{X,Y} = 1$. In this case, we have

$$P(X + Y \geq 900) = 1 - \Phi(1) = 0.1587.$$

Figure 6.6 presents the value of the probability $P(X + Y \geq 900)$, in terms of the correlation coefficient $\rho_{X,Y}$; we observe that, apart from values of $\rho_{X,Y}$ close to -1 (in this case the probability is close to zero), the probability increases almost linearly.

KEY TERMS

approximation of double hypergeometric distribution by trinomial distribution

bivariate normal distribution

double hypergeometric distribution

multinomial distribution

multivariate hypergeometric distribution

trinomial distribution

urn models

CHAPTER 7

GENERATING FUNCTIONS

William Feller (Zagreb 1906 – New York, 1970)

William Feller was born, as Vilibald Srećko Feller, on 7 July 1906, in Zagreb, Croatia (at that time, an autonomous region in the Austro-Hungarian Monarchy) and died on 14 January 1970, in New York, at the age of 63. He studied at University of Zagreb and then at University of Göttingen, and did his doctoral thesis under the supervision of Richard Courant. He started his career at the University of Kiel in 1928, then moved to Scandinavia in 1933 and lectured there at the Universities of Copenhagen, Stockholm and the University of Lund, before moving to the United States in 1939. After teaching at Brown University and Cornell University, he moved to Princeton University in 1950. He published extensively on different topics including functional analysis, differential equations, geometry, probability, and mathematical statistics. He is considered as one of the greatest probabilists of the twentieth century, and his two-volume textbook on probability theory entitled *An Introduction to Probability Theory and Its Applications* is one of the most successful and influential treatises ever written on probability theory. There are many probabilistic topics that bear his name such as Feller Processes, Lindeberg–Feller theorem, Feller operators, and Feller–Brown movement. He was awarded the National Medal of Science in 1969.

7.1   INTRODUCTION

In this chapter, we introduce three types of functions that are related to the distribution of a random variable: moment generating function (m.g.f.), probability generating function (p.g.f.), and characteristic function (known under the collective name *generating functions*). The first two of these are real-valued functions, while the last one is complex-valued (so at least some basic knowledge of complex calculus is needed, otherwise the reader may skip this section).

An important property shared by all three functions is that, under certain conditions, they characterize the distribution of a random variable. So, under some conditions, if we know for example that two variables have the same m.g.f., then they have the same distribution. This property highlights the importance of generating functions in cases in which it is difficult to obtain directly the distribution of a variable of interest. A typical example is when our interest is in the distribution of a sum of independent random variables; if X_1 and X_2 are two independent variables (not necessarily with the same distribution), then the distribution of their sum $Y = X_1 + X_2$ may be hard to obtain explicitly, using convolutions for instance (see Section 4.3). However, using the property that the generating function of Y is simply the product of the generating functions of X_1 and X_2, all one has to do then is to "invert" the generating function found for Y so that its distribution can be identified.

Another important use of them is in providing a proof of a prominent result in probability theory, the "Central Limit Theorem", that is discussed in Chapter 8.

7.2   MOMENT GENERATING FUNCTION

As we know, the expectation and variance of a random variable X provide important information about its probabilistic behavior. This is particularly useful if the underlying distribution of X is not known. For instance, suppose we know that $\mu = E(X)$ and $\sigma^2 = Var(X)$. Then, Chebychev's inequality

$$P(|X - \mu| \geq t) \leq \frac{\sigma^2}{t^2}$$

provides some idea about probabilities associated with X.

However, additional information about the probabilistic structure of a variable may be obtained by using higher order moments. Recall that the central moment of order r is defined as

$$\mu_r = E\left[(X - \mu)^r\right], \qquad r = 2, 3, \ldots$$

(μ_1 is always zero). A typical example where such moments are useful is when one studies skewness (that is, the departure from symmetry) or kurtosis of the distribution of a random variable. To be specific,

- the coefficient of skewness is defined as

$$\gamma_1 = \frac{\mu_3}{\sigma^3},$$

  with a zero value corresponding to symmetry (lack of skewness), while positive and negative values of γ_1 indicate positive and negative asymmetry, respectively, of the associated distribution;

• the coefficient of kurtosis is defined as

$$\gamma_2 = \frac{\mu_4}{\sigma^4}.$$

Roughly speaking, kurtosis measures how peaked (or flat) the probability function (or the density function) of a given distribution is. For the normal distribution, we always have $\gamma_2 = 3$, while, more generally, distributions for which $\gamma_2 = 3$ are called *mesokurtic*. Distributions with $\gamma_2 > 3$ have a more acute peak than the normal and are called *leptokurtic*; distributions with $\gamma_2 < 3$ are more flat and are called *platykurtic*.

The first generating function that is discussed is the moment generating function which, as its name indicates, produces in a rather easy way all the moments of a distribution.

Definition 7.2.1 Let X be a random variable for which $E(e^{tX})$ exists for any t in an interval $(-\delta, \delta)$, for some $\delta > 0$. Then, the function

$$M_X(t) = E(e^{tX}), \qquad\qquad |t| < \delta,$$

is called the **moment generating function** (henceforth abbreviated as m.g.f.) of the variable X.

When there is no potential for any confusion, and it is clear which variable we are referring to, we may omit the subscript and simply write $M(t)$. It is obvious from the definition that $M(0) = 1$.

When X is a discrete variable, then its m.g.f. is given by

$$M_X(t) = \sum_{x \in R_X} e^{tx} f(x), \qquad\qquad |t| < \delta, \qquad\qquad (7.1)$$

while for continuous X we have

$$M_X(t) = \int_{-\infty}^{\infty} e^{tx} f(x)\, dx, \qquad\qquad |t| < \delta. \qquad\qquad (7.2)$$

Readers familiar with advanced calculus may recognize the similarity between the integral in (7.2) and the so-called **Laplace transform** of a function f. As a matter of fact,

$$M_X(-t) = \int_{-\infty}^{\infty} e^{-tx} f(x)\, dx,$$

is the Laplace transform of the function f.

In the ensuing discussion, we assume that the function f is sufficiently smooth so that, when we differentiate $M(t)$, we can interchange the order of derivative and the summation operator or the integral. Moreover, unless otherwise stated, when we refer to an m.g.f. $M(t)$, we shall assume that the argument, t, of this function takes values in an interval $(-\delta, \delta)$ for some $\delta > 0$, such that $M(t)$ is finite in that interval.

As already mentioned, the term "moment generating function" suggests that the function $M(t)$ "generates" the moments of X in some sense. This is stated formally in the following proposition.

Proposition 7.2.1 *Let $M(t)$ denote the m.g.f. of X. Then, we have:*

(i) $\mu'_r = E(X^r) = M^{(r)}(0) = \left[\dfrac{d^r}{dt^r}M(t)\right]_{t=0}, \qquad r = 1, 2, \ldots;$

(ii) $M(t) = \sum\limits_{r=0}^{\infty} \dfrac{E(X^r)}{r!} t^r.$

Proof:

(i) We give the proof for a continuous variable X, and the proof for the discrete case is quite similar. Differentiating both sides of the relation

$$M(t) = E(e^{tX})$$

and assuming, as stated before, that the order of differentiation and integration can be interchanged, we get

$$M'(t) = \frac{d}{dt}\left(\int_{-\infty}^{\infty} e^{tx}f(x)dx\right) = \int_{-\infty}^{\infty}\left(\frac{d}{dt}e^{tx}\right)f(x)dx = \int_{-\infty}^{\infty} xe^{tx}f(x)\,dx.$$

Differentiating once more, we obtain

$$M''(t) = \frac{d}{dt}\left(\int_{-\infty}^{\infty} xe^{tx}f(x)dx\right) = \int_{-\infty}^{\infty} x\left(\frac{d}{dt}e^{tx}\right)f(x)dx = \int_{-\infty}^{\infty} x^2 e^{tx}f(x)\,dx.$$

Proceeding in this way, we find for $r = 1, 2, \ldots$, that

$$M^{(r)}(t) = \frac{d}{dt}\left(\int_{-\infty}^{\infty} x^{r-1}e^{tx}f(x)dx\right) = \int_{-\infty}^{\infty} x^r e^{tx}f(x)\,dx.$$

By setting $t = 0$ in this expression, we see that $M^{(r)}(0) = E(X^r)$, as asserted.

(ii) This is an immediate consequence of Part (i) and McLaurin series expansion which gives

$$M(t) = \sum_{r=0}^{\infty} \frac{M^{(r)}(0)}{r!} t^r.$$

$\square$

Example 7.2.1 (m.g.f. of a binary random variable)

Let X be a random variable taking on only two values, namely, $X = 1$ with probability p and $X = 0$ with probability $q = 1 - p$ (this is a Bernoulli variable). From (7.1), we see that the m.g.f. of X is

$$M(t) = E(e^{tX}) = e^{t \cdot 0}P(X = 0) + e^{t \cdot 1}P(X = 1) = q + pe^t.$$

Example 7.2.2 (Use of the m.g.f. to find moments of a random variable)
Let X be a discrete variable with probability function as

$$f(x) = \frac{7-x}{15}, \quad x = 1, 2, 3.$$

The m.g.f. of X is

$$M(t) = E(e^{tX}) = e^t f(1) + e^{2t} f(2) + e^{3t} f(3)$$

$$= \frac{6}{15} e^t + \frac{5}{15} e^{2t} + \frac{4}{15} e^{3t} = \frac{6e^t + 5e^{2t} + 4e^{3t}}{15}.$$

To find the moments $\mu_r' = E(X^r)$, for $r = 1, 2, \ldots$, we observe that

$$\frac{d}{dt^r} e^{kt} = k^r e^{kt}, \quad k = 1, 2, 3.$$

We then obtain

$$\mu_r' = M^{(r)}(0) = \frac{6 + 5 \cdot 2^r + 4 \cdot 3^r}{15} = \frac{2}{5} + \frac{2^r}{3} + \frac{4 \cdot 3^r}{15}, \quad \text{for } r = 1, 2, \ldots.$$

Of course, we could have also obtained the moments directly as

$$\mu_r' = E(X^r) = \sum_{x \in R_X} x^r f(x) = 1^r f(1) + 2^r f(2) + 3^r f(3).$$

It can be readily verified that this gives the same result.

Example 7.2.3 Consider the following m.g.f. of a random variable X,

$$M(t) = \frac{(3e^t + 2)^4}{625}.$$

The first and second derivatives of M are

$$M'(t) = \frac{12e^t (3e^t + 2)^3}{625}$$

and

$$M''(t) = \frac{12e^t}{625} \left[(3e^t + 2)^3 + 9e^t (3e^t + 2)^2 \right].$$

Upon setting $t = 0$, we then find

$$E(X) = M'(0) = \frac{12 \cdot 5^3}{625} = \frac{12}{5}$$

and

$$E(X^2) = M''(0) = \frac{12}{625} \left(5^3 + 9 \cdot 5^2 \right) = \frac{168}{25}.$$

Consequently, the variance of X is equal to

$$Var(X) = E(X^2) - [E(X)]^2 = M''(0) - [M'(0)]^2 = \frac{24}{25}.$$

In the following example, we derive the m.g.f. of the standard normal distribution.

Example 7.2.4 (m.g.f. of the standard normal distribution)
Let Z be a standard normal variable with density function

$$\phi(z) = \frac{1}{\sqrt{2\pi}} e^{-z^2/2}, \qquad z \in \mathbb{R}.$$

For the m.g.f. of Z, we first observe that

$$M_Z(t) = E(e^{tZ}) = \int_{-\infty}^{\infty} e^{tz} \phi(z) \, dz = \frac{1}{\sqrt{2\pi}} \int_{-\infty}^{\infty} \exp\left(tz - \frac{z^2}{2}\right) dz.$$

Writing the exponent in the form

$$tz - \frac{z^2}{2} = \frac{t^2}{2} - \frac{1}{2}(z - t)^2,$$

we deduce that

$$M_Z(t) = e^{t^2/2} \int_{-\infty}^{\infty} \frac{1}{\sqrt{2\pi}} \exp\left(-\frac{1}{2}(z - t)^2\right) dz = e^{t^2/2}, \qquad t \in \mathbb{R}$$

(the last step above follows from the fact that the integrand is the density function of a normal random variable with $\mu = t$ and $\sigma = 1$, and so the value of the integral is 1).

According to Proposition 7.2.1, knowledge of the m.g.f. of X enables us, at least in theory, to calculate moments of an arbitrary order for X (provided, of course, that such moments exist). A natural question that arises is whether an m.g.f. determines uniquely the distribution of X (that is, the probability function or the density function of X). The answer to this question is given in the following proposition.

Proposition 7.2.2 *Let X and Y be two random variables with m.g.f.s $M_X(t)$, and $M_Y(t)$, respectively. If, for some $\delta > 0$, we have*

$$M_X(t) = M_Y(t), \qquad for\ any\ t \in (-\delta, \delta),$$

then X and Y have the same distribution.

Proof: The proof of this result is omitted.   □

We note that for the result in Proposition 7.2.2 to be valid, the probability functions (respectively density functions) must satisfy some regularity conditions. As these conditions are met in all cases considered in this chapter, we shall make no further comment on this in the discussion to follow.

We now present a first example wherein the form of the m.g.f., in a rather simple form, enables us to identify the underlying distribution.

Example 7.2.5   Obtain the probability function of the random variable associated with the following two m.g.f.s:

(i) $M(t) = \dfrac{5 + 2e^t}{7}$,

(ii) $M(t) = \dfrac{1}{4}e^t + \dfrac{1}{3}e^{2t} + \dfrac{5}{12}e^{3t}$.

SOLUTION

(i) Writing the m.g.f. in the form

$$M(t) = \frac{5}{7} + \frac{2}{7}e^t,$$

we see that this is the same as the m.g.f. of Example 7.2.1 (in the special case when $q = 5/7$, and $p = 2/7$). In view of Proposition 7.2.2, we can then conclude that the distribution of a variable X with this m.g.f. has its probability function as

$$f(x) = P(X = x) = \begin{cases} 5/7, & \text{if } x = 0, \\ 2/7, & \text{if } x = 1. \end{cases}$$

(ii) It is easy to verify that the m.g.f. of a random variable Y with probability function given by

y	1	2	3
$g(y) = P(Y = y)$	1/4	1/3	5/12

is the same as the one given in the statement of Part (ii). Thus, by Proposition 7.2.2 again, we immediately see that the required probability function is the one given above.

We now present a result for the m.g.f. of a linear transformation of a random variable.

Proposition 7.2.3 *Let X be a random variable having m.g.f. $M_X(t)$, and a and b be two real numbers. Then, the m.g.f. of*

$$Y = aX + b$$

is given by

$$M_Y(t) = e^{bt}M_X(at).$$

Proof: From Definition 7.2.1, we can write

$$M_Y(t) = E(e^{tY}) = E\left[\exp(t(aX+b))\right] = E[e^{bt}\ \exp((at)X)]$$

and since the quantity e^{bt} is a constant and $E[\exp((at)X)] = M_X(at)$, we obtain

$$M_Y(t) = e^{bt}E\left[\exp((at)X)\right] = e^{bt}M_X(at),$$

as required.   □

Example 7.2.6 Assume that the continuous random variable X has m.g.f.

$$M_X(t) = \exp(t + 2t^2), \qquad t \in \mathbb{R}.$$

(i) Find the m.g.f. and then the density function of $Y = (X-1)/2$.

(ii) Calculate the probability $P(-1 < X \le 3)$.

SOLUTION

(i) As Y is in the form $Y = aX + b$ with $a = 1/2$ and $b = -1/2$, its m.g.f. is given by

$$M_Y(t) = e^{-t/2}M_X\left(\frac{t}{2}\right) = e^{-t/2}\exp\left[\frac{t}{2} + 2\left(\frac{t}{2}\right)^2\right] = e^{t^2/2}.$$

Comparing this with the result of Example 7.2.4, we conclude (using Proposition 7.2.2) that Y has the standard normal distribution.

(ii) As the variable

$$Z = \frac{X-1}{2}$$

follows the standard normal distribution, we find

$$P(-1 < X \le 3) = P\left(\frac{-1-1}{2} < \frac{X-1}{2} \le \frac{3-1}{2}\right) = P(-1 < Z \le 1)$$

$$= \Phi(1) - \Phi(-1) = 2 \cdot \Phi(1) - 1 = 2 \cdot 0.8413 - 1 = 0.6826.$$

EXERCISES

Group A

1. Calculate the m.g.f. and, using it, find the mean and variance for each of the random variables whose probability functions are as given below:

   (i) $f(x) = \dfrac{1}{3}$, $x = 1, 3, 5$;

   (ii) $f(x) = \dfrac{5-x}{10}$, $x = 1, 2, 3, 4$

(iii) $f(x) = \dfrac{x^2 + 2}{14}$, $\qquad\qquad x = -1, 0, 1, 2;$

(iv) $f(x, y) = p^x(1 - p)^{1-x}$, $\qquad x = 0, 1;$

(v) $f(x) = \left(\dfrac{1}{2}\right)^x$, $\qquad\qquad x = 1, 2, 3, \ldots$

2. The m.g.f. of a random variable X is given by

$$M(t) = \left(\frac{1}{3} + \frac{2}{3} e^t\right)^n.$$

Calculate the expectation and variance of X.

3. For a random variable X, it is known that its m.g.f. is given by

$$M(t) = \exp\left(e^t - 1\right).$$

Calculate the expectation and variance of X.

4. Let the m.g.f. of a variable X be given by

$$M(t) = \exp(e^t - 1), \qquad\qquad t \in \mathbb{R}.$$

Show that the probability function of X is

$$f(x) = P(X = x) = \frac{1}{e\,x!}, \qquad x = 0, 1, 2, \ldots.$$

[Hint: Find the m.g.f. associated with the function $f(x)$ and confirm that this coincides with $M(t)$ given above.]

5. Let X be a discrete random variable having probability function

$$f(x) = c(6 - x), \qquad x = 1, 2, 3, 4, 5.$$

(i) What is the value of the constant c?

(ii) Derive the m.g.f. of X and, using it, obtain the moments $\mu'_r = E(X^r)$, for $r = 1, 2, \ldots$.

6. Suppose that, for a discrete random variable X, it is known that

$$E(X^r) = 0.7, \qquad\qquad r = 1, 2, \ldots.$$

(i) Show that the m.g.f. of X is given by

$$M(t) = 0.3 + 0.7\, e^t, \qquad t \in \mathbb{R}.$$

(ii) Prove that X takes only two values, namely 0 and 1, with probabilities 0.3 and 0.7, respectively.

7. Using the series

$$e^{tx} = \sum_{r=0}^{\infty} \frac{(tx)^r}{r!},$$

along with the properties of expectation, obtain an alternative proof for Part (ii) of Proposition 7.2.1.

8. Let X be a random variable for which the m.g.f. $M_X(t)$ exists in a finite interval $(-\delta, \delta)$, for some $\delta > 0$. Use the results of Propositions 7.2.1 and 7.2.3 to prove the identities

$$E(aX + b) = aE(X) + b, \qquad Var(aX + b) = a^2 Var(X).$$

9. Assume that the m.g.f.s $M_X(t)$ and $M_Y(t)$ of two random variables X and Y are related by $M_Y(t) = [M_X(t)]^a$, where a is a positive integer. Show then that

$$E(Y) = aE(X), \qquad Var(Y) = aVar(X).$$

Group B

10. Consider the following three m.g.f.s:

(i) $M(t) = 0.09 + 0.42e^t + 0.49e^{2t}$;

(ii) $M(t) = \dfrac{0.7e^t}{1 - 0.3e^t}, \qquad t < \ln(0.3)$;

(iii) $M(t) = 0.2 \sum_{x=0}^{4} e^{tx}$.

Assuming, in each case, that a random variable X has m.g.f. $M(t)$, calculate for each of the three cases the following quantities:

$$E(X), \ Var(X), \ P(X = 0), \ P(1 \le X \le 2), \ P(X \ge 2 | X \ge 1).$$

11. A random variable X has m.g.f. given by

$$M(t) = \frac{e^t + e^{-t}}{5} + a, \qquad t \in \mathbb{R}.$$

(i) What is the value of the constant a?

(ii) For $r = 1, 2, \ldots$, calculate the moments $\mu'_r = E(X^r)$ of X.

12. Assume that X has probability function

$$f(x) = c^x, \qquad x = 1, 2, \ldots,$$

for a suitable constant $c > 0$.

(i) Obtain the value of c and thence find the m.g.f. of X.

(ii) Calculate the expectation and variance of X.

(iii) Find the m.g.f. of the random variable $Y = 3X + 4$.

13. A continuous variable X has density function

$$f(x) = \frac{k\theta^k}{x^{k+1}}, \qquad x \geq \theta > 0, \ k > 0$$

(this is the density of Pareto distribution). Show that the m.g.f. of X does not exist (i.e. there does not exist a $\delta > 0$ with $M_X(t) < \infty$ for $|t| < \delta$). What happens with the moments $\mu'_r = E(X^r)$ for $r = 1, 2, \ldots$?

14. Let $M(t)$ be the m.g.f. of a discrete variable X. If the function $G(t) = \ln M(t)$ satisfies

$$G'(t) = G(t) + \lambda$$

for any t such that $M(t) < \infty$, show that $E(X) = Var(X) = \lambda$.

15. The m.g.f. of a variable X is given by

$$M(t) = a \exp\left(\frac{1}{2} t^2\right), \qquad t \in \mathbb{R}.$$

(i) What is the value of a?

(ii) Verify that the moments $\mu'_r = E(X^r)$ are, for $r = 1, 2, \ldots$, given by

$$\mu'_r = \begin{cases} 0, & \text{if } r \text{ is an odd integer,} \\ 1 \cdot 3 \cdot 5 \cdots (2k-1), & \text{if } r = 2k \text{ for some } k = 1, 2, \ldots. \end{cases}$$

7.3 MOMENT GENERATING FUNCTIONS OF SOME IMPORTANT DISTRIBUTIONS

In this section, we find the m.g.f. of some important discrete and continuous distributions.

7.3.1 Binomial Distribution

For a random variable X following the binomial distribution with parameters n and p, the probability function is

$$f(x) = \binom{n}{x} p^x q^{n-x}, \qquad x = 0, 1, 2, \ldots, n,$$

(with $q = 1 - p$), so that

$$M(t) = E(e^{tX}) = \sum_{x=0}^{n} e^{tx} f(x) = \sum_{x=0}^{n} \binom{n}{x} (pe^t)^x q^{n-x}.$$

Using binomial expansion, we then obtain

$$M(t) = (pe^t + q)^n,$$

from which we can obtain any moment of X. In particular, since

$$M'(t) = np\, e^t(pe^t + q)^{n-1},$$

$$M''(t) = np\, e^t(pe^t + q)^{n-1} + n(n-1)(pe^t)^2(pe^t + q)^{n-2},$$

we readily find

$$E(X) = M'(0) = np,$$

$$E(X^2) = M''(0) = np + n(n-1)p^2.$$

From these expressions, we also obtain $Var(X) = E(X^2) - [E(X)]^2 = np(1-p)$. Finally, upon setting $n = 1$ in the m.g.f. of binomial, we arrive at the m.g.f. of Bernoulli distribution (see also Example 7.2.1) as

$$M(t) = pe^t + q.$$

7.3.2 Negative Binomial Distribution

Suppose X has a negative binomial distribution with parameters r and p, so that its probability function is

$$f(x) = \binom{x-1}{r-1} p^r q^{x-r}, \qquad x = r, r+1, \dots,$$

where $q = 1 - p$. The m.g.f. of X is given by

$$M(t) = E(e^{tX}) = \sum_{x=r}^{\infty} e^{tx} f(x) = \sum_{x=r}^{\infty} e^{tx} \binom{x-1}{r-1} p^r q^{x-r}$$

$$= (pe^t)^r \sum_{x=r}^{\infty} \binom{x-1}{r-1} (qe^t)^{x-r}. \tag{7.3}$$

Using the identity

$$(1 - w)^{-r} = \sum_{x=r}^{\infty} \binom{x-1}{r-1} w^{x-r}, \qquad |w| < 1,$$

we conclude from (7.3) that

$$M(t) = (pe^t)^r (1 - qe^t)^{-r} = \left(\frac{pe^t}{1 - qe^t} \right)^r.$$

Note that this formula holds for $|qe^t| < 1$, or equivalently, for $t < -\ln q$ (otherwise, the sum in (7.3) does not converge, in which case we say that the m.g.f. does not exist).

As a special case of the negative binomial distribution, for $r = 1$, we deduce the m.g.f. of the geometric distribution to be

$$M(t) = \frac{pe^t}{1 - qe^t}.$$

7.3.3 Poisson Distribution

A random variable X following the Poisson distribution with parameter λ has probability function as

$$f(x) = e^{-\lambda}\frac{\lambda^x}{x!}, \qquad x = 0, 1, 2, \dots .$$

We then find

$$M(t) = E(e^{tX}) = \sum_{x=0}^{\infty} e^{tx}f(x) = e^{-\lambda}\sum_{x=0}^{\infty}\frac{\left(\lambda e^t\right)^x}{x!} = e^{-\lambda}\exp\left(\lambda e^t\right)$$

$$= \exp\left\{\lambda(e^t - 1)\right\}.$$

Note that the sum here converges for all real t, and so the m.g.f. of Poisson exists in the entire real line.

7.3.4 Uniform Distribution

Suppose that X is a random variable having uniform distribution over the interval $[a, b]$. The density of X is

$$f(x) = \begin{cases} \dfrac{1}{b-a}, & a \leq x \leq b \\[2ex] 0, & \text{otherwise.} \end{cases}$$

As X is a continuous variable, we use (7.2) rather than (7.1) to find the m.g.f. as

$$M(t) = E(e^{tX}) = \int_a^b e^{tx}f(x)\,dx = \frac{1}{b-a}\int_a^b e^{tx}\,dx = \frac{1}{b-a}\left[\frac{e^{tx}}{t}\right]_a^b = \frac{e^{bt} - e^{at}}{t(b-a)}$$

for any real t with $t \neq 0$. Recall that for $t = 0$, we have $M(0) = 1$.

7.3.5 Normal Distribution

Given a random variable X following the normal distribution with parameters μ and σ^2, we know that the random variable

$$Z = \frac{X - \mu}{\sigma}$$

follows standard normal distribution. According to Example 7.2.4, the m.g.f. of Z is given by

$$M_Z(t) = e^{t^2/2}, \qquad t \in \mathbb{R}.$$

Then, by using Proposition 7.2.3, we obtain the m.g.f. of $X = \mu + \sigma Z$ as

$$M_X(t) = e^{\mu t}M_Z(\sigma t), \qquad t \in \mathbb{R},$$

from which we readily obtain

$$M_X(t) = \exp\left(\mu t + \frac{1}{2}\sigma^2 t^2\right), \qquad t \in \mathbb{R}.$$

7.3.6 Gamma Distribution

A random variable X following Gamma distribution with parameters $a > 0$ and $\lambda > 0$ has density function as

$$f(x) = \frac{\lambda^a}{\Gamma(a)}\, x^{a-1} e^{-\lambda x}, \qquad\qquad x > 0.$$

For the m.g.f. of X, we first have

$$M(t) = E(e^{tX}) = \int_{-\infty}^{\infty} e^{tx} f(x)\, dx = \frac{\lambda^a}{\Gamma(a)} \int_0^\infty x^{a-1} e^{-x(\lambda - t)}\, dx.$$

The integral converges if and only if $t < \lambda$. For these values of t, making the change of variable $u = x(\lambda - t)$, we find

$$M(t) = \frac{\lambda^a}{\Gamma(a)} \cdot \frac{1}{(\lambda - t)^a} \int_0^\infty u^{a-1} e^{-u}\, du = \frac{\lambda^a}{\Gamma(a)} \cdot \frac{\Gamma(a)}{(\lambda - t)^a} = \frac{\lambda^a}{(\lambda - t)^a}.$$

We have, thus, shown that the m.g.f. of a $Ga(a, \lambda)$ random variable is

$$M(t) = \frac{\lambda^a}{(\lambda - t)^a}, \qquad\qquad t < \lambda.$$

For the special case when $a = 1$, we deduce the m.g.f. of an exponential random variable with parameter λ as

$$M(t) = \frac{\lambda}{\lambda - t}, \qquad\qquad t < \lambda.$$

The m.g.f.s of the most common probability distributions are presented in Appendix 5.

Example 7.3.1 The m.g.f. of a continuous random variable X is given by

$$M_X(t) = \exp\left[2t(t + 6)\right], \qquad\qquad t \in \mathbb{R}.$$

Calculate the probability $P(9 < X \le 14)$.

SOLUTION   We observe that the given m.g.f. is a special case of the m.g.f. we found for the normal distribution, with $\mu = 12$, and $\sigma^2 = 4$. In view of Proposition 7.2.2, we then have the distribution of X as $N(12, 2^2)$. So, the required probability is

$$P(9 < X \le 14) = P\left(\frac{9 - 12}{2} < \frac{X - 12}{2} \le \frac{14 - 12}{2}\right) = P(-1.5 < Z \le 1)$$

$$= \Phi(1) - \Phi(-1.5) = \Phi(1) + \Phi(1.5) - 1 = 0.8413 + 0.9332 - 1$$

$$= 0.7745 \cong 77.5\%.$$

Example 7.3.2 Let X be a continuous random variable having uniform distribution over $[0, 1]$ and

$$Y = bX + a(1 - X),$$

where $a, b \in \mathbb{R}$ with $a < b$. Find the distribution of Y.

SOLUTION   Writing Y in the form $Y = (b - a)X + a$, Proposition 7.2.3 shows that the m.g.f.s of X and Y are related by

$$M_Y(t) = e^{at}M_X((b - a)t).$$

Using the m.g.f. of X as provided in Section 7.3.4 above, we see that

$$M_X(t) = \frac{e^{1 \cdot t} - e^{0 \cdot t}}{t(1 - 0)} = \frac{e^t - 1}{t}, \qquad t \neq 0,$$

and consequently we find

$$M_Y(t) = \frac{e^{at} \left[e^{(b-a)t} - 1 \right]}{(b - a)t} = \frac{e^{bt} - e^{at}}{(b - a)t}, \qquad t \neq 0.$$

But, as we have seen in Section 7.3.4, this is precisely the m.g.f. of a random variable having the uniform distribution over $[a, b]$, and so this is the distribution of Y.

EXERCISES

Group A

1. Using the formulas for the m.g.f.s of various distributions in this section (negative binomial, Poisson, uniform, exponential, and Gamma), derive the mean and variance of these distributions (and confirm that they agree with known results about these distributions).

2. Let X be a random variable following the $N(\mu, \sigma^2)$ distribution, and let a and b be two real numbers with $a \neq 0$. Show, using m.g.f.s, that the distribution of

$$Y = aX + b$$

   is $N(a\mu + b, a^2\sigma^2)$.

3. Let X be a discrete random variable having m.g.f. $M(t)$. Calculate the probability $P(X < 3)$ for each of the following cases for $M(t)$:

   (i) $M(t) = \dfrac{1}{16}(e^t + 1)^4, \qquad t \in \mathbb{R};$

   (ii) $M(t) = \dfrac{e^{2t}}{(3 - 2e^t)^2}, \qquad t < -\ln(2/3);$

   (iii) $M(t) = \exp\left(5e^t - 5\right), \qquad t \in \mathbb{R}.$

4. The claim amounts, X (in tens of thousands of dollars), paid by an insurance company to its policyholders have a continuous distribution with density

$$f(x) = \frac{1}{3}e^{-2x} + \frac{2}{5}e^{-3x} + \frac{7}{2}e^{-ax}, \qquad x > 0,$$

   for a suitable $a > 0$.

(i) Find the value of a and thence obtain the expected claim amount paid to a policyholder.

(ii) Calculate the m.g.f. of X and, using it, check that the expected claim amount is the same as that found in Part (i).

5. Let X be a continuous random variable having m.g.f. $M(t)$. Calculate the probability $P(X > 1)$ for each of the following $M(t)$:

   (i) $M(t) = \dfrac{e^{4t} - 1}{4t}$, $t \neq 0$;

   (ii) $M(t) = \dfrac{2(e^{2t} - e^{t/2})}{3t}$, $t \neq 0$;

   (iii) $M(t) = \exp\left[t(8t + 6)\right]$, $t \in \mathbb{R}$;

   (iv) $M(t) = \dfrac{1}{1 - 2t}$, $t < \dfrac{1}{2}$;

   (v) $M(t) = \dfrac{1}{(1 - 4t)^2}$, $t < \dfrac{1}{4}$.

6. (i) Calculate the m.g.f. of a random variable having chi-squared distribution with n degrees of freedom.

   (ii) Suppose X has a Gamma distribution with parameters $a = r$ and λ, where r is a positive integer. Using m.g.f.s, show that $Y = 2\lambda X$ has a chi-squared distribution with $n = 2r$ degrees of freedom.

7. A continuous variable X has density function

$$f(x) = \lambda \exp\left[-\lambda(x - \alpha)\right], \qquad\qquad x \geq \alpha,$$

and $f(x) = 0$ for all other values of x.

   (i) Find the m.g.f. of X.

   (ii) Differentiate the m.g.f. found in Part (i) to obtain the mean and variance of X.

   (iii) Calculate the m.g.f. of $Y = \lambda(X - \alpha)$ and thence show that the distribution of Y is standard exponential, i.e. with mean 1.

Group B

8. A random variable taking positive values is said to have **lognormal distribution** with parameters $\mu \in \mathbb{R}$ and $\sigma^2 > 0$ if and only if the random variable

$$Y = \ln X$$

follows normal distribution with parameters μ and σ^2.

   (i) Show that the moments of the lognormal distribution are given by

$$\mu'_r = E(X^r) = \exp\left(r\mu + \frac{1}{2}r^2\sigma^2\right), \qquad\qquad r = 1, 2, \ldots .$$

   (ii) Express the variance of this distribution is terms of its parameters, μ and σ^2.

(iii) Prove that

$$\sigma^2 = \ln \left\{ \frac{Var(X)}{[E(X)]^2 + 1} \right\}.$$

(iv) Suppose $X_1, X_2, \ldots, X_n$ are independent random variables having lognormal distribution with the same parameters. What is the distribution of the product $X_1 X_2 \cdots X_n$?

9. The m.g.f. of a discrete random variable X is given by

$$M_X(t) = e^{3t} \exp \left[5(e^{2t} - 1) \right], \qquad t \in \mathbb{R}.$$

(i) Obtain the mean and variance of X.

(ii) Calculate the probability $P(3 \le X \le 5)$.

[Hint: Calculate the m.g.f. of $(X - 3)/2$ and identify the distribution associated with that m.g.f.]

10. Let X be a discrete random variable and Y be a continuous random variable, and let their m.g.f.s, $M_X(t)$, and $M_Y(t)$ be related by

$$M_X(t) = M_Y(e^t - 1).$$

Assuming that the distribution of Y is exponential with mean 1,

(i) show that the variable $X + 1$ follows geometric distribution and then identify the parameter of that distribution;

(ii) calculate the probabilities $P(X \ge x)$, for $x = 0, 1, 2, \ldots$.

7.4 MOMENT GENERATING FUNCTIONS FOR SUM OF VARIABLES

Many problems in probability theory involve sums of random variables. When we have only two variables, say X and Y, that are independent and their distributions are known, then we can use convolution formulas (see Propositions 4.3.1 and 4.3.2) to calculate the distribution of the sum $X + Y$.

Suppose now we have n variables, $X_1, \ldots, X_n$, and assume further that these are independent. In order to calculate the distribution of the sum $X_1 + \cdots + X_n$, we have to apply the convolution formula repeatedly and this would naturally be cumbersome. Using m.g.f.s instead, things can be much easier as Proposition 7.4.1 shows. In this process, m.g.f.s (and other generating functions discussed in Sections 7.5 and 7.6) play a major role for handling sums of (independent) random variables.

Proposition 7.4.1 *Let $X_1, \ldots, X_n$ be a set of n independent random variables with m.g.f.s $M_{X_1}(t), \ldots, M_{X_n}(t)$, respectively. Then, the m.g.f. of*

$$S_n = X_1 + \cdots + X_n$$

is given by the formula

$$M_{S_n}(t) = M_{X_1}(t)M_{X_2}(t)\cdots M_{X_n}(t) = \prod_{i=1}^{n} M_{X_i}(t). \tag{7.4}$$

Proof: From the definition of m.g.f., we readily have

$$M_{S_n}(t) = E(e^{tS_n}) = E\left[e^{t(X_1+\cdots+X_n)}\right] = E(e^{tX_1}\cdots e^{tX_n}).$$

But, the independence of $X_1, \dots, X_n$ implies that the variables $e^{tX_1}, \dots, e^{tX_n}$ are also independent; consequently,

$$E\left(e^{tX_1}\cdots e^{tX_n}\right) = E(e^{tX_1})\cdots E(e^{tX_n}) = M_{X_1}(t)\cdots M_{X_n}(t),$$

as required. $\qquad\qquad\square$

A particularly useful special case of the above result is when the variables X_i, in addition to being independent, are also identically distributed, so that they form a random sample from a given distribution. Let $M_X(t)$ be the common m.g.f. of X_i's in such a case. It then follows from Proposition 7.4.1 that $S_n = X_1 + X_2 + \cdots + X_n$ has m.g.f.

$$M_{S_n}(t) = \left[M_X(t)\right]^n.$$

As an example, let us consider $Y_1, \dots, Y_n$ to be independent having Bernoulli distribution with parameter p, i.e.

$$P(Y_i = 1) = p, \quad P(Y_i = 0) = q = 1 - p, \quad \text{for } i = 1, \dots, n.$$

We have seen in Example 7.2.1 that the m.g.f. of Y_i, denoted by $M_Y(t)$, is $M_Y(t) = pe^t + q$, and so

$$M_{S_n}(t) = \left(pe^t + q\right)^n.$$

The sum $S_n = Y_1 + \cdots + Y_n$ follows the binomial distribution with parameters n and p, so the quantity on the right-hand side can be identified as the m.g.f. of a random variable having $b(n, p)$ distribution. This result has already been obtained in Section 7.3; but, we can go one step further. Now, let $X_1, \dots, X_r$ be independent random variables with X_i being distributed as binomial with parameters n_i and p for $i = 1, \dots, r$. Then, the m.g.f. of X_i is

$$M_{X_i}(t) = \left(pe^t + q\right)^{n_i}$$

and, by Proposition 7.4.1, the sum $S_r' = X_1 + \cdots + X_r$ has m.g.f.

$$M_{S_r'}(t) = \prod_{i=1}^{r} \left(pe^t + q\right)^{n_i} = \left(pe^t + q\right)^m,$$

where $m = n_1 + \cdots + n_r$. But, the quantity $\left(pe^t + q\right)^m$ is the m.g.f. of binomial distribution with parameters m and p. We have thus proved the following:

Proposition 7.4.2 *Let $X_1, \ldots, X_r$ be independent random variables with $X_i \sim b(n_i, p)$, for $i = 1, 2, \ldots, r$. Then, the distribution of*

$$X_1 + \cdots + X_r$$

is binomial with parameters $m = n_1 + n_2 + \cdots + n_r$ and p.

In Example 4.3.2, we showed that the sum of two Poisson random variables is also Poisson. Following that, we mentioned that the result can be extended to the case when more than two Poisson random variables are involved. We now formally establish this result in the following proposition.

Proposition 7.4.3 *Let $X_1, \ldots, X_n$ be independent random variables following the Poisson distributions with parameters $\lambda_1, \ldots, \lambda_n$, respectively. Then, the distribution of*

$$S_n = X_1 + \cdots + X_n$$

is Poisson with parameter $\lambda = \lambda_1 + \cdots + \lambda_n$.

Proof: Inserting the m.g.f. of a Poisson with parameter λ_i,

$$M_{X_i}(t) = \exp\left(\lambda_i(e^t - 1)\right), \qquad t \in \mathbb{R},$$

into (7.4), we obtain

$$M_{S_n}(t) = \prod_{i=1}^{n} M_{X_i}(t) = \prod_{i=1}^{n} \exp\left(\lambda_i(e^t - 1)\right) = \exp\left\{\left(\sum_{i=1}^{n} \lambda_i\right)(e^t - 1)\right\}$$

$$= \exp\left\{\lambda(e^t - 1)\right\},$$

where $\lambda = \lambda_1 + \cdots + \lambda_n$. As the function above is the m.g.f. of a Poisson random variable with parameter λ, the result then follows. □

The following proposition gives an alternative proof of Corollary 6.4.1 by using m.g.f.s.

Proposition 7.4.4 *Assume that $X_1, \ldots, X_n$ are independent random variables with X_i being normal with mean μ_i and variance σ_i^2, for $i = 1, \ldots, n$. Then, the linear combination*

$$\sum_{i=1}^{n} a_i X_i = a_1 X_1 + \cdots + a_n X_n, \tag{7.5}$$

where $a_1, a_2, \ldots, a_n \in \mathbb{R}$, has a normal distribution with parameters

$$\mu = \sum_{i=1}^{n} a_i \mu_i, \quad \sigma^2 = \sum_{i=1}^{n} a_i^2 \sigma_i^2.$$

Proof: Let $M(t)$ be the m.g.f. of the linear combination in (7.5). Then, for $M(t)$, we find

$$M(t) = E\left[\exp\left(t\sum_{i=1}^{n} a_i X_i\right)\right] = E\left[\exp\left(\sum_{i=1}^{n}(a_i t)X_i\right)\right]$$

$$= E\left[\prod_{i=1}^{n}\exp\left((a_i t)X_i\right)\right] = \prod_{i=1}^{n}E\left[\exp\left((a_i t)X_i\right)\right] = \prod_{i=1}^{n}M_{X_i}(a_i t),$$

using the independence of X_i's. But $X_i \sim N(\mu_i, \sigma_i^2)$, and the associated m.g.f. has already been found in Section 7.3. So, we have

$$M_{X_i}(a_i t) = \exp\left(\mu_i a_i t + \frac{1}{2}\sigma_i^2 a_i^2 t^2\right),$$

which in turn yields

$$M(t) = \prod_{i=1}^{n}\exp\left(\mu_i a_i t + \frac{1}{2}\sigma_i^2 a_i^2 t^2\right) = \exp\left[\left(\sum_{i=1}^{n}a_i\mu_i\right)t + \frac{1}{2}\left(\sum_{i=1}^{n}a_i^2\sigma_i^2\right)t^2\right].$$

Thus, we find $M(t) = \exp(\mu t + \sigma^2 t^2/2)$, with μ and σ^2 as in the statement of the proposition. But, this is the m.g.f. of a $N(\mu, \sigma^2)$ distribution, and we therefore conclude that the distribution of $a_1 X_1 + \cdots + a_n X_n$ is $N(\mu, \sigma^2)$.   □

A result of particular interest that arises immediately from Proposition 7.4.4 concerns the distribution of the sample mean. Specifically, upon setting

$$\mu_1 = \cdots = \mu_n, \quad \sigma_1^2 = \cdots = \sigma_n^2, \quad a_1 = \cdots = a_n = \frac{1}{n},$$

we obtain the following result.

Corollary 7.4.1 *Let $X_1, \ldots, X_n$ be a random sample from normal distribution with parameters μ and σ^2. Then, the distribution of*

$$\bar{X} = \frac{1}{n}\sum_{i=1}^{n}X_i$$

is $N\left(\mu, \dfrac{\sigma^2}{n}\right)$.

Example 7.4.1 The distribution of scores, out of 100, in a psychology test follows the normal distribution with mean 70 and standard deviation 6.

   (i) If 10 persons take the test on a particular day, what is the probability that their average score exceeds 74?

   (ii) On another day, 15 men and 20 women take this test. Suppose the distribution of scores is the same for both men and women. Find the probability that the average scores between the two groups differ by at most 3 units.

SOLUTION

(i) Let $X_1, \ldots, X_{10}$ be the scores of the 10 persons. We require the probability

$$P\left(\overline{X} > 74\right),$$

where $\overline{X}$ is the average score for these 10 persons. But, from Corollary 7.4.1, we have the distribution of $\overline{X}$ to be normal with mean 70 and variance $6^2/10 = 3.6$; that is, $\overline{X} \sim N(70, 3.6)$. This gives

$$P\left(\overline{X} > 74\right) = P\left(\frac{\overline{X} - 70}{\sqrt{3.6}} > \frac{74 - 70}{\sqrt{3.6}}\right) = P(Z > 2.108)$$

$$= 1 - P(Z \le 2.108) = 1 - \Phi(2.108) = 0.018,$$

that is, the required probability is 1.8%.

(ii) Let $\overline{X}_M$ denote the average score achieved in the sample of 15 men and $\overline{X}_W$ denote the average score in the sample of 20 women. Then, Corollary 7.4.1 yields

$$\overline{X}_M \sim N\left(70, \frac{6^2}{15}\right), \quad \overline{X}_W \sim N\left(70, \frac{6^2}{20}\right).$$

Further, since $\overline{X}_M$ is defined only in terms of the scores of males and $\overline{X}_M$ depends only on the scores of females, it follows that $\overline{X}_M$ and $\overline{X}_W$ are independent random variables. As they are also normal, we get from Proposition 7.4.4 that their difference $\overline{X}_M - \overline{X}_W = \overline{X}_M + (-1)\overline{X}_W$ also has a normal distribution with mean 0 and variance

$$\frac{6^2}{15} + \frac{6^2}{20} = 4.2.$$

Therefore, the probability we are seeking is

$$P\left(\left|\overline{X}_M - \overline{X}_W\right| \le 3\right) = P\left(\left|\frac{(\overline{X}_M - \overline{X}_W) - 0}{\sqrt{4.2}}\right| \le \frac{3}{\sqrt{4.2}}\right)$$

$$= \Phi(1.46) - \Phi(-1.46)$$

$$= 2\Phi(1.46) - 1 = 2(0.9279) - 1 = 0.8558.$$

Proposition 7.4.5 gives the m.g.f. for a sum of a random number of random variables.

Proposition 7.4.5 *Let $X_1, X_2, \ldots,$ be a sequence of independent and identically distributed random variables, and N be an integer-valued nonnegative variable independently of X_i's. Let*

$$S_N = X_1 + \cdots + X_N$$

be the sum of $N \ge 1$ of X_i variables (by convention, $S_0 = 0$).

If $M_X(t)$ denotes the common m.g.f. of $X_1, X_2, \ldots$, then the m.g.f. of S_N is given by

$$M_{S_N}(t) = E\left[(M_X(t))^N\right]. \tag{7.6}$$

Proof: First, making use of Proposition 5.4.1, we have

$$M_{S_N}(t) = E(e^{tS_N}) = E\left[E(e^{tS_N}|N)\right] = \sum_{n=0}^{\infty} E(e^{tS_N}|N=n)P(N=n).$$

Observe, that for $n = 0$, we have

$$E(e^{tS_N}|N=0) = 1,$$

while for $n = 1, 2, \ldots$, we have

$$E(e^{tS_N}|N=n) = E(e^{tS_n}|N=n) = E(e^{tS_n}).$$

But, $E(e^{tS_n}) = M_{S_n}(t)$, the m.g.f. of $S_n = X_1 + \cdots + X_n$ (a fixed sum now, rather than a random sum). Thus, we can appeal to (7.4) to obtain

$$E(e^{tS_n}) = M_{S_n}(t) = \prod_{i=1}^{n} M_{X_i}(t) = \left[M_X(t)\right]^n.$$

Consequently, we get

$$M_{S_N}(t) = P(N=0) + \sum_{n=1}^{\infty} \left[M_X(t)\right]^n P(N=n) = \sum_{n=0}^{\infty} \left[M_X(t)\right]^n P(N=n)$$

$$= E\left[(M_X(t))^N\right],$$

as required. □

The last result, suitably reformulated, will be used in some interesting applications in Section 7.5.

Example 7.4.2 Let X be a discrete random variable and Y be a continuous variable, and suppose they satisfy the following conditions:

(a) The conditional distribution of X, given $Y = y$, is binomial with parameters n and y.

(b) The variable Y follows the uniform distribution over $(0, 1)$.

Then, find the m.g.f. of X.

SOLUTION   Clearly, X and Y are not independent variables. Then, we have

$$M_X(t) = E(e^{tX}) = E\left[E(e^{tX}|Y)\right]$$

(note that the first expectation above is considered with respect to Y, while the second one is with respect to X). As the conditional distribution of X, given $Y = y$, is binomial with parameters n and y, the conditional expectation $E(e^{tX}|Y = y)$ will be the same as the m.g.f. of a random variable with $b(n, y)$ distribution; that is,

$$E(e^{tX}|Y = y) = (ye^t + (1 - y))^n = (y(e^t - 1) + 1)^n.$$

Therefore,

$$M_X(t) = E\left[(Y(e^t - 1) + 1)^n\right] = \int_{-\infty}^{\infty} (y(e^t - 1) + 1)^n f_Y(y)\, dy$$

$$= \int_0^1 (y(e^t - 1) + 1)^n\, dy.$$

By making the change of variable $y(e^t - 1) + 1 = w$, we then find

$$M_X(t) = \frac{1}{e^t - 1} \int_1^{e^t} w^n\, dw = \frac{1}{e^t - 1} \left[\frac{w^{n+1}}{n+1}\right]_1^{e^t} = \frac{e^{(n+1)t} - 1}{(n+1)(e^t - 1)}.$$

From the form of the m.g.f. above, we can recognize the distribution of X in this case (see Exercise 7 below).

EXERCISES

Group A

1. Assume that X_1, X_2, and X_3 are independent Bernoulli random variables such that $P(X_i = 1) = p_i = 1 - P(X_i = 0)$, for $i = 1, 2, 3$, with

$$p_1 = \frac{1}{3}, \quad p_2 = \frac{2}{5}, \quad p_3 = \frac{4}{7}.$$

For each of the random variables

$$Y_1 = 3X_1 + 2X_2 + 7X_3, \quad Y_2 = 2X_1 - X_2,$$

(i) calculate the associated m.g.f.;

(ii) obtain its expectation in two ways: first, using the result in Part (i), and then by using the linearity properties of expectation, and check that the two results agree.

2. Let $X_1, \ldots, X_n$ be independent random variables, each having probability function

$$f(x) = pq^{x-1}, \qquad x = 1, 2, 3, \ldots.$$

with $(q = 1 - p)$. Use Proposition 7.4.1 and the results of Section 7.3 to prove that the distribution of $X_1 + \cdots + X_n$ is negative binomial with parameters n and p.

3. Let $X_1, \ldots, X_n$ be independent random variables, each having probability function

$$f(x) = pq^x, \qquad x = 0, 1, 2, \ldots,$$

where $0 < p < 1$ and $q = 1 - p$.

   (i) Obtain the m.g.f. of X_i, for $i = 1, 2, \ldots, n$.

   (ii) Derive the m.g.f. of $S_n = X_1 + \cdots + X_n$.

   (iii) Verify that the distribution of $S_n + n$ is negative binomial with parameters n and p.

4. Let $X_1, \ldots, X_n$ be a set of independent random variables with X_i following negative binomial distribution with parameters r_i and p, for $i = 1, \ldots, n$. Using Proposition 7.4.1 and the results of Section 7.3, show that the distribution of $X_1 + \cdots + X_n$ is negative binomial with parameters $r_1 + \cdots + r_n$ and p.

5. Assume that $X_1, \ldots, X_n$ are independent random variables with X_i following Gamma distribution with parameters a_i and λ, for $i = 1, \ldots, n$. Using Proposition 7.4.1 and the results of Section 7.3, show that the distribution of $X_1 + \cdots + X_n$ is Gamma with parameters $a_1 + \cdots + a_n$ and λ. (Note that this result has been obtained earlier in Example 4.3.1 using a different argument).

6. The measurement errors from two experiments are represented by two independent random variables, X and Y, respectively. Both X and Y are uniformly distributed with a zero mean. Moreover, their variances are

$$Var(X) = \frac{1}{3}, \quad Var(Y) = \frac{27}{25}. \qquad (7.7)$$

Define

$$V = 3X + 2Y, \quad W = 2X - 3Y.$$

   (i) Identify the distributions of V and W, and show that

$$E(V) = E(W) = 0.$$

   Are V and W independent?

   (ii) Write down the m.g.f. for each of V and W.

   (iii) Calculate the variance of V and W in three different ways:

   - using the result in Part (i) of the exercise (easy);
   - using (7.7) and the defining relations for V and W in terms of X and Y (easier!);
   - by differentiating suitably the m.g.f.s found in Part (ii) above (tedious!).

Group B

7. A random variable X has discrete uniform distribution over the set $\{0, 1, 2, \ldots, n\}$ with probability function

$$f(x) = \frac{1}{n+1}, \qquad x = 0, 1, \ldots, n.$$

(i) Show that the m.g.f. of X is given by

$$M_X(t) = E(e^{tX}) = \frac{e^{(n+1)t} - 1}{(n+1)(e^t - 1)}.$$

Compare this result to the one obtained in Example 7.4.2.

(ii) Differentiate the m.g.f. in Part (i) to obtain $E(X)$.

(iii) We ask a computer to generate a decimal number Y from the interval $(0, 1)$. For a given value $Y = y$, we perform $n = 10$ times an experiment with two possible outcomes (success and failure) and success probability y. If X denotes the number of successes in these 10 trials, obtain the m.g.f. of X, and then identify its distribution.

8. (i) A continuous variable X has density function

$$f(x) = \begin{cases} x, & 0 \le x < 1, \\ 2 - x, & 1 \le x < 2, \\ 0, & \text{otherwise.} \end{cases}$$

Obtain the m.g.f. $M(t) = E(e^{tX})$.

(ii) A computer selects at random two numbers, say, X and Y, from the interval $(0, 1)$. Using the form of the m.g.f. of uniform distribution found in Section 7.3 and Proposition 7.4.1, find the m.g.f. and from it, the density function of the sum $X + Y$.

9. Assume that a discrete random pair (X, Y) has joint probability function as follows:

x \ y	-1	0	1
-1	a	b	c
0	c	a	b
1	b	c	a

where $a = 1/9$, $b = 1/9 - s$, $c = 1/9 + s$ for $-1/9 < s < 1/9$.

(i) Prove that X and Y are independent if and only if $s = 0$.

(ii) Show that, for any $s \in (-1/9, 1/9)$, the m.g.f. of the sum $X + Y$ is equal to the product of the m.g.f.s of X and Y. (This is a counterexample showing that the converse of Proposition 7.4.1 fails in general.)

10. Let X be a discrete random variable and Y be a continuous variable such that the m.g.f. of Y is $M_Y(t)$, while the conditional distribution of X, given $Y = y$, is Poisson with parameter y.

    (i) Verify that the m.g.f.s of X and Y are related by

    $$M_X(t) = M_Y(e^t - 1).$$

    (ii) If the distribution of Y is exponential with mean $1/\lambda$, for some $\lambda > 0$, obtain the distribution of X.
    [Hint: Proceed as in Example 7.4.2. You may also find Exercise 10 from Section 7.3 useful.]

11. A continuous variable X has m.g.f. given by

    $$M(t) = \exp\left[2t + 8t^2 - \ln(1 - t)\right], \qquad t < 1.$$

    Show that X has the same distribution as the sum of two independent random variables, one following a normal distribution and the other following an exponential distribution.

7.5 PROBABILITY GENERATING FUNCTION

The m.g.f. discussed so far "generates moments" as we have seen in Proposition 7.2.1; by differentiating m.g.f. successively and setting the argument $t = 0$, we obtain the moments of the distribution. Knowledge of the m.g.f., however, does not lead us, in general, to the form of the distribution itself. Although it is possible in certain cases to identify a distribution if we are given the corresponding m.g.f. (this is true, for example, if the form of the distribution is one of those studied in Section 7.3), this can be hard in many other cases.

In this section, we introduce a function which permits identification of the probability distribution associated with a discrete random variable. Consequently, this is especially useful in cases wherein the probability distribution cannot be obtained in explicit form directly.

Definition 7.5.1 Let X be a random variable taking integer, nonnegative values. Then, the function
$$G_X(t) = E(t^X)$$

is called the **probability generating function** (abbreviated as p.g.f.) of the variable X.

It is apparent from the definition that $G_X(1) = 1$. As in the case of m.g.f.s, when there is no possible confusion, we will simply write $G(t)$ rather than $G_X(t)$. Both m.g.f.s and p.g.f.s are defined in terms of an expectation; note, however, that in the former case existence of that expectation, at least over a finite interval including zero, was *part of the definition*. In contrast, for a variable X taking values over the set $\{0, 1, 2, \ldots, \}$ and $f(x) = P(X = x)$, with

$$G_X(t) = \sum_{x=0}^{\infty} t^x f(x), \tag{7.8}$$

since

$$\sum_{x=0}^{\infty} |t^x f(x)| = \sum_{x=0}^{\infty} |t^x| f(x) \le \sum_{x=0}^{\infty} f(x) = 1 \quad \text{for } t \in [-1, 1],$$

we see that the probability function is always convergent (in fact, absolutely convergent) over the interval $[-1, 1]$.

M.g.f.s produce moments through differentiation; on the other hand, differentiating the p.g.f. of a discrete variable X yields the probability function of X itself, as stated in the following proposition.

Proposition 7.5.1 *Let X be a variable taking values on the nonnegative integers and $G(t)$ be the p.g.f. of X. Then:*

(i) *the probability function of X is given by*

$$P(X = r) = \frac{1}{r!} \left[\frac{d^r}{dt^r} G(t) \right]_{t=0}, \qquad r = 0, 1, 2, \dots$$

(here, by convention, the derivative of order zero is the function itself);

(ii) *the factorial moments of X,*

$$\mu_{(r)} = E[(X)_r] = E[X(X-1)(X-2)\cdots(X-r+1)], \quad r = 1, 2, \dots,$$

are given by

$$\mu_{(r)} = \left[\frac{d^r}{dt^r} G(t) \right]_{t=1}.$$

Proof: The result for $r = 0$ in Part (i) is trivial. Differentiating both sides of (7.8), we obtain

$$G'(t) = \sum_{x=0}^{\infty} (t^x)' f(x) = \sum_{x=1}^{\infty} x t^{x-1} f(x).$$

Differentiating once more gives

$$G''(t) = \sum_{x=1}^{\infty} x (t^{x-1})' f(x) = \sum_{x=2}^{\infty} x(x-1) t^{x-2} f(x) = \sum_{x=2}^{\infty} (x)_2 t^{x-2} f(x).$$

Continuing in this way, it is easy to see (a rigorous proof requires induction, you are encouraged to do this!) that for any positive integer r,

$$G^{(r)}(t) = \sum_{x=r}^{\infty} (x)_r t^{x-r} f(x) = r! f(r) + \sum_{x=r+1}^{\infty} (x)_r t^{x-r} f(x).$$

Putting $t = 0$ and $t = 1$ in the above relation yields, respectively, the first and second parts of the proposition.   □

An immediate deduction from Part (i) of Proposition 7.5.1 is the following "uniqueness result" for p.g.f.s.

Proposition 7.5.2 *Let X and Y be two nonnegative integer-valued random variables, with $G_X(t)$ and $G_Y(t)$ as their p.g.f.s, respectively. Then, if*

$$G_X(t) = G_Y(t), \qquad for \ -1 \le t \le 1,$$

the variables X and Y have the same distribution.

The proof of the following result proceeds in exactly the same way with Propositions 7.2.3 and 7.4.1 and is therefore omitted.

Proposition 7.5.3

(i) *Let X be a discrete random variable, $G_X(t)$ be the associated p.g.f., and $a > 0$ and b be two real numbers. Then, the p.g.f. of*

$$Y = aX + b$$

is given by

$$G_Y(t) = t^b \cdot G_X(t^a).$$

(ii) *Suppose $X_1, \ldots, X_n$ are independent random variables with p.g.f.s $G_{X_1}(t), \ldots, G_{X_n}(t)$ respectively. Then the p.g.f. of*

$$S_n = X_1 + \cdots + X_n$$

is given by

$$G_{S_n}(t) = G_{X_1}(t) \cdots G_{X_n}(t) = \prod_{i=1}^{n} G_{X_i}(t). \tag{7.9}$$

Example 7.5.1 Let X_i, $i = 1, \ldots, n$, be independent binary random variables with

$$P(X_i = 1) = p_i, \quad P(X_i = 0) = q_i = 1 - p_i.$$

Then,

$$G_{X_i}(t) = E(t^{X_i}) = t^0 \cdot P(X_i = 0) + t^1 \cdot P(X_i = 1) = q_i + tp_i,$$

and Part (ii) of Proposition 7.5.3 now implies that the p.g.f. of $S_n = X_1 + \cdots + X_n$ is

$$G_{S_n}(t) = \prod_{i=1}^{n} (q_i + p_i t).$$

In the important special case when $p_i = p = 1 - q$, for all i, we have

$$G_{S_n}(t) = (q + pt)^n.$$

We have thus arrived at the p.g.f. of the binomial distribution with parameters n and p.

Example 7.5.2 *(The sum of two dice outcomes)*
Nick rolls a die twice and suppose X_1 and X_2 are the random variables representing the outcomes of these rolls. We are interested in the distribution of the sum

$$S = X_1 + X_2.$$

To find this, we first note that X_1 and X_2 are independent and identically distributed with probability function (for $i = 1, 2$)

$$f_{X_i}(x) = P(X_i = x) = \frac{1}{6}, \qquad x = 1, 2, \dots, 6.$$

Using this, we find their common generating function to be

$$G_{X_i}(t) = \sum_{x=1}^{6} t^x f_{X_i}(x) = \frac{1}{6} \sum_{x=1}^{6} t^x = \frac{1}{6} \cdot \frac{t(1 - t^6)}{1 - t}, \quad i = 1, 2.$$

Therefore, by the independence of X_1 and X_2, we have

$$G_S(t) = G_{X_1}(t) G_{X_2}(t) = \frac{1}{36} \cdot \frac{t^2 (1 - t^6)^2}{(1 - t)^2}. \tag{7.10}$$

Further, from the definition of p.g.f., we have

$$G_S(t) = \sum_{x=1}^{12} t^x f_S(x). \tag{7.11}$$

We can then obtain the probability function $f_S(x) = P(S = x)$, for $x = 2, 3, \dots, 12$, if we expand the right-hand side of (7.10). More specifically, upon using the identity

$$(1 - t^6) = (1 - t)(1 + t + \cdots + t^5)$$

on the right-hand side of (7.10), we get

$$G_S(t) = \frac{1}{36} t^2 (1 + t + \cdots + t^5)^2.$$

Upon expanding the right-hand side and comparing like terms with the polynomial in (7.11), it is readily checked that we arrive at the simple expression

$$f_S(x) = P(S = x) = \begin{cases} \dfrac{x - 1}{36}, & x = 2, 3, \dots, 7, \\[2mm] \dfrac{13 - x}{36}, & x = 8, 9, \dots, 12, \\[2mm] 0, & \text{otherwise.} \end{cases}$$

This can be written in a more succinct form as

$$f_S(x) = \frac{6 - |7 - x|}{36}, \qquad x = 2, 3, \dots, 12.$$

Now, let X be a discrete random variable with m.g.f. $M(t)$ and p.g.f. $G(t)$. It is then obvious from the definitions of these two functions that

$$M(t) = G(e^t), \quad G(t) = M(\ln t),$$

so that, typically, knowledge of one between the two functions enables us to calculate the other (provided they both exist). This is true in particular for the discrete distributions we have discussed in Section 7.3, where we found their m.g.f.s. In view of the last formula, we see that no extra effort is needed to obtain the associated p.g.f.s. To give an example, we have seen that the m.g.f. associated with $b(n, p)$ distribution is

$$M(t) = (pe^t + q)^n.$$

The associated p.g.f. is simply obtained by replacing e^t by t above, resulting in

$$G(t) = M(\ln t) = (pe^{\ln t} + q)^n = (pt + q)^n,$$

a result we obtained earlier in Example 7.5.1.

The p.g.f.s are quite useful in the study of discrete variables whose probability function is defined by a recursive formula, as is illustrated in Example 7.5.3.

Example 7.5.3 Suppose the probability function $f(x) = P(X = x)$ of a discrete nonnegative random variable X satisfies the recursive formula

$$f(x) = \frac{2}{3}f(x-1) + \frac{4}{27}f(x-2), \qquad x = 2, 3, \ldots,$$

with initial conditions

$$f(0) = \frac{1}{9}, \quad f(1) = \frac{4}{27}.$$

Then, writing the p.g.f.

$$G(t) = E(t^X) = \sum_{x=0}^{\infty} t^x f(x)$$

in the form

$$G(t) = f(0) + tf(1) + \sum_{x=2}^{\infty} t^x f(x),$$

and replacing $f(0)$, $f(1)$, and $f(x)$ by the given relations, we obtain

$$G(t) = \frac{1}{9} + \frac{4}{27}t + \sum_{x=2}^{\infty} \left(\frac{2}{3}f(x-1) + \frac{4}{27}f(x-2) \right) t^x$$

$$= \frac{1}{9} + \frac{4}{27}t + \frac{2}{3}t \cdot \sum_{x=2}^{\infty} f(x-1)t^{x-1} + \frac{4}{27}t^2 \cdot \sum_{x=2}^{\infty} f(x-2)t^{x-2}. \qquad (7.12)$$

Noting now that

$$\sum_{x=2}^{\infty} f(x-1)t^{x-1} = \sum_{r=1}^{\infty} f(r)t^r = G(t) - f(0)t^0 = G(t) - \frac{1}{9}$$

and

$$\sum_{x=2}^{\infty} f(x-2)t^{x-2} = \sum_{r=0}^{\infty} f(r)t^r = G(t),$$

and substituting these in (7.12), we get

$$G(t) = \frac{1}{9} + \frac{4}{27}t + \frac{2}{3}t\left(G(t) - \frac{1}{9}\right) + \frac{4}{27}t^2 G(t).$$

Upon solving for $G(t)$, we obtain

$$G(t) = \frac{\dfrac{1}{9} + \dfrac{2}{27}t}{1 - \dfrac{2}{3}t - \dfrac{4}{27}t^2}.$$

With this p.g.f., we may then obtain the expectation of X by differentiating $G(t)$ and then setting $t = 1$; more explicitly, we find that

$$E(X) = G'(1) = \frac{28}{5}.$$

The following proposition deals with p.g.f. for a sum of random number of random variables.

Proposition 7.5.4 *Let $X_1, X_2, \ldots$, be a sequence of independent and identically distributed random variables, and N be a discrete nonnegative random variable independently of X_is. Consider the random sum*

$$S_N = X_1 + \cdots + X_N.$$

Let $M_X(t)$ be the common m.g.f. of the variables X_i and $G_N(t)$ be the p.g.f. of N. Then, the m.g.f. of S_N is given by

$$M_{S_N}(t) = G_N(M_X(t)). \qquad (7.13)$$

If, in addition, the variables $X_1, X_2, \ldots$, are nonnegative and take integer values with common p.g.f. $G_X(t)$, then the p.g.f. of S_N is given by the formula

$$G_{S_N}(t) = G_N(G_X(t)).$$

Proof: From Proposition 7.4.5, we have

$$M_{S_N}(t) = E\left[(M_X(t))^N\right]$$

and, since $G_N(s) = E(s^N)$, the first result follows immediately. The second result is proved using arguments similar to those in the proof of Proposition 7.4.5. □

Upon differentiating (7.13) with respect to t, we obtain

$$M'_{S_N}(t) = G'_N(M_X(t))M'_X(t),$$

and setting $t = 0$, we get

$$M'_{S_N}(0) = G'_N(M_X(0))M'_X(0) = G'_N(1)M'_X(0) \quad \text{(since } M_X(0) = 1\text{)}$$

which is the same as
$$E(S_N) = E(N)E(X).$$

We have seen the last result earlier in the first part of Proposition 5.4.4. The second part of that proposition may also be obtained from above, by differentiating (7.13) once more and then setting $t = 0$.

Proposition 7.5.4 has found numerous applications in practice, in situations involving random sums of random variables. The following two simple examples demonstrate this:

1. Suppose the amount of daily customers at a supermarket is a random variable N. Each customer buys a certain product with probability p. For $i = 1, 2, \ldots$, let us consider the variables

$$X_i = \begin{cases} 1, & \text{if the } i\text{th customer buys the product,} \\ 0, & \text{if the } i\text{th customer does not buy the product.} \end{cases}$$

Then, it is evident that the number of customers buying the product during a day is given by the random sum
$$S_N = X_1 + \cdots + X_N.$$

According to Proposition 7.5.4 (see also Example 7.5.1), the p.g.f. of S_N is given by

$$G_{S_N}(t) = G_N(q + pt),$$

where $q = 1 - p$ and $G_N(t)$ is the p.g.f. of N.

2. The number of claims arriving at an insurance company during a month is a discrete random variable N. Assuming that the sizes of the individual claim amounts $X_1, X_2, \ldots$ are independent and identically distributed random variables, the total amount paid by the company is represented by the random sum

$$S_N = X_1 + \cdots + X_N.$$

Note that if the common distribution of the X_i's is discrete, then S_N also has a discrete distribution, while if the variables X_i have a continuous distribution, the distribution of S_N is also continuous (provided we are certain that at least one claim arrives at the company, otherwise $S_N = 0$ with positive probability). In either case, the m.g.f. (and, in the former case, also the p.g.f.) of S_N can be studied using Proposition 7.5.4, assuming that the distributions (or the respective generating functions) of N and X_i are known.

Note that the model discussed here is known in insurance mathematics as the **collective model of risk theory**. In general, distributions that arise as the sum of a random number of independent and identically distributed random variables are called **compound distributions**.

Example 7.5.4 Suppose the number of claims arriving at an insurance company during a month is a variable N with probability function

$$f_N(n) = P(N = n) = q^n p, \qquad n = 0, 1, 2, \ldots.$$

Then, for the random variable $Y = N + 1$, we have

$$P(Y = y) = P(N + 1 = y) = P(N = y - 1) = q^{n-1} p, \qquad n = 1, 2, \ldots,$$

that is, Y follows the (usual) geometric distribution with parameter p. The m.g.f. of Y is (see Section 7.3)

$$M_Y(t) = \frac{pe^t}{1 - qe^t}.$$

Therefore, the m.g.f. of N is given by

$$M_N(t) = e^{-t} M_Y(t) = \frac{p}{1 - qe^t},$$

while the p.g.f. of N is

$$G_N(t) = M_N(\ln t) = \frac{p}{1 - qt}.$$

As a result, using Proposition 7.5.4, we observe that the random sum

$$S_N = X_1 + \cdots + X_N$$

has its p.g.f. as

$$G_{S_N}(t) = G_N(G_X(t)) = \frac{p}{1 - qG_X(t)},$$

while the corresponding m.g.f. is

$$M_{S_N}(t) = G_N(M_X(t)) = \frac{p}{1 - qM_X(t)}.$$

Suppose, for example, the sizes of the claims are exponentially distributed with parameter λ. Then, we have

$$M_X(t) = \frac{\lambda}{\lambda - t}, \qquad t < \lambda,$$

and consequently

$$M_{S_N}(t) = \frac{p}{1 - q\dfrac{\lambda}{\lambda - t}} = \frac{p(\lambda - t)}{\lambda - t - q\lambda} = \frac{p(\lambda - t)}{\lambda p - t}.$$

From this, we can find easily (by differentiation) the moments $E(S_N^r)$, for $r = 1, 2, \ldots$.

EXERCISES

Group A

1. A discrete random variable X has p.g.f.

$$G(t) = \alpha + \frac{3}{7} \cdot t^2 + \frac{1}{3} \cdot t^3,$$

   for a suitable $\alpha \in \mathbb{R}$.

   (i) Find the value of α, and thence show that

   $$P(X = 2 | X > 0) = \frac{9}{16}.$$

   (ii) Calculate the mean and variance of X.

2. A random variable X has probability function

   $$f(x) = c \cdot 2^{-x+1}, \qquad x = 0, 1, 2, 3, 4.$$

   (i) What is the value of the constant c?

   (ii) Calculate the p.g.f. of X.

   (iii) Find the expected value and the factorial moment of second order, $E[X(X - 1)]$, of X.

   (iv) Calculate the variance of X.

3. The probability function of a discrete variable X is given by

   $$f(x) = \frac{1}{ex!}, \qquad x = 0, 1, 2, \ldots.$$

   (i) Obtain the p.g.f. of X.

   (ii) Find the expected value and the factorial moment of second order, $E[X(X - 1)]$, of X.

   (iii) Calculate the variance of X.

4. A discrete variable X has probability function

$$f(x) = \binom{n}{x} \cdot \frac{3^x 7^{n-x}}{10^n}, \qquad x = 0, 1, 2, \ldots, n.$$

(i) Obtain the p.g.f. of X.

(ii) Using this p.g.f., verify that

$$E(X) = 0.3n, \quad Var(X) = 0.21n.$$

5. The m.g.f. of a discrete variable X is

$$M(t) = a \left(\frac{8 - e^{3t}}{2 - e^t} \right)^2$$

for a suitable constant a.

(i) Obtain the value of a.

(ii) Find the p.g.f. of X. Using this, calculate the mean and variance of X.

(iii) Derive the probability function $f(x) = P(X = x)$, for $x = 0, 1, 2, \ldots$.

6. The p.g.f. of a discrete variable X is

$$G(t) = \frac{3t^3 - t^4}{a - 81t + 2t^4},$$

where a is a suitable real constant.

(i) What is the value of a?

(ii) Calculate the mean and variance of X.

7. Give an alternative proof for the second part of Proposition 7.5.4, using the first part of that proposition and the relation

$$G(t) = M(\ln t).$$

Group B

8. The probability function $f(x) = P(X = x)$, $x = 0, 1, 2, \ldots$, of a discrete random variable X satisfies the recursive relation

$$f(x) = \frac{1}{2}f(x - 1) + \frac{1}{8}f(x - 2), \qquad x = 2, 3, \ldots,$$

with initial conditions
$$f(0) = \frac{1}{4}, \quad f(1) = \frac{1}{4}.$$

(i) Find the p.g.f. of X.

(ii) Calculate the expectation $E(X)$ and the factorial moment of second order, $E[X(X-1)]$ of X.

(iii) Calculate the variance of X.

9. The p.g.f. $G(t)$ of a discrete random variable X is

$$G(t) = \frac{2t^2 - t^3}{8 - 8t + t^3}.$$

Show that the associated probability function

$$f(x) = P(X = x), \qquad x = 0, 1, 2, \ldots$$

satisfies the recursive relationship

$$f(x) = f(x-1) - \frac{1}{8}f(x-3), \qquad x = 4, 5, 6, \ldots,$$

subject to the initial conditions

$$f(0) = f(1) = 0, \quad f(2) = \frac{1}{4}, \quad f(3) = \frac{1}{8}.$$

[Hint: From the given form of $G(t)$, deduce that

$$8G(t) - 8tG(t) + t^3G(t) = 2t^2 - t^3.$$

Then, substitute $G(t)$ from the expression

$$G(t) = \sum_{k=0}^{\infty} f(k)t^k$$

and equate the coefficients of the same powers of t on the two sides.]

10. The discrete variable X takes values on the set of nonnegative integers and has cumulative distribution function $F(x) = P(X \leq x)$. Let $G(t)$ be the p.g.f. of X. Then, show that

$$\sum_{x=0}^{\infty} F(x)t^x = \frac{G(t)}{1-t}, \qquad -1 < t < 1,$$

and

$$\sum_{x=0}^{\infty} (1 - F(x))t^x = \frac{1 - G(t)}{1-t}, \qquad -1 < t < 1.$$

(The above two functions above are called generating functions of the cumulative distribution function and survival function $1 - F(x)$.)

11. For a discrete random variable X, let $G(t) = E(t^X)$. Let us define the function $R(t)$ as

$$R(t) = G(t+1),$$

and write this, in powers of t, as

$$R(t) = \sum_{r=0}^{\infty} a_r \frac{t^r}{r!}.$$

Show that the coefficients a_r above are given by

$$a_r = E\left[(X)_r\right] = E\left[X(X-1)(X-2)\cdots(X-r+1)\right], \qquad r = 1, 2, \ldots.$$

(The function $R(t)$ is called the **factorial moment generating function** of X.)

Application: Let X be a random variable with probability function

$$f(x) = pq^x, \qquad x = 0, 1, 2, \ldots,$$

where $0 < p < 1$ and $q = 1 - p$.

(i) Obtain the p.g.f. of X.

(ii) Write down the corresponding factorial moment generating function $R(t)$.

(iii) Expanding $R(t)$ in powers of t, verify that

$$E\left[(X)_r\right] = r!\left(\frac{q}{p}\right)^r, \qquad r = 1, 2, \ldots.$$

12. The number of persons entering a shop during a day is a discrete random variable following Poisson distribution with mean 70. The percentage of persons, among those who visit the shop, that buy a particular product is 10%.

   (i) Derive the probability function for the number of persons, S, who buy this product during the day.

   (ii) Obtain the mean and variance of S.

13. Suppose the total amount of claims arriving at an insurance company during a week, denoted by S, is given by

$$S = \begin{cases} X_1 + \cdots + X_N, & \text{if } N = 1, 2, \ldots, \\ 0, & \text{if } N = 0, \end{cases}$$

where $X_1, X_2, \ldots$ are the sizes of individual claim amounts and N is the number of claims arriving at the company during a week. We assume that

- the variables X_i are independent and identically distributed having a mean μ, variance σ^2, while the third moment around their mean is

$$E[(X - \mu)^3] = \mu_3;$$

- the distribution of N is Poisson with parameter λ;

- N is independent of X_i, for $i = 1, 2, \ldots$.

Using the above information,

(i) find the mean and variance of S, the aggregate claim amount;

(ii) show that the coefficient of skewness associated with S,

$$\gamma_{1,S} = \frac{E[(S - E(S))^3]}{Var(S)^{3/2}}$$

is given by

$$\gamma_{1,S} = \frac{\mu_3}{\sqrt{\lambda \mu_2^3}}.$$

Application: Under the above assumptions, suppose $\lambda = 4$ and that the distribution of X_i is exponential with $\mu = 3$ (in thousands of dollars). Calculate the mean, variance, and the coefficient of skewness of the total claim amount S.

7.6   CHARACTERISTIC FUNCTION

When we use m.g.f.s for the study of a random variable X, it is necessary to assume the existence of $E(e^{tX})$ in an interval $(-\delta, \delta)$. Although this is satisfied by most of the discrete and continuous distributions we have met, there are even simple cases in which this condition fails. For example, suppose X has density function

$$f(x) = \frac{1}{x^2}, \qquad x \geq 1.$$

We then have

$$E(e^{tX}) = \int_{-\infty}^{\infty} e^{tx} f(x)\, dx = \int_1^{\infty} \frac{e^{tx}}{x^2}\, dx$$

which does not converge for any $t > 0$.

For this reason, another type of generating function can be considered. The members of this class are generating functions that exist always (i.e. for any distribution) and they are called **characteristic functions**. These functions are in fact complex-valued functions of a single variable.

Let i denote the imaginary unit (with $i^2 = -1$), and X_1 and X_2 be two variables. Then, we say that

$$X = X_1 + iX_2$$

is a **complex random variable** whose expectation equals

$$E(X) = E(X_1) + iE(X_2).$$

Next, using the formula

$$e^{itX} = \cos tX + i \sin tX,$$

we can also write

$$E(e^{itX}) = E(\cos tX) + iE(\sin tX).$$

This last quantity, considered as a function of t, is called the characteristic function of X.

Definition 7.6.1 Let X be a random variable. Then, its characteristic function, $\varphi_X(t)$, is defined as
$$\varphi_X(t) = E(e^{itX}) = E(\cos tX) + iE(\sin tX).$$

As usual, when there is no possibility of confusion, we drop the subscript and simply write $\varphi(t)$ instead of $\varphi_X(t)$.

First, it is clear from the definition that any characteristic function φ satisfies $\varphi(0) = 1$. Next, it is important to note that the characteristic function of a random variable X exists for all real t. To see this, assume that X is discrete with probability function $f(x), x = 0, 1, 2, \ldots$. Then, its characteristic function is given by

$$\varphi(t) = \sum_{x=0}^{\infty} e^{itx} f(x),$$

and the series converges for any $t \in \mathbb{R}$, since

$$\sum_{x=0}^{\infty} |e^{itx} f(x)| = \sum_{x=0}^{\infty} |e^{itx}| f(x) = \sum_{x=0}^{\infty} f(x) = 1.$$

We have used above the well-known fact that

$$|e^{itx}| = |\cos tx + i\cos tx| = \sqrt{\cos^2(tx) + \sin^2(tx)} = 1$$

for any $t \in \mathbb{R}$.

For a continuous variable X with density function $f(x)$, its characteristic function is given by

$$\varphi(t) = \int_{-\infty}^{\infty} e^{itx} f(x) \, dx.$$

The function $\varphi(t)$ defined above is called the **Fourier transform** of the function f, and we then see that

$$|\varphi(t)| = \left| \int_{-\infty}^{\infty} e^{itx} f(x) \, dx \right| \le \int_{-\infty}^{\infty} |e^{itx}| f(x) dx = \int_{-\infty}^{\infty} f(x) dx = 1,$$

implying that the characteristic function exists for any real t in this case as well.

From Definition 7.6.1, it is apparent that

$$\varphi(-t) = E(\cos tX) - E(\sin tX) = \overline{\varphi(t)},$$

where $\bar{z}$ denotes the conjugate of a complex number z. Moreover, if $M(t)$ is the m.g.f. of X, then it is easy to see that

$$\varphi(t) = M(it), \quad M(t) = \varphi\left(\frac{t}{i}\right) = \varphi(-it)$$

for all $t \in \mathbb{R}$ for which the m.g.f. exists. Using this we find, for example:

- if X follows binomial distribution with parameters n and p, its characteristic function is

$$\varphi(t) = (pe^{it} + q)^n;$$

- if X follows $N(\mu, \sigma^2)$ distribution, its characteristic function is

$$\varphi(t) = \exp\left(i\mu t - \frac{1}{2}\sigma^2 t^2\right), \qquad t \in \mathbb{R},$$

and so on.

The next result, given without proof, gives some main properties of a characteristic function, with analogous properties of the m.g.f. having been seen earlier.

Proposition 7.6.1 *Let X be a random variable with characteristic function $\varphi(t)$. Then, we have:*

(i) *If the moment of order r of X exists, it is given by*

$$\mu'_r = E(X^r) = i^{-r}\left[\frac{d^r}{dt^r}\varphi(t)\right]_{t=0}, \qquad r = 1, 2, \ldots.$$

(ii) *For any real a and b, the characteristic function of $Y = aX + b$ is given by*

$$\varphi_Y(t) = e^{ibt}\varphi(at), \qquad t \in \mathbb{R}.$$

As the name indicates, characteristic functions determine probability distributions in a unique manner, so that there is a one-to-one correspondence between characteristic functions and distribution functions over the real line. The following result enables us to obtain the distribution of a variable X, given the characteristic function of X, and it is known as **inversion formula** for a characteristic function.

Proposition 7.6.2 *The characteristic function, $\varphi(t)$, of a random variable X determines uniquely the distribution function $F(t)$ of X. Specifically, for any $a, b \in \mathbb{R}$ such that $a < b$ and F is continuous at a and b, we have*

$$P(a < X \leq b) = F(b) - F(a) = \lim_{s \to \infty} \frac{1}{2\pi} \int_{-s}^{s} \frac{e^{-ibt} - e^{-iat}}{t} i\varphi(t)\, dt.$$

If, in addition, the integral

$$\int_{-\infty}^{\infty} |\varphi(t)|\, dt$$

is finite, then X is a continuous variable and its density function f is given by

$$f(x) = \frac{1}{2\pi} \int_{-\infty}^{\infty} e^{-itx}\varphi(t)\, dt.$$

The following result is regarding the characteristic function of a sum of random variables.

Proposition 7.6.3 *Suppose $X_1, \ldots, X_n$ are independent random variables having characteristic functions $\varphi_{X_1}(t), \ldots, \varphi_{X_n}(t)$, respectively. Then, the characteristic function of*

$$S_n = X_1 + \cdots + X_n$$

is given by

$$\varphi_{S_n}(t) = \varphi_{X_1}(t) \cdots \varphi_{X_n}(t) = \prod_{r=1}^{n} \varphi_{X_r}(t).$$

Example 7.6.1 (Cauchy distribution) Suppose X is a continuous variable with density function

$$f(x) = \frac{1}{\pi(1 + x^2)}, \qquad\qquad x \in \mathbb{R}. \tag{7.14}$$

The corresponding distribution is known as the **Cauchy distribution** (a more general form is given in Exercise 1). The characteristic function of X is

$$\varphi(t) = E(e^{itX}) = \int_{-\infty}^{\infty} e^{itx} \frac{1}{\pi(1 + x^2)} \, dx.$$

Employing the following known result from calculus

$$\int_{-\infty}^{\infty} \frac{e^{itx}}{1 + x^2} \, dx = \pi e^{-|t|}, \qquad\qquad t \in \mathbb{R},$$

we get the expression

$$\varphi(t) = e^{-|t|}, \qquad\qquad t \in \mathbb{R}.$$

Suppose now we have n independent random variables $X_1, \ldots, X_n$, each having a Cauchy distribution with density (7.14). Then, Proposition 7.6.3 yields the characteristic function of

$$S_n = X_1 + \cdots + X_n$$

to be

$$\varphi_{S_n}(t) = \prod_{r=1}^{n} \varphi_{X_r}(t) = \prod_{r=1}^{n} e^{-|t|} = e^{-n|t|}.$$

Finally, considering the average of the variables X_i,

$$\bar{X} = \frac{1}{n} \sum_{i=1}^{n} X_i = \frac{1}{n} S_n = aS_n + b$$

(with $a = 1/n$, and $b = 0$), we obtain its characteristic function to be

$$\varphi_{\bar{X}}(t) = e^{itb} \varphi_{S_n}(at) = 1 \cdot \varphi_{S_n}(t/n) = e^{-|t|}.$$

This shows that, when we have a sample from the Cauchy distribution, the distribution of the sample mean is also Cauchy. Note that this result could not have been obtained via m.g.f.s, since the m.g.f. does not exist for the Cauchy distribution.

EXERCISES

Group A

1. Let $X_1, \dots, X_n$ be a set of independent random variables, each having Cauchy distribution (in its general form) with density function

$$f(x) = \frac{1}{\pi} \cdot \frac{\lambda}{\lambda^2 + (x - \mu)^2}, \qquad x \in \mathbb{R},$$

where $\lambda > 0$ and $\mu \in \mathbb{R}$ are given parameters of the distribution.

(i) Show that the characteristic function of X_i, for $i = 1, \dots, n$, is given by

$$\varphi(t) = \exp(i\mu t - \lambda|t|), \qquad t \in \mathbb{R}.$$

(ii) Show that the average of X_i's,

$$\overline{X} = \frac{1}{n} \sum_{i=1}^{n} X_i,$$

has exactly the same distribution as each of $X_1, \dots, X_n$.

[Hint: Use the formula

$$\int_{-\infty}^{\infty} \frac{e^{itx}}{1 + x^2} \, dx = \pi \exp(-|t|), \qquad t \in \mathbb{R}. \ ]$$

2. (i) Let X be a continuous random variable with characteristic function

$$\varphi(t) = E(e^{itX}) = \frac{it\mu}{1 + \lambda^2 t^2}, \qquad t \in \mathbb{R},$$

where $\lambda > 0$ and $\mu \in \mathbb{R}$. Using the inversion formula for characteristic functions (Proposition 7.6.2) and the formula in the hint to Exercise 1, show that X has density function

$$f(x) = \frac{1}{2\lambda} \exp\left\{ -\frac{|x - \mu|}{\lambda} \right\}, \qquad x \in \mathbb{R}.$$

(ii) Let X_1 and X_2 be two independent random variables with the same density function, which is given by

$$f(x) = \frac{1}{\lambda} \exp\left(-\frac{x}{\lambda} \right), \qquad x > 0.$$

Using characteristics functions, derive the density function of the random variable $Y = (X_1 - X_2) + \mu$, where μ is a given real number.

3. A computer generates at random three (decimal) numbers from the interval $(0, 1)$. Denote by X_1, X_2, X_3 the three numbers generated.

   (i) Verify that the characteristic function of X_j, for $j = 1, 2, 3$, is given by

   $$\varphi(t) = E(e^{itX_j}) = \frac{i(1 - e^{it})}{t}.$$

   (ii) Obtain the characteristic function of the sum $S_3 = X_1 + X_2 + X_3$ and hence find its first two moments, $E(S_3)$ and $E(S_3^2)$.

4. Let $Z_1, \ldots, Z_n$ be a random sample from standard normal distribution.

   (i) Obtain the m.g.f. of the variables $Z_j^2, j = 1, \ldots, n$, and then find their characteristic functions.

   (ii) Show that the characteristic function of

   $$S = Z_1^2 + \cdots + Z_n^2$$

   is given by

   $$\varphi_S(t) = E(e^{itS}) = (1 - 2it)^{-n/2}, \qquad t \in \mathbb{R}.$$

   (iii) Use the formula for the m.g.f. of a Gamma distribution to conclude that $\varphi_S(t)$ coincides with the characteristic function of a Gamma random variable with parameters $\lambda = 1/2, a = n/2$.

   (iv) Write down the density function of S.

   (This exercise provides an alternate derivation for the density function of the chi-squared distribution.)

5. Assume that the characteristic function φ of a random variable X is an even function, that is, $\varphi(-t) = \varphi(t)$, for any real t. Then, show that the variables X and $-X$ have exactly the same distribution.

   Application: Check that, in each of the following cases, the variables X and $-X$ have the same distribution:

   (i) X has normal distribution $N(0, \sigma^2)$;

   (ii) X has Cauchy distribution with density as in (7.14).

7.7 GENERATING FUNCTIONS FOR MULTIVARIATE CASE

In the following definition, we extend the concept of generating functions to the case of multivariate distributions.

Definition 7.7.1 Let $(X_1, \ldots, X_n)$ be an n-dimensional random variable.

(i) The m.g.f. of $(X_1, \ldots, X_n)$ is a function of n variables defined by

$$M_{X_1, \ldots, X_n}(t_1, \ldots, t_n) = E\left[\exp(t_1 X_1 + \cdots + t_n X_n)\right].$$

(ii) If the variables $X_1, \ldots, X_n$ are discrete, the p.g.f. of $(X_1, \ldots, X_n)$ is a function of n variables defined as

$$G_{X_1, \ldots, X_n}(t_1, \ldots, t_n) = E\left(t_1^{X_1} \cdots t_n^{X_n}\right).$$

(iii) The characteristic function of $(X_1, \ldots, X_n)$ is the function defined for any $t_1, \ldots, t_n \in \mathbb{R}$ as

$$\varphi_{X_1, \ldots, X_n}(t_1, \ldots, t_n) = E\left[\exp(i(t_1 X_1 + \cdots + t_n X_n))\right].$$

In analogy to the univariate case, the characteristic function always exists, the probability function is defined at least for $-1 \leq t_j \leq 1, j = 1, \ldots, n$, while for the m.g.f. one has to check again that the expectation appearing there is finite in certain open intervals including zero.

In the rest of this section, for ease in notation, we drop the subscripts $X_1, \ldots, X_n$ in the multivariate generating functions.

Looking at the defining relations for the three generating functions, it is clear that these functions are related to each other just as in the univariate case. For example,

$$M(t_1, \ldots, t_n) = E\left[(e^{t_1})^{X_1} \ldots (e^{t_n})^{X_n}\right] = G\left(e^{t_1}, \ldots, e^{t_n}\right)$$

$$\varphi(t_1, \ldots, t_n) = M(it_1, \ldots, it_n),$$

and so on.

For ease of presentation, in what follows we only present some results for the case of a two-dimensional random variable, while the interested reader may try to confirm the validity of these results when $n > 2$ variables are involved. To begin with, let (X, Y) be a random pair with probability function (in the discrete case) or probability density function (continuous case) $f(x, y)$. The m.g.f. of (X, Y) is then given by

$$M(t, s) = \begin{cases} \displaystyle\sum_{(x,y) \in R_{X,Y}} e^{tx+sy} f(x, y), & \text{(discrete case)}, \\ \displaystyle\int_{-\infty}^{\infty} \int_{-\infty}^{\infty} e^{tx+sy} f(x, y) \, dx \, dy, & \text{(continuous case)}. \end{cases}$$

For $s = 0$, this immediately gives

$$M(t, 0) = E(e^{tX+0 \cdot Y}) = E(e^{tX}) = M_X(t),$$

where $M_X(t)$ is the m.g.f. of X (specifically, the m.g.f. corresponding to the marginal distribution of X). Similarly, for $t = 0$, we obtain $M(0, s) = M_Y(s)$, where $M_Y(t)$ is the m.g.f. of Y.

If X and Y are independent random variables, then the same is true for the variables e^{tX} and e^{sY}, for any $t, s \in \mathbb{R}$, and so we have in this case

$$M(t, s) = E(e^{tX+sY}) = E(e^{tX}e^{sY}) = E(e^{tX})E(e^{sY}) = M_X(t)M_Y(s).$$

It can be shown that the converse implication also holds, as stated in the following result.

Proposition 7.7.1 *Two random variables X and Y are independent if and only if the m.g.f. of the pair (X, Y) is the product of the m.g.f.s of X and Y.*

Next, the m.g.f. of the sum $Z = X + Y$ arises immediately from the function $M(t, s)$ upon setting $s = t$. To be specific, from Definition 7.7.1 we obtain

$$M(t, t) = E(e^{tX+tY}) = E\left[e^{t(X+Y)}\right] = E(e^{tZ}) = M_Z(t).$$

Further, assuming that the regularity conditions needed to interchange derivatives, sums, and integrals are satisfied, we deduce the following relations:

$$\frac{\partial}{\partial t}M(t, s) = \frac{\partial}{\partial t}E(e^{tX+sY}) = E\left(\frac{\partial}{\partial t}e^{tX+sY}\right) = E(Xe^{tX+sY}),$$

$$\frac{\partial}{\partial s}M(t, s) = \frac{\partial}{\partial s}E(e^{tX+sY}) = E\left(\frac{\partial}{\partial s}e^{tX+sY}\right) = E(Ye^{tX+sY})$$

$$\frac{\partial^2}{\partial t \partial s}M(t, s) = E\left(\frac{\partial^2}{\partial t \partial s}e^{tX+sY}\right) = E(XYe^{tX+sY}).$$

Now, by setting $t = s = 0$, we get from the above expressions that

$$E(X) = \left[\frac{\partial}{\partial t}M(t, s)\right]_{t=s=0}, \qquad E(Y) = \left[\frac{\partial}{\partial s}M(t, s)\right]_{t=s=0}$$

and

$$E(XY) = \left[\frac{\partial^2}{\partial t \partial s}M(t, s)\right]_{t=s=0}.$$

Finally, if we consider the linear combinations

$$U = aX + bY, \quad V = cX + dY$$

for any given $a, b, c, d, \in \mathbb{R}$, for the m.g.f. of (U, V) we obtain

$$M_{U,V}(t, s) = E\left[\exp(tU + sV)\right] = E\left[\exp(t(aX + bY) + s(cX + dY))\right]$$
$$= E\left[\exp((at + cs)X + (bt + ds)Y)\right] = M(at + cs, bt + ds).$$

The above formula enables us to study the joint, and also the individual, behavior of the variables U and V. For example, the m.g.f. of U is given by

$$M_{U,V}(t, 0) = M(at, bt),$$

while that of V is given by

$$M_{U,V}(0, s) = M(bs, ds).$$

Moreover, if we have

$$M(at + cs, bt + ds) = M(at, bt)M(cs, ds)$$

for all t, s in the domain of M, then the variables U and V are independent.

It needs to be mentioned that analogous results also hold for both the p.g.f. of a discrete bivariate random variable and the characteristic function of a two-dimensional variable, either discrete or continuous.

Example 7.7.1 (Moment generating function of trinomial distribution)
Let (X_1, X_2) be a random pair which follows the trinomial distribution with parameters n and p_1, p_2. Then, its probability function is given by

$$f(x_1, x_2) = P(X_1 = x_1, X_2 = x_2) = \frac{n!}{x_1! x_2! (n - x_1 - x_2)!} p_1^{x_1} p_2^{x_2} q^{n - x_1 - x_2}$$

for $0 \leq x_1, x_2 \leq n$ and $x_1 + x_2 \leq n$, where $q = 1 - p_1 - p_2$. The m.g.f. of the pair (X_1, X_2) is given by

$$M(t_1, t_2) = E\left[e^{t_1 X_1 + t_2 X_2}\right] = \sum \frac{n!}{x_1! x_2! (n - x_1 - x_2)!} (p_1 \, e^{t_1})^{x_1} (p_2 \, e^{t_2})^{x_2} q^{n - x_1 - x_2},$$

where the summation is over all nonnegative integers x_1, x_2 with $x_1 + x_2 \leq n$. Making use of the polynomial expansion (the extension of the binomial expansion, when we have three rather than two summands), we get

$$M(t_1, t_2) = \left(p_1 e^{t_1} + p_2 e^{t_2} + q\right)^n.$$

Setting $t_2 = 0$, we find the m.g.f. of X_1 to be

$$M(t_1, 0) = \left(p_1 e^{t_1} + p_2 e^0 + q\right)^n = \left(p_1 e^{t_1} + (p_2 + q)\right)^n,$$

from which we see that X_1 follows the binomial distribution with parameters n and p_1. Similarly, we find that the (marginal) distribution of X_2 is binomial with parameters n and p_2. Thus, we have obtained an alternative proof for the result in Part (i) of Proposition 6.2.2.

By differentiating the function $M(t_1, t_2)$, first with respect to t_1 and then with respect to t_2, we obtain

$$\frac{\partial^2}{\partial t_1 \partial t_2} M(t_1, t_2) = n(n - 1) p_1 e^{t_1} p_2 e^{t_2} \left(p_1 e^{t_1} + p_2 e^{t_2} + q\right)^{n-2},$$

and this yields

$$E(X_1 X_2) = \left[\frac{\partial^2}{\partial t_1 \partial t_2} M(t_1, t_2)\right]_{t_1 = t_2 = 0} = n(n-1)p_1 p_2.$$

Finally, we obtain the covariance between X_1 and X_2 as

$$Cov(X_1, X_2) = E(X_1 X_2) - E(X_1)E(X_2) = n(n-1)p_1 p_2 - (np_1)(np_2) = -np_1 p_2,$$

a result we have already seen in Proposition 6.2.3.

Example 7.7.2 (*Moment generating function of bivariate normal distribution*)
Let (X, Y) follow bivariate normal distribution with parameters

$$\mu_X = \mu_Y = 0, \quad \sigma_X^2 = \sigma_Y^2 = 1, \quad \text{and } \rho \in (-1, 1).$$

Then, the joint density, $f(x, y)$, of X and Y can be expressed as

$$f(x, y) = f_{Y|X}(y|x)f(x),$$

where the first density on the right-hand side above corresponds to $N(\mu, \sigma^2)$ distribution, with $\mu = \rho x$ and $\sigma^2 = 1 - \rho^2$, while the second density is just the standard normal distribution.

Then, the m.g.f. of the pair (X, Y) is

$$M(t, s) = E(e^{tX + sY}) = \int_{-\infty}^{\infty} \int_{-\infty}^{\infty} e^{tx + sy} f(x, y) \, dy \, dx$$

$$= \int_{-\infty}^{\infty} e^{tx} \left(\int_{-\infty}^{\infty} e^{sy} f_{Y|X}(y|x) \, dy \right) f_X(x) \, dx.$$

Since the interior integral above is the m.g.f., at point s, of a normal distribution with parameters $\mu = \rho x$ and $\sigma^2 = 1 - \rho^2$, we simply have

$$M(t, s) = \int_{-\infty}^{\infty} e^{tx} \exp\left(\mu s + \frac{1}{2}\sigma^2 s^2\right) f_X(x) dx$$

$$= \int_{-\infty}^{\infty} \exp\left[tx + \rho x s + \frac{1}{2}(1 - \rho^2)s^2\right] f_X(x) \, dx$$

$$= \exp\left(\frac{1}{2}(1 - \rho^2)s^2\right) \int_{-\infty}^{\infty} e^{(t + rs)x} f_X(x) \, dx.$$

But the last integral is the m.g.f. of a standard normal distribution at the point $t + \rho s$, and so we have

$$M(t, s) = \exp\left(\frac{1}{2}(1 - \rho^2)s^2\right) \exp\left(\frac{1}{2}(t + \rho s)^2\right).$$

Thus, we finally obtain the m.g.f. of bivariate normal distribution with stated parameters to be

$$M(t,s) = E(e^{tX+sY}) = \exp\left[\frac{1}{2}(t^2 + 2\rho ts + s^2)\right].$$

EXERCISES

Group A

1. Nick tosses a biased coin n times. Let X be the number of times that *Heads* appear and Y be the number of *Tails*. We assume that in a single toss, the probability of *Heads* is p, for some $0 < p < 1$.

   (i) Show that the m.g.f. of (X, Y) is given by

   $$M(t,s) = E(e^{tX+sY}) = (pe^t + qe^s)^n.$$

   (ii) Find the covariance between X and Y.

   [Hint: Observe that, since $Y = n - X$, the problem can essentially be tackled with univariate methods.]

2. Let (X_1, X_2) be a random pair following trinomial distribution with parameters n and p_1, p_2. Use the m.g.f. found in Example 7.7.1 to show that

   (i) the variables X_1 and X_2 are not independent;
   (ii) the distribution of $X_1 + X_2$ is binomial with parameters n and $p_1 + p_2$.

3. Let $M(t, s)$ be the m.g.f. of (X_1, Y_1) and define the variables X and Y by

   $$X = aX_1 + b, \quad Y = cY_1 + d,$$

   for some given real numbers a, b, c, d. Prove that the m.g.f. of (X, Y) is given by

   $$M_{X,Y}(t, s) = e^{bt+ds} M(at, cs).$$

   Application: Assume that (X, Y) follows a bivariate normal distribution with parameters

   $$\mu_X, \mu_Y, \sigma_X^2, \sigma_Y^2, \text{ and } \rho_{X,Y} = \rho \in (-1, 1).$$

   (i) Show that the m.g.f. of (X, Y) is given by

   $$E(e^{tX+sY}) = \exp\left[\mu_X t + \mu_Y s + \frac{1}{2}\left(\sigma_X^2 t^2 + 2\rho\sigma_X\sigma_Y ts + \sigma_Y^2 s^2\right)\right].$$

   (ii) Obtain the m.g.f. of X, and use it to identify its distribution.
   (iii) Obtain the m.g.f. of Y, and use it to identify its distribution.

(iv) Calculate the m.g.f. of $U = aX + bY$, for $a, b \in \mathbb{R}$, and use it to derive the distribution of U.

4. Let (X, Y) follow a bivariate normal distribution with parameters

$$\mu_X = 0, \mu_Y = 0, \sigma_X^2 = 1, \sigma_Y^2 = 1, \text{ and } \rho = 1/2.$$

(i) Calculate the m.g.f. of the pair $(X, 2Y - X)$.

(ii) Verify that the variables X and $2Y - X$ are independent and identify their distributions.

(iii) Calculate the probability

$$P\left(-1 \leq X \leq 1 \text{ and } \frac{X - \sqrt{3}}{2} \leq Y \leq \frac{X + \sqrt{3}}{2}\right).$$

5. Two discrete variables X and Y have joint probability function as

$$f(x, y) = c \exp(-2y - 1)\frac{y^x}{x!}, \qquad x = 0, 1, 2, \ldots \text{ and } y = 0, 1, 2, \ldots,$$

for a suitable constant $c \in \mathbb{R}$.

(i) Show that $c = e - 1$.

(ii) Prove that the p.g.f. of (X, Y) is

$$G(t, s) = \frac{e - 1}{e(1 - se^{t-2})}.$$

What restrictions are needed for the values of t and s in the above expression?

(iii) Calculate the p.g.f. for each of the variables X and Y and examine whether these variables are independent.

(iv) Establish that

$$Cov(X, Y) = \frac{e}{(e - 1)^2}.$$

Group B

6. Assume that the k-dimensional random variable $(X_1, \ldots, X_k)$ $(k \geq 2)$ follows multinomial distribution with parameters n and $p_1, p_2, \ldots, p_k$.

(i) Show that the m.g.f. of $(X_1, \ldots, X_k)$ is given by

$$M(t_1, \ldots, t_k) = \left(p_1 e^{t_1} + \cdots + p_k e^{t_k} + q\right)^n,$$

where $q = 1 - p_1 - \cdots - p_k$.

(ii) Verify that the marginal distribution of X_i, for $i = 1, \ldots, k$, is binomial with parameters n and p_i.

(iii) Show that, for any i, j with $i \neq j$, we have

$$Cov(X_i, X_j) = -np_i p_j.$$

(iv) Show that the distribution of $X_1 + \cdots + X_k$ is binomial with parameters n and $p = p_1 + \cdots + p_k$.

[This exercise provides alternate proofs, using m.g.f.s, for the results of Proposition 6.2.4.]

7. Let (X, Y) be a discrete, two-dimensional random variable, with probability function

$$f(x, y) = P(X = x, Y = y), \qquad x, y = 0, 1, 2, \ldots,$$

and assume that $G(t, s)$ is the corresponding p.g.f.

(i) Differentiate the function $G(t, s)$, k times with respect to t and r times with respect to s, to establish that

$$f(k, r) = \frac{1}{k! r!} \left[\frac{\partial^{k+r}}{\partial t^k \partial s^r} G(t, s) \right]_{t=0, s=0}.$$

(ii) Show that the p.g.f.s of X and Y are given by

$$G_X(t) = G(t, 1), \quad G_Y(s) = G(1, s).$$

(iii) Using Parts (i) and (ii), prove that if

$$G(t, s) = G_X(t) G_Y(s) \qquad \text{for any } t, s,$$

then

$$f(k, r) = f_X(k) f_Y(r) \qquad \text{for any } k, r,$$

i.e. the variables X and Y are independent.

Application: Let (X, Y) be a discrete bivariate random variable and assume its p.g.f. is

$$G(t, s) = \frac{1}{9}(2ts + t + 4s + 2).$$

Examine whether X and Y are independent.

8. Let X_1, X_2, X_3, X_4 be four discrete random variables with the corresponding p.g.f. of the four-dimensional variable (X_1, X_2, X_3, X_4) as

$$G(t_1, t_2, t_3, t_4) = \frac{1}{8}\left(t_1 t_2 t_3 t_4 + t_1 t_2 + t_1 t_3 + t_1 t_4 + t_2 t_3 + t_2 t_4 + t_3 t_4 + 1 \right).$$

Prove that

(i) the variables X_1, X_2, X_3, X_4 are pairwise independent;

(ii) any three among the variables X_1, X_2, X_3, X_4 are independent;

(iii) X_1, X_2, X_3, X_4 are *not* independent.

9. The joint probability function of the discrete random variables X and Y is given by

$$f(x, y) = (1 - a)(b - a)a^x b^{y-x-1}, \qquad 0 \leq y \leq x$$

   (x, y are integers), where a, b are given real numbers with $0 < a < 1$ and $a < b$.

   (i) Show that the p.g.f. of (X, Y) is given by

$$G(t, s) = \frac{(1 - a)(b - a)}{(1 - ast)(b - at)}, \qquad \text{for } |st| \leq \frac{1}{a}.$$

   (ii) Derive the p.g.f. of X and that of Y. Are X and Y independent?

   (iii) Show that the covariance between X and Y is equal to $a(1 - a)^{-2}$.

7.8 BASIC CONCEPTS AND FORMULAS

Moment generating function (m.g.f.)	$M(t) = E(e^{tX}) = \begin{cases} \displaystyle\sum_{x \in R_X} e^{tx} f(x), & \text{if } X \text{ is discrete} \\[2ex] \displaystyle\int_{-\infty}^{\infty} e^{tx} f(x) \, dx, & \text{if } X \text{ is continuous} \end{cases}$
Properties of the m.g.f.	$\bullet$ $\mu_r' = E(X^r) = M^{(r)}(0) = \left[\dfrac{d^r}{dt^r} M(t) \right]_{t=0}$, for $r = 1, 2, \ldots$ $\bullet$ $M(t) = \displaystyle\sum_{r=0}^{\infty} \frac{E(X^r)}{r!} t^r$. $\bullet$ If $M_X(t) = M_Y(t)$ for any $t \in (-\delta, \delta)$, then the variables X and Y have the same distribution. $\bullet$ $M_{aX+b}(t) = e^{bt} M_X(at)$ $\bullet$ If $X_1, \ldots, X_n$ are independent random variables and $$S_n = X_1 + \cdots + X_n,$$ then $$M_{S_n}(t) = \prod_{i=1}^{n} M_{X_i}(t) = M_{X_1}(t) \cdots M_{X_n}(t).$$

Probability generating function (p.g.f.) of a nonnegative integer-valued variable X	$$G_X(t) = G(t) = E(t^X) = \sum_{x \in R_X} t^x f(x), \quad |t| \leq 1.$$
Properties of the p.g.f.	• $f(r) = \dfrac{1}{r!} \left[\dfrac{d^r}{dt^r} G(t) \right]_{t=0}, \quad$ for $r = 0, 1, 2, \ldots$ • $\mu_{(r)} = E\left[(X)_r\right] = \left[\dfrac{d^r}{dt^r} G(t) \right]_{t=1}, \quad$ for $r = 1, 2, \ldots$ • If $G_X(t) = G_Y(t)$ for any $t \in [-1, 1]$, then the variables X and Y have the same distribution. • $G_{aX+b}(t) = t^b G_X(t^a)$, for $a > 0$ • If $X_1, \ldots, X_n$ are independent random variables and $$S_n = X_1 + \cdots + X_n,$$ then $$G_{S_n}(t) = \prod_{i=1}^{n} G_{X_i}(t) = G_{X_1}(t) \cdots G_{X_n}(t).$$
Characteristic function	$\varphi_X(t) = \varphi(t) = E(e^{itX}) = E(\cos tX) + iE(\sin tX)$, $t \in \mathbb{R}$
Properties of a characteristic function	• $\mu_r' = E(X^r) = i^{-r} \left[\dfrac{d^r}{dt^r} \varphi(t) \right]_{t=0}$, for $r = 1, 2, \ldots$ • $\varphi_{aX+b}(t) = e^{ibt} \varphi_X(at)$ • $P(a < X \leq b) = \lim_{s \to \infty} \dfrac{1}{2\pi} \displaystyle\int_{-s}^{s} \dfrac{e^{-ibt} - e^{-iat}}{t} i\varphi(t)\, dt$ • $f(x) = \dfrac{1}{2\pi} \displaystyle\int_{-\infty}^{\infty} e^{-itx} \varphi(t)\, dt$ (for continuous random variables) provided $\int_{-\infty}^{\infty} |\varphi(t)|\, dt < \infty$. • If $\varphi_X(t) = \varphi_Y(t)$ for any t, then the variables X, Y have the same distribution • If $X_1, \ldots, X_n$ are independent random variables and $$S_n = X_1 + \cdots + X_n,$$ then $$\varphi_{S_n}(t) = \prod_{i=1}^{n} \varphi_{X_i}(t) = \varphi_{X_1}(t) \cdots \varphi_{X_n}(t).$$

Moment generating function of a random sum (the sum of a random number of random variables)	If $S_N = X_1 + \cdots + X_N$, where $X_1, X_2, \ldots$, are independent and identically distributed random variables and N is a nonnegative integer-valued variable independent of X_i's, then $$M_{S_N}(t) = G_N(M_X(t)), \quad G_{S_N}(t) = G_N(G_X(t)).$$
Relations between m.g.f., p.g.f., and characteristic function	$$M(t) = G(e^t) = \varphi(-it)$$ $$G(t) = M(\ln t) = \varphi(-i \ln t)$$ $$\varphi(t) = M(it) = G(e^{it}).$$
Generating functions of a random pair (X, Y)	$$M_{X,Y}(t, s) = E\left[\exp(tX + sY)\right]$$ $$G_{X,Y}(t, s) = E\left(t^X s^Y\right)$$ $$\varphi_{X,Y}(t, s) = E\left[\exp(i(tX + sY))\right].$$
Properties of the m.g.f. of a random pair	• $M_{X,Y}(t, 0) = M_X(t)$ • $M_{X,Y}(0, s) = M_Y(s)$ • $M_{X,Y}(t, t) = M_{X+Y}(t)$ • $E(X) = \left[\dfrac{\partial}{\partial t}M(t, s)\right]_{t=s=0}$, $E(Y) = \left[\dfrac{\partial}{\partial s}M(t, s)\right]_{t=s=0}$ • $E(XY) = \left[\dfrac{\partial^2}{\partial t \partial s}M(t, s)\right]_{t=s=0}$ • If $U = aX + bY$, and $V = cX + dY$, then $$M_{U,V}(t, s) = M_{X,Y}(at + cs, bt + ds).$$

7.9 COMPUTATIONAL EXERCISES

With the following commands, we calculate the m.g.f., mean, and variance of a discrete random variable with probability function

$$f(x) = \frac{7 - x}{15}, \quad x = 1, 2, 3$$

(see Example 7.2.2).

```
In[1]: f[x_] := (7 - x)/15;
M[t_] := Sum[f[x]*Exp[t*x], {x, 1, 3}]
```

```
Print["THE MGF OF THE DISTRIBUTION WITH PROBABILITY FUNCTION "]
Print["f(x)= ", f[x] ]
Print["IS   "]
Print["M(t)= ", M[t]]
m := M'[0]
var := M''[0] - (M'[0])^2
Print["THE MEAN IS ", m]
Print["THE VARIANCE IS ", var]
```

```
Out[1]= THE MGF OF THE DISTRIBUTION WITH PROBABILITY FUNCTION
f(x)= (7-x)/15
IS
M(t)= (2 Exp[t])/5+Exp[2 t]/3+(4 Exp[3 t])/15
THE MEAN IS 28/15
THE VARIANCE IS 146/225
```

Next, the following program is used to find the m.g.f. of a random variable X following standard normal distribution, and then the m.g.f. of $Y = aX + b$ (see Example 7.2.4 and Proposition 7.2.3.)

```
In[2]: f[x_] := (1/Sqrt[2*Pi])*Exp[-x^2/2]
M[s_] := Integrate[f[x]*Exp[s*x], {x, -Infinity, Infinity}];
Print["THE MGF OF THE DISTRIBUTION WITH DENSITY "]
Print["f(x)= ", f[x] ]
Print["IS   "]
Print["M(t)= ", M[t]]
MY[t_] := Exp[b*t]*M[a*t]
Print["THE MOMENT GENERATING FUNCTION OF THE VARIABLE Y=aX+b IS" ]
Print["MY(t)= ", MY[t]]
```

```
Out[2]= THE MGF OF THE DISTRIBUTION WITH DENSITY
```

$$f(x)=\frac{e^{-\frac{x^2}{2}}}{\sqrt{2\pi}}$$

```
IS
M(t)= Exp[t^2/2]
THE MOMENT GENERATING FUNCTION OF THE VARIABLE Y=aX+b IS
MY(t)= Exp[b t+a^2 t^2/2]
```

1. Use Mathematica to find the m.g.f. of the following discrete distributions. Then, differentiate the m.g.f. to derive the first two moments of the distribution and check that they are what you expect:

    (i) binomial with $n = 20$, $p = 1/3$;

    (ii) geometric with $p = 2/5$;

    (iii) negative binomial with $r = 10$, $p = 1/10$.

2. Let X be a discrete random variable with probability function f. For each of the following cases for f, calculate

    • the m.g.f. associated with f,

- the mean and variance of X,

- the coefficient of skewness of the distribution of X

(recall that the coefficient of skewness is defined as $\mu_3/(\mu_2)^{3/2}$).

(i) $f(x) = \dfrac{1}{n}$, $\qquad\qquad x = 1, \ldots, n$;

(ii) $f(x) = \dfrac{20(2x+3)}{229(x+1)}$, $\qquad x = 1, 2, 3, 4, 5$;

(iii) $f(x) = \dfrac{1}{165}\, x^2$, $\qquad x = 1, 3, 5, 7, 9$;

(iv) $f(x) = \dfrac{3}{10} \cdot \dfrac{2^x}{x!(4-x)!}$, $\quad x = 1, 2, 3, 4$;

(v) $f(x) = \dfrac{16}{15} \cdot \dfrac{1}{2^x}$, $\qquad x = 1, 2, 3, 4$;

(vi) $f(x) = \dfrac{9}{428} \cdot \dfrac{4^x}{x!}$, $\qquad x = 1, 2, 3, 4, 5, 6$.

3. Repeat Exercise 2 when X is now a continuous random variable having density f, in each of the following cases:

(i) $f(x) = \begin{cases} x, & 0 \le x < 1, \\ 2-x, & 1 \le x < 2, \\ 0, & \text{otherwise}; \end{cases}$

(ii) $f(x) = 60x^3(1-x)^2$, $\qquad 0 < x < 1$;

(iii) $f(x) = \dfrac{1}{8196}(x^7 + 1)$, $\qquad 0 < x < 4$;

(iv) $f(x) = \dfrac{3}{32}x^3(4 - x^2)^2$, $\qquad 0 < x < 2$;

(v) $f(x) = \dfrac{625}{6}x^3 e^{-5x}$, $\qquad x > 0$.

4. The m.g.f. of X is given by each of the following formulas. Determine the m.g.f. of $Y = 3X - 2$ and use this it calculate the mean and variance of Y:

(i) $M(t) = \dfrac{e^{6t} - 1}{6(e^t - 1)}$;

(ii) $M(t) = \exp\left[3t + 7t^2 - \ln(1-t)\right]$, $\quad t < 1$;

(iii) $M(t) = \dfrac{1}{8}e^{2t} + \dfrac{1}{2}e^{4t} + \dfrac{3}{8}e^{7t}$;

(iv) $M(t) = \left(\dfrac{3}{5}e^{5t} + \dfrac{3}{10}e^{7t} + \dfrac{1}{10}e^{10t}\right)^9$;

(v) $M(t) = \dfrac{2e^{3t^2} + 3e^{5t} + 4e^{t+3t^2}}{9}$;

(vi) $M(t) = \dfrac{7e^t}{10 - 3e^t}$, $t < -\ln(3/10)$.

5. Write a program in Mathematica which, for a given discrete random variable X with probability function f, calculates the p.g.f. of X and by a suitable differentiation,

the mean and variance of X. Then, use it to find the p.g.f., the mean and variance of a discrete variable X having probability function f, for each of the probability functions presented in Exercise 2.

6. Write a program in Mathematica which, for a given continuous random variable X with density function f, calculates the characteristic function of X and by a suitable differentiation, the mean and the variance of X. Then, use it to find the characteristic function, the mean and variance of a continuous variable X having density function f, for each of the density functions presented in Exercise 3.

7.10   SELF-ASSESSMENT EXERCISES

7.10.1   True–False Questions

1. In a single toss of a coin, let X be the number of heads that appear. The m.g.f. of X is
$$M(t) = (e^t + 1)/2.$$

2. In two rolls of a die, let Y be the number of sixes that appear. The m.g.f. of Y is
$$M(t) = \left(\frac{5e^t + 1}{6} \right)^2.$$

3. Assume that X has m.g.f. as
$$M(t) = \frac{9 + 8e^{2t} + 7e^{4t}}{24}, \qquad t \in \mathbb{R}.$$

Then, $P(X > 3) = 7/24$.

4. The number of accidents per week in two roads A and B, in a city, are two independent random variables, denoted by X and Y, which follow the Poisson distributions with parameters 3 and 5, respectively. Let $W = X + Y$ be the total number of accidents in the two roads during a week. Then, the p.g.f. of W is $G(t) = e^{8-8t}$.

5. The service times (in minutes) of customers by a cashier at a bank is denoted by a continuous variable X. If it is known that
$$P(X > y) = e^{-y/2} \qquad \text{for any } y \geq 0,$$

then the characteristic function of X is
$$\varphi(t) = \frac{1}{1 - 2it}, \qquad t \in \mathbb{R}.$$

6. The service time, in minutes, at a gas station is a continuous random variable X with density function
$$f(x) = 4x^2 e^{-2x}, \qquad x \geq 0.$$

The characteristic function of X is given by

$$\varphi(t) = \left(\frac{2}{2 - it}\right)^3, \qquad t \in \mathbb{R}.$$

7. A variable X has m.g.f.

$$M(t) = \exp[4(e^t - 1)], \qquad t \in \mathbb{R}.$$

Then,

$$P(X > 1) = 4e^{-4}.$$

8. Let X and Y be two independent random variables whose m.g.f.s are given by

$$M_X(t) = \frac{e^t + 1}{2}, \qquad M_Y(t) = \frac{3e^t + 1}{4}, \qquad t \in \mathbb{R}.$$

Then, the distribution of $X + Y$ is binomial.

9. The measurement error, X, from an experiment, expressed in a suitable unit, has m.g.f $M(t) = e^{8t^2}$. Then, the probability that X lies in the interval $[-1, 1]$ is equal to 0.383.

10. Suppose we roll a die 5 times, and let X and Y denote, respectively, the number of aces and sixes that appear. The joint p.g.f. of X and Y is

$$G_{X,Y}(t, s) = \left(\frac{t + s + 4}{6}\right)^5, \qquad t, s \in \mathbb{R}.$$

11. The m.g.f. of a variable X is

$$M(t) = \left(\frac{2}{2 - t}\right)^3, \qquad t < 2.$$

Then, we have $E(X^2) = E(X)$.

12. The number of fish caught by a fisherman during a day follows negative binomial distribution with parameters $r = 3$ and $p = 1/4$. The weight of a fish caught, in kilos, is assumed to follow $N(3, 1)$ distribution. Then, the expected total weight of fish caught by the fisherman during a day is 2.25 kilograms.

13. The p.g.f. of a random variable X takes always values in the interval $[0, 1]$.

14. The characteristic function of a random variable X always takes values in the interval $[-1, 1]$.

15. The characteristic function $\varphi(t)$ of a continuous random variable X is

$$\varphi(t) = \exp\left(5it - 2t^2\right), \qquad t \in \mathbb{R}.$$

Then, $X \sim N(2, 5)$.

16. The characteristic function of a continuous variable X is given by

$$\varphi(t) = \frac{5}{5 - it}.$$

Then, the mean of X is equal to its standard deviation, i.e. $E(X) = \sqrt{Var(X)}$.

17. Assume that (X, Y) has a bivariate discrete distribution with joint p.g.f. $G_{X,Y}(t, s)$. Then, the p.g.f. of the variable $W = X + Y$, denoted by $G_W(v)$, is given by

$$G_W(v) = G_{X,Y}(v, 0).$$

18. A discrete variable X has characteristic function

$$\varphi(t) = \frac{1}{6} + \frac{2}{9}e^{4it} + c \cdot e^{10it}, \qquad t \in \mathbb{R},$$

for a suitable $c \in \mathbb{R}$.

The expected value of X is $E(X) = 7/3$.

19. Let X and Y be two continuous random variables with joint m.g.f. $M_{X,Y}(t, s)$, and set $L_{X,Y}(t, s) = \ln M_{X,Y}(t, s)$. Then, we have

$$Cov(X, Y) = \left[\frac{\partial^2}{\partial t \partial s} L(t, s) \right]_{t=s=0}.$$

7.10.2 Multiple Choice Questions

1. The variable X has uniform distribution over $[0, 1]$. The m.g.f. of $Y = 3X - 2$ is given by

(a) $M(t) = \dfrac{e^t - 1}{t}$ (b) $M(t) = \dfrac{e^{3t} - e^t}{3t}$ (c) $M(t) = \dfrac{e^{2t} - e^t}{t}$

(d) $M(t) = \dfrac{e^t - e^{-2t}}{3t}$ (e) $M(t) = \dfrac{e^{2t} - e^{-t}}{3t}$

2. The internal diameter, in inches, of copper tubes manufactured by a factory is a random variable X having m.g.f.

$$M(t) = \exp \left[9t \left(\frac{t}{50} + 1 \right) \right], \qquad t \in \mathbb{R}.$$

Let μ and σ denote, respectively, the mean and standard deviation of the tube diameter, in inches. The values of μ and σ are

(a) $\mu = 1, \sigma = \dfrac{1}{5}$ (b) $\mu = 1, \sigma = \dfrac{3}{\sqrt{50}}$ (c) $\mu = 9, \sigma = \dfrac{1}{5}$

(d) $\mu = 3, \sigma = \dfrac{3}{5}$ (e) $\mu = 9, \sigma = \dfrac{3}{5}$

3. The discrete variable X has m.g.f. as

$$M(t) = \left(\frac{1}{6} e^t + \frac{5}{6} \right)^3.$$

The value of $P(3X - 1 = 5)$ is equal to

(a) $\dfrac{5}{216}$ (b) $\dfrac{1}{36}$ (c) $\dfrac{5}{72}$ (d) $\dfrac{25}{72}$ (e) $\dfrac{13}{36}$

4. Let X be a discrete random variable with p.g.f.

$$G(t) = \left(\frac{2e^t}{5 - 3e^t} \right)^4.$$

Then, $P(X = 3)$ is equal to

(a) 0   (b) $\dfrac{2}{5}$ (c) $\dfrac{3}{5}$ (d) $\left(\dfrac{3}{5}\right)^2 \cdot \dfrac{2}{5}$ (e) $\left(\dfrac{3}{5}\right)^3 \cdot \dfrac{2}{5}$

5. The p.g.f., $G(t)$, of a discrete random variable X is

$$G(t) = \frac{2t + 5t^2 + 4t^3}{20} + c,$$

for a suitable real constant c. Then, $P(X = 2 | X \le 3)$ is equal to

(a) $\dfrac{1}{4}$ (b) $\dfrac{5}{11}$ (c) $\dfrac{1}{5}$ (d) $\dfrac{9}{20}$ (e) $\dfrac{1}{3}$

6. The m.g.f. of a variable X is

$$M(t) = \frac{3}{2(2 - t)} + \frac{3}{4(3 - t)}, \qquad t < 2.$$

The distribution of X is

(a) binomial   (b) normal   (c) Poisson

(d) exponential   (e) none of the above

7. The number of books a customer, entering a bookstore, purchases follows Poisson distribution with parameter $\lambda = 3/2$. Suppose, on a Thursday afternoon, 60 customers entered the bookstore, and that

$$\overline{X} = \frac{1}{60} \sum_{i=1}^{60} X_i$$

denotes the average number of books bought by them (here, X_i is the number of books bought by customer i). Then, the p.g.f. of $\overline{X}$, denoted by $G(t)$, is

(a) $G(t) = \dfrac{1}{60} \exp[3(t - 1)/2]$ (b) $G(t) = \left(\exp[3(e^t - 1)/2] \right)^{1/60}$

(c) $G(t) = \dfrac{1}{60} \exp[90(t^{60} - 1)]$ (d) $G(t) = \exp[90(t^{1/60} - 1)]$

(e) $G(t) = \exp[90(e^{t/60} - 1)]$

8. The number of claims, N, arriving at an insurance company during a day has probability function

$$P(N = n) = 5^{n-1}6^{-n}, \qquad n = 1, 2, \dots .$$

The sizes $X_1, X_2, \dots$ of these claims (in thousands of dollars) are i.i.d. random variables, independently of N, with density function

$$f(x) = \frac{2}{3} e^{-2x/3}, \qquad x > 0.$$

Then, the expected total claim amount, $E(S_N)$, where $S_N = X_1 + \cdots + X_N$, in thousands of dollars, is equal to

(a) $\dfrac{5}{9}$ (b) $\dfrac{5}{4}$ (c) $\dfrac{15}{4}$ (d) $\dfrac{10}{3}$ (e) 9

9. A discrete variable X has characteristic function

$$\varphi(t) = \frac{1}{4} + \frac{2}{3} e^{2it} + \frac{1}{12} e^{4it}.$$

The conditional probability $P(X > 2 | X > 0)$ is equal to

(a) $\dfrac{1}{4}$ (b) $\dfrac{1}{8}$ (c) $\dfrac{1}{9}$ (d) $\dfrac{1}{12}$ (e) $\dfrac{3}{4}$

10. A discrete variable X has characteristic function

$$\varphi(t) = \frac{e^{it}}{3 - 2e^{it}}, \qquad t \in \mathbb{R}.$$

Then, the probability $P(X \le 4)$ is equal to

(a) $\dfrac{1}{9}$ (b) $\dfrac{16}{27}$ (c) $\dfrac{65}{81}$ (d) $\dfrac{8}{9}$ (e) $\dfrac{80}{81}$

11. Consider the following m.g.f. of a discrete random variable X,

$$M(t) = \left(\frac{e^t + 3}{4} \right)^4.$$

Then, the mean and variance of $Y = 2X + 1$ are given by

(a) $E(Y) = 7, Var(Y) = \dfrac{3}{4}$ (b) $E(Y) = 3, Var(Y) = \dfrac{3}{4}$

(c) $E(Y) = 3, Var(Y) = 3$ (d) $E(Y) = 7, Var(Y) = 3$

(e) $E(Y) = 3, Var(Y) = \dfrac{7}{4}$

12. Let X_1 and X_2 be two independent random variables having exponential distribution with the same parameter $\lambda = 1$, and let $Y = \min\{X_1, X_2\}$. Then, the m.g.f. of Y, for the range of values t it is defined, is given by

(a) $\left(\dfrac{2}{2-t} \right)^2$ (b) $\left(\dfrac{1}{1-t} \right)^2$ (c) $\dfrac{1}{1-t}$

(d) $\dfrac{2}{2-t}$ (e) $\dfrac{1}{1-2t}$

13. Suppose we select five digits from 1 to 9 with replacement. Let X and Y be the number of times the digits 1 and 4, respectively, have been selected. The p.g.f. of (X, Y), at the point (t_1, t_2), is given by

    (a) $\left(\dfrac{t_1 + t_2}{9}\right)^5$
    (b) $\left(e^{t_1} + e^{t_2}\right)^5$
    (c) $\left(\dfrac{e^{t_1} + e^{t_2} + 7}{9}\right)^5$

    (d) $\left(\dfrac{\ln t_1 + 4 \ln t_2}{9}\right)^5$
    (e) $\left(\dfrac{t_1 + t_2 + 7}{9}\right)^5$

14. Consider the following m.g.f. of a random variable X,

    $$M(t) = \frac{e^t}{2 - e^t}, \qquad t < \ln 2.$$

    The distribution of X is

    (a) geometric   (b) binomial   (c) exponential

    (d) normal   (e) gamma

15. Let X_1 and X_2 be two independent discrete random variables, each having the following m.g.f.

    $$M(t) = \frac{3e^t + 5e^{3t} + 4e^{5t}}{12}.$$

    Then, $P(X_1 = X_2)$ is equal to

    (a) $\dfrac{1}{16}$ (b) $\dfrac{1}{8}$ (c) $\dfrac{25}{144}$ (d) $\dfrac{35}{144}$ (e) $\dfrac{25}{72}$.

16. A continuous random variable X has characteristic function

    $$\varphi(t) = \left(\frac{3}{3 - it}\right)^\alpha, \qquad t \in \mathbb{R}.$$

    If it is known that $E(X) = 1$, then the value of α is

    (a) $\alpha = 1$ (b) $\alpha = 2$ (c) $\alpha = 3$ (d) $\alpha = 9$ (e) $\alpha = \dfrac{1}{2}$.

17. Assume that the m.g.f. of a continuous variable X is

    $$M(t) = \exp\left[2t + \frac{9t^2}{2}\right].$$

    Then, the distribution of

    $$Y = \frac{(X - 2)^2}{9}$$

    is

    (a) normal   (b) χ^2 (c) exponential

    (d) Cauchy   (e) none of the above

18. Consider the random sum $S_N = X_1 + \cdots + X_N$, where X_i's are i.i.d. random variables having geometric distribution with parameter p, and assume that N is

a discrete random variable, independent of X_i's, following geometric distribution with parameter r. Then, the p.g.f. of S_N, denoted by $G(t)$, is given by

(a) $G(t) = \dfrac{(1-pr)t}{1-prt}$ 　　　(b) $G(t) = \dfrac{(pr)^t}{1-(1-pr)t}$

(c) $G(t) = \dfrac{prt}{1-(1-pr)t}$ 　　　(d) $G(t) = \dfrac{prt}{(1-pr)t}$

(e) $G(t) = \dfrac{(1-p)(1-r)t}{1+prt}$

19. The joint m.g.f. of (X, Y) is given by

$$M(t, s) = \exp\left[\frac{1}{6}(3t^2 + 3s^2 + 2ts)\right].$$

Then, the conditional expectation $E(Y|X = 1)$ is equal to

(a) 0 　　(b) $\dfrac{1}{3}$ 　　(c) $\dfrac{2}{3}$ 　　(d) $\dfrac{\sqrt{\pi}}{2}$ 　　(e) 1

7.11 REVIEW PROBLEMS

1. The change in the price of the stock of Company A during a day follows uniform distribution over $[-0.8, 1.4]$. The daily change in the price of another stock of Company B also has uniform distribution over $[-0.5, 1.3]$. An investor has 300 stocks of Company A and 200 stocks of Company B.

   Let Y be the total daily change in the price of stocks the investor has. Obtain the m.g.f. of Y, and thence calculate the mean and variance of Y.

2. For a discrete random variable X, it is known that

$$\mu'_k = E(X^k) = a^k, \qquad k = 1, 2, \dots (a > 0).$$

   Obtain the m.g.f. of X. What do you conclude about the distribution of X?

3. Let $M(t)$ denote the m.g.f. of a discrete random variable X, and let $L(t) = \ln M(t)$. Verify that

$$L'(t) = \frac{M'(t)}{M(t)}, \qquad L''(t) = \frac{M'(t)}{M(t)} - [L'(t)]^2,$$

   and thence show that

$$E(X) = L'(0), \qquad Var(X) = L''(0).$$

   Application: Use the derivatives of L, introduced above, to find the expectation and variance of X whose m.g.f. is given by

   (i) $M(t) = \exp\left[\lambda(e^t - 1)\right]$ 　　　　(Poisson distribution);

   (ii) $M(t) = \exp\left(\mu t + \dfrac{1}{2}\sigma^2 t^2\right)$ 　　　(Normal distribution);

   (iii) $M(t) = \left(\dfrac{\lambda}{\lambda - t}\right)^a,$ 　for $t < \lambda$ 　　(Gamma distribution).

4. Let X be a random variable following exponential distribution with density function

$$f(x) = \begin{cases} \lambda e^{-\lambda x}, & x \geq 0, \\ 0, & \text{otherwise.} \end{cases}$$

Using the m.g.f. of X, show that

$$\mu'_r = E(X^r) = \frac{r!}{\lambda^r}, \qquad r = 1, 2, \ldots.$$

5. The m.g.f. of a discrete random variable X is given by

$$M(t) = c \sum_{k=1}^{n} k e^{kt},$$

where c is a real constant and n is a given positive integer.

(i) Obtain the value of c, in terms of n.

(ii) Write down the probability function of X.

Application: Derive the probability function of a discrete variable X if it is known that its m.g.f. is
$$M(t) = \frac{e^t + 2e^{2t} + 3e^{3t} + 4e^{4t}}{10}.$$

6. Nick rolls a red die and a black die. Let X be the outcome of the red die and Y be the outcome of the black one. Calculate the p.g.f. for each of U and V, where

$$U = |X - Y|, \quad V = 2Y^2 + 1.$$

7. The density function of a continuous variable X is an even function, that is,

$$f(x) = f(-x), \qquad \text{for any } x \in \mathbb{R}.$$

(i) Show that the variables X and $-X$ have the same distribution.

(ii) Show that the characteristic function of X is also an even function.

8. For a continuous variable X, it is known that its moments are given by

$$\mu'_r = E(X^r) = r! a^r, \qquad r = 1, 2, \ldots$$

for a given constant a.

(i) Find the moment generating function of X.

(ii) Determine the distribution of X.

9. Assume that X is a continuous random variable with density function

$$f(x) = \frac{1}{2} e^{-|x|}, \quad x \in \mathbb{R},$$

(this distribution is known as **Laplace distribution**).

(i) Obtain the m.g.f. and the characteristic function of X.

(ii) Use the result from Part (i) to compute the moments $E(X^n)$, for $n = 1, 2, \dots$.

10. Recall from the beginning of Section 7.2 that the coefficient of kurtosis associated with the distribution of a random variable X is given by

$$\gamma_2 = \frac{\mu_4}{\sigma^4},$$

where μ and σ are the mean and standard deviation of X, and $\mu_4 = E[(X - \mu)^4]$.

A random variable X follows binomial distribution with parameters $n = 3$ and p, where $0 < p < 1$.

(i) Find, in terms of p, the coefficient of kurtosis of X.

(ii) For which value(s) of p is the distribution of X mesokurtic (i.e. we have $\gamma_2 = 3$)?

11. Let $X_1, \dots, X_n$ be independent random variables, each following $N(0, 1)$ distribution. Use characteristics functions to show that the distribution of

$$W = \sum_{i=1}^{n} X_i^2$$

is chi-squared with n degrees of freedom.

[Hint: Use the uniqueness property of characteristic functions and the fact that chi-squared distribution is a special case of Gamma distribution.]

12. An insurance company has two motor insurance portfolios. Let S_1 and S_2 be the total claim amounts from these two portfolios during a month. Each of S_1 and S_2 can be written as a random sum as

$$S_1 = X_1 + \cdots + X_{N_1}, \quad S_2 = Y_1 + \cdots + Y_{N_2},$$

where X_i and Y_i denote the individual claim amounts in the two portfolios, respectively, and N_1 and N_2 are the numbers of claims for the first and second portfolio during a month (by convention, $S_i = 0$ if $N_i = 0$ for $i = 1, 2$). The company makes the assumption that

$$N_1 \sim P(\lambda_1), \quad N_2 \sim P(\lambda_2),$$

and

$$X_i \sim \mathcal{E}(\theta_1), \quad Y_j \sim \mathcal{E}(\theta_2), \quad \text{for } i, j = 1, 2, \dots.$$

Let $S = S_1 + S_2$ denote the combined total claim amount in the two portfolios.

Assuming that $\{X_i: i = 1, 2 \dots\}$, $\{Y_j: j = 1, 2, \dots\}$ are sequences of i.i.d. random variables, independent of one another, answer the following:

(i) Find the m.g.f for each of S_1 and S_2. Differentiate these functions to obtain $E(S_i)$ and $Var(S_i)$, for $i = 1, 2$.

(ii) Calculate the m.g.f. of S, and thence find its mean and variance.

(iii) Show that, under the above assumptions, S can also be expressed as a random sum; that is, S can be written in the form

$$S = Z_1 + \cdots + Z_N, \quad (S = 0 \text{ if } N = 0),$$

where N has a Poisson distribution with parameter $\lambda = \lambda_1 + \lambda_2$, i.e. $S \sim \mathcal{P}(\lambda)$ and $Z_1, Z_2, \ldots$, is a sequence of independent and identically distributed random variables. What is the distribution function of Z_i? How do you interpret this result intuitively?

13. The p.g.f. of a discrete nonnegative variable X is

$$G(t) = \frac{(pt)^k(1 - pt)}{1 - t + qp^k t^{k+1}},$$

where k is a positive integer, $0 < p < 1$ and $q = 1 - p$.

(i) Obtain the mean and variance of X.

(ii) Show that the probability function of X,

$$f(x) = P(X = x), \qquad x = 0, 1, 2, \ldots,$$

satisfies the recursive relation

$$f(x) = f(x - 1) - qp^k f(x - k - 1), \qquad x \geq k + 2,$$

with initial conditions

$$f(x) = 0, \text{ for } x = 0, 1, 2, \ldots, k - 1$$

and

$$f(k) = p^k, \quad f(k + 1) = qp^k.$$

14. Assume that $X_1, \ldots, X_n$ are independent random variables and that each of them follows exponential distribution with mean 1.

(i) Determine the density function of

$$U = X_{(n)} = \max \{X_1, \ldots, X_n\},$$

and use it to show that the m.g.f. of U is

$$E(e^{tU}) = \frac{n!}{(1 - t)(2 - t) \cdots (n - t)}, \qquad \text{for } t < 1.$$

(ii) Derive the m.g.f. of the variables X_i/i, for $i = 1, \ldots, n$, and use them to obtain the m.g.f. of

$$V = \sum_{i=1}^{n} \frac{X_i}{i}.$$

(iii) Show that U and V have exactly the same distribution.

[Hint: For Part (i), you may use the formula

$$\int_0^1 t^{a-1}(1-t)^{b-1} \, dt = B(a,b) = \frac{\Gamma(a)\Gamma(b)}{\Gamma(a+b)}. \ ]$$

15. A random variable X takes integer nonnegative values and has probability function $f(x)$. It is known that

$$f(0) = f(1) = 0, \ f(2) = p^2$$

and

$$p^2 P(X > x) = pf(x+1) + f(x+2), \qquad x = 1, 2, \ldots,$$

where $0 < p < 1$, and $q = 1 - p$.

   (i) Verify that the p.g.f., $G(t)$, of X satisfies the equation

$$\frac{1 - G(t)}{1 - t} p^2 t^2 = (1 + pt)G(t).$$

   (ii) Solve this equation to obtain and expression for $G(t)$.

   (iii) Show that the probability function $f(x)$ satisfies the recursion

$$f(x) = qf(x-1) + pqf(x-2), \qquad x = 3, 4, \ldots .$$

[Hint: Multiply both sides of the given equation by t^{x+1} and sum over x. Then, express the sums arising in terms of $G(t)$.]

16. The number of children born by a female animal is a discrete variable N with probability function

$$P(N = 1) = 0.5, \quad P(N = 2) = 0.3, \quad P(N = 3) = 0.2.$$

The weight of a newborn animal, in lbs, has normal distribution with $\mu = 20$ and $\sigma = 3$.

Let W be the variable representing the *total weight*, at birth, of the animals being born by the animal.

   (i) Compute the m.g.f. of W.

   (ii) Use the result from Part (i) to calculate the mean and variance of the total weight for the newborns.

17. Let X be a discrete random variable with p.g.f. $G(t)$. Use the fact

$$\frac{1}{x+1} = \int_0^1 t^x \, dt$$

to show that the expectation of $1/(X + 1)$ can be found from the formula

$$E\left(\frac{1}{X+1}\right) = \int_0^1 G(t) \, dt.$$

Applications:

(i) If X has binomial distribution with parameters n and p, show that

$$E\left(\frac{1}{X+1}\right) = \frac{1-(1-p)^{n+1}}{(n+1)p}.$$

(ii) Assuming that X has a Poisson distribution with parameter $\lambda > 0$, show that

$$E\left(\frac{1}{X+1}\right) = \frac{1-e^{-\lambda}}{\lambda}.$$

18. A discrete variable X is said to follow **logarithmic distribution** with parameter p $(0 < p < 1)$ if its probability function is given by

$$f(x) = P(X = x) = \frac{1}{\ln(1/p)} \cdot \frac{q^x}{x}, \qquad x = 1, 2, \ldots,$$

where $q = 1 - p$.

(i) Verify that the probability generating function of X is

$$G(t) = E(t^X) = \frac{\ln(1-qt)}{\ln p}, \qquad |t| < 1/q$$

and thence obtain the expectation and variance of X.

(ii) Assume that $X_1, X_2, \ldots$ is a sequence of independent and identically distributed random variables following logarithmic distribution with parameter p. Consider the random sum

$$S_N = X_1 + \cdots + X_N,$$

where the variable N is independent of the X_i's and follows Poisson distribution with parameter $\lambda > 0$. Prove that the p.g.f. of S_N is given by

$$E(t^{S_N}) = \left(\frac{p}{1-qt}\right)^s,$$

where $s = -\lambda/\ln p$.

Application: The number of CDs bought by a customer entering a record store is a random variable following logarithmic distribution with parameter $p = 0.5$. Assuming that the number of persons entering the store during a day has Poisson distribution with parameter $\lambda = 60$, find the p.g.f. of the number, S, of CDs sold during a day. Calculate also the expected value and variance of S.

[Hint: Use the formula

$$-\ln(1-x) = \sum_{k=1}^{\infty} \frac{x^k}{k}, \qquad |x| < 1.]$$

19. Let Y be a nonnegative integer-valued random variable with p.g.f. $G_Y(t)$, and X be another discrete variable such that, given that $Y = y$, the conditional distribution of X is Poisson with parameter y.

(i) Show that the p.g.f. of $Z = X + Y$ is given by

$$G_Z(t) = E(t^Z) = G_Y(te^{t-1}).$$

(ii) Assuming in addition that the distribution of Y is Poisson with parameter λ, determine the p.g.f. of Z, and thence find its expectation and variance.

[Hint: Proceed as in Example 7.4.2.]

20. The m.g.f. of the discrete random pair (X, Y) is given by

$$M_{X,Y}(t, s) = E(e^{tX+sY}) = \exp\left[\lambda_1(e^t - 1) + \lambda_2(e^s - 1) + \lambda(e^{t+s} - 1)\right],$$

for $t, s \in \mathbb{R}$, where $\lambda_1, \lambda_2, \lambda > 0$ are given parameters.

(i) Show that the marginal distributions of both X and Y are Poisson.

(ii) Show that the variable $Z = X + Y$ does not follow a Poisson distribution unless $\lambda = 0$.

21. The m.g.f. of the two-dimensional discrete random variable (X, Y) is given by

$$M_{X,Y}(t, s) = E(e^{tX+sY}) = \left[\frac{1 - (p_1 + p_2)}{1 - (p_1 e^t + p_2 e^s)}\right]^r,$$

where $p_1, p_2 \in (0, 1)$ and r is a positive integer.

(i) Determine the m.g.f. of X and verify that the distribution of $X_1 = X + r$ is negative binomial.

(ii) Determine the m.g.f. of Y and verify that the distribution of $Y_1 = Y + r$ is negative binomial.

(iii) Obtain the m.g.f. of $X + Y$ and show that

$$X + Y + r$$

follows a negative binomial distribution.

(iv) Calculate the covariance between X and Y in two different ways:
(a) directly from the m.g.f. of (X, Y), and
(b) using the identity

$$Var(X + Y) = Var(X) + Var(Y) + 2Cov(X, Y)$$

and the result for the variance of a negative binomial distribution.

22. A discrete variable X defined on the set of nonnegative integers is said to follow **power series distribution** if its probability function has the form

$$f(x) = P(X = x) = \frac{\alpha_x \lambda^x}{A(\lambda)}, \quad x = 0, 1, \dots,$$

where λ is a positive constant, $\{\alpha_x: x = 0, 1, 2, \ldots\}$ is a sequence of nonnegative real numbers, and

$$A(\lambda) = \sum_{x=0}^{\infty} \alpha_x \lambda^x.$$

(i) Express the m.g.f. of X in terms of the function $A(\cdot)$.

(ii) Establish that

$$E(X) = \lambda \frac{d}{d\lambda} [\ln A(\lambda)]$$

and

$$Var(X) = \lambda \frac{d}{d\lambda} [\ln A(\lambda)] + \lambda^2 \frac{d^2}{d\lambda^2} [\ln A(\lambda)].$$

(iii) Verify that binomial, $b(n, p)$, Poisson, $\mathcal{P}(\lambda)$, and the negative binomial, $Nb(r, p)$, distributions all belong to the family of power series distributions. Hence, use the result in Part (ii) to find the mean and variance of these distributions, and ensure that they are what you expect.

23. Assume that X is a continuous variable with density function

$$f(x) = \frac{e^{-x/\lambda}}{\lambda(1 + e^{-x/\lambda})^2}, \qquad x \in \mathbb{R}$$

where $\lambda > 0$ is a real parameter (the associated distribution is known as **logistic distribution**).

(i) Show that the m.g.f. of X is

$$M(t) = \Gamma(1 - \lambda t)\Gamma(1 + \lambda t), \qquad |t| < \frac{1}{\lambda}.$$

(ii) Let X_1 and X_2 be two independent random variables, each having an exponential distribution with mean 1. Compute the m.g.f. of the random variable

$$Y = -\ln(X_2/X_1),$$

and thence identify the distribution of Y.

[Hint: For Part (i), make a change of variable so that you have an integral of the form

$$\int_0^1 t^{a-1}(1 - t)^{b-1} dt = \frac{\Gamma(a)\Gamma(b)}{\Gamma(a + b)}. \qquad]$$

24. In the collective model of risk theory (see Example 7.5.4), we assume that the total claim amount an insurance company pays to its policyholders is expressed by the random sum

$$S_N = X_1 + \cdots + X_N \quad (S = 0 \text{ if } N = 0),$$

where N takes integer values and is independent of the i.i.d. random variables $X_1, X_2, \ldots$, representing the individual claim amounts. In the case when N has probability function

$$f_N(n) = P(N = n) = q^n p, \quad n = 0, 1, 2, \ldots,$$

$0 < p < 1$ and $q = 1 - p$, we found in Example 7.5.4 that S_N has m.g.f. as

$$M_{S_N}(t) = \frac{p}{1 - qM_X(t)}.$$

(i) Show that $M_{S_N}(t)$ can be written in the form

$$M_{S_N}(t) = p + qM^*(t),$$

where

$$M^*(t) = \frac{pM_X(t)}{1 - qM_X(t)}.$$

(ii) Assuming in addition that the variables $X_i, i = 1, 2, \ldots$, are exponentially distributed with parameter $\lambda > 0$, verify that the function $M^*(t)$ above coincides with the m.g.f. of the exponential distribution with parameter $p\lambda$. Use the above results to show that S_N has a distribution of *mixed type* (see, for example, Section 6.5 in Volume I); in particular, $P(S_N = 0) = p$ (discrete part), while on the interval $(0, \infty)$, S_N has a density given by

$$f_{S_N}(x) = q(p\lambda)e^{-p\lambda x}, \qquad x > 0.$$

Thence, show that, for $x > 0$,

$$P(S_N > x) = qe^{-p\lambda x}.$$

7.12 APPLICATIONS

7.12.1 Random Walks

We have seen several instances in this chapter where generating functions play a key role in studying sums of independent random variables. There is a particular model in probability theory which concerns the sequence of partial sums of independent and identically distributed random variables; more precisely, this is called a random walk and has found numerous applications ranging from economics and finance to physics, ecology, and more recently, computer science. For example, a recent paper by Xia et al. (2020) surveys various applications of random walks in engineering and computer science.

Here we shall only sketch the simplest and prototype model of a random walk, called simple random walk and we shall illustrate the use of generating functions in studying various quantities associated with such a probability model.

To begin with, let $\{X_1, X_2, \ldots\}$ be a sequence of independent and identically distributed random variables and, for $n = 1, 2, \ldots$, define their partial sums

$$S_n = X_1 + X_2 + \cdots + X_n.$$

The sequence $\{S_n\}$ constitutes then a **random walk**; the use of this term may be justified by the fact that this sequence offers a tool to model the successive movements of a particle along a one-dimensional axis. More specifically, the variable X_i represents the movement of the particle at the ith step and, typically, X_i can take both positive and negative values, so that the particle may move at either direction.

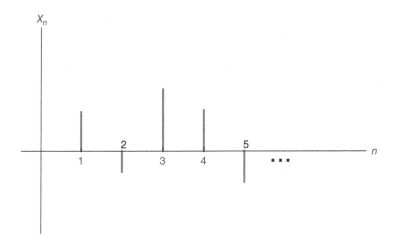

From the definition of S_n above, the variable S_n stands for the total movement of the particle (positive, zero or negative) from its starting position. It is customary to define $S_0 = 0$ so that the walk starts from the origin. With this notation, the sequence of pairs $\{(n, S_n): n = 0, 1, \ldots\}$ depicts on a Cartesian plane the way that the particle moves through time, as shown in the following graph.

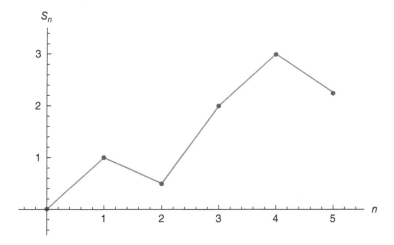

The above definition is very general; no assumptions are made about the common distribution of the variables X_i (called the **step distribution** of the random walk). For an arbitrary distribution, finding the distribution of S_n can be very hard. We shall confine ourselves to the case where each X_i assumes only two values, namely $+1$ and -1. More explicitly, suppose that for any i and for some $0 < p < 1$,

$$P(X_i = +1) = p, \quad P(X_i = -1) = q = 1 - p.$$

In such a case, the sequence $\{S_n: n = 0, 1, 2, \ldots\}$ is called a **simple random walk**. (If, in addition, $p = 1/2$, the random walk is called symmetric, however we shall not restrict ourselves to this case.)

There are various questions that arise naturally, concerning the movement of the random walk: for instance, starting from the origin, how long will it take to return there? More precisely, what is the distribution of the time until the first return? Moreover, is it always (that is, for all values of p) certain that a return will occur? Let us try to answer some of these questions with the help of generating functions.

Since the variables S_n assume discrete values, we may use p.g.f.s, however for the problems raised above we shall need to use this notion in a more general sense. More precisely, so far a p.g.f. has been associated with a discrete variable X or with a discrete probability distribution, say with probability function $p_n = P(X = n)$. Let now $\{a_n: n = 0, 1, 2, \ldots\}$ be a real sequence (i.e. not necessarily a probability function). Then we say that the function

$$A(s) = \sum_{n=0}^{\infty} a_n s^n$$

is the *generating function associated with the sequence* $\{a_n: n = 0, 1, 2, \ldots\}$.

For any nonnegative integer n, let $g_0(n)$ be the probability that the random walk will be at zero after n steps; that is, $g_0(n) = P(S_n = 0)$. For the simple random walk that we consider here, it is easy to find this probability (in the general case, however, it is not!). A first observation is that the random walk may return to the origin after n steps only if n is an even integer; that is, $P(S_{2k+1} = 0) = 0$. If $n = 2k$, then for the event $S_n = 0$ to occur, we must have exactly k steps of size $+1$ and k steps of size -1 in the first n movements. Any such realization has probability $p^k q^k$ and since there are precisely

$$\binom{n}{k} = \binom{2k}{k}$$

realizations of this form, we deduce that

$$g_0(n) = P(S_n = 0) = P(S_{2k} = 0) = \binom{2k}{k}(pq)^k, \qquad k = 1, 2, 3, \ldots . \qquad (7.15)$$

Moreover, we define for $n = 1, 2, \ldots$,

$$h_0(n) = P(S_1 \neq 0, S_2 \neq 0, \ldots, S_{n-1} \neq 0, S_n = 0)$$

so that $h_0(n)$ is the probability that the random walk returns *for the first time* to zero after n steps.

We denote the generating functions associated with these two sequences by

$$G(s) = \sum_{n=0}^{\infty} g_0(n)s^n, \qquad H(s) = \sum_{n=1}^{\infty} h_0(n)s^n.$$

There is a nice and simple formula that links these two functions so that, knowing one of them, we can immediately find the other. To see this, let A be the event that $S_n = 0$ and B_k be the event that the first return to the origin happens after k steps. Since the events B_k are pairwise disjoint, the law of the total probability yields that

$$P(A) = \sum_{k=1}^{n} P(A|B_k)P(B_k).$$

But $P(B_k) = h_0(k)$, while it is easy to see that $P(A|B_k) = g_0(n-k)$, so that we obtain the formula

$$g_0(n) = \sum_{k=1}^{n} g_0(n-k)h_0(k), \quad n = 1, 2, \dots.$$

On the right-hand side, we recognize the "convolution" between the two sequences $\{h_0(n)\}, \{g_0(n)\}$; we use the term in quotes since these two sequences are not probability functions. Now if we multiply both sides in the last equation by s_n and sum for $n = 1, 2, \dots$, on the left-hand side we get $G(s) - 1$ (since the term for $n = 0$ is missing); similarly, on the right we identify the generating function associated with the convolution between $\{h_0(n)\}$ and $\{g_0(n)\}$. We have seen that the generating function associated with a convolution is the product of the generating functions (see e.g. Proposition 7.4.1; this holds also for the more general case of generating functions that we consider here). We thus arrive at the formula

$$G(s) = 1 + G(s)H(s), \tag{7.16}$$

which gives, if we solve for $G(s)$,

$$G(s) = [1 - H(s)]^{-1}.$$

Solving (7.16) for $H(s)$, we also see that

$$H(s) = \frac{G(s) - 1}{G(s)} = 1 - [G(s)]^{-1}. \tag{7.17}$$

The last two formulas link the generating functions for the two sequences $\{h_0(n)\}, \{g_0(n)\}$. Since we have obtained the probabilities $g_0(n)$ in (7.15), we get first the generating function $G(s)$. Recalling that $g_0(n) = 0$ if n is odd, we get that

$$G(s) = \sum_{n=0}^{\infty} g_0(n)s^n = \sum_{k=0}^{\infty} g_0(2k)s^{2k}$$

$$= \sum_{k=0}^{\infty} \binom{2k}{k} (pqs^2)^k.$$

By a standard combinatorial formula, the last sum equals $(1 - 4pqs^2)^{-1/2}$, thus we have shown that

$$G(s) = (1 - 4pqs^2)^{-1/2}.$$

In view of (7.17), we now find the generating function $H(s)$ to be

$$H(s) = 1 - \sqrt{1 - 4pqs^2}. \tag{7.18}$$

The above results enable us to answer some of the questions we posed earlier. First, are we certain that a return to the origin will occur? The probability that a return *ever occurs* is

$$\sum_{n=1}^{\infty} h_0(n) = H(1)$$

and, in view of (7.18), this is

$$H(1) = 1 - \sqrt{1 - 4pq}.$$

But making use of the fact that $p + q = 1$, we get $1 - 4pq = (p + q)^2 - 4pq = (p - q)^2$, and combining the above results we see that

$$\sum_{n=1}^{\infty} h_0(n) = 1 - |p - q|.$$

Thus, a return to the origin is certain only if $p = q$, that is, $p = 1/2$. Intuitively, this can be explained by the fact that for $p > 1/2$, there is a possibility that the random walk may only assume positive values, while for $p < 1/2$, it may stay in the negative half-axis forever. A random walk is called **persistent** (or **recurrent**) if a return to the origin is certain; otherwise, it is called **transient**.

For a persistent random walk, a natural question is how long one has to wait until the first return to the origin. Since we are certain that a return will happen, we may define a random variable T representing the time until the first return. Then, $P(T = n) = h_0(n)$, so that T has p.g.f. (in the usual sense) $H(s) = 1 - \sqrt{1 - s^2}$. This gives that

$$E(T) = \sum_{n=1}^{\infty} n h_0(n) = H'(1).$$

A simple differentiation of the function $H(s)$ gives $H'(s) = s/\sqrt{1 - s^2}$, therefore $H'(1) = \infty$ which shows that even if a return to the origin is certain, one has to wait infinitely long on average for this return to occur!

Figure 7.1 shows a simulated path of a random walk with $p = 3/4$ (a) and $p = 1/2$ (b). In particular, the graphs depict the pairs (n, S_n) for $1 \le n \le 200$. We see that in the former case, the random walk which is transient, increases unboundedly while in the latter case, the persistent random walk fluctuates around zero.

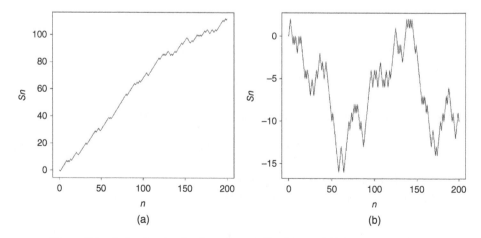

Figure 7.1   A simulated path of a random walk with $p = 3/4$ (a) and $p = 1/2$ (b).

KEY TERMS

Cauchy distribution

characteristic function

coefficient of kurtosis

coefficient of skewness

collective risk model

complex random variable

inversion formulas (for characteristic functions)

compound distribution

factorial moment generating function

Fourier transform

Laplace transform

moment generating function (m.g.f.)

probability geberating function (p.g.f.)

random walk

recursive relation

CHAPTER 8

LIMIT THEOREMS

Harald Cramér (Stockholm, Sweden 1893 – Stockholm 1985)

Harald Cramér was born on 25 September 1893, in Stockholm, Sweden, and died on 5 October 1985, at the age of 92, in Stockholm, Sweden. After completing his degree in Mathematics and Chemistry at the University of Stockholm, he finished his PhD in Mathematics in 1917 under the supervision of Professor Marcel Riesz at the University of Stockholm. Following that, he continued working at the University of Stockholm in the Department of Mathematics and produced many fine works in analytic number theory, probability, and probabilistic number theory. Then, in 1929, he was appointed in Stockholm University as a chair professor of Actuarial Mathematics and Mathematical Statistics. His main focus of work then became actuarial science and mathematical statistics, resulting in several well-known results such as Cramér-Rao bound, Cramér's decomposition theorem, and Cramér's theorem on large deviations. In 1946, he published his book *Mathematical Methods of Statistics*, which became a highly influential book. He went on to become President of Stockholm University in 1950, and then Chancellor of the University in 1958.

Introduction to Probability: Multivariate Models and Applications, Volume 2, First Edition.
N. Balakrishnan, Markos V. Koutras, and Konstadinos G. Politis.
© 2022 John Wiley & Sons, Inc. Published 2022 by John Wiley & Sons, Inc.

8.1 INTRODUCTION

In this final chapter, we discuss one of the cornerstones of probability theory, namely, the Central Limit Theorem (CLT). Before that, in Section 8.2, we state two other important limiting results, known as laws of large numbers. Apart from their theoretical importance, these results provide a justification for the common use of probability as (the limit of an appropriate) relative frequency. Specifically, during the eighteenth and nineteenth centuries, when the first attempts were made to formalize the concept of probability, many mathematicians of that time observed that, when we perform an experiment a large number of times, the relative frequency of an event approaches a certain value. A notable achievement in this respect was the *statistical definition of probability*, given by Richard von Mises (1883–1953): if an event A associated with an experiment occurs n_A times during the first n repetitions of this experiment, von Mises defined the probability of A by the formula

$$P(A) = \lim_{n \to \infty} \frac{n_A}{n}.$$

Although the results of this chapter are asymptotic in nature (limit theorems), their value in practice should not be underestimated. The limit theorems of Section 8.2, on one hand, are mathematically sound and precise statements of what we commonly understand as "the law of averages". The CLT, on the other hand, displays the great importance of normal distribution in all applications of probability theory and statistics. Specifically, this theorem states that if we consider the sum (or average) of i.i.d. random variables, its distribution approaches the normal distribution as the number of variables becomes large, regardless of the original distribution of the variables.

It is important to mention that there are versions of the results in this chapter for the case when the underlying variables are not identically distributed (or they are not independent). However, we restrict our attention to the i.i.d. case which is far easier to deal with, in addition to being sufficient for many practical applications.

8.2 LAWS OF LARGE NUMBERS

We are all familiar from calculus with the concept of convergence for a real sequence. If $\{a_n : n = 1, 2, \ldots\}$ is such a sequence, we say that a_n converges to some value a when the values of a_n become arbitrarily close to a as n becomes large. The notation we use in mathematics for this is $a_n \to a$ as $n \to \infty$, or

$$\lim_{n \to \infty} a_n = a.$$

When we consider a sequence of random variables instead, $\{X_n : n = 1, 2, \ldots\}$, it is not clear in which way the sequence may converge, since the values of the variables X_i are not known with certainty. In fact, there are several types of convergence for a sequence of random variables; in other words, there are several ways in which a random sequence may approach its limit (which may not always be the same!).

The following result is known as the weak law of large numbers (WLLN).

Proposition 8.2.1 (Weak law of large numbers)   Let $X_1, X_2, \ldots$ be a sequence of i.i.d. random variables with $E(X_i) = \mu$ and a finite variance, $Var(X_i) = \sigma^2$, for $i = 1, 2, \ldots$.

Then, for the sequence of sample averages

$$\overline{X}_n = \frac{1}{n}\sum_{i=1}^{n} X_i,$$

we have, for any positive ϵ,

$$\lim_{n\to\infty} P\left(\left|\overline{X}_n - \mu\right| > \epsilon\right) = 0.$$

Proof: Let $\epsilon > 0$ and n be an arbitrary positive integer. Then, by applying Chebyshev's inequality (see, e.g. Sections 4.6 and 6.4 in Volume I) to $\overline{X}_n$, we have

$$P\left(\left|\overline{X}_n - E\left(\overline{X}_n\right)\right| > \epsilon\right) \leq \frac{Var\left(\overline{X}_n\right)}{\epsilon^2}.$$

As

$$E\left(\overline{X}_n\right) = E\left(\frac{1}{n}\sum_{i=1}^{n} X_i\right) = \frac{1}{n}\sum_{i=1}^{n} E(X_i) = \mu$$

and (by the independence of the X_i's, see Example 5.2.4)

$$Var\left(\overline{X}_n\right) = \frac{\sigma^2}{n},$$

we obtain

$$P\left(\left|\overline{X}_n - \mu\right| > \epsilon\right) \leq \frac{\sigma^2}{n\epsilon^2}.$$

Letting $n \to \infty$, the right-hand side tends to zero, and so it must be true for the left-hand side, too. This concludes the proof of the proposition.   □

We note that Proposition 8.2.1 can be proved under the weaker assumption that $X_1, X_2, \ldots$ are pairwise uncorrelated (not necessarily independent), since that condition suffices to get $Var(\overline{X}_n) = \sigma^2/n$. Furthermore, the assumption that the variables X_i have the same distribution can also be relaxed and, provided that for $i = 1, 2, \ldots$, each variable X_i has a finite mean and variance, μ_i and σ_i^2, respectively, a more general result can be proved (see Exercise 8).

Proposition 8.2.1 asserts that the sample mean $\overline{X}_n$ "converges" to the theoretical mean; however, it is clear that the type of convergence given here is different from the usual notion in mathematical analysis. Next, we introduce a notion of convergence called *convergence in probability*.

Definition 8.2.1 Let $X_1, X_2, \ldots$ be a sequence of random variables defined on a common probability space Ω, and X be another variable, also defined on Ω. Then, we say that the sequence X_n converges in probability to X if, for any $\epsilon > 0$,

$$\lim_{n\to\infty} P\left(\left|X_n - X\right| > \epsilon\right) = 0.$$

In view of the definition above, the WLLN can now be restated as follows:

Let $X_1, X_2, \ldots$ be a sequence of i.i.d. random variables with mean μ and finite variance. Then, as $n \to \infty$, the sequence of sample means $\overline{X}_n$ converges in probability to μ.

Notice that, in the WLLN, the sequence $\overline{X}_n$ converges to a constant, although in Definition 8.2.1 the limit can, at least in principle, be any (continuous or discrete) random variable.

Next, we discuss how the WLLN may be used to justify theoretically the statistical definition of probability, as mentioned in Section 8.1.

Let A be an event associated with a random experiment, and let n_A denote the number of times A occurs in n repetitions of this experiment. Define now the variables

$$X_i = \begin{cases} 1, & \text{if } A \text{ occurs in the } i\text{th repetition of the experiment,} \\ 0, & \text{otherwise.} \end{cases}$$

It is then clear that

$$n_A = \sum_{i=1}^{n} X_i, \quad \frac{n_A}{n} = \frac{1}{n} \sum_{i=1}^{n} X_i = \overline{X}_n,$$

and also that

$$\mu = E(X_i) = 1 \cdot P(X_i = 1) + 0 \cdot P(X_i = 0) = P(X_i = 1) = P(A), \quad i = 1, 2, \ldots.$$

Consequently, according to the WLLN, we have

$$\lim_{n \to \infty} P\left(\left| \frac{n_A}{n} - P(A) \right| > \epsilon \right) = 0$$

or, equivalently,

$$\lim_{n \to \infty} P\left(\left| \frac{n_A}{n} - P(A) \right| \leq \epsilon \right) = 1.$$

This equation shows that, for a large number of repetitions ($n \to \infty$), the ratio n_A/n is expected to be arbitrarily close to the quantity $P(A)$, with high probability.

Another immediate application of WLLN is given in the following corollary.

Corollary 8.2.1 (Bernoulli's law of large numbers)   *Consider an infinite sequence of (independent) Bernoulli trials with identical success probability p, and let R_n be the proportion of successes in the first n trials. Then, the sequence R_n converges in probability to p.*

Proof: For $i = 1, 2, \ldots$, define the sequence of variables

$$X_i = \begin{cases} 1, & \text{if the } i\text{th trial results in success,} \\ 0, & \text{otherwise.} \end{cases}$$

Then, $R_n = \overline{X}_n$ and the result follows from WLLN applied to the sequence $\{X_i : i = 1, 2, \ldots\}$ and observing that

$$E(X_i) = P(X_i = 1) = p, \qquad Var(X_i) = p(1 - p) < \infty, \qquad i = 1, 2, \ldots. \qquad \square$$

The corollary is named after Jacob Bernoulli (1654–1705), who was the first to prove it. The proof that Bernoulli gave was obviously different than the one given here, since WLLN was not known in his day; the proof appeared in Part IV of Bernoulli's book *Ars Conjectandi*, published in 1713, eight years after his death. Siméon Poisson was the first to use the term "Law of Large Numbers" (*la Loi des Grandes Nombres*) for Bernoulli's result. The WLLN in its general form in Proposition 8.2.1 appeared at the end of nineteenth century, following the development of Chebyshev's inequality. An even more general result, which does not assume a finite variance for the variables $X_1, X_2, \ldots$ (and so cannot be proved using Chebyshev's inequality) was given by another Russian mathematician, A. Khintchine, in 1929.

Example 8.2.1 Let $X_1, X_2, \ldots$ be a sequence of i.i.d. random variables with density function

$$f(x) = \begin{cases} 12x^2(1 - x), & \text{if } 0 \le x \le 1, \\ 0, & \text{otherwise.} \end{cases}$$

Show that the average, $\overline{X}_n$, of the X_i's satisfies

$$\lim_{n \to \infty} P\left(\frac{2}{5} \le \overline{X}_n \le \frac{4}{5}\right) = \lim_{n \to \infty} P\left(\frac{1}{2} \le \overline{X}_n \le \frac{7}{10}\right) = 1.$$

SOLUTION The common mean of the variables X_i is

$$E(X_i) = \int_{-\infty}^{\infty} xf(x)\,dx = \int_0^1 12x^3(1 - x)\,dx = 12\left[\frac{x^4}{4}\right]_0^1 - 12\left[\frac{x^5}{5}\right]_0^1 = \frac{3}{5}.$$

Moreover, the variance of X_i's is finite (in fact, one can see easily that $Var(X_i) = 1/25$). Therefore, for any $\epsilon > 0$, we have from WLLN that

$$\lim_{n \to \infty} P\left(\left|\overline{X}_n - \frac{3}{5}\right| > \epsilon\right) = 0.$$

This yields immediately that

$$\lim_{n \to \infty} P\left(\frac{3}{5} - \epsilon \le \overline{X}_n \le \frac{3}{5} + \epsilon\right) = \lim_{n \to \infty} P\left(\left|\overline{X}_n - \frac{3}{5}\right| \le \epsilon\right) = 1.$$

The required results now follow upon taking $\epsilon = 1/5$ and $\epsilon = 1/10$, respectively.

Since the law of large numbers given in Proposition 8.2.1 is called "weak", there must be a strong one! In order to state this "strong law", we need a stronger form of convergence than the one considered in Definition 8.2.1.

Let $X_1, X_2, \ldots$ be a sequence of random variables defined on the same probability space Ω and assume that X is another random variable defined on Ω. Then, the notation

$$P\left(\lim_{n\to\infty} X_n = X\right)$$

denotes the probability of the event

$$A = \left\{\omega \in \Omega \; : \; \lim_{n\to\infty} X_n(\omega) = X(\omega)\right\}.$$

Definition 8.2.2 A sequence $X_1, X_2, \ldots$ of random variables is said to converge to the random variable X **almost surely** or **with probability 1** if

$$P\left(\lim_{n\to\infty} X_n = X\right) = 1.$$

This is the same as the requirement that, for any $\epsilon > 0$,

$$\lim_{n\to\infty} P\left(|X_r - X| > \epsilon \text{ for at least one } r \text{ such that } r \geq n\right) = 0.$$

From the above definition, we see that almost sure convergence means that, for a set A with $P(A) = 1$, the difference $X_n(\omega) - X(\omega)$ can be made arbitrarily small for every ω in A as n goes to infinity. Recall that convergence in probability implies that the probability of the event $\{X_n(\omega) - X(\omega) > \epsilon\}$ can be made sufficiently close to zero by letting $n \to \infty$.

Further, the obvious inequality

$$P\left(|X_n - X| > \epsilon\right) \leq P\left(|X_r - X| > \epsilon \text{ for at least one } r \text{ such that } r \geq n\right)$$

shows that almost sure convergence implies convergence in probability. The converse is not true in general.

In order to distinguish between the two types of convergence considered so far, we use the following notation. If a sequence $X_1, X_2, \ldots$ of random variables converges in probability to a variable X, we write $X_n \overset{P}{\longrightarrow} X$, while if $X_1, X_2, \ldots$ converge to X almost surely, we write $X_n \overset{a.s.}{\longrightarrow} X$. In view of this, the argument above shows that

$$X_n \overset{a.s.}{\longrightarrow} X \Rightarrow X_n \overset{P}{\longrightarrow} X. \tag{8.1}$$

The following proposition, known as the *strong law of large numbers* (SLLN) shows that, under the same requirements as those of WLLN, the convergence of the sample mean to the theoretical mean μ is almost sure (with probability 1). The proof of this result is not presented here, however. It can be found in more advanced textbooks, such as Feller (1971) or Grimmett and Stirzaker (2001a).

Proposition 8.2.2 (Strong law of large numbers, SLLN) *Suppose $X_1, X_2, \ldots$ is a sequence of independent and identically distributed random variables with $E(X_i) = \mu$*

and Var$(X_i) = \sigma^2 < \infty$, for $i = 1, 2, \dots$. Then, the sequence of sample means

$$\overline{X}_n = \frac{1}{n}\sum_{i=1}^{n} X_i$$

converges to μ with probability one, that is,

$$P\left(\lim_{n\to\infty} \overline{X}_n = \mu\right) = 1.$$

If we repeat the arguments following Definition 8.2.1, and denote again by n_A the number of appearances of event A in n repetitions of the experiment, we obtain

$$P\left(\lim_{n\to\infty} \frac{n_A}{n} = P(A)\right) = 1.$$

This reaffirms the reasoning underlying the statistical definition of probability, now expressed in terms of "almost sure convergence".

To clarify the distinction between weak and strong laws of large numbers, we note the following. As n becomes large, WLLN asserts that $\overline{X}_n$ gets close to the mean μ. However, this is not guaranteed for the differences $|\overline{X}_r - \mu|$ for *all* values $r \geq n$. It does not, therefore, preclude the possibility that large values of the differences $|\overline{X}_r - \mu|$ occur (rarely but) infinitely often. This possibility, however, is ruled out by SLLN.

Further, it is worth noting that there are examples of sequences of random variables obeying WLLN, but not SLLN. Cases where this may happen are quite rare and technical in nature; they are therefore not discussed here.

Example 8.2.2 *(The empirical distribution function)*

Suppose we measure a certain characteristic (*a random variable*) X, and we are interested in estimating the cumulative distribution function, $F(x)$, of X at a *specific* point $x \in \mathbb{R}$. For this purpose, we take a random sample $X_1, \dots, X_n$ (so that these are n independent variables, each having distribution function F) and, subsequently, for $i = 1, \dots, n$, we define new variables

$$Y_i = \begin{cases} 1, & \text{if } X_i \leq x, \\ 0, & \text{otherwise.} \end{cases}$$

Viewed as a function of x, a quantity widely used in statistics is the so-called empirical distribution function, defined by

$$\hat{F}_n(x) = \frac{1}{n}\sum_{i=1}^{n} Y_i = \overline{Y}_n.$$

We then see that, for each real x, $\hat{F}_n(x)$ takes values in the interval $[0, 1]$ and represents the proportion of values in the sample that are less than or equal to x.

Next, since $X_1, \dots, X_n$ are independent variables, the same must be true for $Y_1, \dots, Y_n$. Moreover, we have

$$\mu = E(Y_i) = 1 \cdot P(Y_i = 1) + 0 \cdot P(Y_i = 0) = P(Y_i = 1) = P(X_i \leq x) = F(x)$$

and

$$E\left(Y_i^2\right) = 1^2 \cdot P(Y_i = 1) + 0^2 \cdot P(Y_i = 0) = P(Y_i = 1) = F(x).$$

Therefore, the variance of Y_i, for $i = 1, \dots, n$, is finite; in particular,

$$\sigma^2 = Var(Y_i) = E\left(Y_i^2\right) - \left[E(Y_i)\right]^2 = F(x) - [F(x)]^2 = F(x)(1 - F(x)) < \infty.$$

We thus see that the SLLN is in force and yields

$$\lim_{n\to\infty} \hat{F}_n(x) = \lim_{n\to\infty} \overline{Y}_n = F(x)$$

with probability 1. This result implies that the empirical distribution function is a suitable estimator of the distribution function $F(x)$ (*at least for large samples*), in cases when F is not known, but a sample of values from it is available.

Exercises
Group A

1. Consider the following three density functions:

   (i) $f(x) = \begin{cases} 6x(1 - x), & \text{if } 0 \le x \le 1, \\ 0, & \text{otherwise;} \end{cases}$

   (ii) $f(x) = \begin{cases} 16xe^{-4x}, & x > 0, \\ 0, & \text{otherwise;} \end{cases}$

   (iii) $f(x) = \begin{cases} 1, & 0 \le x \le 1, \\ 0, & \text{otherwise.} \end{cases}$

   Show that, in each of the three cases where $X_1, X_2, \dots$ is a sequence of i.i.d. random variables with density function f, the sequence of sample means converges, with probability 1, to the same constant.

2. Let $X_1, X_2, \dots$ be a sequence of independent continuous random variables with common density function

   $$f(x) = cx^3(2 - x), \qquad 0 \le x \le 2,$$

   for a suitable real constant c.

   (i) Identify the limit of the sequence

   $$\overline{X}_n = \frac{X_1 + \cdots + X_n}{n}$$

   in terms of almost sure convergence.

   (ii) Show that

   $$\lim_{n\to\infty} P\left(1 \le \overline{X}_n \le \frac{5}{3}\right) = \lim_{n\to\infty} P\left(\frac{7}{6} \le \overline{X}_n \le \frac{3}{2}\right) = 1.$$

(iii) Verify that the following holds:

$$\lim_{n\to\infty} P\left(\overline{X}_n < 1\right) = \lim_{n\to\infty} P\left(\overline{X}_n > \frac{5}{3}\right) = 0.$$

3. Let X be a nonnegative random variable, and $Y_1, Y_2, \ldots$ be a sequence of variables defined as

$$Y_n = \begin{cases} 1, & \text{if } X > n, \\ 0, & \text{otherwise.} \end{cases}$$

Prove that the sequence $Y_1, Y_2, \ldots$ converges in probability to 0. Does it converge with probability 1?

4. Let $\{X_n : n = 1, 2, \ldots\}$, and $\{Y_n : n = 1, 2, \ldots\}$ be two independent sequences of random variables such that

$$X_n \xrightarrow{a.s.} X, \quad Y_n \xrightarrow{a.s.} Y,$$

for two variables X and Y. Establish that $X_n + Y_n \xrightarrow{a.s.} X + Y$.

5. Assume that $X_n \xrightarrow{P} X$ for a variable X such that $P(X \neq 0) = 1$. Then, show that $X_n^{-1} \xrightarrow{P} X^{-1}$.

Group B

6. Let $X_1, X_2, \ldots$ be a sequence of i.i.d. random variables with a positive mean and a finite variance. Show that, for any $a > 0$, we have

$$\lim_{n\to\infty} P(X_1 + \cdots + X_n > a) = 1.$$

7. Let $X_1, X_2, \ldots$ be independent variables with a common density function

$$f(x) = 2x, \qquad 0 < x < 1.$$

(i) Verify that the distribution function of the random variable

$$Y_n = \sqrt{n} \min_{1 \leq i \leq n} X_i, \qquad n = 1, 2, \ldots$$

is given by

$$F_{Y_n}(y) = 1 - \left(1 - \frac{y^2}{n}\right)^n, \qquad 0 < y < \sqrt{n}.$$

(ii) Show that the sequence of distribution functions F_{Y_n} converges to a continuous distribution with density function

$$f_Y(y) = 2y \exp(-y^2), \qquad y > 0.$$

8. **(Chebyshev's law of large numbers)** Let $X_1, X_2, \ldots$ be independent random variables with

$$E(X_i) = \mu_i, \qquad Var(X_i) = \sigma_i^2, \qquad i = 1, 2, \ldots.$$

Set

$$\overline{X}_n = \frac{1}{n}\sum_{i=1}^{n} X_i, \quad \overline{\mu}_n = \frac{1}{n}\sum_{i=1}^{n} \mu_i, \qquad n = 1, 2, \dots .$$

Then, provided

$$\lim_{n\to\infty} \frac{1}{n^2}\sum_{i=1}^{n} \sigma_i^2 = 0,$$

demonstrate that the sequence $\overline{X}_n - \overline{\mu}_n$ converges to zero in probability.

Examine whether the result remains true under the weaker assumption that the variables $X_1, X_2, \dots$ are pairwise uncorrelated (rather than independent).

[Hint: Work as in the proof of Proposition 8.2.1.]

9. (**Poisson's law of large numbers**) Consider an infinite sequence of Bernoulli trials, such that the probability of a "success" outcome at the ith trial is p_i, and define a sequence of random variables $X_1, X_2, \dots$ as $X_i = 1$ if the outcome in the ith trial is a success and $X_i = 0$ otherwise. Let

$$\overline{X}_n = \frac{1}{n}\sum_{i=1}^{n} X_i$$

be the proportion of successes during the first n trials.

(i) Prove that

$$\overline{p}_n = E\left(\overline{X}_n\right) = \frac{1}{n}\sum_{i=1}^{n} p_i, \quad Var\left(\overline{X}_n\right) \le \frac{1}{4n}.$$

(ii) Use the result in Part (i) and Chebyshev's inequality to show that the sequence $\overline{X}_n - \overline{p}_n$ converges to zero in probability.

8.3 CENTRAL LIMIT THEOREM

This section is devoted to one of the most renowned and profound results in probability theory. This result, the *CLT*, is also of major importance in statistics, and plays a vital role in many applications.

In Volume I, we discussed normal approximation to binomial distribution (De Moivre–Laplace theorem). Suppose X has binomial distribution with parameters n, p. Then, for any $a, b \in \mathbb{R}$ with $a < b$, we have

$$\lim_{n\to\infty} P\left(a < \frac{X - np}{\sqrt{np(1-p)}} < b\right) = \frac{1}{\sqrt{2\pi}}\int_a^b e^{-t^2/2}\, dt.$$

Abraham De Moivre was the first to establish this for $p = 1/2$, when he was searching for satisfactory approximations to the cumulative distribution of binomial distribution; in 1812, Pierre Simon Laplace showed that the result remains true for any $p \in (0, 1)$.

The De Moivre–Laplace theorem shows that the distribution function of $b(n, p)$ distribution can be approximated by that of normal distribution, when n is large. It is not easy to imagine how the result can be generalized to offer approximations for distributions other than binomial.

However, a key observation is made by noting that, if $X \sim b(n, p)$, then X can be written in the form

$$X = X_1 + \cdots + X_n = S_n,$$

where $X_i, i = 1, \ldots, n$, are independent, binary (or Bernoulli) random variables with $P(X_i = 1) = p$, $P(X_i = 0) = 1 - p$. For such an X_i, we have

$$\mu = E(X_i) = p, \quad \sigma^2 = Var(X_i) = p(1 - p).$$

We thus see that De Moivre–Laplace theorem yields the following approximation for the cumulative distribution function associated with a normalized version of a binomial random variable. Specifically, for $b = x$ and letting $a \to -\infty$, we obtain

$$\lim_{n \to \infty} P\left(\frac{S_n - n\mu}{\sigma\sqrt{n}} \le x\right) = \frac{1}{\sqrt{2\pi}} \int_{-\infty}^{x} e^{-t^2/2} \, dt = \Phi(x),$$

where $\Phi(\cdot)$ is the c.d.f. of the standard normal distribution. It is this last expression that is amenable to generalizations. The first observation in this regard was made by P. Chebyshev in 1887; then A. Markov, one of his students, gave a more rigorous (but also simpler) proof to Chebyshev's result. A few years later, in 1901, A. Lyapounov, another Russian mathematician and also a student of Chebyshev, presented the following result, which is known as the **Central Limit Theorem**, and it is one of the cornerstones of probability theory.

Proposition 8.3.1 (Central Limit Theorem, CLT) *Let $X_1, X_2, \ldots$ be a sequence of independent and identically distributed random variables with mean μ and variance σ^2. Then, the distribution function of*

$$Z_n = \frac{S_n - n\mu}{\sigma\sqrt{n}} = \frac{X_1 + \cdots + X_n - n\mu}{\sigma\sqrt{n}}$$

converges, as $n \to \infty$, to the distribution function of the standard normal distribution, $N(0, 1)$; that is,

$$\lim_{n \to \infty} P(Z_n \le x) = \frac{1}{\sqrt{2\pi}} \int_{-\infty}^{x} e^{-t^2/2} \, dt = \Phi(x).$$

Proof: By employing a result known as the *Lévy–Cramer continuity theorem*, in order to show that a sequence of distribution functions converges to another distribution function, it is sufficient to establish convergence of the corresponding moment generating functions. (The Lévy–Cramer continuity theorem is actually stronger, as it links convergence of distributions with pointwise convergence of characteristic functions – which always exist, while a moment generating function may not exist.) Among the numerous proofs that can be found in the literature, we present here a simple proof that does not use characteristic functions, but simply requires that, in the Lévy–Cramer theorem, the involved moment generating functions are twice differentiable.

We first express the variable Z_n as

$$Z_n = \frac{(X_1 - \mu) + \cdots + (X_n - \mu)}{\sigma\sqrt{n}} = \frac{Y_1 + \cdots + Y_n}{\sigma\sqrt{n}},$$

where the variables $Y_i = X_i - \mu$ are independent and identically distributed with

$$E(Y_i) = 0, \qquad Var(Y_i) = Var(X_i) = \sigma^2, \quad i = 1, 2, \dots.$$

Then, the moment generating function of Z_n is given by

$$M_{Z_n}(t) = \prod_{i=1}^{n} M_{Y_i/(\sigma\sqrt{n})}(t) = \prod_{i=1}^{n} M_{Y_i}\left(\frac{t}{\sigma\sqrt{n}}\right) = \left[M\left(\frac{t}{\sigma\sqrt{n}}\right)\right]^n,$$

where $M(t)$ denotes the moment generating function of Y_i, for $i = 1, 2, \dots$. According to the continuity theorem mentioned earlier, it is now sufficient to show that

$$\lim_{n\to\infty} M_{Z_n}(t) = \exp(t^2/2),$$

which is the same as

$$\lim_{n\to\infty} \left(\ln M_{Z_n}(t)\right) = \frac{t^2}{2}. \tag{8.2}$$

But,

$$\ln M_{Z_n}(t) = \ln \left[M\left(\frac{t}{\sigma\sqrt{n}}\right)\right]^n = n \ln M\left(\frac{t}{\sigma\sqrt{n}}\right).$$

If we now set, for convenience, $h = t/\left(\sigma n^{1/2}\right)$ (so that $n = t^2/\left(\sigma^2 h^2\right)$), we see that

$$\ln M_{Z_n}(t) = \frac{t^2}{\sigma^2 h^2} \ln M(h) = \frac{t^2}{\sigma^2} \cdot \frac{\ln M(h)}{h^2}, \qquad h \neq 0,$$

which, in turn, yields

$$\lim_{n\to\infty} (\ln M_{Z_n}(t)) = \frac{t^2}{\sigma^2} \lim_{h\to 0} \frac{\ln M(h)}{h^2}.$$

As $M(0) = 1$, in order to obtain the limit on the right-hand side, we employ l'Hôpital's rule. Upon applying this rule twice (since $M'(0) = E(Y_i) = 0$), we get

$$\lim_{h\to 0} \frac{\ln M(h)}{h^2} = \lim_{h\to 0} \frac{[\ln M(h)]'}{(h^2)'} = \frac{1}{2}\lim_{h\to 0} \frac{M'(h)}{hM(h)}$$

$$= \frac{1}{2}\lim_{h\to 0} \frac{M''(h)}{M(h) + hM'(h)} = \frac{1}{2} \cdot \frac{M''(0)}{M(0) + 0 \cdot M'(0)} = \frac{1}{2}M''(0).$$

Further, we note that

$$M''(0) = E\left(Y_i^2\right) = E\left[(X_i - \mu)^2\right] = Var(X_i) = \sigma^2,$$

so that we obtain

$$\lim_{h \to 0} \frac{\ln M(h)}{h^2} = \frac{\sigma^2}{2}.$$

This, in turn, implies that

$$\lim_{n \to \infty} \left(\ln M_{Z_n}(t) \right) = \frac{t^2}{\sigma^2} \cdot \frac{\sigma^2}{2} = \frac{t^2}{2},$$

as we set out to prove. This completes the proof of the proposition. □

We observe that since $E(S_n) = n\mu$ and $Var(S_n) = n\sigma^2$, the random variables Z_n in the CLT form a standardized sequence of partial sums of the sequence $\{X_n : n = 1, 2, \ldots\}$. Moreover, writing Z_n in the form

$$Z_n = \frac{\dfrac{S_n - n\mu}{n}}{\dfrac{\sigma\sqrt{n}}{n}} = \frac{\dfrac{S_n}{n} - \mu}{\sqrt{\dfrac{\sigma^2}{n}}} = \frac{\overline{X}_n - E\left(\overline{X}_n\right)}{\sqrt{Var\left(\overline{X}_n\right)}},$$

we deduce that $\{Z_n : n = 1, 2, \ldots\}$ is also a sequence of standardized *sample means* of $X_1, X_2, \ldots$.

The result of the CLT is often stated in the following way that is easier to remember:

The sample mean $\overline{X}_n$ follows asymptotically (i.e. for large values of n) normal distribution, $N(\mu, \sigma^2/n)$. Equivalently, the asymptotic distribution of the sum S_n is $N(n\mu, n\sigma^2)$.

We note that the type of convergence we have seen in the CLT differs from the other two types seen in Section 8.2 (viz convergence in probability and almost sure convergence). The convergence seen in CLT is known as *convergence in distribution*. The formal definition of this concept is given next.

Definition 8.3.1 Let $X_1, \ldots, X_n, \ldots$, be a sequence of random variables defined on the same probability space Ω, and X be another variable defined on Ω. Then, we say that the sequence $\{X_n : n = 1, 2, \ldots\}$ converges in distribution to X if the sequence of distribution functions, $F_{X_n}(x)$, of X_n converges pointwise to the distribution function $F(x)$ of X, that is,

$$\lim_{n \to \infty} F_{X_n}(x) = F(x)$$

for all points $x \in \mathbb{R}$ at which F is continuous.

As with convergence in probability and almost sure convergence, there is a common notation for convergence in distribution. Specifically, if a sequence $X_1, X_2, \ldots$ of random variables converges in distribution to X, this is denoted by $X_n \overset{d}{\longrightarrow} X$.

We have seen in Section 8.2 that almost sure convergence implies convergence in probability, and so it is natural to wonder how convergence in distribution is related to the other two types of convergence. To this end, we state the following result without a proof.

> **Proposition 8.3.2** *Let $X_1, X_2, \ldots$ be a sequence of variables that converges in probability to variable X. Then, $\{X_n : n = 1, 2, \ldots\}$ converges also in distribution to X.*
> *When X is constant, i.e. $P(X = c) = 1$ for some real c, the converse is also true.*

We thus see that convergence in distribution is the weakest among the three types of convergence we have discussed so far. In fact, it is easy to construct sequences that converge in distribution, but fail to converge in terms of the other two forms. Grimmett and Stirzaker (2001a) have presented Example 8.3.1 for illustrating this aspect.

> ***Example 8.3.1*** Let X be a Bernoulli random variable with $P(X = 1) = P(X = 0) = 1/2$, and define a sequence $\{X_n : n = 1, 2, \ldots\}$ in which $X_n = X$ for all n, so that each X_n is identical to X. It is then immediate that $X_n \xrightarrow{d} X$. Now, let $Y = 1 - X$ so that Y has the same distribution as X, and so $X_n \xrightarrow{d} Y$ also. However, we have $|X_n - Y| = 1$ for all n, and it is therefore clear that X_n cannot converge to Y in any other form of convergence.

Combining Proposition 8.3.2 with (8.1), we have the following relation between the three types of convergence:

$$X_n \xrightarrow{a.s.} X \Rightarrow X_n \xrightarrow{P} X \Rightarrow X_n \xrightarrow{d} X.$$

Now, in view of Definition 8.3.1 and the above notation, we may restate the result of CLT as follows:

> For a sequence $X_1, X_2, \ldots$ of i.i.d. random variables with mean μ and finite variance σ^2, the sequence of standardized sample means
> $$Z_n = \frac{\overline{X}_n - \mu}{\sigma/\sqrt{n}} = \frac{\sqrt{n}\left(\overline{X}_n - \mu\right)}{\sigma}$$
> converges in distribution to a standard normal random variable. In short, $Z_n \xrightarrow{d} Z$, where $Z \sim N(0, 1)$.

The CLT is one of the primary reasons for the normal distribution to be widely used in many areas. In physical and social sciences, for example, in order to describe a particular characteristic of interest, we use a random variable that can often be expressed as a sum of "smaller" random effects. Although each of these small effects may in itself be negligible, adding a large number of these produces a quantity which is no longer negligible; and, what is more, according to CLT, this quantity follows (approximately) normal distribution. Examples of the use of normal distribution in this context include (among many others) a person's weight or IQ score, monthly or yearly salaries, the position and the velocity of a gas molecule, while also measurement errors in physical experiments are invariably modeled by normal distribution.

Since CLT is an asymptotic result, in order to use it as an approximation in practice, one has to ensure that a "large" number of variables are summed or "averaged". One may wonder what is actually meant by "large" here; it obviously depends on the level of accuracy we seek in the approximation, but $n \geq 30$ is the rule of thumb that is typically used and, in most cases, the resulting approximation works sufficiently well.

Example 8.3.2 Suppose the measurement error of an instrument follows uniform distribution over the interval $(-0.05, 0.05)$. Using CLT, find an approximation for the probability that

(i) the cumulative error (i.e. *the sum*) of 100 independent measurements has a magnitude less than 0.2;

(ii) the sum of the squares of 100 independent measurements

   (a) does not exceed 1/12;

   (b) lies between 0.1 and 0.15.

SOLUTION   Let us denote by X_i the error in the ith measurement, for $i = 1, \dots, 100$. Then, it is clear that, for any i, we have

$$\mu = E(X_i) = \frac{0.05 + (-0.05)}{2} = 0, \quad Var(X_i) = \frac{[0.05 - (-0.05)]^2}{12} = \frac{1}{1200}.$$

We now use CLT to calculate probabilities related to the cumulative error, represented by

$$S_{100} = X_1 + \cdots + X_{100}.$$

Specifically, the CLT yields the approximate distribution of S_{100} to be $N(100\mu, 100\sigma^2)$ and, in view of the above, it is $N(0, 1/12)$.

(i) The required probability is

$$P\left(|S_{100}| < 0.2\right) = P\left(-0.2 < S_{100} < 0.2\right)$$

$$= P\left(\frac{-0.2 - 0}{\sqrt{1/12}} < \frac{S_{100} - 0}{\sqrt{1/12}} < \frac{0.2 - 0}{\sqrt{1/12}}\right)$$

$$\cong \Phi(0.69) - \Phi(-0.69) = 2\Phi(0.69) - 1 = 0.51.$$

(ii) For the sum

$$S = X_1^2 + \cdots + X_{100}^2,$$

we can also use a normal approximation, since we have added a large number ($n = 100$) of i.i.d. random variables. In this case, we have

$$E\left(X_i^2\right) = Var(X_i) + [E(X_i)]^2 = \frac{1}{1200} + 0 = \frac{1}{1200},$$

while the associated variance is

$$Var\left(X_i^2\right) = E\left(X_i^4\right) - [E\left(X_i^2\right)]^2,$$

with

$$E\left(X_i^4\right) = \int_{-\infty}^{\infty} x^4 f(x)\, dx = \frac{1}{0.1}\int_{-0.05}^{0.05} x^4\, dx = \frac{1}{0.1}\left[\frac{x^5}{5}\right]_{-0.05}^{0.05} = \frac{1}{8\cdot 10^5}.$$

Thus, we have

$$Var\left(X_i^2\right) = \frac{1}{8\cdot 10^5} - \left(\frac{1}{1200}\right)^2 = \frac{1}{18\cdot 10^5}.$$

Now, due to CLT,

$$Z_{100} = \frac{S - \dfrac{100}{1200}}{\sqrt{\dfrac{100}{18\cdot 10^5}}} = 60\sqrt{5}\left(S - \frac{1}{12}\right)$$

follows (approximately) standard normal distribution.

(a) Using the above results we find that

$$P\left(S \le \frac{1}{12}\right) = P\left[60\sqrt{5}\left(S - \frac{1}{12}\right) \le 60\sqrt{5}\left(\frac{1}{12} - \frac{1}{12}\right)\right] = P(Z_{100} \le 0)$$
$$\cong \Phi(0) = 0.5.$$

(b) The required probability is

$$P(0.10 < S < 0.15) = P\left(\frac{1}{10} < S < \frac{3}{20}\right)$$
$$= P\left[60\sqrt{5}\left(\frac{1}{10} - \frac{1}{12}\right) < 60\sqrt{5}\left(S - \frac{1}{12}\right)\right.$$
$$\left. < 60\sqrt{5}\left(\frac{3}{20} - \frac{1}{12}\right)\right]$$
$$= P\left(\sqrt{5} < Z_{100} < 4\sqrt{5}\right) \cong \Phi\left(4\sqrt{5}\right) - \Phi\left(\sqrt{5}\right)$$
$$= 1 - \Phi(2.236) = 0.013.$$

Following the discussion on the three types of convergence for a sequence of random variables seen thus far, a natural question that arises is: if a sequence $\{X_n : n = 1, 2, \ldots\}$ converges to X in one form or another, does it follow that the moments of X_n converge to those of X? For example, if $X_n \xrightarrow{P} X$ (in which case we also have $X_n \xrightarrow{d} X$), does $E(X_n)$ converges to $E(X)$ as $n \to \infty$? The answer is negative in general (unless extra conditions are imposed), as shown in Example 8.3.3.

Example 8.3.3 Consider a sequence of variables $\{X_n : n = 1, 2, \ldots\}$, defined for $n = 1, 2, \ldots$ as follows

$$P(X_n = 0) = 1 - \frac{1}{n}, \quad P(X_n = n) = \frac{1}{n}.$$

Thus, each X_n assumes only two values, 0 and n, and the probabilities $P(X_n = 0)$ converge to one as $n \to \infty$. This shows that $X_n \xrightarrow{P} 0$, i.e. the sequence converges in probability to a random variable X which is 0 with probability 1. This yields, in particular, that the distribution of X_n converges to a distribution degenerate at zero. However, for $n = 1, 2, \ldots$, we have

$$E(X_n) = n \cdot \frac{1}{n} = 1,$$

so that the limit of the real sequence $E(X_n)$ is 1 as $n \to \infty$. Moreover, it is easy to verify that $\lim_{n \to \infty} Var(X_n) = \infty$.

In view of Example 8.3.3 and the discussion preceding it, we note that there is yet another type of convergence that guarantees convergence of the respective means.

Definition 8.3.2 Let $r \geq 1$. Then a sequence $X_1, \ldots, X_n, \ldots$ of random variables is said to converge in r-th mean to a random variable X if

$$\lim_{n \to \infty} E\left(|X_n - X|^r\right) = 0. \tag{8.3}$$

In this case, we denote it by $X_n \xrightarrow{r\text{-mean}} X$.

Although we shall not consider this type of convergence in great detail, we present a few remarks and results below. The interested reader may refer to Grimmett and Stirzaker (2001a, Chapter 7) for an elaborate discussion on this concept.

First, note in the above definition that r need not be an integer, but just a positive number not less than 1. Of course, integer values of r are generally easier to deal with and are usually sufficient for applications. In particular, the cases $r = 1$ and $r = 2$ are frequently encountered. Specifically, if (8.3) holds when $r = 1$, we say that X_n *converges in mean* to X, while if (8.3) holds for $r = 2$, we say that X_n *converges in mean square* (or **in quadratic mean**) to X.

Proposition 8.3.3 *Let $r \geq 1$ and assume that a sequence $X_1, X_2, \ldots$ of random variables converges in r-th mean to a variable X. Then, the following hold:*

(i) *the r-th absolute moment of X_n converges to the respective moment of X, that is,*

$$\lim_{n \to \infty} E\left|X_n^r\right| = E|X^r|;$$

(ii) *the sequence* $\{X_n : n = 1, 2, ...\}$ *converges in s-th mean to X for any s such that* $1 \leq s < r;$

(iii) *the sequence* $\{X_n : n = 1, 2, ...\}$ *converges to X in probability (and so also in distribution).*

Part (i) is easy to prove and it is in fact one of the exercises at the end of this section. The proof of Part (ii) is harder and is therefore omitted. Finally, for the statement in Part (iii), it follows from Part (ii) that if $X_n \xrightarrow{r\text{-mean}} X$ for some $r \geq 1$, the sequence converges *in* mean (i.e. with $r = 1$) to X. The result in (iii) is then obvious from Markov's inequality, which yields

$$P\left(|X_n - X| \geq \epsilon\right) \leq \frac{E|X_n - X|}{\epsilon} \quad \text{for all } \epsilon > 0$$

(simply take the limit as $n \to \infty$ on both sides).

Finally, we note that convergence in r-th mean neither implies nor is implied by almost sure convergence; this means that we can find sequences that converge in r-th mean, but not almost surely, and vice versa.

The following diagram depicts the relationship between the four types of convergence that we have discussed so far:

$$X_n \xrightarrow{a.s.} X$$
$$X_n \xrightarrow{p} X \implies X_n \xrightarrow{d} X$$
$$X_n \xrightarrow{r\text{-mean}} X$$

Example 8.3.4 (Sample size needed for estimating the mean)
The number of "likes" that Sophia receives on her Facebook account each day is a sequence of independent random variables with the same distribution. The expected number of likes per day, denoted by μ, is not known, but suppose the variance of this number is known to be $\sigma^2 = 256$. Further, suppose Sophia wants to estimate the mean number of daily likes and that she wants to be 99% sure that her estimate does not differ from the true mean by three likes or more. What is the minimum number of days she needs to record the number of likes in order to achieve this desired level of accuracy?

SOLUTION   Let X_i be the number of likes on day i. Then, $X_1, X_2, ...$ are independent and identically distributed with

$$E(X_i) = \mu, \qquad Var(X_i) = 256,$$

so that the standard deviation is $\sigma = 16$. For a period of n days, the total number of likes is simply

$$S_n = X_1 + \cdots + X_n,$$

and so Sophia needs to find the minimum value of n satisfying

$$P\left(-3 < \frac{S_n}{n} - \mu < 3\right) \geq 0.99.$$

Assuming that n is sufficiently large, we can use the CLT to obtain

$$P\left(-3 < \frac{S_n}{n} - \mu < 3\right) = P(-3n < S_n - n\mu < 3n)$$

$$= P\left(-\frac{3\sqrt{n}}{16} < \frac{S_n - n\mu}{16\sqrt{n}} < \frac{3\sqrt{n}}{16}\right)$$

$$= P\left(-\frac{3\sqrt{n}}{16} < Z_n < \frac{3\sqrt{n}}{16}\right)$$

$$\cong \Phi\left(\frac{3\sqrt{n}}{16}\right) - \Phi\left(-\frac{3\sqrt{n}}{16}\right) = 2\Phi\left(\frac{3\sqrt{n}}{16}\right) - 1,$$

so that the required condition becomes

$$2\Phi\left(\frac{3\sqrt{n}}{16}\right) - 1 \geq 0.99,$$

which in turn gives

$$\Phi\left(\frac{3\sqrt{n}}{16}\right) \geq 0.995.$$

From the tables of standard normal distribution, we have

$$\Phi(2.57) = 0.9949, \quad \Phi(2.58) = 0.9951.$$

As Φ is an increasing function, we note that a sufficient requirement is that

$$\frac{3\sqrt{n}}{16} \geq 2.58,$$

and therefore

$$n \geq \left[\frac{16 \cdot (2.58)}{3}\right]^2 = 189.3.$$

This shows that the minimum number of days required so that Sophia achieves the desired level of accuracy is 190.

Note that we have used CLT in the above derivation *under the proviso* that n is sufficiently large. The value of n found, which is 190, ascertains that CLT can in fact be used here.

It needs to be mentioned that Chebyshev's inequality may also be employed in problems of this kind. For example, for the sample mean $\overline{X}_n = S_n/n$, we have

$$E\left(\overline{X}_n\right) = \mu, \quad Var\left(\overline{X}_n\right) = \frac{\sigma^2}{n} = \frac{256}{n}.$$

Then, Chebyshev's inequality yields

$$P\left(\left|\overline{X}_n - \mu\right| \geq t\right) \leq \frac{256}{nt^2}$$

for any real t. Thus, for $t = 3$, we get

$$P\left(-3 < \frac{S_n}{n} - \mu < 3\right) = 1 - P\left(\left|\overline{X}_n - \mu\right| \geq 3\right) \geq 1 - \frac{256}{9n}.$$

Because we require

$$P\left(-3 < \frac{S_n}{n} - \mu < 3\right) \geq 0.99,$$

a sufficient condition becomes

$$1 - \frac{256}{9n} \geq 0.99,$$

which yields

$$n \geq \frac{256}{9(1 - 0.99)} = 2844.4.$$

Notice the much larger sample size compared to the previous answer (Sophia would now have to wait almost eight years until she finds out how many persons "like" her announcements!). Generally, if one is certain that the sample size required in a specific problem is at least moderately large (say, larger than 30), CLT is preferable to Chebyshev's inequality (which, on the other hand, can be used for any value of n).

The CLT is not only used for approximating continuous quantities, but it can also be employed to approximate the *distribution of the sum of discrete random variables*. In such cases, however, to compensate for the discrete nature of the original quantities, a modification known as **continuity correction** is typically used. To be specific, let X_i (for $i = 1, 2, \ldots$) be a sequence of i.i.d. variables with a discrete distribution, having mean μ and variance σ^2, respectively, and let i and j be two integers with $i \leq j$. Then,

$$P(i \leq S_n \leq j) \cong \Phi\left(\frac{j + 0.5 - n\mu}{\sigma\sqrt{n}}\right) - \Phi\left(\frac{i - 0.5 - n\mu}{\sigma\sqrt{n}}\right) \tag{8.4}$$

is in general more accurate than the approximation (without continuity correction)

$$P(i \leq S_n \leq j) = P\left(\frac{i - n\mu}{\sqrt{n\sigma^2}} \leq \frac{S_n - n\mu}{\sqrt{n\sigma^2}} \leq \frac{j - n\mu}{\sqrt{n\sigma^2}}\right)$$

$$\cong \Phi\left(\frac{j - n\mu}{\sigma\sqrt{n}}\right) - \Phi\left(\frac{i - n\mu}{\sigma\sqrt{n}}\right)$$

that arises directly from CLT; more details can be found in Volume I.

The continuity correction may also be used to give an approximation for probabilities of the form $P(S_n = i)$, where i is an integer (such probabilities are nonzero when X_i's are discrete). In particular, applying (8.4) when $i = j$, we immediately obtain

$$P(S_n = i) \cong \Phi\left(\frac{i + 0.5 - n\mu}{\sigma\sqrt{n}}\right) - \Phi\left(\frac{i - 0.5 - n\mu}{\sigma\sqrt{n}}\right)$$

(whereas the approximation without the continuity correction would yield zero for the right-hand side).

Example 8.3.5 (Normal approximation to a discrete distribution using continuity correction)
Let $X_1, X_2, \ldots$ be a sequence of independent variables, each having a Poisson distribution with mean $\mu = 1/5$. Suppose we want to use the normal approximation for the sum

$$S_{40} = X_1 + \cdots + X_{40}.$$

As $n = 40$ is sufficiently large, and $E(X_i) = Var(X_i) = 1/5$ for $i = 1, 2, \ldots$, according to CLT we may write for any nonnegative integers i and j (with $i < j$)

$$P(i \le S_{40} \le j) = P\left(\frac{i - 40 \cdot 1/5}{\sqrt{40 \cdot 1/5}} \le \frac{S_{40} - 40 \cdot 1/5}{\sqrt{40 \cdot 1/5}} \le \frac{j - 40 \cdot 1/5}{\sqrt{40 \cdot 1/5}}\right)$$

$$\cong \Phi\left(\frac{j - 8}{\sqrt{8}}\right) - \Phi\left(\frac{i - 8}{\sqrt{8}}\right) := p_1(i, j).$$

On the other hand, upon using continuity correction, we have

$$P(i \le S_{40} \le j) = \Phi\left(\frac{j - 8 + 0.5}{\sqrt{8}}\right) - \Phi\left(\frac{i - 8 - 0.5}{\sqrt{8}}\right) := p_2(i, j).$$

Note in this case that the exact distribution of S_{40} is known since sum of independent Poisson variables is again Poisson. Thus, in particular, we have

$$P(i \le S_{40} \le j) = \sum_{x=i}^{j} e^{-8} \frac{8^x}{x!} := p(i, j).$$

Table 8.1 presents the exact probabilities, $p(i, j)$, along with the two approximations $p_1(i, j)$ and $p_2(i, j)$ above, for different choices of i and j. Neither approximation is very good (explain why!), but the use of continuity correction does result in significant improvement.

i	j	$p(i,j)$	$p_1(i,j)$	$p_2(i,j)$
0	4	0.0996	0.0763	0.1066
5	8	0.4929	0.3556	0.4622
9	12	0.3437	0.2832	0.3740
13	∞	0.0638	0.0385	0.0558

Table 8.1 Normal approximation to Poisson probabilities, with $(p_2(i,j))$ and without $(p_1(i,j))$ continuity correction.

We close this section by a presentation of a well-known result, used often in combinatorics, known as *Stirling's formula*. This states that, for large n, $n!$ can be approximated by the quantity $\sqrt{2\pi n}\, n^n e^{-n}$ (recall that $n!$ can be extremely large even for $n = 200$). Precisely, we have the asymptotic relation

$$\lim_{n\to\infty} \frac{n!}{\sqrt{2\pi n}\, n^n e^{-n}} = 1. \tag{8.5}$$

In Example 8.3.6, which can be found in Ghahramani (2005), we use CLT to provide an (informal) proof of the above Stirling's formula.

Example 8.3.6 Let $X_1, X_2, \ldots$ be a sequence of independent and identically distributed random variables, having Poisson distribution with parameter 1. Then,

$$\mu = E(X_i) = 1, \quad \sigma^2 = Var(X_i) = 1,$$

for $i = 1, 2, \ldots$, so that CLT gives, for $S_n = X_1 + \cdots + X_n$, that

$$P(S_n = n) = P(n - 1 < S_n \leq n) = P\left(\frac{(n-1) - n\cdot 1}{1\cdot\sqrt{n}} < \frac{S_n - n\mu}{\sigma\sqrt{n}} \leq \frac{n - n\cdot 1}{1\cdot\sqrt{n}}\right)$$

$$= P\left(-n^{-1/2} < Z_n \leq 0)\right) \cong \Phi(0) - \Phi(-n^{-1/2}).$$

But, the Mean Value Theorem of calculus, implies that there exists a $t_n \in (-n^{-1/2}, 0)$ such that

$$\Phi(0) - \Phi(-n^{-1/2}) = \Phi'(t_n)\left[0 - (-n^{-1/2})\right] = \phi(t_n)\frac{1}{\sqrt{n}}$$

$$= \frac{1}{\sqrt{n}} \cdot \frac{1}{\sqrt{2\pi}} \cdot \exp\left(-\frac{1}{2}t_n^2\right).$$

As $t_n \in (n^{-1/2}, 0)$, we have $\lim_{n\to\infty} t_n = 0$, so that for large n, we can write

$$P(S_n = n) \cong \frac{1}{\sqrt{2\pi n}}.$$

Note now that the distribution of S_n is Poisson with parameter n, and so

$$P(S_n = n) = e^{-n} \cdot \frac{n^n}{n!}.$$

The last two results immediately yield

$$e^{-n} \cdot \frac{n^n}{n!} \cong \frac{1}{\sqrt{2\pi n}},$$

and we thus arrive at Stirling's approximation formula which asserts that

$$n! \cong \sqrt{2\pi n} \cdot n^n e^{-n}.$$

This result, in a more precise mathematical form, is as given in (8.5).

Exercises

Group A

1. A fair coin is tossed 20 000 times. Find the probability that the number of times the coin lands heads is between 9900 and 10 100.

2. The time, X (in seconds), that a PC takes to execute an algorithm has exponential distribution with parameter $\lambda = 3\,\text{s}^{-1}$. Peter wants to produce 2500 simulations (runs) using that algorithm, but he only has 10 minutes to perform that task. What is the probability that he will manage to complete it?

3. An insurance company has a portfolio with 150 000 customers. The company has estimated that, during a year, the amount X claimed by a customer chosen at random has expected value US$1200 and standard deviation US$350. What is the probability that, in a given year, the total amount claimed by the customers for this portfolio will exceed US$250 million?

4. The proportion of time an automatic cash machine is used by customers outside a bank, during a day, has uniform distribution over the interval $(0, 1)$. What is the probability that

   (i) the average daily proportion of time the ATM is in use over a 30-day period exceeds 60%?

   (ii) the average proportion of time the ATM is used by customers over a year (365 days) is between 50% and 53%?

5. For a certain brand of cigarettes, it has been suggested that the amount of nicotine entering the body of the smoker, per cigarette, is 0.7 mg with a standard deviation of 0.1 mg. If a person smokes 10 cigarettes per day, what is the probability that the total amount of nicotine entering his body over a six-month period will be more than 150 mg? (Assume that a month has 30 days.)

6. The number of accidents at a certain junction follows Poisson distribution with a mean of two accidents per week. Assuming that a year has 52 weeks,

   (i) give an approximation for the probability that, during a given year, the number of accidents will be between 95 and 110;

   (ii) calculate the exact probability that there will be at least 100, but no more than 105 accidents, in a given year. Then, compare this exact value with the approximations, both with and without continuity correction.

   [Hint: For the exact probability in Part (ii), use the result of Proposition 7.4.3.]

7. Suppose the number of defective pixels on a TFT screen having an area of $t\,\text{cm}^2$ has the Poisson distribution with mean $t/150$. A store has 100 screens, each with an area of $300\,\text{cm}^2$. What is the probability that the total number of defective pixels in these screens

   (i) does not exceed 180?

   (ii) is between 190 and 230?

8. The number of calories a person receives during breakfast, lunch, and dinner is a random variable with mean 600, 800, and 1100, respectively, and with corresponding standard deviations of 50, 80, and 100 calories. Assuming that the three variables associated with the three meals of the day are independent, what is the probability that the person receives more than 2700 calories per day on average during a period of 365 days?

9. Sophie has estimated that, among the books she has at home, the average thickness is 3 cm per book with a standard deviation of 1 cm. If she selects 40 books at random, what is the probability that they will fit in a bookshelf having a width of 1 m?

10. Jim and Zoe play the following game: they roll a die in turn; if the outcome of a roll is 1, 3, or 5, Jim receives from Zoe US$1, US$2, or US$3, respectively. If the outcome is 2, 4, or 6, then Jim gives Zoe US$3, US$2, or US$1, respectively. Find the probability that after 60 rolls of the die,

    (i) Zoe's earnings from the game will be at least US$15;

    (ii) Zoe's earnings will be between US$5 and US$15.

11. Prove the first part of Proposition 8.3.3; that is, if $X_n \xrightarrow{r\text{-mean}} X$ for some $r \geq 1$, then

$$\lim_{n\to\infty} E|X_n^r| = E|X^r|.$$

12. Show that if a sequence $\{X_n : n = 1, 2, \ldots\}$ of variables converges to X in mean, then $E(X_n) \to E(X)$ as $n \to \infty$. Is the converse of this statement also true?

Group B

13. Let $X_1, X_2, \ldots$ be a sequence of i.i.d. random variables following uniform distribution over the interval $[0, 1]$. For $n = 1, 2, \ldots$, define

$$Y_n = n \cdot \min_{1 \leq i \leq n} X_i.$$

Show that $Y_n \xrightarrow{d} Y$, where Y follows exponential distribution with parameter $\lambda = 1$.

14. Let $X_1, X_2, \ldots$ be a sequence of independent random variables, each having Poisson distribution with mean r, where r is a positive integer.

(i) Using CLT, prove that, for any real x,

$$\lim_{n \to \infty} P\left(X_1 + \cdots + X_n \le rn + x\sqrt{rn} \right) = \Phi(x).$$

Here, Φ denotes the distribution function of the standard normal distribution.

(ii) By an appeal to the result of Example 4.3.2, show that

$$\lim_{n \to \infty} e^{-rn} \sum_{k=0}^{rn} \frac{(rn)^k}{k!} = \frac{1}{2}.$$

As a special case, deduce that

$$\lim_{n \to \infty} \left[e^{-n} \left(1 + \frac{n}{1!} + \frac{n^2}{2!} + \cdots + \frac{n^n}{n!} \right) \right] = \frac{1}{2}.$$

15. A call operator handles phone calls whose duration (in minutes), denoted by X, has exponential distribution with mean two minutes.

(i) Find the probability that the next phone call she receives will last for less than two minutes.

(ii) Suppose we know that for a certain day the operator handled 200 telephone calls. Calculate the probability that she was busy for at least seven hours on that day.

(iii) For these 200 calls, we need to calculate the probability that at least 30 of them lasted for more than four minutes. Find an approximation for this probability using the normal distribution, first without and then with the continuity correction. How do these values compare with the exact value of 0.301?

16. Suppose the life length, in days, of a certain type of light bulbs has exponential distribution with parameter $\lambda = 0.002$. A drugstore uses 49 bulbs of that type for lighting.

(i) Find the probability that the average life length of the 49 bulbs currently in use will be more than 540 days.

(ii) Give an approximation for the probability that, among the 49 bulbs, at least 25 will have a life length less than 480 days.

17. A supermarket has four brands of cereals, labeled A, B, C, and D. If it is assumed that, during a day, the supermarket sells a total of 600 packs of cereals from these brands, and that all four brands are equally likely for a customer, find how many packs from each brand should be in stock so that the supermarket meets its daily demand (for all brands) with probability at least 99%.

8.4   BASIC CONCEPTS AND FORMULAS

Convergence in probability	$X_n \xrightarrow{P} X \Leftrightarrow \lim_{n\to\infty} P\left(|X_n - X| > \epsilon\right) = 0$ for any $\epsilon > 0$
Convergence almost surely (with probability 1)	$X_n \xrightarrow{a.s.} X \Leftrightarrow P\left(\lim_{n\to\infty} X_n = X\right) = 1$
Convergence in distribution	$X_n \xrightarrow{d} X \Leftrightarrow \lim_{n\to\infty} F_n(x) = F(x),$ for every x at which F is continuous. Here, F is the c.d.f. of X and $F_1, F_2, \ldots$ are the distribution functions of $X_1, X_2, \ldots$, respectively.
Convergence in r-th mean (for $r \geq 1$)	$X_n \xrightarrow{r\text{-mean}} X \Leftrightarrow \lim_{n\to\infty} E\left(|X_n - X|^r\right) = 0.$
Relationship among the four forms of convergence	$X_n \xrightarrow{a.s.} X \Rightarrow X_n \xrightarrow{P} X \Rightarrow X_n \xrightarrow{d} X$ and $X_n \xrightarrow{r\text{-mean}} X \Rightarrow X_n \xrightarrow{P} X \Rightarrow X_n \xrightarrow{d} X,$ but (in general) $X_n \xrightarrow{a.s.} X \not\Rightarrow X_n \xrightarrow{r\text{-mean}} X$ and $X_n \xrightarrow{r\text{-mean}} X \not\Rightarrow X_n \xrightarrow{a.s.} X$
Weak law of large numbers (WLLN)	If $X_1, X_2, \ldots$ are independent and identically distributed random variables with $E(X_i) = \mu$ and $Var(X_i) = \sigma^2$, for $i = 1, 2, \ldots$, then $\overline{X}_n \xrightarrow{p} \mu$.
Strong law of large numbers (SLLN)	If $X_1, X_2, \ldots$ are independent and identically distributed random variables with $E(X_i) = \mu$ and $Var(X_i) = \sigma^2$, for $i = 1, 2, \ldots$, then $\overline{X}_n \xrightarrow{a.s.} \mu$.
Central limit theorem (CLT)	Let $X_1, X_2, \ldots$ be a sequence of independent and identically distributed random variables with $E(X_i) = \mu, \quad Var(X_i) = \sigma^2, \quad i = 1, 2, \ldots.$ Then, the standardized sequence of means $Z_n = \dfrac{X_1 + X_2 + \cdots + X_n - n\mu}{\sigma\sqrt{n}} = \dfrac{\overline{X}_n - \mu}{\sqrt{\sigma^2/n}} = \dfrac{\sqrt{n}\left(\overline{X}_n - \mu\right)}{\sigma}$ converges in distribution to a standard normal random variable; that is, $\lim_{n\to\infty} P(Z_n \leq x) = \dfrac{1}{\sqrt{2\pi}} \displaystyle\int_{-\infty}^{x} e^{-t^2/2}\, dt = \Phi(x).$

8.5 COMPUTATIONAL EXERCISES

The following command sequence in Mathematica is used to calculate the probability $P(|\overline{X}_n - \mu| > \epsilon)$ for a sequence of i.i.d. variables $X_1, X_2, \ldots$ with mean μ and variance σ^2. The value of ϵ is taken to be 0.01. Specifically, we first produce random variables $X_1, X_2, \ldots$, from the exponential distribution with parameter $\lambda = 3$; then, we simply find the proportion of times (in 3000 repetitions of the experiment) that the inequality $|\overline{X}_n - \mu| > \epsilon$ is satisfied. The graph below illustrates the result (WLLN, Proposition 8.2.1)

$$\lim_{n \to \infty} P(|\overline{X}_n - \mu| > \epsilon) = 0$$

for $\epsilon = 0.01$.

```
In[1]:= ndist = ExponentialDistribution[3]
m = Mean[ndist]
epsilon = 0.01;
nr = 3000; (*Number of repetitions to estimate the probability
P(|Xbar-m|>epsilon)*)
prob = {};
Do[S[n] = 0;
(*n is the sample size*)
Do[sample = RandomReal[ndist, n];
xbar = Mean[sample];
S[n] = S[n] + If[Abs[xbar - m] > epsilon, 1, 0];
, {j, 1, nr}];
AppendTo[prob, {n, S[n]/(nr)}], {n, 10, 3000, 10}]
Print["Graph of the probabilities P(|Xbar-m|>", epsilon, ")"]
Print["Distribution of Xi :" , ndist]
Print["epsilon=", epsilon]
ListPlot[prob]

Out[10]= Graph of the probabilities P(|Xbar-m|>0.01)
Distribution of Xi :ExponentialDistribution[3]
epsilon=0.01
```

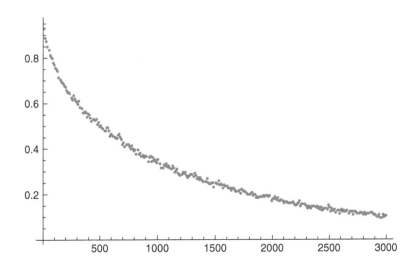

1. Working in a similar way, draw a graph to estimate the probabilities $P(|\overline{X}_n - \mu| > \epsilon)$ for the following pairs of values of n and ϵ:

   - $n = 30, \epsilon = 0.02$,

   - $n = 80, \epsilon = 0.005$,

   - $n = 150, \epsilon = 0.002$,

   when $X_1, \ldots, X_n$ is a random sample taken from each of the following distributions:

   (i) standard normal;

   (ii) uniform over the interval $(-1, 1)$;

   (iii) geometric with $p = 1/4$;

   (iv) Gamma with parameters $\alpha = 3, \beta = 2$.

   In each case, first use $B = 5000$ random samples and then increase this to $B = 15\,000$. What conclusions can you draw overall about the importance of the different parameters used?

2. Continuing from Exercise 1, let $X_i, i = 1, 2, \ldots$, be a sequence of i.i.d. random variables with mean μ and variance σ^2. Write a program to compare the probabilities $P(|\overline{X}_n - \mu| > \epsilon)$ with the quantity $\sigma^2/(n\epsilon^2)$. Do this for each of the cases in the previous exercise. What do you observe this time?

   The following program produces a histogram for the frequencies of the standardized means, based on a sequence of i.i.d. variables $X_1, X_2, \ldots$ from exponential distribution with $\lambda = 3$. The graphs illustrate the "effect of CLT"; as the sample size increases, the shape of the histogram approaches that of the standard normal density function.

```
In[1]:= nr = 10000;  (*Number of random samples taken*)
ndist = ExponentialDistribution[3];
m = Mean[ndist];
var := Variance[ndist];
stdmeanslist = {};
n = 100;   (*This is the size for each sample*)
Do[sample = RandomReal[ndist, n];
xbar = Mean[sample];
stdmean = (xbar - m)/Sqrt[var/n];
AppendTo[stdmeanslist, stdmean], {i, 1, nr} ]
Print["HISTOGRAM FOR STANDARDIZED MEANS"]
Print["Sample Size n=", n]
Print["Distribution of Xi :" , ndist]
Histogram[stdmeanslist, n=40]

HISTOGRAM FOR STANDARDIZED MEANS
Sample Size n=10
Distribution of Xi :ExponentialDistribution[3]
```

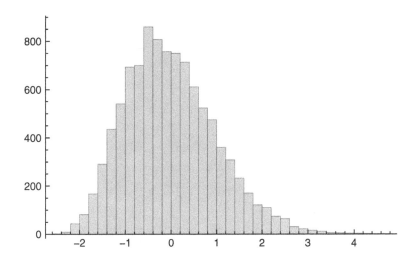

HISTOGRAM FOR STANDARDIZED MEANS
Sample Size n=100
Distribution of Xi :ExponentialDistribution[3]

HISTOGRAM FOR STANDARDIZED MEANS
Sample Size n=1000
Distribution of Xi :ExponentialDistribution[3]

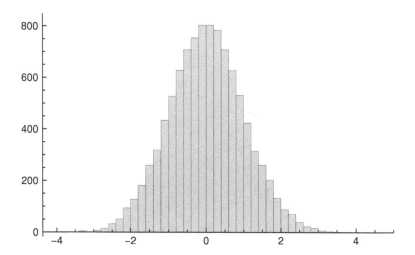

3. Produce similar histograms for the four distributions in Exercise 1, for each of the following sample sizes: 10, 100, and 1000. Note that two of the four distributions in Exercise 1 are symmetric (normal and uniform), the other two are not. What do you observe?

4. If X has a χ_n^2 (chi-squared with n degrees of freedom) distribution, then we know from Chapter 4 that

$$E(X) = n, \quad Var(X) = 2n.$$

   Draw, on the same graph, a plot of the χ_n^2 density, along with the density of $N(n, 2n)$ distribution for the following values of n: 10, 20, 30, 50, 100. What do you observe?

5. We know that, as $n \to \infty$, t_n distribution approaches standard normal distribution. Compare (by plotting them on the same graph) the densities of t_n and $N(0, 1)$ for the same values of n as in the previous exercise.

6. The duration, in minutes, of a phone call on Diane's cell phone is represented by a random variable whose density function is

$$f(x) = cx^4 e^{-3x}, \qquad x > 0.$$

   After identifying the value of the constant c in the above density, answer the following questions:

   (i) What is the probability that the duration of the next phone call that Diane receives will be more than two minutes?

   (ii) Suppose we know that, in a certain week, Diane used her cell phone to make or receive 120 calls. Assuming the durations of these phone calls are independent, each having a density f as above, calculate the probability that she spent at least four hours talking over her cell phone during that week.

(iii) For these 120 calls, we need to calculate the probability that at least 40 of them lasted for more than two minutes. Calculate both the exact value for this probability, as well as an approximation using the normal distribution, first without and then with continuity correction. How do these results compare?

7. Suppose we want to assess the approximation in Stirling's formula (Example 8.3.6) for calculating factorials. Use Mathematica to calculate $n!$ for $n = 10, 20, 30, 40, \dots, 150$; for each value of n, find both the exact value and the approximation from Example 8.3.6. Make a table in Mathematica that presents the relative percentage errors, that is, the values of

$$\frac{n_{St} - n!}{n!} \cdot 100,$$

where n_{St} is Stirling's approximation for $n!$. What do you observe?

8.6 SELF-ASSESSMENT EXERCISES

In the following, $X_1, X_2, \dots$ denotes a sequence of independent and identically distributed random variables with mean μ and variance $\sigma^2 < \infty$, while $\overline{X}_n = (X_1 + X_2 + \cdots + X_n)/n$.

8.6.1 True–False Questions

1. Nick tosses a pair of dice repeatedly and records the relative frequency for the occurrence of event A: *the sum of the outcomes in the two dice is 7*. As the number of throws grows large, this relative frequency will converge to $7/36$.

2. Let the common density of the variables X_i be

$$f(x) = \begin{cases} x^{-2}, & \text{if } x \geq 1, \\ 0, & \text{otherwise.} \end{cases}$$

Then, neither SLLN nor CLT can be applied to the sequence $\{X_i : i = 1, 2, \dots\}$ because the variables X_i have an infinite mean.

3. We toss a fair die n times, and let

$$X_i = \begin{cases} 3, & \text{if the outcome in the } i\text{th trial is 5 or 6,} \\ 0, & \text{otherwise.} \end{cases}$$

Then, as $n \to \infty$, the sequence $\{\overline{X}_n\}$ converges to 1 almost surely.

4. Suppose X_i have a geometric distribution with probability function

$$f(x) = \frac{3 \cdot 2^{x-1}}{5^x}, \qquad x = 1, 2, \dots .$$

Then, the following holds:

$$\lim_{n \to \infty} P\left(\overline{X}_n > \frac{8}{5}\right) = 1.$$

5. The empirical distribution function always takes values in the interval $[0, 1]$.

6. If the sequence $\{X_n\}$ satisfies $X_n \xrightarrow{P} X$ for some variable X, then we also have $X_n \xrightarrow{d} X$.

7. If the sequence $\{X_n\}$ is such that $X_n \xrightarrow{P} 0$, it follows that $E(X_n) \to 0$.

8. Suppose we take a random sample of size $n = 3000$ from an unknown distribution F, and observe that 1800 values in that sample are less than or equal to 5. Then, assuming that X has distribution F, a suitable estimate for the probability $P(X > 5)$ is $2/5$.

9. Assume that the sequence $\{X_n\}$ converges in quadratic mean to a random variable X. Then,

$$\lim_{n \to \infty} Var(X_n) = Var(X).$$

10. The Central Limit Theorem states that $\overline{X}_n$ follows normal distribution, no matter what the original distribution of the variables X_i is.

11. Assume that the variables $X_1, X_2, \dots$ have uniform distribution over the interval $[-1, 1]$. Then, the asymptotic distribution of $\sqrt{n} \cdot \overline{X}_n$ is $N(0, 1)$.

12. Assume that the variables $X_1, X_2, \dots$ have a uniform distribution over the interval $[0, 100]$. Then,

$$\lim_{n \to \infty} P\left(\max_{1 \leq k \leq n} X_k \leq 99\right) = 1.$$

8.6.2 Multiple Choice Questions

1. The sequence $\left\{\overline{X}_n : n = 1, 2, \dots\right\}$ converges to a random variable X with $P(X = \mu) = 1$

   a. almost surely

   b. in quadratic mean

   c. in probability

   d. in distribution

   e. in all of the above ways.

2. Which of the following statements is **not correct**?

   a. $\lim_{n \to \infty} P\left(\left|\overline{X}_n - \mu\right| > \epsilon\right) = 0$ for any $\epsilon > 0$

   b. $P\left(\left|\overline{X}_n - \mu\right| > \epsilon\right) \leq \dfrac{2\sigma^2}{n\epsilon^2}$ for any $\epsilon > 0$

   c. $\overline{X}_n \xrightarrow{d} \mu$

   d. $\overline{X}_n \xrightarrow{a.s.} \mu$

   e. $\lim_{n \to \infty} P\left(\left|\overline{X}_n - \mu\right| = 0\right) = 1.$

3. Assume that the variables $X_1, X_2, \ldots$ have Poisson distribution with mean $\mu = 3$. Then the asymptotic distribution of $\overline{X}_n$ is

   a. $N(0, 1)$

   b. $N\left(0, \dfrac{3}{\sqrt{n}}\right)$

   c. $N(3, 3)$

   d. $N\left(3, \dfrac{3}{n}\right)$

   e. $N\left(3, \dfrac{3}{\sqrt{n}}\right)$

4. Assume that the variables $X_1, X_2, \ldots$ are continuous with density function

$$f(x) = \frac{3x(2 - x)}{4}, \qquad 0 \le x \le 2.$$

   Then, which of the following statements **is correct?**

   a. $P(\overline{X}_n = 1) = 1$ for all n

   b. $\overline{X}_n \xrightarrow{a.s.} 1$

   c. $\overline{X}_n \xrightarrow{d} \dfrac{1}{2}$

   d. $\overline{X}_n \xrightarrow{\text{2-mean}} \dfrac{6}{5}$

   e. $P\left(\left|\overline{X}_n - \dfrac{1}{2}\right| > \epsilon\right) \le \dfrac{1}{n\epsilon^2}$ for any $\epsilon > 0$.

5. Assume that the variables $X_1, X_2, \ldots$ have uniform distribution over the interval $[0, 2]$. Which of the following statements is **not correct?**

   a. $\overline{X}_n \xrightarrow{P} 1$

   b. $\overline{X}_n \xrightarrow{\text{mean}} 1$

   c. $\lim_{n \to \infty} P\left(\overline{X}_n > 1\right) = 1$

   d. $\lim_{n \to \infty} P\left(\overline{X}_n > \dfrac{3}{4}\right) = 1$

   e. $\lim_{n \to \infty} P\left(\overline{X}_n > \dfrac{5}{4}\right) = 0$.

6. The measurement error of an instrument (in inches) is uniformly distributed over the interval $(-0.3, 0.3)$. Using CLT, we see that the probability the **cumulative error of 48 such measurements** does not exceed 0.6 in. in absolute value is

   a. 0.38   b. 0.44   c. 0.51   d. 0.68   e. 0.76.

7. Assume that the variables $X_1, X_2, \ldots$ have uniform distribution over the interval $[0, 1]$ and define a new sequence $\{Y_n : n = 1, 2, \ldots\}$ as

$$Y_n = 3X_n - 2.$$

Which of the following statements is correct?

a. $\lim_{n \to \infty} P\left(\left| \overline{Y}_n - \frac{1}{2} \right| > \epsilon \right) = 0$ for any $\epsilon > 0$

b. $\{\overline{Y}_n : n = 1, 2, \ldots\}$ converges in quadratic mean to 0

c. $\{\overline{Y}_n : n = 1, 2, \ldots\}$ does not converge in mean

d. The sequences $\{\overline{X}_n\}, \{\overline{Y}_n\}$ converge almost surely to the same constant

e. $\lim_{n \to \infty} P\left(\left| \overline{Y}_r + \frac{1}{2} \right| > \epsilon \text{ for at least one } r \text{ with } r \geq n \right) = 0$, for any $\epsilon > 0$.

8. Let X be an exponential random variable with mean $\mu = 1/2$, and define a sequence of variables $\{Y_n : n = 1, 2, \ldots\}$ as

$$Y_n = \begin{cases} 1, & \text{if } X > n \\ 0, & \text{otherwise.} \end{cases}$$

Then,

a. $\{Y_n\}$ converges in probability, but not in distribution

b. $\{Y_n\}$ converges in distribution, but not in probability

c. $\{Y_n\}$ converges both in mean and in probability to $1/2$

d. $\{Y_n\}$ converges both in mean and in probability to 0

e. $P(Y_n > \epsilon) = 0$ for any $\epsilon > 0$ and any $n = 1, 2, \ldots$.

9. Suppose we have taken a random sample of size $n = 8$ from a distribution F and the values we have observed are

$$2, 4, 4, 5, 7, 10, 11, 14.$$

The value of the empirical distribution $\hat{F}_8(5)$ is equal to

a. $1/8$ b. $1/7$ c. $3/8$ d. $1/2$ e. 1

10. Assume that the common density function of $X_1, X_2, \ldots$ is given by

$$f(x) = \frac{x^3 + 1}{6}, \quad 0 \leq x \leq 2.$$

Which of the following statements is correct?

a. The sequence $\{\overline{X}_n\}$ does not converge almost surely

b. The sequence $\{\overline{X}_n\}$ converges in quadratic mean to $5/6$

c. $\lim_{n \to \infty} P(1 \leq \overline{X}_n \leq 8/5) = 1$

d. $\lim_{n \to \infty} E(X_n) = 7/2$

e. The sequence $\{\overline{X}_n\}$ does not converge in probability

11. The number of visitors, per minute, at a web site has Poisson distribution with mean 1. Using the normal approximation with continuity correction, the probability that there will be at least 30 visitors between 10 : 20 a.m. and 11 a.m. on a particular day is

    a. $\Phi(1.66)$ b. $1 - \Phi(1.66)$ c. $\Phi(1.83)$ d. $1 - \Phi(0.26)$ e. $\Phi(1.92)$

12. If the common distribution of $X_1, X_2, \ldots$ is uniform over the interval $[2, 3]$, then the distribution of

$$S_{60} = \sum_{i=1}^{60} X_i$$

is approximately

a. $N(150, 150)$ b. $N\left(150, \dfrac{25}{2}\right)$ c. $N(150, 25)$ d. $N(150, 45)$

e. none of the above

8.7   REVIEW PROBLEMS

1. Prove that, if a sequence $\{X_n : n = 1, 2, \ldots\}$ satisfies

$$X_n \xrightarrow{P} X \quad \text{and} \quad X_n \xrightarrow{P} Y$$

for two variables X, Y, then we have $P(X = Y) = 1$. Does the same conclusion hold if $\xrightarrow{P}$ above is replaced by $\xrightarrow{d}$?

2. Suppose $X_n \xrightarrow{d} X$ and $Y_n \xrightarrow{P} 1$. Then, show that $X_n Y_n \xrightarrow{d} X$.

3. Let $X_1, X_2, \ldots$ be a sequence of i.i.d. random variables with density function

$$f(x) = \frac{\Gamma(\alpha + \beta)}{\Gamma(\alpha)\Gamma(\beta)\lambda^{\alpha+\beta-1}}\, x^{\alpha-1}(\lambda - x)^{\beta-1}, \qquad 0 \le x \le \lambda,$$

where α, β, λ are positive parameters. Investigate whether the limit

$$\lim_{n\to\infty} \frac{X_1 + \cdots + X_n}{n}$$

exists, in terms of almost sure convergence.

4. Suppose we select at random a number from the interval $[0, 1)$, and denote by $X_1, X_2, \ldots$ the decimal places of that number, so that $X_i \in \{0, 1, 2, \ldots, 9\}$, for $i = 1, 2, \ldots$. For a fixed integer k in the set $\{0, 1, 2, \ldots, 9\}$, we define a sequence of variables $Y_i, i = 1, 2, \ldots$, as follows:

$$Y_i = \begin{cases} 1, & \text{if } X_i = k, \\ 0 & \text{otherwise.} \end{cases}$$

(i) Verify that the variables $Y_1, Y_2, \ldots$ are independent and have the same distribution.

(ii) Show that, with probability 1,

$$\lim_{n\to\infty} \frac{Y_1 + \cdots + Y_n}{n} = \frac{1}{10}.$$

This result shows that, when representing a number in $[0, 1)$ as a decimal with infinite decimal places, each digit k in $\{0, 1, 2, \ldots, 9\}$ has a (limiting) relating frequency of appearance equal to $1/10$.

5. Nick rolls a die 75 times. Find an approximation for the probability that the sum of outcomes in these rolls is

   (i) at least 225;

   (ii) between 210 and 240;

   (iii) equal to 225.

6. Assume that a chemical solution has a true value for pH (which measures its acidity) equal to μ, which is not known. Using an instrument, we take n independent measurements of pH. We assume that the value of each measurement is a random variable having mean μ and variance σ^2. How many measurements need to be taken so that with probability a at least (for $0 < a < 1$), the average of these measurements differs from μ, in absolute value by σ/k at most? As an application, find the number of measurements needed for $k = 4, 5$, and 10 and $a = 90\%, 95\%$, and 99%.

7. Steve keeps a record every time he uses his credit card. However, for each transaction, he rounds the exact amount to the nearest integer, in dollars. Suppose in a given year, Steve has had n transactions with his credit card.

   What is the probability that the total amount of these transactions differs (in absolute value) from the amount in his records by a dollars at least? As an illustration, give a numerical answer to the question for $n = 300$ and $a = 10$.

8. We want to estimate the proportion p of defective items produced in a factory unit. What is the minimum number of items that we need to inspect so that the proportion of defective items in the sample differs from p by 1% at most, with a probability at least 0.95? Calculate the sample size required for this if

   (i) p is completely unknown;

   (ii) it is known that $p < 0.1$.

9. (i) Use CLT to prove that χ^2 distribution with n degrees of freedom converges (after an appropriate normalization) to standard normal distribution, as $n \to \infty$.

   (ii) Let $X_1, X_2, \ldots$ be a sequence of independent standard normal random variables. Calculate the value of

   $$\lim_{n\to\infty} P\left(X_1^2 + X_2^2 + \cdots + X_n^2 \le n + k\sqrt{2n}\right),$$

   for $k = 0, 1, 2, 3$.

10. A factory produces X_n machines during day n, where $X_1, X_2, \ldots$, are i.i.d. random variables with a mean value of 10 and standard deviation of four machines.

(i) Find an approximation, with and without the continuity correction, for the probability the total number of machines produced in 100 days is less than 900.

(ii) Use CLT to find (approximately) the largest value of n such that

$$P(X_1 + X_2 + \cdots + X_n \geq 450 + 5n) \leq 0.02.$$

11. Assume that, for $n = 1, 2, \ldots$, the variable X_n has binomial distribution with parameters n and p_n $(0 < p_n < 1)$ in such a way that

$$\lim_{n \to \infty} np_n = \lambda > 0.$$

Let X be a Poisson random variable with mean λ.

(i) Write down the moment generating function of X and X_n, for $n = 1, 2, \ldots$.

(ii) Use the Lévy–Cramer continuity theorem to show that X_n converges in distribution to X.

12. Let X be a random variable with Poisson distribution with mean $\lambda > 0$.

(i) Establish that the moment generating function of

$$Z = \frac{X - \lambda}{\sqrt{\lambda}}$$

is given by

$$M_Z(t) = \exp\left[\lambda\left(\exp\left(\frac{t}{\sqrt{\lambda}}\right) - 1 - \frac{t}{\sqrt{\lambda}}\right)\right].$$

(ii) By making the change of parameter $\lambda = 1/h^2$, deduce that

$$\lim_{\lambda \to \infty}\left[\lambda\left(\exp\left(\frac{t}{\sqrt{\lambda}}\right) - 1 - \frac{t}{\sqrt{\lambda}}\right)\right] = \lim_{h \to 0}\frac{e^{ht} - 1 - ht}{h^2},$$

and use l' Hôpital's rule to find the limit on the right-hand side.

(iii) Show that

$$\lim_{\lambda \to \infty} M_Z(t) = \exp\left(\frac{1}{2}t^2\right).$$

What do you conclude about the Poisson distribution, as the parameter λ becomes large?

13. The number of persons who visit an emergency clinic per day follows Poisson distribution with mean 50. What is the probability that

(i) the number of patients visiting the clinic on a given day is between 35 and 70?

(ii) the number of patients visiting the clinic during a six-month period (180 days) will be at least 8800 but no more than 9100?

[Hint: Use the result of Exercise 12.]

14. Let $X_1, X_2, \ldots$ be a sequence of variables with $E(X_i) = \mu_i, i = 1, 2, \ldots$. Assuming that there exist two constants c_1, c_2 such that for any $n \geq 1$,

$$\sum_{i=1}^{n} Var(X_i) \leq nc_1 \quad \text{and} \quad \sum_{i=1}^{n-1} \sum_{j=i+1}^{n} Cov(X_i, X_j) \leq nc_2,$$

prove that the sequence $\overline{X}_n - \overline{\mu}_n, n = 1, 2, \ldots$ where

$$\overline{X}_n = \frac{1}{n} \sum_{i=1}^{n} X_i, \quad \overline{\mu}_n = \frac{1}{n} \sum_{i=1}^{n} \mu_i,$$

converges to zero in probability.

[Hint: First check that

$$E\left(\overline{X}_n\right) = \overline{\mu}_n, \quad \lim_{n \to \infty} Var\left(\overline{X}_n\right) = 0,$$

and then apply Chebyshev's inequality.]

8.8 APPLICATIONS

8.8.1 Use of the CLT for Capacity Planning

Capacity planning is the process of determining the capacity (in terms of products, services available, etc.) needed by an organization to meet changing demands for its products and services. We shall discuss an application of the CLT with reference to a simple example for designing the capacity of a hospital so that it meets its daily needs with a probability close to 1.

We have seen that the normal distribution may serve as an approximation, via the CLT, not only for sums of continuous random variables but also of discrete ones, provided that the number of variables is large. Consider now the following problem. A new hospital is going to be built in a city with 80 000 inhabitants. It has been estimated that the proportion of persons who need to spend the night at the emergency unit of the hospital is $3 \cdot 10^{-4}$. How many beds need to be put into the emergency unit so that the daily needs of the unit are met with probability 98%?

In this case, the number of persons X who, at a given day, stay overnight at the emergency unit follows the binomial distribution. To see this, define for $1 \leq i \leq 80\,000$ a random variable X_i such that X_i takes the value 1 if the ith person spends the night at the unit and 0 otherwise. Then the number of persons who spend the night at the unit can be expressed as

$$X = \sum_{i=1}^{80\,000} X_i$$

and, since the distribution of each X_i is Bernoulli with success probability $p = 3 \cdot 10^{-4}$ (note that here spending the night at emergencies is a "success"!), the distribution of X is binomial with parameters $n = 80\,000$ and p.

As the value of n is very large, we would expect the normal approximation to be very good in this case. To calculate the required (minimum) number of beds, we need to find the smallest value of k such that $P(X > k) \leq 0.02$, or

$$P(X \leq k) \geq 0.98. \tag{8.6}$$

Using the formula for binomial probabilities, this is written as

$$\sum_{i=0}^{k} P(X = i) = \sum_{i=0}^{k} \binom{80\ 000}{i} p^i (1-p)^{80\ 000-i} \leq 0.98$$

with $p = 3 \cdot 10^{-4}$.

Using the CLT, we may approximate the distribution of X by the normal distribution with mean

$$\mu = n \cdot p = 80\ 000 \cdot 3 \cdot 10^{-4} = 24$$

and variance

$$\sigma^2 = n \cdot p \cdot (1-p) = 23.9928,$$

so that the associated standard deviation is

$$\sigma = \sqrt{23.9928} = 4.8982.$$

Therefore, using the CLT, condition (8.6) reduces to

$$P(X \leq k) = P\left(\frac{X - \mu}{\sigma} \leq \frac{k - \mu}{\sigma} \right) \geq 0.98. \tag{8.7}$$

Replacing the values of μ, σ we find

$$P(X \leq k) = P\left(\frac{X - 24}{4.8982} \leq \frac{k - 24}{4.8982} \right) = \Phi\left(\frac{k - 24}{4.8982} \right)$$

and using the tables of the standard normal distribution we see that

$$\Phi(2.0537) = 0.98.$$

Hence, (8.7) can be restated as

$$\Phi\left(\frac{k - 24}{4.8982} \right) \geq \Phi(2.0537)$$

and taking into account that $\Phi(\cdot)$ is an increasing function we deduce that

$$\frac{k - 24}{4.8982} \geq 2.0537.$$

Solving this for k, we see that

$$k \geq 24 + (2.0537) \cdot (4.8982) = 34.06.$$

This means that the hospital should have at least 35 beds available in the unit in order to meet its needs with probability 98%.

If we use a continuity correction, we replace k by $k + 1/2$ in (8.7) and the condition becomes

$$P(X \leq k + 1/2) = \Phi\left(\frac{k + 1/2 - 24}{4.8982}\right) \geq 0.98.$$

Therefore, we obtain in this case

$$k \geq 23.5 + (2.0537) \cdot (4.8982) = 33.56,$$

so that $k = 34$ is sufficient.

The following table provides the required number of beds (with and without continuity correction) so that the hospital meets its daily needs with probability $1 - \alpha$, for six different choices of α.

α	Number of beds	Number of beds (with continuity correction)
0.05	33	32
0.02	35	34
0.01	36	35
0.005	37	37
0.001	40	39
0.0001	44	43

It is perhaps interesting to note that a relatively small increase to the numbers of beds has a drastic impact on the probability that the daily needs exceed the bed capacity; in particular, by increasing the number of beds by a third the value of α decreases from 5% to 0.01%.

KEY TERMS

Central Limit Theorem (CLT)
convergence almost surely (or with probability 1)
convergence in distribution
convergence in mean
convergence in probability
convergence in r-th mean
convergence in quadratic mean (or in mean square)
empirical distribution function
Lévy–Cramer continuity theorem
Stirling formula
strong law of large numbers (SLLN)
weak law of large numbers (WLLN)

APPENDIX A

TAIL PROBABILITY UNDER STANDARD NORMAL DISTRIBUTION[a]

z	0.00	0.01	0.02	0.03	0.04	0.05	0.06	0.07	0.08	0.09
0.0	0.5000	0.4960	0.4920	0.4880	0.4840	0.4801	0.4761	0.4721	0.4681	0.4641
0.1	0.4602	0.4562	0.4522	0.4483	0.4443	0.4404	0.4364	0.4325	0.4286	0.4247
0.2	0.4207	0.4168	0.4129	0.4090	0.4052	0.4013	0.3974	0.3936	0.3897	0.3859
0.3	0.3821	0.3783	0.3745	0.3707	0.3669	0.3632	0.3594	0.3557	0.3520	0.3483
0.4	0.3446	0.3409	0.3372	0.3336	0.3300	0.3264	0.3228	0.3192	0.3156	0.3121
0.5	0.3085	0.3050	0.3015	0.2981	0.2946	0.2912	0.2877	0.2843	0.2810	0.2776
0.6	0.2743	0.2709	0.2676	0.2643	0.2611	0.2578	0.2546	0.2514	0.2483	0.2451
0.7	0.2420	0.2389	0.2358	0.2327	0.2297	0.2266	0.2231	0.2206	0.2177	0.2148
0.8	0.2119	0.2090	0.2061	0.2033	0.2005	0.1977	0.1949	0.1922	0.1984	0.1867
0.9	0.1841	0.1814	0.1788	0.1762	0.1736	0.1711	0.1685	0.1660	0.1635	0.1611
1.0	0.1587	0.1562	0.1539	0.1515	0.1492	0.1469	0.1446	0.1423	0.1401	0.1379
1.1	0.1357	0.1335	0.1314	0.1292	0.1271	0.1251	0.1230	0.1210	0.1190	0.1170
1.2	0.1151	0.1131	0.1112	0.1093	0.1075	0.1056	0.1038	0.1020	0.1003	0.0985
1.3	0.0968	0.0951	0.0934	0.0918	0.0901	0.0885	0.0869	0.0853	0.0838	0.0823
1.4	0.0808	0.0793	0.0778	0.0764	0.0749	0.0735	0.0721	0.0708	0.0694	0.0681
1.5	0.0668	0.0655	0.0643	0.0630	0.0618	0.0606	0.0594	0.0582	0.0571	0.0559
1.6	0.0548	0.0537	0.0526	0.0516	0.0505	0.0495	0.0485	0.0475	0.0465	0.0455
1.7	0.0446	0.0436	0.0427	0.0418	0.0409	0.0401	0.0392	0.0384	0.0375	0.0367

z	0.00	0.01	0.02	0.03	0.04	0.05	0.06	0.07	0.08	0.09
1.8	0.0359	0.0351	0.0344	0.0336	0.0329	0.0322	0.0314	0.0307	0.0301	0.0294
1.9	0.0287	0.0281	0.0274	0.0268	0.0262	0.0256	0.0250	0.0244	0.0239	0.0233
2.0	0.0228	0.0222	0.0217	0.0212	0.0207	0.0202	0.0197	0.0192	0.0188	0.0183
2.1	0.0179	0.0174	0.0170	0.0166	0.0162	0.0158	0.0154	0.0150	0.0146	0.0143
2.2	0.0139	0.0136	0.0132	0.0129	0.0125	0.0122	0.0119	0.0116	0.0113	0.0110
2.3	0.0107	0.0104	0.0102	0.0099	0.0096	0.0094	0.0091	0.0089	0.0087	0.0084
2.4	0.0082	0.0080	0.0078	0.0075	0.0073	0.0017	0.0069	0.0068	0.0066	0.0064
2.5	0.0062	0.0060	0.0059	0.0057	0.0055	0.0054	0.0052	0.0051	0.0049	0.0048
2.6	0.0047	0.0045	0.0044	0.0043	0.0041	0.0040	0.0039	0.0038	0.0037	0.0036
2.7	0.0035	0.0034	0.0033	0.0032	0.0031	0.0030	0.0029	0.0028	0.0027	0.0026
2.8	0.0026	0.0025	0.0024	0.0023	0.0023	0.0022	0.0021	0.0021	0.0020	0.0019
2.9	0.0019	0.0018	0.0018	0.0017	0.0016	0.0016	0.0015	0.0015	0.0014	0.0014
3.0	0.0013	0.0013	0.0013	0.0012	0.0012	0.0011	0.0011	0.0011	0.0010	0.0010

[a]This table gives the probability that the standard normal variable Z will exceed a given positive value z, that is, $P\{Z > z_\alpha\} = \alpha$. The probabilities for negative values of z are obtained by symmetry.

APPENDIX B

CRITICAL VALUES UNDER CHI-SQUARE DISTRIBUTION[a]

Degrees of Freedom	α												
	0.99	0.98	0.95	0.90	0.80	0.70	0.50	0.30	0.20	0.10	0.05	0.02	0.01
1	0.000157	0.000628	0.00393	0.0158	0.0642	0.148	0.455	1.074	1.642	2.706	3.841	5.412	6.635
2	0.0201	0.0404	0.103	0.211	0.446	0.713	1.386	2.408	3.219	4.605	5.991	7.824	9.210
3	0.115	0.185	0.352	0.584	1.005	1.424	2.366	3.665	4.642	6.251	7.815	9.837	11.341
4	0.297	0.429	0.711	1.064	1.649	2.195	3.357	4.878	5.989	7.779	9.488	11.668	13.277
5	0.554	0.752	1.145	1.610	2.343	3.000	4.351	6.064	7.289	9.236	11.070	13.388	15.086
6	0.872	1.134	1.635	2.204	3.070	3.828	5.348	7.231	8.558	10.645	12.592	15.033	16.812
7	1.239	1.564	2.167	2.833	3.822	4.671	6.346	8.383	9.803	12.017	14.067	16.622	18.475
8	1.646	2.032	2.733	3.490	4.594	5.527	7.344	9.524	11.030	13.362	15.507	18.168	20.090
9	2.088	2.532	3.325	4.168	5.380	6.393	8.343	10.656	12.242	14.684	16.919	19.679	21.666
10	2.558	3.059	3.940	4.865	6.179	7.267	9.342	11.781	13.442	15.987	18.307	21.161	23.209
11	3.053	3.609	4.575	5.578	6.989	8.148	10.341	12.899	14.631	17.275	19.675	22.618	24.725
12	3.571	4.178	5.226	6.304	7.807	9.034	11.340	14.011	15.812	18.549	21.026	24.054	26.217
13	4.107	4.765	5.892	7.042	8.634	9.926	12.340	15.119	16.985	19.812	22.362	25.472	27.688
14	4.660	5.368	6.571	7.790	9.467	10.821	13.339	16.222	18.151	21.064	23.685	26.873	29.141
15	5.229	5.985	7.261	8.547	10.307	11.721	14.339	17.322	19.311	22.307	24.996	28.259	30.578
16	5.812	6.614	7.962	9.312	11.152	12.624	15.338	18.418	20.465	23.542	26.296	29.633	32.000
17	6.408	7.255	8.672	10.085	12.002	13.531	16.338	19.511	21.615	24.669	27.587	30.995	33.409
18	7.015	7.906	9.390	10.865	12.857	14.440	17.338	20.601	22.760	25.989	28.869	32.346	34.805
19	7.633	8.567	10.117	11.651	13.716	15.352	18.338	21.689	23.900	27.204	30.144	33.687	36.191
20	8.260	9.237	10.851	12.443	14.578	16.266	19.337	22.775	25.038	28.412	31.410	35.020	37.566
21	8.897	9.915	11.591	13.240	15.445	17.182	20.337	23.858	26.171	29.615	32.671	36.343	38.932
22	9.542	10.600	12.338	14.041	16.314	18.101	21.337	24.939	27.301	30.813	33.924	37.659	40.289

23	10.196	11.293	13.091	14.848	17.187	19.021	22.337	26.018	28.429	32.007	35.172	38.968	41.638
24	10.856	11.992	13.848	15.659	18.062	19.943	23.337	27.096	29.553	33.196	36.415	40.270	42.980
25	11.524	12.697	14.611	16.473	18.940	20.867	24.337	28.172	30.675	34.382	37.652	41.566	44.314
26	12.198	13.409	15.379	17.292	19.820	21.792	25.336	29.246	31.795	35.563	38.885	42.856	45.642
27	12.879	14.125	16.151	18.114	20.703	22.719	26.336	30.319	32.912	36.741	40.113	44.140	46.963
28	13.565	14.847	16.928	18.939	21.588	23.647	27.336	31.391	34.027	37.916	41.337	45.419	48.278
29	14.256	15.574	17.708	19.768	22.475	24.577	28.336	32.461	35.139	39.087	42.557	46.693	49.588
30	14.953	16.306	18.493	20.599	23.364	25.508	29.336	33.530	36.250	40.256	43.773	47.962	50.892

[a]For degrees of freedom greater than 30, the expression $\sqrt{2\chi^2} - \sqrt{2n-1}$ may be used as a normal deviate with unit variance, where n is the number of degrees of freedom.

APPENDIX C

STUDENT'S *t*-DISTRIBUTION[a]

	α				
n	0.10	0.05	0.025	0.01	0.005
1	3.078	6.314	12.706	31.821	63.657
2	1.886	2.920	4.303	6.965	9.925
3	1.638	2.353	3.182	4.541	5.841
4	1.533	2.132	2.776	3.747	4.604
5	1.476	2.015	2.571	3.365	4.032
6	1.440	1.943	2.447	3.143	3.707
7	1.415	1.895	2.365	2.998	3.499
8	1.397	1.860	2.306	2.896	3.355
9	1.383	1.833	2.262	2.821	3.250
10	1.372	1.812	2.228	2.764	3.169
11	1.363	1.796	2.201	2.718	3.106
12	1.356	1.782	2.179	2.681	3.055
13	1.350	1.771	2.160	2.650	3.012
14	1.345	1.761	2.145	2.624	2.977
15	1.341	1.753	2.131	2.602	2.947

	α				
n	0.10	0.05	0.025	0.01	0.005
16	1.337	1.746	2.120	2.583	2.921
17	1.333	1.740	2.110	2.567	2.898
18	1.330	1.734	2.101	2.552	2.878
19	1.328	1.729	2.093	2.539	2.861
20	1.325	1.725	2.086	2.528	2.845
21	1.323	1.721	2.080	2.518	2.831
22	1.321	1.717	2.074	2.508	2.819
23	1.319	1.714	2.069	2.500	2.807
24	1.318	1.711	2.064	2.492	2.797
25	1.316	1.708	2.060	2.485	2.787
26	1.315	1.706	2.056	2.479	2.779
27	1.314	1.703	2.052	2.473	2.771
28	1.313	1.701	2.048	2.467	2.763
29	1.311	1.699	2.045	2.462	2.756
30	1.310	1.697	2.042	2.457	2.750
40	1.303	1.684	2.021	2.423	2.704
60	1.296	1.671	2.000	2.390	2.660
120	1.289	1.658	1.980	2.358	2.617
∞	1.282	1.645	1.960	2.326	2.576

[a]The first column lists the number of degrees of freedom (n). The headings of the other columns give probabilities (α) for t to exceed the entry value. Use symmetry for negative t values.

APPENDIX D

F-DISTRIBUTION: 5% (LIGHTFACE TYPE) AND 1% (BOLDFACE TYPE) POINTS FOR THE F-DISTRIBUTION

Introduction to Probability: Multivariate Models and Applications, Volume 2, First Edition.
N. Balakrishnan, Markos V. Koutras, and Konstadinos G. Politis.
© 2022 John Wiley & Sons, Inc. Published 2022 by John Wiley & Sons, Inc.

Degrees of Freedom for Numerator (*m*)

Degrees of Freedom for Denominator (*n*)	1	2	3	4	5	6	7	8	9	10	11	12	14	16	20	24	30	40	50	75	100	200	500	∞
1	161	200	216	225	230	234	237	239	241	242	243	244	245	246	248	249	250	251	252	253	253	254	254	254
	4052	**4999**	**5403**	**5625**	**5764**	**5859**	**5928**	**5981**	**6022**	**6056**	**6082**	**6106**	**6142**	**6169**	**6208**	**6234**	**6258**	**6286**	**6302**	**6323**	**6334**	**6352**	**6361**	**6366**
2	18.51	19.00	19.16	19.25	19.30	19.33	19.36	19.37	19.38	19.39	19.40	19.41	19.42	19.43	19.44	19.45	19.46	19.47	19.47	19.48	19.49	19.49	19.50	19.50
	98.49	**99.01**	**99.17**	**99.25**	**99.30**	**99.33**	**99.34**	**99.36**	**99.38**	**99.40**	**99.41**	**99.42**	**99.43**	**99.44**	**99.45**	**99.46**	**99.47**	**99.48**	**99.48**	**99.49**	**99.49**	**99.49**	**99.50**	**99.50**
3	10.13	9.55	9.28	9.12	9.01	8.94	8.88	8.84	8.81	8.78	8.76	8.74	8.71	8.69	8.66	8.64	8.62	8.60	8.58	8.57	8.56	8.54	8.54	8.53
	34.12	**30.81**	**29.46**	**28.71**	**28.24**	**27.91**	**27.67**	**27.49**	**27.34**	**27.23**	**27.13**	**27.05**	**26.92**	**26.83**	**26.69**	**26.60**	**26.50**	**26.41**	**26.30**	**26.27**	**26.23**	**26.18**	**26.14**	**26.12**
4	7.71	6.94	6.59	6.39	6.26	6.16	6.09	6.04	6.00	5.96	5.93	5.91	5.87	5.84	5.80	5.77	5.74	5.71	5.70	5.68	5.66	5.65	5.64	5.63
	21.20	**18.00**	**16.69**	**15.98**	**15.52**	**15.21**	**14.98**	**14.80**	**14.66**	**14.54**	**14.45**	**14.37**	**14.24**	**14.15**	**14.02**	**13.93**	**13.83**	**13.74**	**13.69**	**13.61**	**13.57**	**13.52**	**13.48**	**13.46**
5	6.61	5.79	5.41	5.19	5.05	4.95	4.88	4.82	4.78	4.74	4.70	4.68	4.64	4.60	4.56	4.53	4.50	4.46	4.44	4.42	4.40	4.38	4.37	4.36
	16.26	**13.27**	**12.06**	**11.39**	**10.97**	**10.67**	**10.45**	**10.27**	**10.15**	**10.05**	**9.96**	**9.89**	**9.77**	**9.68**	**9.55**	**9.47**	**9.38**	**9.29**	**9.24**	**9.17**	**9.13**	**9.07**	**9.04**	**9.02**
6	5.99	5.14	4.76	4.53	4.39	4.28	4.21	4.15	4.10	4.06	4.03	4.00	3.96	3.92	3.87	3.84	3.81	3.77	3.75	3.72	3.71	3.69	3.68	3.67
	13.74	**10.92**	**9.78**	**9.15**	**8.75**	**8.47**	**8.26**	**8.10**	**7.98**	**7.87**	**7.79**	**7.72**	**7.60**	**7.52**	**7.39**	**7.31**	**7.23**	**7.14**	**7.09**	**7.02**	**6.99**	**6.94**	**6.90**	**6.88**
7	5.59	4.74	4.35	4.12	3.97	3.87	3.79	3.73	3.68	3.63	3.60	3.57	3.52	3.49	3.44	3.41	3.38	3.34	3.32	3.29	3.28	3.25	3.24	3.23
	12.25	**9.55**	**8.45**	**7.85**	**7.46**	**7.19**	**7.00**	**6.84**	**6.71**	**6.62**	**6.54**	**6.47**	**6.35**	**6.27**	**6.15**	**6.07**	**5.98**	**5.90**	**5.85**	**5.78**	**5.75**	**5.70**	**5.67**	**5.65**
8	5.32	4.46	4.07	3.84	3.69	3.58	3.50	3.44	3.39	3.34	3.31	3.28	3.23	3.20	3.15	3.12	3.08	3.05	3.03	3.00	2.98	2.96	2.94	2.93
	11.26	**8.65**	**7.59**	**7.01**	**6.63**	**6.37**	**6.19**	**6.03**	**5.91**	**5.82**	**5.74**	**5.67**	**5.56**	**5.48**	**5.36**	**5.28**	**5.20**	**5.11**	**5.06**	**5.00**	**4.96**	**4.91**	**4.88**	**4.86**
9	5.12	4.26	3.86	3.63	3.48	3.37	3.29	3.23	3.18	3.13	3.10	3.07	3.02	2.98	2.93	2.90	2.86	2.82	2.80	2.77	2.76	2.73	2.72	2.71
	10.56	**8.02**	**6.99**	**6.42**	**6.06**	**5.80**	**5.62**	**5.47**	**5.35**	**5.26**	**5.18**	**5.11**	**5.00**	**4.92**	**4.80**	**4.73**	**4.64**	**4.56**	**4.51**	**4.45**	**4.41**	**4.36**	**4.33**	**4.31**
10	4.96	4.10	3.71	3.48	3.33	3.22	3.14	3.07	3.02	2.97	2.94	2.91	2.86	2.82	2.77	2.74	2.70	2.67	2.64	2.61	2.59	2.56	2.55	2.54
	10.04	**7.56**	**6.55**	**5.99**	**5.64**	**5.39**	**5.21**	**5.06**	**4.95**	**4.85**	**4.78**	**4.71**	**4.60**	**4.52**	**4.41**	**4.33**	**4.25**	**4.17**	**4.12**	**4.05**	**4.01**	**3.96**	**3.93**	**3.91**
11	4.84	3.98	3.59	3.36	3.20	3.09	3.01	2.95	2.90	2.86	2.82	2.79	2.74	2.70	2.65	2.61	2.57	2.53	2.50	2.47	2.45	2.42	2.41	2.40
	9.65	**7.20**	**6.22**	**5.67**	**5.32**	**5.07**	**4.88**	**4.74**	**4.63**	**4.54**	**4.46**	**4.40**	**4.29**	**4.21**	**4.10**	**4.02**	**3.94**	**3.86**	**3.80**	**3.74**	**3.70**	**3.66**	**3.62**	**3.60**
12	4.75	3.88	3.49	3.26	3.11	3.00	2.92	2.85	2.80	2.76	2.72	2.69	2.64	2.60	2.54	2.50	2.46	2.42	2.40	2.36	2.35	2.32	2.31	2.30
	9.33	**6.93**	**5.95**	**5.41**	**5.06**	**4.82**	**4.65**	**4.50**	**4.39**	**4.30**	**4.22**	**4.16**	**4.05**	**3.98**	**3.86**	**3.78**	**3.70**	**3.61**	**3.56**	**3.49**	**3.46**	**3.41**	**3.38**	**3.36**
13	4.67	3.80	3.41	3.18	3.02	2.92	2.84	2.77	2.72	2.67	2.63	2.60	2.55	2.51	2.46	2.42	2.38	2.34	2.32	2.28	2.26	2.24	2.22	2.21
	9.07	**6.70**	**5.74**	**5.20**	**4.86**	**4.62**	**4.44**	**4.30**	**4.19**	**4.10**	**4.02**	**3.96**	**3.85**	**3.78**	**3.67**	**3.59**	**3.51**	**3.42**	**3.37**	**3.30**	**3.27**	**3.21**	**3.18**	**3.16**
14	4.60	3.74	3.34	3.11	2.96	2.85	2.77	2.70	2.65	2.60	2.56	2.53	2.48	2.44	2.39	2.35	2.31	2.27	2.24	2.21	2.19	2.16	2.14	2.13
	8.86	**6.51**	**5.56**	**5.03**	**4.69**	**4.46**	**4.28**	**4.14**	**4.03**	**3.94**	**3.86**	**3.80**	**3.70**	**3.62**	**3.51**	**3.43**	**3.34**	**3.26**	**3.21**	**3.14**	**3.11**	**3.06**	**3.02**	**3.00**
15	4.54	3.68	3.29	3.06	2.90	2.79	2.70	2.64	2.59	2.55	2.51	2.48	2.43	2.39	2.33	2.29	2.25	2.21	2.18	2.15	2.12	2.10	2.08	2.07
	8.68	**6.36**	**5.42**	**4.89**	**4.56**	**4.32**	**4.14**	**4.00**	**3.89**	**3.80**	**3.73**	**3.67**	**3.56**	**3.48**	**3.36**	**3.29**	**3.20**	**3.12**	**3.07**	**3.00**	**2.97**	**2.92**	**2.89**	**2.87**
16	4.49	3.63	3.24	3.01	2.85	2.74	2.66	2.59	2.54	2.49	2.45	2.42	2.37	2.33	2.28	2.24	2.20	2.16	2.13	2.09	2.07	2.04	2.02	2.01
	8.53	**6.23**	**5.29**	**4.77**	**4.44**	**4.20**	**4.03**	**3.89**	**3.78**	**3.69**	**3.61**	**3.55**	**3.45**	**3.37**	**3.25**	**3.18**	**3.10**	**3.01**	**2.96**	**2.89**	**2.86**	**2.80**	**2.77**	**2.75**
17	4.45	3.59	3.20	2.96	2.81	2.70	2.62	2.55	2.50	2.45	2.41	2.38	2.33	2.29	2.23	2.19	2.15	2.11	2.08	2.04	2.02	1.99	1.97	1.96
	8.40	**6.11**	**5.18**	**4.67**	**4.34**	**4.10**	**3.93**	**3.79**	**3.68**	**3.59**	**3.52**	**3.45**	**3.35**	**3.27**	**3.16**	**3.08**	**3.00**	**2.92**	**2.86**	**2.79**	**2.76**	**2.70**	**2.67**	**2.65**

df																								
18	4.41	3.55	3.16	2.93	2.77	2.66	2.58	2.51	2.46	2.41	2.37	2.34	2.29	2.25	2.19	2.15	2.11	2.07	2.04	2.00	1.98	1.95	1.93	1.92
	8.28	**6.01**	**5.09**	**4.58**	**4.25**	**4.01**	**3.85**	**3.71**	**3.60**	**3.51**	**3.44**	**3.37**	**3.27**	**3.19**	**3.07**	**3.00**	**2.91**	**2.83**	**2.78**	**2.71**	**2.68**	**2.62**	**2.59**	**2.57**
19	4.38	3.52	3.13	2.90	2.74	2.63	2.55	2.48	2.43	2.38	2.34	2.31	2.26	2.21	2.15	2.11	2.07	2.02	2.00	1.96	1.94	1.91	1.90	1.88
	8.18	**5.93**	**5.01**	**4.50**	**4.17**	**3.94**	**3.77**	**3.63**	**3.52**	**3.43**	**3.36**	**3.30**	**3.19**	**3.12**	**3.00**	**2.92**	**2.84**	**2.76**	**2.70**	**2.63**	**2.60**	**2.54**	**2.51**	**2.49**
20	4.35	3.49	3.10	2.87	2.71	2.60	2.52	2.45	2.40	2.35	2.31	2.28	2.23	2.18	2.12	2.08	2.04	1.99	1.96	1.92	1.90	1.87	1.85	1.84
	8.10	**5.85**	**4.94**	**4.43**	**4.10**	**3.87**	**3.71**	**3.56**	**3.45**	**3.37**	**3.30**	**3.23**	**3.13**	**3.05**	**2.94**	**2.86**	**2.77**	**2.69**	**2.63**	**2.56**	**2.53**	**2.47**	**2.44**	**2.42**
21	4.32	3.47	3.07	2.84	2.68	2.57	2.49	2.42	2.37	2.32	2.28	2.25	2.20	2.15	2.09	2.05	2.00	1.96	1.93	1.89	1.87	1.84	1.82	1.81
	8.02	**5.78**	**4.87**	**4.37**	**4.04**	**3.81**	**3.65**	**3.51**	**3.40**	**3.31**	**3.24**	**3.17**	**3.07**	**2.99**	**2.88**	**2.80**	**2.72**	**2.63**	**2.58**	**2.51**	**2.47**	**2.42**	**2.38**	**2.36**
22	4.30	3.44	3.05	2.82	2.66	2.55	2.47	2.40	2.35	2.30	2.26	2.23	2.18	2.13	2.07	2.03	1.98	1.93	1.91	1.87	1.84	1.81	1.80	1.78
	7.94	**5.72**	**4.82**	**4.31**	**3.99**	**3.76**	**3.59**	**3.45**	**3.35**	**3.26**	**3.18**	**3.12**	**3.02**	**2.94**	**2.83**	**2.75**	**2.67**	**2.58**	**2.53**	**2.46**	**2.42**	**2.37**	**2.33**	**2.31**
23	4.28	3.42	3.03	2.80	2.64	2.53	2.45	2.38	2.32	2.28	2.24	2.20	2.14	2.10	2.04	2.00	1.96	1.91	1.88	1.84	1.82	1.79	1.77	1.76
	7.88	**5.66**	**4.76**	**4.26**	**3.94**	**3.71**	**3.54**	**3.41**	**3.30**	**3.21**	**3.14**	**3.07**	**2.97**	**2.89**	**2.78**	**2.70**	**2.62**	**2.53**	**2.48**	**2.41**	**2.37**	**2.32**	**2.28**	**2.26**
24	4.26	3.40	3.01	2.78	2.62	2.51	2.43	2.36	2.30	2.26	2.22	2.18	2.13	2.09	2.02	1.98	1.94	1.89	1.86	1.82	1.80	1.76	1.74	1.73
	7.82	**5.61**	**4.72**	**4.22**	**3.90**	**3.67**	**3.50**	**3.36**	**3.25**	**3.17**	**3.09**	**3.03**	**2.93**	**2.85**	**2.74**	**2.66**	**2.58**	**2.49**	**2.44**	**2.36**	**2.33**	**2.27**	**2.23**	**2.21**
25	4.24	3.38	2.99	2.76	2.60	2.49	2.41	2.34	2.28	2.24	2.20	2.16	2.11	2.06	2.00	1.96	1.92	1.87	1.84	1.80	1.77	1.74	1.72	1.71
	7.77	**5.57**	**4.68**	**4.18**	**3.86**	**3.63**	**3.46**	**3.32**	**3.21**	**3.13**	**3.05**	**2.99**	**2.89**	**2.81**	**2.70**	**2.62**	**2.54**	**2.45**	**2.40**	**2.32**	**2.29**	**2.23**	**2.19**	**2.17**
26	4.22	3.37	2.98	2.74	2.59	2.47	2.39	2.32	2.27	2.22	2.18	2.15	2.10	2.05	1.99	1.95	1.90	1.85	1.82	1.78	1.76	1.72	1.70	1.69
	7.72	**5.53**	**4.64**	**4.14**	**3.82**	**3.59**	**3.42**	**3.29**	**3.17**	**3.09**	**3.02**	**2.96**	**2.86**	**2.77**	**2.66**	**2.58**	**2.50**	**2.41**	**2.36**	**2.28**	**2.25**	**2.19**	**2.15**	**2.13**
27	4.21	3.35	2.96	2.73	2.57	2.46	2.37	2.30	2.25	2.20	2.16	2.13	2.08	2.03	1.97	1.93	1.88	1.84	1.80	1.76	1.74	1.71	1.68	1.67
	7.68	**5.49**	**4.60**	**4.11**	**3.79**	**3.56**	**3.39**	**3.26**	**3.14**	**3.06**	**2.98**	**2.93**	**2.83**	**2.74**	**2.63**	**2.55**	**2.47**	**2.38**	**2.33**	**2.25**	**2.21**	**2.16**	**2.12**	**2.10**
28	4.20	3.34	2.95	2.71	2.56	2.44	2.36	2.29	2.24	2.19	2.15	2.12	2.06	2.02	1.96	1.91	1.87	1.81	1.78	1.75	1.72	1.69	1.67	1.65
	7.64	**5.45**	**4.57**	**4.07**	**3.76**	**3.53**	**3.36**	**3.23**	**3.11**	**3.03**	**2.95**	**2.90**	**2.80**	**2.71**	**2.60**	**2.52**	**2.44**	**2.35**	**2.30**	**2.22**	**2.18**	**2.13**	**2.09**	**2.06**
29	4.18	3.33	2.93	2.70	2.54	2.43	2.35	2.28	2.22	2.18	2.14	2.10	2.05	2.00	1.94	1.90	1.85	1.80	1.77	1.73	1.71	1.68	1.65	1.64
	7.60	**5.52**	**4.54**	**4.04**	**3.73**	**3.50**	**3.33**	**3.20**	**3.08**	**3.00**	**2.92**	**2.87**	**2.77**	**2.68**	**2.57**	**2.49**	**2.41**	**2.32**	**2.27**	**2.19**	**2.15**	**2.10**	**2.06**	**2.03**
30	4.17	3.32	2.92	2.69	2.53	2.42	2.34	2.27	2.21	2.16	2.12	2.09	2.04	1.99	1.93	1.89	1.84	1.79	1.76	1.72	1.69	1.66	1.64	1.62
	7.56	**5.39**	**4.51**	**4.02**	**3.70**	**3.47**	**3.30**	**3.17**	**3.06**	**2.98**	**2.90**	**2.84**	**2.74**	**2.66**	**2.55**	**2.47**	**2.38**	**2.29**	**2.24**	**2.16**	**2.13**	**2.07**	**2.03**	**2.01**
32	4.15	3.30	2.90	2.67	2.51	2.40	2.32	2.25	2.19	2.14	2.10	2.07	2.02	1.97	1.91	1.86	1.82	1.76	1.74	1.69	1.67	1.64	1.61	1.59
	7.50	**5.34**	**4.46**	**3.97**	**3.66**	**3.42**	**3.25**	**3.12**	**3.01**	**2.94**	**2.86**	**2.80**	**2.70**	**2.62**	**2.51**	**2.42**	**2.34**	**2.25**	**2.20**	**2.12**	**2.08**	**2.02**	**1.98**	**1.96**
34	4.13	3.28	2.88	2.65	2.49	2.38	2.30	2.23	2.17	2.12	2.08	2.05	2.00	1.95	1.89	1.84	1.80	1.74	1.71	1.67	1.64	1.61	1.59	1.57
	7.44	**5.29**	**4.42**	**3.93**	**3.61**	**3.38**	**3.21**	**3.08**	**2.97**	**2.89**	**2.82**	**2.76**	**2.66**	**2.58**	**2.47**	**2.38**	**2.30**	**2.21**	**2.15**	**2.08**	**2.04**	**1.98**	**1.94**	**1.91**
36	4.11	3.26	2.86	2.63	2.48	2.36	2.28	2.21	2.15	2.10	2.06	2.03	1.98	1.93	1.87	1.82	1.78	1.72	1.69	1.65	1.62	1.59	1.56	1.55
	7.39	**5.25**	**4.38**	**3.89**	**3.58**	**3.35**	**3.18**	**3.04**	**2.94**	**2.86**	**2.78**	**2.72**	**2.62**	**2.54**	**2.43**	**2.35**	**2.26**	**2.17**	**2.12**	**2.04**	**2.00**	**1.94**	**1.90**	**1.87**
38	4.10	3.25	2.85	2.62	2.46	2.35	2.26	2.19	2.14	2.09	2.05	2.02	1.96	1.92	1.85	1.80	1.76	1.71	1.67	1.63	1.60	1.57	1.54	1.53
	7.35	**5.21**	**4.34**	**3.86**	**3.54**	**3.32**	**3.15**	**3.02**	**2.91**	**2.82**	**2.75**	**2.69**	**2.59**	**2.51**	**2.40**	**2.32**	**2.22**	**2.14**	**2.08**	**2.00**	**1.97**	**1.90**	**1.86**	**1.84**
40	4.08	3.23	2.84	2.61	2.45	2.34	2.25	2.18	2.12	2.07	2.04	2.00	1.95	1.90	1.84	1.79	1.74	1.69	1.66	1.61	1.59	1.55	1.53	1.51
	7.31	**5.18**	**4.31**	**3.83**	**3.51**	**3.29**	**3.12**	**2.99**	**2.88**	**2.80**	**2.73**	**2.66**	**2.56**	**2.49**	**2.37**	**2.29**	**2.20**	**2.11**	**2.05**	**1.97**	**1.94**	**1.88**	**1.84**	**1.81**
42	4.07	3.22	2.83	2.59	2.44	2.32	2.24	2.17	2.11	2.06	2.02	1.99	1.94	1.89	1.82	1.78	1.73	1.68	1.64	1.60	1.57	1.54	1.51	1.49

(Continued)

Degrees of Freedom for Numerator (*m*)

Degrees of Freedom for Denominator (*n*)	1	2	3	4	5	6	7	8	9	10	11	12	14	16	20	24	30	40	50	75	100	200	500	∞
44	**7.27**	**5.15**	**4.29**	**3.80**	**3.49**	**3.26**	**3.10**	**2.96**	**2.86**	**2.77**	**2.70**	**2.64**	**2.54**	**2.46**	**2.35**	**2.26**	**2.17**	**2.08**	**2.02**	**1.94**	**1.91**	**1.85**	**1.80**	**1.78**
	4.06	3.21	2.82	2.58	2.43	2.31	2.23	2.16	2.10	2.05	2.01	1.98	1.92	1.88	1.81	1.76	1.72	1.66	1.63	1.58	1.56	1.52	1.50	1.48
46	**7.24**	**5.12**	**4.26**	**3.78**	**3.46**	**3.24**	**3.07**	**2.94**	**2.84**	**2.75**	**2.68**	**2.62**	**2.52**	**2.44**	**2.32**	**2.24**	**2.15**	**2.06**	**2.00**	**1.92**	**1.88**	**1.82**	**1.78**	**1.75**
	4.05	3.20	2.81	2.57	2.42	2.30	2.22	2.14	2.09	2.04	2.00	1.97	1.91	1.87	1.80	1.75	1.71	1.65	1.62	1.57	1.54	1.51	1.48	1.46
48	**7.21**	**5.10**	**4.24**	**3.76**	**3.44**	**3.22**	**3.05**	**2.92**	**2.82**	**2.73**	**2.66**	**2.60**	**2.50**	**2.42**	**2.40**	**2.22**	**2.13**	**2.04**	**1.98**	**1.90**	**1.86**	**1.80**	**1.76**	**1.72**
	4.04	3.19	2.80	2.56	2.41	2.30	2.21	2.14	2.08	2.03	1.99	1.96	1.90	1.86	1.79	1.74	1.70	1.64	1.61	1.56	1.53	1.50	1.47	1.45
50	**7.19**	**5.08**	**4.22**	**3.74**	**3.42**	**3.20**	**3.04**	**2.90**	**2.80**	**2.71**	**2.64**	**2.58**	**2.48**	**2.40**	**2.28**	**2.20**	**2.11**	**2.02**	**1.96**	**1.88**	**1.84**	**1.78**	**1.73**	**1.70**
	4.03	3.18	2.79	2.56	2.40	2.29	2.20	2.13	2.07	2.02	1.98	1.95	1.90	1.85	1.78	1.74	1.69	1.63	1.60	1.55	1.52	1.48	1.46	1.44
55	**7.17**	**5.06**	**4.20**	**3.72**	**3.41**	**3.18**	**3.02**	**2.88**	**2.78**	**2.70**	**2.62**	**2.56**	**2.46**	**2.39**	**2.26**	**2.18**	**2.10**	**2.00**	**1.94**	**1.86**	**1.82**	**1.76**	**1.71**	**1.68**
	4.02	3.17	2.78	2.54	2.38	2.27	2.18	2.11	2.05	2.00	1.97	1.93	1.88	1.83	1.76	1.72	1.67	1.61	1.58	1.52	1.50	1.46	1.43	1.41
60	**7.12**	**5.01**	**4.16**	**3.68**	**3.37**	**3.15**	**2.98**	**2.85**	**2.75**	**2.66**	**2.59**	**2.53**	**2.43**	**2.35**	**2.23**	**2.15**	**2.06**	**1.96**	**1.90**	**1.82**	**1.78**	**1.71**	**1.66**	**1.64**
	4.00	3.15	2.76	2.52	2.37	2.25	2.17	2.10	2.04	1.99	1.95	1.92	1.86	1.81	1.75	1.70	1.65	1.59	1.56	1.50	1.48	1.44	1.41	1.39
65	**7.08**	**4.98**	**4.13**	**3.65**	**3.34**	**3.12**	**2.95**	**2.82**	**2.72**	**2.63**	**2.56**	**2.50**	**2.40**	**2.32**	**2.20**	**2.12**	**2.03**	**1.93**	**1.87**	**1.79**	**1.74**	**1.68**	**1.63**	**1.60**
	3.99	3.14	2.75	2.51	2.36	2.24	2.15	2.08	2.02	1.98	1.94	1.90	1.85	1.80	1.73	1.68	1.63	1.57	1.54	1.49	1.46	1.42	1.39	1.37
70	**7.04**	**4.95**	**4.10**	**3.62**	**3.31**	**3.09**	**2.93**	**2.79**	**2.70**	**2.61**	**2.54**	**2.47**	**2.30**	**2.37**	**2.18**	**2.09**	**2.00**	**1.90**	**1.84**	**1.76**	**1.71**	**1.64**	**1.60**	**1.56**
	3.98	3.13	2.74	2.50	2.35	2.23	2.14	2.07	2.01	1.97	1.93	1.89	1.84	1.79	1.72	1.67	1.62	1.56	1.53	1.47	1.45	1.40	1.37	1.35
80	**7.01**	**4.92**	**4.08**	**3.60**	**3.29**	**3.07**	**2.91**	**2.77**	**2.67**	**2.59**	**2.51**	**2.45**	**2.35**	**2.28**	**2.15**	**2.07**	**1.98**	**1.88**	**1.82**	**1.74**	**1.69**	**1.63**	**1.56**	**1.53**
	3.96	3.11	2.72	2.48	2.33	2.21	2.12	2.05	1.99	1.95	1.91	1.88	1.82	1.77	1.70	1.65	1.60	1.54	1.51	1.45	1.42	1.38	1.35	1.32
100	**6.96**	**4.88**	**4.04**	**3.56**	**3.25**	**3.04**	**2.87**	**2.74**	**2.64**	**2.55**	**2.48**	**2.41**	**2.32**	**2.24**	**2.11**	**2.03**	**1.94**	**1.84**	**1.78**	**1.70**	**1.65**	**1.57**	**1.52**	**1.49**
	3.94	3.09	2.70	2.46	2.30	2.19	2.10	2.03	1.90	1.92	1.88	1.85	1.79	1.75	1.68	1.63	1.57	1.51	1.48	1.42	1.69	1.34	1.30	1.28
125	**6.90**	**4.82**	**3.98**	**3.51**	**3.20**	**2.99**	**2.82**	**2.69**	**2.59**	**2.51**	**2.43**	**2.36**	**2.26**	**2.19**	**2.06**	**1.98**	**1.89**	**1.79**	**1.73**	**1.64**	**1.59**	**1.51**	**1.46**	**1.43**
	3.92	3.07	2.68	2.44	2.29	2.17	2.08	2.01	1.95	1.90	1.86	1.83	1.77	1.72	1.65	1.60	1.55	1.49	1.45	1.39	1.36	1.31	1.27	1.25
150	**6.84**	**4.78**	**3.94**	**3.47**	**3.17**	**2.95**	**2.79**	**2.65**	**2.56**	**2.47**	**2.40**	**2.33**	**2.23**	**2.15**	**2.03**	**1.94**	**1.85**	**1.75**	**1.68**	**1.59**	**1.54**	**1.46**	**1.40**	**1.37**
	3.91	3.06	2.67	2.43	2.27	2.16	2.07	2.00	1.94	1.89	1.85	1.82	1.76	1.71	1.64	1.57	1.54	1.47	1.44	1.37	1.34	1.29	1.25	1.22
200	**6.81**	**4.75**	**3.91**	**3.44**	**3.13**	**2.92**	**2.76**	**2.62**	**2.53**	**2.44**	**2.37**	**2.30**	**2.20**	**2.12**	**2.00**	**1.91**	**1.83**	**1.72**	**1.66**	**1.56**	**1.51**	**1.43**	**1.37**	**1.33**
	3.89	3.04	2.65	2.41	2.26	2.14	2.05	1.98	1.92	1.87	1.83	1.80	1.74	1.69	1.62	1.57	1.52	1.45	1.42	1.35	1.32	1.26	1.22	1.19
400	**6.76**	**4.71**	**3.88**	**3.41**	**3.11**	**2.90**	**2.73**	**2.60**	**2.50**	**2.41**	**2.34**	**2.28**	**2.17**	**2.09**	**1.97**	**1.88**	**1.79**	**1.69**	**1.62**	**1.53**	**1.48**	**1.39**	**1.33**	**1.28**
	3.86	3.02	2.62	2.39	2.23	2.12	2.03	1.96	1.89	1.85	1.81	1.78	1.72	1.67	1.60	1.54	1.49	1.42	1.38	1.32	1.28	1.22	1.16	1.13
1000	**6.70**	**4.66**	**3.83**	**3.36**	**3.06**	**2.85**	**2.69**	**2.55**	**2.46**	**2.37**	**2.29**	**2.23**	**2.12**	**2.04**	**1.92**	**1.84**	**1.74**	**1.64**	**1.57**	**1.47**	**1.42**	**1.32**	**1.24**	**1.19**
	3.85	3.00	2.61	2.38	2.22	2.10	2.02	1.95	1.89	1.84	1.80	1.76	1.70	1.65	1.58	1.53	1.47	1.41	1.36	1.30	1.26	1.19	1.13	1.08
∞	**6.66**	**4.62**	**3.80**	**3.34**	**3.04**	**2.82**	**2.66**	**2.53**	**2.43**	**2.34**	**2.26**	**2.20**	**2.09**	**2.01**	**1.89**	**1.81**	**1.71**	**1.61**	**1.54**	**1.44**	**1.38**	**1.28**	**1.19**	**1.11**
	3.84	2.99	2.60	2.37	2.21	2.09	2.01	1.94	1.88	1.83	1.79	1.75	1.69	1.64	1.57	1.52	1.46	1.40	1.35	1.28	1.24	1.17	1.11	1.00
	6.64	**4.60**	**3.78**	**3.32**	**3.02**	**2.80**	**2.64**	**2.51**	**2.41**	**2.32**	**2.24**	**2.18**	**2.07**	**1.99**	**1.87**	**1.79**	**1.69**	**1.59**	**1.52**	**1.41**	**1.36**	**1.25**	**1.15**	**1.00**

APPENDIX E

GENERATING FUNCTIONS

Introduction to Probability: Multivariate Models and Applications, Volume 2, First Edition.
N. Balakrishnan, Markos V. Koutras, and Konstadinos G. Politis.
© 2022 John Wiley & Sons, Inc. Published 2022 by John Wiley & Sons, Inc.

DISCRETE DISTRIBUTIONS

Distribution	Probability function	Range (R_X)	$E(X)$	$V(X)$	mgf	pgf	cf
Bernoulli	$f(x) = \begin{cases} p, & x=1 \\ q=1-p, & x=0 \end{cases}$	$R_X = \{0,1\}$	p	pq	$q + pe^t$	$q + pt$	$q + pe^{it}$
Binomial	$f(x) = \binom{n}{x} p^x q^{n-x}$	$R_X = \{0,1,2,\ldots,n\}$	np	npq	$(q+pe^t)^n$	$(q+pt)^n$	$(q+pe^{it})^n$
Geometric	$f(x) = q^{x-1}p$	$R_X = \{1,2,\ldots\}$	$\dfrac{1}{p}$	$\dfrac{q}{p^2}$	$\dfrac{pe^t}{1-qe^t}$	$\dfrac{pt}{1-qt}$	$\dfrac{pe^{it}}{1-qe^{it}}$
Negative binomial	$f(x) = \binom{x-1}{r-1} p^r q^{x-r}$	$R_X = \{r, r+1, r+2,\ldots\}$	$\dfrac{r}{p}$	$\dfrac{rq}{p^2}$	$\left(\dfrac{pe^t}{1-qe^t}\right)^r$	$\left(\dfrac{pt}{1-qt}\right)^r$	$\left(\dfrac{pe^{it}}{1-qe^{it}}\right)^r$
Poisson	$f(x) = e^{-\lambda}\dfrac{\lambda^x}{x!}$	$R_X = \{0,1,2,\ldots\}$	λ	λ	$e^{\lambda(e^t-1)}$	$e^{\lambda(t-1)}$	$e^{\lambda(e^{it}-1)}$
Discrete uniform	$f(x) = \dfrac{1}{n+1}$	$R_X = \{0,1,2,\ldots,n\}$	$\dfrac{n}{2}$	$\dfrac{n(n+2)}{12}$	$\dfrac{e^{n+1}-1}{(n+1)e^t-1}$	$\dfrac{e^{n+1}-1}{(n+1)t-1}$	$\dfrac{e^{n+1}-1}{(n+1)e^{it}-1}$

CONTINUOUS DISTRIBUTIONS

Distribution	Density function	Range (R_X)	$E(X)$	$V(X)$	mgf	cf
Uniform	$f(x) = \dfrac{1}{b-a}$	$R_X = [a,b]$	$\dfrac{a+b}{2}$	$\dfrac{(b-a)^2}{12}$	$\dfrac{e^{bt} - e^{at}}{t(b-a)}$	$\dfrac{e^{bit} - e^{ait}}{it(b-a)}$
Normal	$f(x) = \dfrac{1}{\sqrt{2\pi}\sigma}\, e^{-(x-\mu)^2/(2\sigma^2)}$	$R_X = (-\infty, \infty)$	μ	σ	$e^{\mu t + \sigma^2 t^2/2}$	$e^{i\mu t - \sigma^2 t^2/2}$
Exponential	$\lambda e^{-\lambda x}$	$R_X = [0, \infty)$	$\dfrac{1}{\lambda}$	$\dfrac{1}{\lambda^2}$	$\dfrac{\lambda}{\lambda - t}$	$\dfrac{\lambda}{\lambda - it}$
Gamma	$f(x) = \dfrac{\lambda^a x^{a-1} e^{-\lambda x}}{\Gamma(a)}$	$R_X = [0, \infty)$	$\dfrac{a}{\lambda}$	$\dfrac{a}{\lambda^2}$	$\left(\dfrac{\lambda}{\lambda - t}\right)^a$	$\left(\dfrac{\lambda}{\lambda - it}\right)^a$

BIBLIOGRAPHY

Balakrishnan, N. and Rao, C.R. (eds.) (1998a). *Order Statistics: Theory and Methods, Handbook of Statistics-16*. Amsterdam: North-Holland.

Balakrishnan, N. and Rao, C.R. (eds.) (1998b). *Order Statistics: Applications, Handbook of Statistics-17*. Amsterdam: North-Holland.

Bassett, E.E., Bremner, J.M., Jolliffe, I.T. et al. (2000). *Statistics: Problems and Solutions*. Singapore: World Scientific.

Bean, M.A. (2001). *Probability: The Science of Uncertainty, Brooks/Cole Series in Advanced Mathematics*. Pacific Grove, California, USA: Brooks/Cole.

Bertsekas, D.P. and Tsitsiklis, J.N. (2013). *Introduction to Probability*, 3e. Belmont, Massachusetts, USA: Athena Scientific.

Blom, G., Holst, S., and Sandell, D. (1994). *Problems and Snapshots from the World of Probability*. New York: Springer-Verlag.

Devore, J.L. (2012). *Probability and Statistics for Engineering and the Sciences*, 8e. Brooks/Cole. Cengage Learning.

Dorogovtsev, A.Y., Silvestrov, D.S., Skorohod, A.V., and Yadrenko, M.I. (1997). *Probability Theory: Collection of Problems, American Mathematical Society: Translations of Mathematical Monographs*. Providence, Rhode Island, USA: American Mathematical Society.

Durrett, R. (2009). *Elementary Probability for Applications*. Cambridge, New York: Cambridge University press.

Feller, W. (1968). *An Introduction to Probability Theory and its Applications*, 3e. New York: Wiley.

Feller, W. (1971). *An Introduction to Probability Theory and its Applications*, vol. II, 2e. New York: Wiley.

Ghahramani, S. (2005). *Fundamentals of Probability with Stochastic Processes*, 3e. London: Pearson Education Ltd.

Grimmett, G. and Stirzaker, D.R. (2001a). *Probability and Random Processes*, 3e. Oxford: Oxford University Press.

Grimmett, G. and Stirzaker, D.R. (2001b). *One Thousand Exercises in Probability*. Oxford: Oxford University Press.

Grimmett, G. and Welsh, D. (2014). *Probability: An Introduction*, 2e. Oxford: Oxford University Press.

Hald, A. (1998). *A History of Mathematical Statistics from 1750 to 1930*. New York: Wiley.

Kotz, S., Balakrishnan, N., and Johnson, N.L. (2000). *Continuous Multivariate Distributions*, vol. 1, 2e. New York: Wiley.

Koutras, M.V. (2005). *Introduction to Probability: Theory and Applications*, vol. II. Athens, Greece: Stamoulis Publications (in Greek).

Marshall, A.W. and Olkin, I. (2007). *Life Distributions: Structure of Nonparametric, Semiparametric, and Parametric Families, Springer Series in Statistics*. New York, USA: Springer.

Markowitz, H. (1952). Portfolio selection. *The Journal of Finance* 7 77–91.

Pitman, J. (1993). *Probability*. New York: Springer–Verlag.

Ross, S.M. (2006). *A First Course in Probability*, 7e. London, NJ: Pearson Prentice Hall (International Edition).

Ross, S.M. (2019). *An Introduction to Probability Models*, 12e. Boston, Massachusetts, USA: Academic Press.

Weiss, N.A. (2006). *A Course in Probability*. London, England: Pearson Education.

Xia, F., Liu, J. Nie, H. et al. (2020). Random walks: a review of algorithms and applications. *IEEE Transactions on Emerging Topics in Computational Intelligence* 4 (2): 95–107.

INDEX